Springer Tracts in Advanced Robotics 104

Editors

Prof. Bruno Siciliano
Dipartimento di Ingegneria Elettrica
e Tecnologie dell'Informazione
Università degli Studi di Napoli
Federico II
Via Claudio 21, 80125 Napoli
Italy
E-mail: siciliano@unina.it

Prof. Oussama Khatib
Artificial Intelligence Laboratory
Department of Computer Science
Stanford University
Stanford, CA 94305-9010
USA
E-mail: khatib@cs.stanford.edu

For further volumes:
http://www.springer.com/series/5208

STAR (Springer Tracts in Advanced Robotics) has been promoted under the auspices of EURON (European Robotics Research Network)

M. Ani Hsieh · Gregory Chirikjian
Editors

Distributed Autonomous Robotic Systems

The 11th International Symposium

Springer

Editors
M. Ani Hsieh
Mechanical Engineering and Mechanics
Drexel University
3141 Chestnut St.
Randell 115
Philadelphia, PA 19104
USA
e-mail: mhsieh1@drexel.edu

Gregory Chirikjian
Department of Mechanical Engineering
Johns Hopkins University
3400 N. Charles St.
223 Latrobe Hall
Baltimore, MD 21218
USA
e-mail: gchirik1@jhu.edu

ISSN 1610-7438 ISSN 1610-742X (electronic)
ISBN 978-3-662-52332-2 ISBN 978-3-642-55146-8 (eBook)
DOI 10.1007/978-3-642-55146-8
Springer Heidelberg New York Dordrecht London

Printed on acid-free paper

Springer is part of Springer Science+Business Media (www.springer.com)

Foreword

Robotics is undergoing a major transformation in scope and dimension. From a largely dominant industrial focus, robotics is rapidly expanding into human environments and vigorously engaged in its new challenges. Interacting with, assisting, serving, and exploring with humans, the emerging robots will increasingly touch people and their lives.

Beyond its impact on physical robots, the body of knowledge robotics has produced is revealing a much wider range of applications reaching across diverse research areas and scientific disciplines, such as: biomechanics, haptics, neurosciences, virtual simulation, animation, surgery, and sensor networks among others. In return, the challenges of the new emerging areas are proving an abundant source of stimulation and insights for the field of robotics. It is indeed at the intersection of disciplines that the most striking advances happen.

The *Springer Tracts in Advanced Robotics (STAR)* is devoted to bringing to the research community the latest advances in the robotics field on the basis of their significance and quality. Through a wide and timely dissemination of critical research developments in robotics, our objective with this series is to promote more exchanges and collaborations among the researchers in the community and contribute to further advancements in this rapidly growing field.

DARS is a well-established single-track conference that gathers every two years the main researchers in *Distributed Autonomous Robotic Systems.* Since the Tenth edition in 2010, STAR has welcomed DARS among the volumes resulting from thematic symposia devoted to excellence in robotics research.

The volume edited by Ani Hsieh and Gregory Chirikjian offers in its thirty-one chapters an interdisciplinary collection of technologies, algorithms, system architectures, and applications of advanced distributed robotic systems. The contents are effectively grouped into five thematic sections: coordination for perception, coverage, and tracking; task allocation and coordination strategies; modular robots and novel mechanisms and sensors; formation control and planning for robot teams; learning, adaptation, and cognition for robot teams.

Rich by topics and authoritative contributors, the Eleventh edition of DARS in 2012 culminates with this unique reference on the current developments and new directions in the field of distributed autonomous robotic systems. A very fine addition to the series!

Naples, Italy
March 2014

Bruno Siciliano
STAR Editor

Preface

The goal of the Symposium on Distributed Autonomous Robotic Systems (DARS) is to exchange and stimulate research ideas to realize advanced distributed robotic systems. Distributed robotics is a rapidly growing, interdisciplinary research area lying at the intersection of computer science, communication and control systems, and electrical and mechanical engineering. Technologies, algorithms, system architectures, and applications were presented and discussed during the symposium. The 11th edition of DARS took place at Johns Hopkins University in Baltimore, Maryland in the United States of America.

As in previous years, this 11th edition of DARS boasts an extremely strong technical program. We received a total of 73 submissions and 31 of which were accepted for oral presentations. Following the precedence established by DARS 2010, each paper was reviewed by at least three reviewers and a Senior Program Committee member. The Program Chair and the Senior Program Committee coordinated the review process with the help of 132 reviewers. We are very grateful to all the reviewers and the senior program committee for their thoroughness and thoughtfulness in reviewing the papers. All accepted papers were included in the digital proceedings distributed at the symposium. Authors were then encouraged to submit a revised version after the conference before being accepted for inclusion in this STAR volume. The final collection consists of 31 original contributions. As in previous years, networked robots, robot swarms, motion planning, and modular robots continue their strong presence at DARS. For this 11th edition of DARS we saw an increase in contributions in the areas of multi-robot perception and machine learning. The papers are organized into five sections: Coordination for Perception, Coverage, and Tracking; Task Allocation and Coordination Strategies; Modular Robots and Novel Mechanisms and Sensors; Formation Control and Planning for Robot Teams; and Learning,

Adaptation, and Cognition for Robot Teams. Additionally, we also added a poster session that highlights novel preliminary results. Authors and poster presenters were then invited to submit more mature versions of their contributions to a Robotica Special Issue on Distributed Autonomous Robotic Systems.

The DARS 2012 program also included four invited keynote talks by world-renowned speakers covering emerging scientific discoveries in and application areas for distributed autonomous systems: Francesco Bullo, University of California, Santa Barbara on optimization and control of distributed robotic networks, Vijay Kumar, University of Pennsylvania on coordinating swarms of micro aerial vehicles, David Gracias, Johns Hopkins University on 3D assembly with micro/nano robots, and Luis Mier-y-Teran-Romero, Naval Research Laboratory on the effects of noise and delays in large coupled multi-agent systems. We have included the abstracts and bio-sketches for each invited talk and speaker in this volume.

DARS 2012 presented a total of five awards, one for overall Best Paper, one for overall Best Paper Runner-Up, one for Best Student Paper, one for Best Student Paper Runner-Up, and one for Best Poster. These awards were co-sponsored by the DARS 2010 organizing committee, represented by Alcherio Martinoli at the symposium. The award panel was chaired by Alcherio Martinoli (École Polytechnique Fédérale de Lausanne) and included Nikolaus Correll (University of Colorado), Roderich Groß (University of Sheffield), and Brian Sadler (Army Research Laboratory). The award selection process took into account various factors, including the reviewers' score, the revised contribution included in the digital proceedings, the oral presentation, and related discussion at the symposium. Best Paper was awarded to Brian Coltin and Manuela Veloso (Carnegie Mellon University) for their paper "Optimizing for Transfers in a Multi-Vehicle Collection and Delivery Problem." Best Paper Runner-Up was awarded to Nils Napp and Radhika Nagpal (Harvard University) for their paper "Distributed Amorphous Ramp Construction in Unstructured Environments." Best Student Paper was awarded to Ross Anderson and Dejan Milutinović (University of California, Santa Cruz) for their paper "A Stochastic Optimal Enhancement of Feedback Control for Unicycle Formations." Best Student Paper Runner-Up was awarded to Lantao Liu and Dylan Shell (Texas A & M University) for their paper "Multi-robot Formation Morphing through Matching Graph." Best Poster was awarded to Nisar Ahmed, Tsung-Lin Yang, Eric Sample, and Mark Campbell for their poster "Bayesian Sketch and Share: Enhanced Information Fusion for Large Scale Mixed Robot-Human Search Teams."

Last but not least, we would like to thank the National Science Foundation (NSF) who provide travel support for qualifying student authors pursuing their degrees in U.S. institutions. The NSF sponsored travel awards were facilitated by Timothy Chung (Naval Postgraduate School). We would also like to thank Thomas Ditzinger and Holger Schäpe, representatives from Springer Verlag who are responsible for coordinating the series, for giving us the opportunity to continue publishing the DARS proceedings in such a prestigious venue. Finally, we would like to thank the local organization team, which consists of administrative assistants, PhD students, particularly Mr. M. Kendal Ackerman, and research collaborators. Their hard work was instrumental ensuring the success of the symposium.

Baltimore, MD
February 10, 2014

M. Ani Hsieh
Gregory Chirikjian

Program Committee

Senior Program Committee

Hajime Asada	University of Tokyo
Tucker Balch	Georgia Institute of Technology
Nikolaus Correll	University of Colorado
Roderich Groß	University of Sheffield
Francesco Mondada	École Polytechnique Fédérale de Lausanne
Derek Paley	University of Marlyand
Brian Sadler	Army Research Laboratory
Wei-Min Shen	University of Southern California
Herbert Tanner	University of Delaware
Mark Yim	University of Pennsylvania

Paper Awards Committee

Nikolaus Correll	University of Colorado
Roderich Groß	University of Sheffield
Alcherio Martinoli	École Polytechnique Fédérale de Lausanne
Brian Sadler	Army Research Laboratory

Reviewers

Pramod Abichandani
Bruno Adorno
Francesco Amigoni
Ross Anderson
Marcelo Ang
Thomas Apker
Adrian Arfire
Federico Augugliaro
Spring Berman
Nicola Bezzo
Subhrajit Bhattacharya
Andreas Breitenmoser
Zack Butler
Mario Campos
Luiz Chaimowicz
Young-Jo Cho

Nak Young Chong
Anders Christensen
Timothy Chung
Igor Cizelj
Christopher Clark
Thomas Collins
Nikolaus Correll
Hugo Costelha
Anthony Cowley
Yanzhe Cui
Alexander Cunningham
Sangeeta Daingade
Gautham Das
Jnaneshwar Das
Prithviraj Dasgupta
Levi DeVries
Ezequiel Di Mario
Ayan Dutta
Younes El Hamdi
Chris Evans
Pooyan Fazli
Jonathan Fiene
Rafael Fierro
Jonathan Fink
Robert Fitch
Lutz Frommberger
Jie Fu
Toyomi Fujita
Dean Galarowicz
Melvin Gauci
Asish Ghoshal
Stephanie Gil
Sven Gowal
Teawon Han
David Herrero-Perez
Bobby Hodgkinson
Geoff Hollinger
Feili Hou
Brian Hrolenok
M. Ani Hsieh
Fumiya Iida
Hyunja Im
Minsu Jang
Seohyun Jeon
Sertac Karaman
James Keller
Jeffrey Khan
Young-Ho Kim
Christoph Kirsch
David Kriesel
Venkat Krovi
Haruhisa Kurokawa
Hyukseong Kwon
Rolf Lakaemper
Somchaya Liemhetcharat
Quentin Lindsey
Lantao Liu
Wenguo Liu
Savvas Loizou
Tyler MacCready
Stéphane Magnenat
Kenneth Mallory
Gian Luca Mariottini
Eric Martinson
Ana Medina
Yan Meng
Luis Mier-y-Teran-Romero
Dejan Milutinovic
Marco Montes de Oca
Nils Napp
Daniele Nardi
Michael Novitzky
Kazuhiro Ohkura
Jun Ota
Moises Pacheco
Jose Pereira
Cammy Peterson
Cody Phillips
Alyssa Pierson
Luciano Pimenta
David Portugal
Amanda Prorok
Nadeesha Ranasinghe
Dustin Reishus
Shai Revzen
Rui Rocha
Michael Rubenstein
Brian Sadler
Erol Sahin
Bharath Sankaran

Sanem Sariel-Talay
Kosuke Sekiyama
Rajnikant Sharma
Dylan Shell
Stephen Smith
Huan Tan
Yew Teck Tan
Matt Taylor
Kohji Tomita
Vito Trianni
Elio Tuci
Jeffrey Twigg
Alphan Ulusoy
Luis Valbuena
Andrew Vardy
Gregory Vorobyev
Briana Wellman
Michael West
Alan Winfield
Kevin Wolfe
Eric Wolff
James Worcester
Atsushi Yamashita
Gen'ichi Yasuda
Toshiyuki Yasuda
Dingjiang Zhou

Invited Keynote Presentations

Optimization and Control in Robotic Coordination

Francesco Bullo — *University of California, Santa Barbara*

Abstract. Motion coordination is an extraordinary phenomenon in biological systems and a powerful tool in man-made systems; although individual agents have no global system knowledge, complex behaviors emerge from local interactions. This talks focuses on robotic networks, that is, group of robots that communicate, compute, and coordinate their motions to perform useful tasks. I will propose adaptive and distributed algorithms based on concepts from stochastic analysis, geometric optimization, and stability theory. Time permitting, I will discuss recent developments on stochastic surveillance and attention allocation for mixed teams.

Biosketch. Professor Francesco Bullo is a professor with the Mechanical Engineering Department at the University of California, Santa Barbara. His main research interests lie in multi-agent networks with application to robotic coordination, distributed computing and power networks.

Control and Planning for Groups of Micro Aerial Vehicles

Vijay Kumar — *University of Pennsylvania*

Abstract. This talk addresses the development of autonomous quadrotors, which are on the order of 0.1 - 0.5 meters in length and 0.1 - 0.5 kilograms. The small size/inertia of flying robots enables agile, three-dimensional movement. However, the control of these vehicles is challenging because of underactuation and uncertainty in the dynamics. I will first discuss the modeling and control of quadrotors, and the sequential composition of controllers for aggressive flight. I will then show how lower-dimensional abstractions can be used to develop real-time planning algorithms allowing the robots to navigate in three-dimensional environments. Finally, I will discuss the challenges of extending the control and planning algorithms for groups of vehicles.

Biosketch. Professor Vijay Kumar is the UPS Foundation Professor in the School of Engineering and Applied Science at the University of Pennsylvania, and on sabbatical leave at White House Office of Science and Technology Policy where he serves as the assistant director for robotics and cyber physical systems. His research interests, covering mechanical, electrical and computer engineering, are manifest in his work on the design, control and coordination of aerial robots.

Stimuli-responsive Sub-millimeter Scale 3D Assembly and Robotics

David Gracias *Johns Hopkins University*

Abstract. There are significant challenges in miniaturizing assembly and robotic operations to sub-millimeter length scales. These challenges are especially pronounced off-chip and include, for example, (a) the challenge of powering large numbers of tiny devices in either a wired or wireless manner using electromechanical or pneumatic actuators, and (b) limits on the information that can be retrieved and the positional real-time control that can be achieved in 3D microenvironments and especially under in vivo conditions. Hence, fabrication, guidance, tracking and controlling decisions can all be prohibitively difficult to implement at sub-mm length scales.

In this talk, I will discuss the use of stimuli responsive strategies to enable 3D assembly and robotic functions at the sub-millimeter to 100 nanometer length scales. I will present three examples of such an approach. These include, (a) the self-assembly of complex 3D structures such as polyhedra; (b) the guidance and realization of functional tasks such as pick-and-place operations or drilling using bacterial-backpacking and self-propulsion; and (c) the in vivo surgical biopsy with large numbers of tiny stimuli-responsive tools. These studies suggest the need for robotic paradigms wherein structures assemble on their own, tiny devices respond autonomously to their microenvironments, and function is achieved using statistical, defect-tolerant approaches.

Biosketch. Professor David Gracias is an associate professor in the Department of Chemical and Biomolecular Engineering at Johns Hopkins University. His diverse research interests center around the fabrication, self-assembly and characterization of biotic/abiotic systems, devices and materials at very small size scales.

The Dynamics of Large Numbers of Interacting Autonomous Agents

Presented by *Luis Mier-y-Teran* in place of *Ira Schwartz* *Naval Research Lab*

Abstract. Recently, much attention has been given to the study of interacting multi-agent, particle or swarming systems in various natural and engineering fields. Interestingly, these multi-agent swarms can self-organize and form complex spatio-temporal patterns even when the coupling between agents is weak. Many of these investigations have been motivated by a multitude of biological systems such as schooling fish, swarming locusts, flocking birds, bacterial colonies, ant movement, etc. In the engineering world, the results of these studies have been successfully applied in the design of systems of autonomous, inter-communicating robotic systems to solve specific goals, such as robotic motion planning, spatio-temporal formation, obstacle avoidance, and boundary tracking.

Two topics of ongoing research in interacting particle systems, and, in particular, in the dynamics of swarms, are the effects of system noise and time delays. The inclusion of noise in such studies has revealed noise-induced transitions between different coherent patterns, and to the discovery of different phase transitions. On the other hand, it is well known that time delays can have profound dynamical consequences, such as destabilization and synchronization. Initially, such studies focused on the case of one or a few discrete time delays. More recently, however, the complex situation of several and random time delays has been considered. An additional important case is that of distributed time delays, when the dynamics of the system depends on a continuous interval in its past instead of a discrete instant.

In this talk I will review some of the results on the interaction of delay and noise in particle systems and, in particular, I will introduce new results on the effects of random delays on swarm dynamics.

This work is supported by the Office of Naval Research.

Biosketch. Luis Mier-y-Teran did his undergraduate studies at UNAM, in Mexico city and then obtained a PhD in Physics at Northwestern University in 2010 working with professors, Mary Silber and Vassily Hatzimanikatis. His thesis research was on the mathematical modeling of protein synthesis from a dynamical systems perspective. He currently holds a postdoctoral position at Johns Hopkins University/Naval Reseach Lab and is doing research on time-delayed models of swarming agents and on mathematical epidemiology.

Biosketch. Dr. Ira Schwartz is a theoretical physicist/applied Mathematician and the head of the Nonlinear Dynamical Systems Section at the Naval Research Laboratory. The main themes of his work have been mathematical and numerical techniques of nonlinear dynamics and chaos, and most recently, nonlinear stochastic analysis and control of natural and artificial swarms and other coupled networked systems.

Contents

Part V: Learning, Adaptation, and Cognition in Many Robot Systems

Part I
Coordination for Perception, Coverage and Tracking

Adaptive Multi–Robot Coverage of Curved Surfaces

Andreas Breitenmoser, Hannes Sommer, and Roland Siegwart

Abstract. This paper presents two adaptive coverage algorithms for the deployment of multiple robots into discrete partitions over curved surfaces. The algorithms compute a metric tensor field locally on the surface in order to shape the robots' partitions in position, size, orientation and aspect ratio according to the present anisotropy. The coverage algorithms are further incorporated into a hybrid coverage method for the complete coverage of surface areas. Each robot iteratively deploys and adapts the partition, then subsequently sweeps its assigned area. The algorithms are demonstrated in simulations on different mesh models, including meshes reconstructed from real laser point cloud data.

1 Introduction

Coverage of work space is an elementary task that arises in multi–robot systems. Multi–robot coverage methods describe how multiple robots coordinate and partition their work load and work space among each other. Depending on the tasks a group of robots has to accomplish, robot coverage can have different meaning: coverage equally involves static and dynamic robot distributions, and notions of sensing as well as actuation. Robot locations in mobile sensor networks for instance are optimized to provide good communication and sensor coverage of the environment [1, 2]. Similar approaches lead to robot deployments that guarantee fair distribution of work loads among robots, or short response times when providing services to allocated sites [3]. Another class of coverage tasks requires more permanent movements. In or-

Andreas Breitenmoser · Hannes Sommer · Roland Siegwart
Autonomous Systems Laboratory (ASL), ETH Zurich, Tannenstrasse 3, 8092 Zurich, Switzerland
e-mail: {andreas.breitenmoser,hannes.sommer}@mavt.ethz.ch, rsiegwart@ethz.ch

M.A. Hsieh and G. Chirikjian (eds.), *Distributed Autonomous Robotic Systems*, Springer Tracts in Advanced Robotics 104,
DOI: 10.1007/978-3-642-55146-8_1,

der to continuously monitor an environment, inspect complex structures, or apply tools to a surface in manufacturing, the robots perform coverage by sweeping through their work spaces [4, 5].

We are interested in robot coverage of curved surfaces as they are typically found in inspection applications for tanks and tubes of industrial plants. In our work, we consider robots that move directly on the surface [6, 7]. Spaces in industrial structures are often narrow and thus forbid robots to work together in close side by side formations. Therefore, the robots need to deploy and partition their work space in a reasonable way. However, robot deployment and visual inspection are often not sufficient, and robots are required to sweep their sensor probes in close contact over the surface areas.

Hybrid coverage combines deployment and sweeping motion by performing one after the other [6]. Hybrid coverage methods build on and combine aforementioned coverage concepts. The variant of hybrid coverage we are looking at in this paper first spreads the robots into disjoint regions by constructing a centroidal Voronoi configuration [1]. This is further related to the classical concepts of cell decomposition, just that now the cell decomposition depends on the robot locations rather than on the environment boundaries. Once deployed, each robot sweeps its assigned Voronoi region along a covering path. Here, basically any space filling curve can be applied as long as the kinematic constraints of the robots are met. Examples are swaths similar to those used in the Boustrophedon decomposition on planar and curved surfaces [5].

The main focus of this paper is on the first stage of the two–tired hybrid coverage method. It is important to provide an effective deployment and partitioning of space in the first stage, as each robot needs to cover the entire surface of its assigned Voronoi region in the subsequent second stage. The proposed coverage algorithms make use of anisotropic centroidal Voronoi tessellations [8], and extend our previous work [7] with enhanced adaptivity of the Voronoi regions. The Voronoi regions adapt to local anisotropy, which is defined by a tensor field on the curved surface. The tensor field allows for controlling shape, densitiy and size, as well as orientation and aspect ratio of the Voronoi regions. This new adaptivity may improve multi–robot coverage in several ways. First, adapting the size and orientation of the Voronoi regions according to environment characteristics, such as surface curvature, salient features or representation uncertainty, makes robot movements during operation (e.g., sweeping) in a region on the surface safer and more efficient. Second, adapting the density, orientation or aspect ratio of the Voronoi regions by an input tensor field enables user guidance of the robot configuration. Finally, adapting the shape and size of the Voronoi regions allows to match the region to a sweeping pattern, which is executed by a robot in the region during the second stage of the hybrid coverage method.

Similar to [7, 9, 10, 11], we perform the robot deployment based on a discrete representation and model the environment as a graph. In our previous work [7], we suggested two fundamentally different approaches for covering a surface, based on discrete geodesic and approximative Euclidean distance computation. The algorithms in this paper follow up on both approaches in presence of anisotropy. Discrete geodesic alternatives for covering a graph can be found in [9, 10, 11]. Furthermore, geodesic distances under the influence of anisotropy can be computed by Fast Marching methods [12]. The adaptive remeshing of triangular meshes [13], in contrast, applies similar techniques of approximating geodesics by Euclidean distances. The partitioning is driven by an iterative process of pairwise optimization between adjacent regions, which shares similarities with the generation of pairwise optimal partitions as defined in [10]. Anisotropy has been previously used in the context of Voronoi coverage for modeling anisotropic sensors [14]; however, the anisotropy is formulated with respect to the robot positions and not with respect to points in the environment. The concept of adjusting a density function, defined over the areas of the Voronoi regions, is applied in [1] for formation control and robot guidance. Finally, in [3], fat– and skinny–shaped equitable partitions are created using power diagrams, with applications to vehicle routing and optimal workload sharing.

The remainder of the paper is organized as follows. A formal definition of the problem is given in Section 2. Section 3 describes the adaptive coverage algorithms for the deployment of multiple robots on curved surfaces. In Section 4, the hybrid coverage method for area coverage of curved surfaces is presented, which relies on the algorithms of Section 3 in the first stage. Simulation experiments in Section 5 verify the coverage method on synthetic and real data gathered in an inspection scenario. Section 6 concludes the paper.

2 Problem Definition

Given a curved connected orientable surface $\mathcal{S} \subset \mathbb{R}^3$ with the Riemannian metric induced by the Euclidean scalar product, a group of n robots r_i, $i \in \{1, ..., n\}$, with start positions $\mathbf{q}_i \in \mathcal{S}$, is to be distributed over a fair partition $\mathcal{V} = \{V_i\}_{i=1}^n$ of $\mathcal{S}$ into regions $V_i \subset \mathcal{S}$, each containing one robot at end position $\mathbf{z}_i \in V_i$. In the partitioning process, it is desirable to adjust the meaning of fair as well as the shapes of the regions V_i as much as possible.

First, we drop the constraint of being located on the surface $\mathcal{S}$ and consider the distribution of n points $\mathbf{Z} = [\mathbf{z}_i]_{i=1}^n \in \mathbb{R}^{3n}$ into a partition of a convex three–dimensional submanifold $\Omega \subset \mathbb{R}^3$. The resulting simplified problem is typically formulated as an optimization problem, where the following cost functional within the regions V_i is minimized,

$$\underset{\mathbf{Z},\,\mathcal{V}}{\operatorname{argmin}}\ \mathcal{H}_{\text{cont, iso}}(\mathbf{Z},\,\mathcal{V}) = \underset{\mathbf{Z},\,\mathcal{V}}{\operatorname{argmin}} \sum_{i=1}^{n} \int_{V_i} \rho(\mathbf{x})\ \|\mathbf{z}_i - \mathbf{x}\|_2^2\ dV(\mathbf{x})\ , \tag{1}$$

where we use the integral with volume element dV, and $\rho(\mathbf{x})$ describes a non–negative weight (a mass density) at every point $\mathbf{x}$ in Ω. The points $\mathbf{Z}$ and the partition $\mathcal{V}$ are dependent. The partition $\mathcal{V}$ that minimizes the cost functional $\mathcal{H}_{\text{cont, iso}}(\mathbf{Z},\, \mathcal{V})$ for fixed points $\mathbf{Z}$ is the Voronoi tessellation. If the points $\mathbf{Z}$ are taken as the generators of the Voronoi tessellation and are moved iteratively to the mass centers of the Voronoi regions, the cost converges to a local minimum and a *centroidal Voronoi tessellation (CVT)* results.

While choosing ρ gives good control over the fairness of the resulting partition, there is no possibility to influence the regions' shapes.

Therefore, we follow the idea of generalizing the approach given by Equation (1) and replace the Euclidean distance with an anisotropic distance, with which we hope to influence the shapes of the regions,

$$\operatorname*{argmin}_{\mathbf{Z},\, \mathcal{V}} \mathcal{H}_{\text{cont, aniso}}(\mathbf{Z},\, \mathcal{V}) \;=\; \operatorname*{argmin}_{\mathbf{Z},\, \mathcal{V}} \sum_{i=1}^{n} \int_{V_i} \rho(\mathbf{x})\;\, f\left(d\left(\mathbf{x},\, \mathbf{z}_i\right)\right)\, dV(\mathbf{x})\;, \quad (2)$$

where f is a smooth and strictly increasing function and d denotes a general anisotropic distance metric. Similar to the cost in Equation (1), $\mathcal{H}_{\text{cont, aniso}}(\mathbf{Z},\, \mathcal{V})$ is minimized when the points $\mathbf{Z}$ are placed at the anisotropic mass centers of $\mathcal{V}$.

In particular, we consider functions $f(d(\mathbf{x},\, \mathbf{z})) = d^{\,p}_{\mathbf{x},\mathbf{z}}$ and distances

$$d_{\mathbf{x},\, \mathbf{z}} \;=\; \inf_{\substack{\alpha:\; [0,\, 1] \to \Omega \\ \alpha(0) = x,\, \alpha(1) = z}} \int_0^1 \left\| F_{\alpha(t)}\; \dot{\alpha}(t) \right\|_p dt\;, \quad (3)$$

where $\|\mathbf{x}\|_p = \sqrt[p]{|x_1|^p + |x_2|^p + |x_3|^p}$ denotes the p–norm for $p \in \mathbb{N} \setminus \{0\}$ and $F_{\mathbf{x}}$ is a symmetric positive definite matrix smoothly depending on $\mathbf{x} \in \Omega$, which describes the local anisotropy. $F_{\mathbf{x}}$ is called the *Finsler tensor field* in the following[1].

We are now interested in formulating the minimization problem of Equation (2) over the surface $\mathcal{S}$. To do so, we follow the same approach as used in [7], and minimize the discrete version of the cost functional defined over the triangle mesh M, which is a graph–based representation of the surface $\mathcal{S}$, defined by the graph G_M. The discretization gives

$$\mathcal{H}_{\text{disc, aniso}}(\mathbf{Z},\, \mathcal{G}) \;=\; \sum_{i=1}^{n} \sum_{v \in G_i} \int_{A_v} \rho(\mathbf{x})\, d^{\,p}_{\mathbf{x},\mathbf{z}_i}\, dS(\mathbf{x})\;, \quad (4)$$

where dS denotes the surface element. The partition is represented by $\mathcal{G} = \{G_i\}_{i=1}^{n}$, with the subgraphs $G_i \subset G_M$ given by sets of connected vertices v. A_v denotes the area of the dual cell of each vertex. In the case of a planar mesh, this corresponds to the Voronoi region around v obtained from the Voronoi diagram with the mesh vertices as generators.

---

[1] This yields the usual distance of the Finsler manifold, given by Ω together with the Finsler function $(\mathbf{x}, \mathbf{v}) \mapsto \|F_{\mathbf{x}}\, \mathbf{v}\|_p$.

To determine the distance $d_{\mathbf{x},\mathbf{z}}$, the integral in Equation (3) is approximated with a sum of lengths $\left\|\widetilde{F}_{\mathbf{x}_l}\left(\mathbf{x}_{l+1}-\mathbf{x}_l\right)\right\|_p$ of segments given by a sequence of nodes $\mathbf{x}_l$, $l \in \mathbb{N}$, in the graph G_M, connecting $\mathbf{x}$ and $\mathbf{z}$.

Although $\mathcal{S}$ is a two–dimensional manifold, it is not sufficient in the discrete version to define $F_{\mathbf{x}}$ on the two–dimensional tangent spaces of S because the distance vectors $(\mathbf{x}_{l+1}-\mathbf{x}_l)$ are not restricted to the tangent spaces of $\mathcal{S}$ at $\mathbf{x}_l$. So, however we choose $F_{\mathbf{x}}$ on the tangent spaces, it must be given as a three–by–three matrix $\widetilde{F}_{\mathbf{x}}$. This matrix can be constructed from the desired matrix $F_{\mathbf{x}}$ given in a basis of $T_{\mathbf{x}}\mathcal{S}$, e.g., such that in the local basis extended by the normal vector of the oriented surface, it looks like: $\widetilde{F}_{\mathbf{x}} = \left(\begin{smallmatrix} F_{\mathbf{x}} & \mathbf{0} \\ \mathbf{0} & 1 \end{smallmatrix}\right)$.

In the following, we pursue two different approaches of choosing the sequence of nodes $\mathbf{x}_l$ for the distance computation. In the first approach, $d_{\mathbf{x},\mathbf{z}}$ is computed for arbitrary p–norms as the discrete geodesic distance. The distance segments correspond to the edges along a shortest path from $\mathbf{x}$ to $\mathbf{z}$ on the graph G_M. In the second approach, the path distance $d_{\mathbf{x},\mathbf{z}}$ is approximated by the length of a single segment, which connects $\mathbf{x}$ and $\mathbf{z}$ along the direct shortcut through $\mathbb{R}^3$, using the 2–norm. We arrive this way at the *directional distance* as defined in [8][2]. That yields $d_{\mathbf{x},\mathbf{z}} \simeq \left\|\widetilde{F}_{\mathbf{x}}\left(\mathbf{z}-\mathbf{x}\right)\right\|_2 = \sqrt{(\mathbf{z}-\mathbf{x})^T\, \widetilde{K}_{\mathbf{x}}\, (\mathbf{z}-\mathbf{x})}$, where $\widetilde{K}_{\mathbf{x}} := \widetilde{F}_{\mathbf{x}}^T\, \widetilde{F}_{\mathbf{x}}$, which is a usual anisotropic quadratic distance in $\mathbb{R}^3$.

For the second approach, under the approximation that $\widetilde{K}_{\mathbf{x}_v}$ (in short $\widetilde{K}_v$) is constant over a given vertex area A_v, the cost functional $\mathcal{H}_{\text{disc, aniso}}$ can be rewritten with $d^2_{\mathbf{x},\mathbf{z}_i} = (\mathbf{z}_i-\mathbf{x})^T\, \widetilde{K}_{\mathbf{x}}\, (\mathbf{z}_i-\mathbf{x})$ as

$$\mathcal{H}_{\text{disc, aniso}}(\mathbf{Z},\mathcal{G}) = \sum_{i=1}^{n} \sum_{v\in G_i} \left[W_v + (\mathbf{z}_i-\gamma_v)^T\, \widetilde{K}_v^*\, (\mathbf{z}_i-\gamma_v)\right] =$$

$$\sum_{i=1}^{n} \left[\sum_{v\in G_i} W_v + \sum_{v\in G_i} \left(\gamma_v^T\, \widetilde{K}_v^*\, \gamma_v\right) + {\mathbf{z}_i}^T \left(\sum_{v\in G_i} \widetilde{K}_v^*\right) \mathbf{z}_i - 2\,\mathbf{z}_i^T \left(\sum_{v\in G_i} \widetilde{K}_v^* \gamma_v\right)\right], \tag{5}$$

where γ_v is the centroid of A_v, $W_v = \int_{A_v} \rho(\mathbf{x})\, (\gamma_v-\mathbf{x})^T\, \widetilde{K}_{\mathbf{x}}\, (\gamma_v-\mathbf{x})\, dS(\mathbf{x})$, and $\widetilde{K}_v^* = m_v\, \widetilde{K}_v$, with $m_v = \int_{A_v} \rho(\mathbf{x})\, dS(\mathbf{x})$, the concentrated weight over vertex area A_v. From Equation (5) follows that it is sufficient to minimize for the last two terms, since the first two terms do not depend on a particular choice of $\{\mathbf{Z},\mathcal{G}\}$. The locations of the points $\mathbf{Z}$ can be chosen freely. However, if we select the generators $\mathbf{Z}$ in Equation (5) to be the mass centers of the Voronoi regions, which is in accordance with [1, 8] and the notion of CVT,

[2] Note that this distance measure is not symmetric and thus no proper distance metric as such; however, it is consistent with the definition of the standard Voronoi diagram in the isotropic metric and straightforward to compute. See [8] for further discussions.

the last two terms in Equation (5) can be further simplified. The anisotropic mass centers are obtained as $\mathbf{z}_i = \text{inv}\left(\sum_{v \in G_i} \widetilde{K}_v^*\right) \sum_{v \in G_i} \widetilde{K}_v^* \gamma_v$, which are the minimizers of the cost functional in Equation (4). Substituting $\mathbf{z}_i$ for the last two terms of Equation (5) leads to the partial cost

$$\mathcal{H}^*_{\text{disc, aniso}}(\mathbf{Z}, \mathcal{G}) = \sum_{i=1}^{n} \left[- \left(\sum_{v \in G_i} \widetilde{K}_v^* \gamma_v \right)^T \text{inv}\left(\sum_{v \in G_i} \widetilde{K}_v^* \right) \left(\sum_{v \in G_i} \widetilde{K}_v^* \gamma_v \right) \right], \tag{6}$$

which does not depend on the generators $\mathbf{z}_i$ anymore. In the isotropic case, Equation (6) becomes

$$\mathcal{H}^*_{\text{disc, iso}}(\mathbf{Z}, \mathcal{G}) = \sum_{i=1}^{n} \left[- \frac{\left\| \sum_{v \in G_i} m_v \gamma_v \right\|_2^2}{\sum_{v \in G_i} m_v} \right]. \tag{7}$$

The original problem of distributing n robots r_i over the surface S can now be solved for the second approach by simply minimizing Equation (6), or Equation (7) respectively, and letting $\mathbf{Q} = [\mathbf{q}_i]_{i=1}^n$ approach $\mathbf{Z} = [\mathbf{z}_i]_{i=1}^n$ over time. The partial cost $\mathcal{H}^*_{\text{disc, aniso}}(\mathbf{Z}, \mathcal{G})$ and $\mathcal{H}^*_{\text{disc, iso}}(\mathbf{Z}, \mathcal{G})$ are minimized in an efficient way by exchanging vertices between neighboring Voronoi regions, which only requires updates of the sums $\sum \widetilde{K}_v^* \gamma_v$ and $\sum \widetilde{K}_v^*$, or $\sum m_v \gamma_v$ and $\sum m_v$ respectively.

For the first approach, the original cost in Equation (4) must be minimized; gradient descent methods provide a practical solution [1, 7, 9]. This involves finding the direction of fastest decline of distance. We use the negated tangent vector of the minimizing geodesic, which differs from its gradient [9][3].

The Finsler tensor given by $F_{\mathbf{x}}$ offers several ways to adapt the partition over the surface $\mathcal{S}$ (see Figure 1). The orientation of the Voronoi regions is influenced by the directions of the eigenvectors of $F_{\mathbf{x}}$. The regions' aspect ratio is given by the ratios of the eigenvalues of $F_{\mathbf{x}}$, measuring the strength of directionality. As mentioned at the beginning of the section, the regions' size can be changed by the weighting factor or mass density ρ. In addition, the distance in the cost functionals can assume the general p–norm. Depending on p, the p–norm results in a more circular or square–shaped distance field, which leads to modifications in the shapes of the regions.

Here, a remark is in place: the p–norm only applies to the first approach, which computes discrete geodesic distances. Note that the second approach with approximate distance computation, as presented in here, requires $p = 2$. This is inherent to above derivation (from Equation (4) to Equation (5)), which is based on the parallel axis theorem. The parallel axis theorem as such, however, relies on specific properties of the 2–norm.

[3] In fact, they even differ in their mathematical type (tangent vector vs. tangent covector). Their value's relation can be found for $p = 2$ in [9]. This must be generalized for $p \neq 2$. However, this is not included here for the sake of brevity.

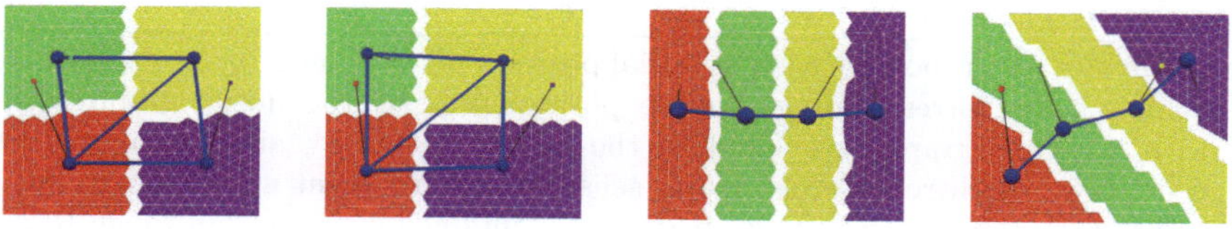

Fig. 1 Adaptive deployment of four robots with uniform, isotropic (to the left) and anisotropic (upward and diagonal) metric tensor fields provided by the user guidance

3 Adaptive Surface Coverage

The iterative construction of a Voronoi tessellation on a graph G_M, representing a triangle mesh M, forms the basis for our adaptive surface coverage algorithms. Each Voronoi region is owned by a robot r_i and represented by a subgraph $G_i \subset G_M$. As the partitioning process evolves, the Voronoi regions—and together with the regions the robots—are distributed over the surface. The subgraphs G_i result for the first approach from minimizing the cost with discrete geodesic distance in Equation (4), and for the second approach from minimizing the cost with approximative Euclidean distance in Equation (6) or Equation (7). The basic algorithms have been introduced in [7] and are restated in the following. Algorithm 1 summarizes the approaches' main function, and Algorithm 2 and Algorithm 3 describe the coordination functions of the two approaches.

The first approach computes a discrete geodesic distance by propagating a wavefront over the graph. Each robot computes a distance field from its current position $\mathbf{v}_{\mathbf{p}_i}$, which is the robot's position $\mathbf{p}_i \in \mathcal{S}$ projected onto the graph G_M, to positions of neighboring robots on the graph. As soon as the wavefront reaches a neighbor's vertex, a wavefront is propagated back in order to find the boundary vertices with equal discrete geodesic distance between the two robots. This procedure is continued by each robot r_i until the complete Voronoi partition $\mathcal{G}$ is constructed. Once the Voronoi regions are computed, the robots move toward the regions' mass centers. The regions are updated continuously, and it is sufficent to determine the direction toward the mass center instead of its absolute position. As the direction, we use the negated tangent vector of the minimizing geodesic (see Section 2).

The second approach minimizes the cost by exchanging boundary vertices iteratively across the boundaries of adjacent regions. For each vertex $v_A \in G_i$ located at a region boundary, following local vertex exchanges are possible. If a vertex at the boundary next to v_A has not yet been assigned to any region, it is directly added to G_i. If there exists at least one neighbor vertex $v_B \in G_j$, $G_i \cap G_j = \emptyset$, the change in the cost of Equation (6) or Equation (7) is evaluated for the following three cases: 1) the vertex v_A is added to G_j, 2) the vertex v_B is added to G_i, or 3) no vertex is exchanged across the boundary. The case resulting in the highest reduction of cost is used for

Algorithm 1. Adaptive Surface Coverage

Require: Set of n robots r_i, each at initial position $\mathbf{p}_i^0 = \mathbf{q}_i$ on S in the continuous domain. The corresponding discrete position $\mathbf{v}_{\mathbf{p}_i}^0$ on the initial graph $G_{M_i}^0$, with $G_{M_i}^0$ the graph associated to the initial mesh M_i^0, and initial regions $G_i^0 \subset G_{M_i}^0$. Each robot r_i is able to sense the environment within range $R_{\mathbf{sens}}$ and to communicate with robot neighbors, contained in N_i, within range R_{com}.

while state == `DEPLOY` **do**
  if action == `COORDINATE` **then**
    $\mathbf{v}_{\mathbf{g}_i} \leftarrow$ `Coordinate`(algorithm) // "Algorithm 2" or "Algorithm 3"
  end if
  if action == `MOVE` **then**
    $\{\mathbf{p}_i, M_i\} \leftarrow$ `UpdateNeighborhood(SENS)`
    $T_i \leftarrow$ `PlanPath`$(\mathbf{v}_{\mathbf{p}_i}, \mathbf{v}_{\mathbf{g}_i}, G_{M_i}, M_i)$
    $\mathbf{p}_i \leftarrow$ `MoveToGoal`$(\mathbf{p}_i, T_i)$
    `UpdateState()`
  end if
end while

Algorithm 2. Coordination By Wavefront Propagation

Require: Voronoi region G_i, graph G_{M_i} and mesh M_i.

$N_i \leftarrow$ `UpdateNeighborhood(COM)`
$G_i \leftarrow$ `PropagateWavefront`(N_i, G_i, G_{M_i}, M_i)
$\mathbf{v}_{\mathbf{g}_i} \leftarrow$ `UpdateGoalDirection`(G_i)
`UpdateState()`

Algorithm 3. Coordination By Vertex Exchange

Require: Voronoi region G_i, graph G_{M_i} and mesh M_i; k rounds of vertex exchange.

$G_i \leftarrow$ `AssignFreeVertices`(G_i, G_{M_i})
$N_i \leftarrow$ `UpdateNeighborhood(COM)`
for all r_j contained in N_i **do**
  `ExchangeVertices`(G_i, G_j, k)
end for
$\mathbf{v}_{\mathbf{g}_i} \leftarrow$ `UpdateGoal`(G_i)
`UpdateState()`

the local update of the two Voronoi regions. If a vertex exchange leads to a disconnected region, the last case applies and the respective vertex will not be exchanged. Note that the last case does not affect the convergence of the overall algorithm; the cost is not minimized but remains unchanged in this iteration step. However, over subsequent iterations the pairwise optimization by exchanging vertices among adjacent regions minimizes the overall cost.

Each robot is assumed to be equipped with communication and sensing devices. The information collected from the environment is stored in the triangle mesh maps $M_i \subset M$, and each mesh map is associated with a graph $G_{M_i} \subset G_M$. For applications, where the algorithms are required to adapt to characteristics of the surface (e.g., surface curvature), the matrix $\widetilde{F}_{\mathbf{x}}$ is computed from the mesh map M_i. A Voronoi region is defined as the subgraph $G_i \subset G_{M_i}$. The neighborhood N_i of a robot r_i includes all the neighboring robots r_j within communication range R_{com}, together with the additional information of their positions, Voronoi regions and mesh maps. The `UpdateNeighborhood` function updates N_i; for argument `SENS`, the robot r_i updates its own position and mesh map, for argument `COM`, the robot detects other robots and updates its mesh map from communicated information. If the meshes of two neighboring robots overlap, the meshes are merged, $M_i \leftarrow M_i \cup M_j$, $M_j \leftarrow M_i \cup M_j$. A matching among two neighboring robots is only established if the robots' Voronoi regions are adjacent, i.e., if two vertices in G_i and G_j are adjacent in the graph G_M.

The adaptive surface coverage algorithms are divided into a *coordination* and a *moving* action. The coordination action implements the previously described cost minimization process. In the coordination action of the second approach, k rounds of vertex exchange are performed by the `ExchangeVertices` function. With k chosen large enough, a robot will not enter a local vertex exchange with another robot neighbor before a partition with optimal cost between its own and the current neighboring region is obtained. Provided that a vertex exchange procedure can find the global optimum for the two regions, this leads to pairwise optimal partitions similar to [10]. The coordination action updates the goal direction and goal position $\mathbf{v}_{\mathbf{g}_i}$ on the graph for each region and robot, which is then handed over to the moving action of the algorithms in order to deploy the robots on the surface. During the moving action the robots navigate on the surface. Each robot perceives the environment and computes the shortest path to its current goal. The `PlanPath` function updates the path T_i, based on the discrete positions $\mathbf{v}_{\mathbf{p}_i}$, whereas the `MoveToGoal` function controls the robot along the path.

In the Voronoi coverage method of [1], the robots act as the generators and converge to the mass centers of their Voronoi regions. However, this is no longer a necessary condition in our case, as the Voronoi regions are generated either by wavefront propagation between vertices or by local vertex exchanges of boundary vertices on the underlying graphs. The coordination and moving actions are loosely coupled[4]. The decoupling of the two actions allows to place the robots' goal positions at different locations that may improve the overall performance of the algorithms, with benefits such as lower cost paths, better accessibility of the goal positions, or better sensor coverage. Placing

[4] The dependence is only given through the communication and the environment update; the robot positions influence what the robots will communicate, perceive from their environment and incorporate into the mesh maps.

goals closer to the boundary of a Voronoi region, for example, could increase the explorative behavior of the basic coverage algorithms.

4 Application to Hybrid Coverage Control

This section explains how the adaptive surface coverage algorithms can be applied within the hybrid coverage framework to achieve area coverage on a surface. The hybrid area coverage method is outlined in Figure 2 and Algorithm 4. The idea behind hybrid coverage is to combine robot deployment and sweeping motions [6]. The method starts with the robots spreading out on the surface. The robots cooperatively partition the surface and get their assigned areas of operation. In this first stage, the adaptive surface coverage algorithms from Section 3 are applied to realize an effective deployment and adaptive decomposition of the surface. The provided adaptivity can be actively used to shape the Voronoi regions and simplify the coverage of the area, e.g., leading to compact or more elongated shapes that support circular spiralling or rectangular back–and–forth sweeping patterns. Upon convergence, the robots switch to the second stage, where each robot sweeps its assigned region locally to search the area. Depending on the size of the environment to be covered and the range of coverage of the single robots, the two stages of deployment and sweeping (state `DEPLOY` and `SWEEP` in Algorithm 4) are iterated. The robots relocate and redistribute outside the already covered area by applying the hybrid coverage subroutines `AdaptiveSurfaceCoverage` and `SweepSurfaceCoverage` again. By iterating the process, the complete surface is finally covered by the robots.

Once a Voronoi region is covered, it is marked as covered in the robot's mesh map M_i and is locked; the robot communicates the status to its neighbors, which update their mesh maps accordingly. As the adaptive surface coverage algorithms deploy the robots by generating a Voronoi tessellation, a dual Delaunay graph is established simultaneously (see Figures 1 – 4). The graph connects the Voronoi regions, represents the surface topology and gives a simplified low resolution representation of the environment. This representation remains valid and may serve as roadmap for future relocation and redistribution phases of the robots. The covered regions are known to the robots as they have swept and explored these areas before. Hence, paths transferring from one location on the surface to another are preferably planned through the known and safe area of the regions along the Delaunay graph.

Besides covering and locking of regions, many more operations on the regions are possible. A robot can leave its region uncovered and reassign it to other robots for coverage, or ask other robots for support. Regions can be deleted cooperatively if a robot fails or relocates. Furthermore, a new region is instantiated whenever a robot joins during deployment, or a region is created inside already covered area in order to initiate redundant coverage.

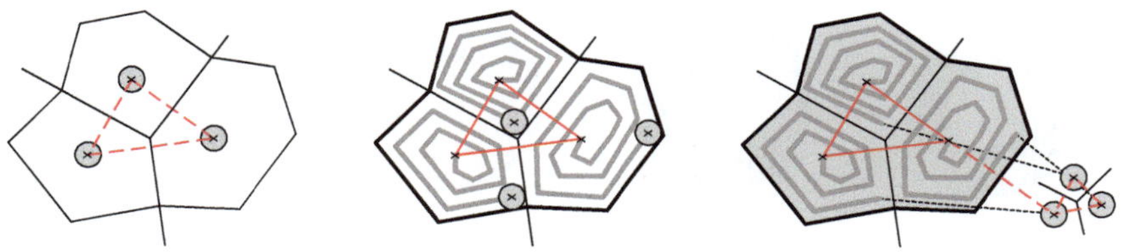

Fig. 2 Hybrid area coverage. Left: Robot deployment. Center: Sweeping motions. Right: Relocation and redistribution. The dual Delaunay graph is marked in red.

Algorithm 4. Hybrid Area Coverage

Require: Set of n robots r_i, with sensing and communication capabilities. Subroutines for the two stages of deployment and sweeping motion.

```
1: loop
2:    AdaptiveSurfaceCoverage(state) // state == DEPLOY
3:    SweepSurfaceCoverage(state) // state == SWEEP
4:    RelocateAndRestart()
5: end loop
```

5 Simulation Experiments

The adaptive surface coverage and the hybrid area coverage algorithms are tested on different surfaces. The surface models vary in complexity and consist of synthetic meshes, and meshes obtained by surface reconstruction from real laser point clouds. In the following simulations we assumed that the robots always remain connected (i.e., $R_{\text{com}} = \infty$).

The first experiment demonstrates the adaptivity of the Voronoi coverage. Figure 1 shows how four robots are deployed on a planar mesh under varying user guidance. The metric tensor field is specified directly by the user. The first deployment is uniform, which sets equal weights to all vertices and directions. The second deployment is isotropic and directs the robots to the left of the mesh. The metric tensor field for the first two deployments is of the form $\widetilde{K}_v^* = m_v\, I_3$, where m_v is constant for the uniform deployment and varies as function of the location for the isotropic deployment. The third and fourth deployments use an anisotropic metric. The Voronoi partitions point along the principal directions of 90° upwards and 45° to the left.

In the second and third experiment, 10 and 50 robots are distributed on the synthetic mesh model of a torus (Figure 3) and a cactus (Figure 4). The deployment on the torus is isotropic and the deployment on the cactus is anisotropic. In both cases, the matrix $\widetilde{F}_v$ is constructed from the local curvature of the surface, given by weights $\rho(\mathbf{v}) = 1 + \sqrt{k_{v,1}^2 + k_{v,2}^2}$, the principal curvatures $k_{v,1}$ and $k_{v,2}$, and the principal directions. The principal curvatures and principal directions are estimated from the mesh model following [15], and are shown in Figure 5 on the left and in the center.

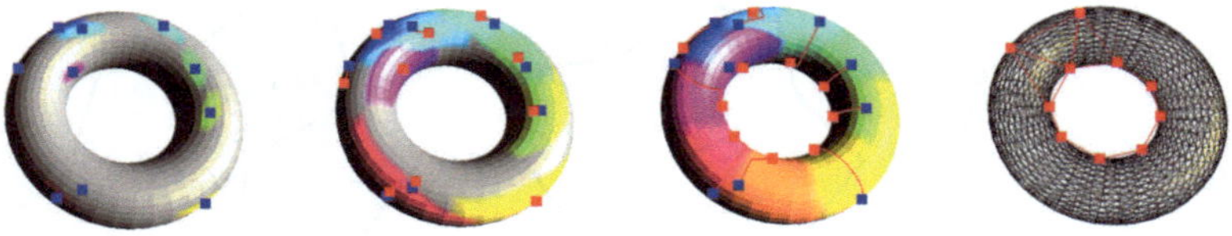

Fig. 3 Adaptive deployment of 10 robots (red) from start (blue), with isotropic curvature metric. The dual Delaunay graph at convergence is shown on the right.

Fig. 4 Adaptive deployment of 50 robots (red) from start (blue), with anisotropic curvature metric. The dual Delaunay graph at convergence is shown on the right.

Fig. 5 Left, center: Principal curvature directions estimated for the torus and cactus models. Right: Close–up of the planned triangle strip path and the vector field generated along the strip for robot control toward the mass center of the Voronoi region.

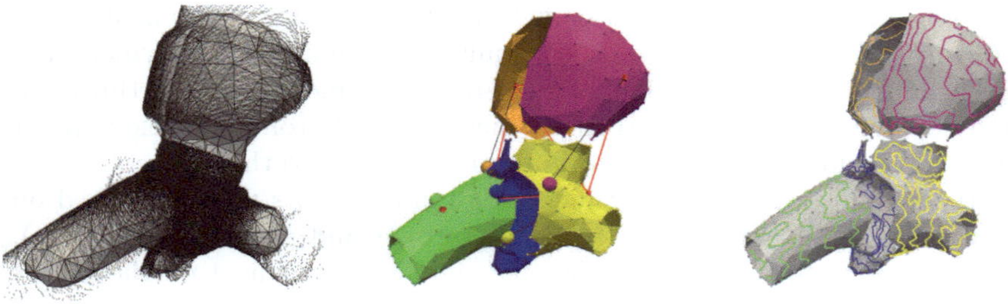

Fig. 6 Left: 3D laser point cloud and reconstructed triangle mesh of a steam chest tube. Center: Partition resulting from the deployment of five robots. Right: Covering triangle strip paths defining a sweeping pattern for each of the Voronoi regions.

In the fourth experiment, the inner surface of an industrial tube structure is covered by five robots. Figure 6 shows one iteration of the hybrid area coverage method. The input is a 3D point cloud, which was recorded with a rotating Hokuyo URG–04LX laser scanner (Figure 6, left). From the point cloud, a triangle mesh is reconstructed. Five robots are deployed over the mesh in a first stage, using the second adaptive surface coverage algorithm (Algorithm 3) with $k = 1$ rounds and uniform weights (Figure 6, center). In a second stage, each robot covers its Voronoi region with a sweeping motion (Figure 6, right). The path planner computes transferring as well as sweeping paths. Paths to transfer a robot to a goal position are planned by an A* search, which generates a triangle strip path on the mesh. The sweeping paths to cover the surface area result from a covering strip planner, which is based on [16]. The robots are controlled by smooth vector fields generated along the strip, using concepts similar to [17]. Figure 5 on the right gives an example for a generated vector field that steers a robot toward the mass center of its Voronoi region.

6 Conclusion

In this paper, we have presented two adaptive surface coverage algorithms to deploy multiple robots on curved surfaces. The algorithms operate on triangle meshes and distribute the robots into discrete regions on the mesh. The size, positions and shapes of the regions are adaptive; they are influenced by a metric tensor field defined on the surface. The adaptive surface coverage algorithms are applied under the hybrid coverage concept for surface area coverage. The algorithms implement the adaptive deployment of the robots. In a second stage, the robots cover the areas of their assigned regions by a sweeping motion. Depending on the environment and the coverage task, this procedure is iterated, which finally leads to complete coverage of the surface. The algorithms are successfully tested in several simulation experiments on different synthetic as well as more realistic mesh models.

In our future work, we plan to test the algorithms on real robots, with added localization and mapping functionalities. Regarding adaptive surface coverage, we will look at the optimization of pairwise partitions and the adaption of regions by user guidance. Concerning hybrid area coverage, it is interesting to further study schemes of robot relocation and redistribution of regions and tasks among robots, particularly for deployment, exploration and search scenarios in heterogeneous robot teams.

References

1. Cortés, J., Martínez, S., Karatas, T., Bullo, F.: Coverage Control for Mobile Sensing Networks. IEEE Transactions on Robotics and Automation 20(2), 243–255 (2004)

2. Meguerdichian, S., Koushanfar, F., Potkonjak, M., Srivastava, M.B.: Coverage Problems in Wireless Ad-hoc Sensor Networks. In: Proc. of the 20th Annual Joint Conf. of the IEEE Computer and Communications Societies (INFOCOM), pp. 1380–1387 (2001)
3. Pavone, M., Savla, K., Frazzoli, E.: Sharing the Load. IEEE Robotics and Automation Magazine 16(2), 52–61 (2009)
4. Smith, S.L., Schwager, M., Rus, D.: Persistent Robotic Tasks: Monitoring and Sweeping in Changing Environments. IEEE Transactions on Robotics 28(2), 410–426 (2012)
5. Atkar, P.N., Choset, H., Rizzi, A.A., Acar, E.U.: Exact Cellular Decomposition of Closed Orientable Surfaces Embedded in $\mathfrak{R}^3$. In: Proc. of the Int. Conf. on Robotics and Automation (ICRA), pp. 699–704 (2001)
6. Breitenmoser, A., Tâche, F., Caprari, G., Siegwart, R., Moser, R.: MagneBike—Toward Multi Climbing Robots for Power Plant Inspection. In: Proc. of the 9th Int. Conf. on Autonomous Agents and Multiagent Systems, AAMAS (2010)
7. Breitenmoser, A., Metzger, J.-C., Siegwart, R., Rus, D.: Distributed Coverage Control on Surfaces in 3D Space. In: Proc. of the IEEE/RSJ Int. Conf. on Intelligent Robots and Systems (IROS), pp. 5569–5576 (2010)
8. Du, Q., Wang, D.: Anisotropic Centroidal Voronoi Tessellations and Their Applications. SIAM Journal on Scientific Computing 26, 737–761 (2005)
9. Bhattacharya, S., Ghrist, R., Kumar, V.: Multi–Robot Coverage and Exploration in Non–Euclidean Metric Spaces. In: Proc. of the 10th Int. Workshop on the Algorithmic Foundations of Robotics, WAFR (2012)
10. Durham, J.W., Carli, R., Frasca, P., Bullo, F.: Discrete Partitioning and Coverage Control for Gossiping Robots. IEEE Transactions on Robotics 28(2), 364–378 (2012)
11. Yun, S., Rus, D.: Distributed Coverage with Mobile Robots on a Graph: Locational Optimization. In: Proc. of the IEEE Int. Conf. on Robotics and Automation (ICRA), pp. 634–641 (2012)
12. Bougleux, S., Peyré, G., Cohen, L.: Anisotropic Geodesics for Perceptual Grouping and Domain Meshing. In: Forsyth, D., Torr, P., Zisserman, A. (eds.) ECCV 2008, Part II. LNCS, vol. 5303, pp. 129–142. Springer, Heidelberg (2008)
13. Valette, S., Chassery, J.-M., Prost, R.: Generic Remeshing of 3D Triangular Meshes with Metric–Dependent Discrete Voronoi Diagrams. IEEE Transactions on Visualization and Computer Graphics 14(2), 369–381 (2008)
14. Gusrialdi, A., Hirche, S., Hatanaka, T., Fujita, M.: Voronoi Based Coverage Control with Anisotropic Sensors. In: Proc. of the Am. Control Conf (ACC), pp. 736–741 (2008)
15. Meyer, M., Desbrun, M., Schröder, P., Barr, A.H.: Discrete Differential–Geometry Operators for Triangulated 2–Manifolds. In: Proc. of VisMath, pp. 35–57 (2002)
16. Gopi, M.: Controllable Single–Strip Generation for Triangulated Surfaces. In: Pacific Conf. on Computer Graphics and Applications, pp. 61–69 (2004)
17. Pereira, G.A.S., Pimenta, L.C.A., Fonseca, A.R., de, L., Corrêa, Q., Mesquita, R.C., Chaimowicz, L., de Almeida, D.S.C., Campos, M.F.M.: Robot Navigation in Multi–Terrain Outdoor Environments. Int. Journal of Robotic Research 28(6), 685–700 (2009)

Nonlinear Cyclic Pursuit Based Cooperative Target Monitoring

Sangeeta Daingade and Arpita Sinha

Abstract. This paper presents a nonlinear cyclic pursuit based target monitoring strategy for a group of autonomous vehicles. The vehicles are modeled as unicycles and are assumed to be heterogeneous. Each vehicle follows its next neighbor as well as the target. Detailed analysis is done for a stationary target and the effectiveness of the proposed strategy with a moving target is shown in simulation. At equilibrium the vehicles capture the target and move along concentric circles in a rigid polygonal formation around the target with equal angular speeds. A necessary condition for the existence of equilibrium formation is derived. Local stability analysis is carried out for homogeneous agents. Simulation results illustrate the objective of the proposed method and demonstrates the derived results.

1 Introduction

In various military and civil applications such as surveillance, security systems, space and underwater exploration, it is often required to monitor a target continuously from all the directions. In such situations cooperative target tracking would be an attractive solution rather than employing a single, intelligent and sophisticated vehicle. In this paper we address the cooperative control problem for monitoring a target with multiple autonomous vehicles. In cooperative target monitoring, the objective is to coordinate the motion of vehicles in such a way that the vehicles reach the desired relative positions and orientations with respect to the target and keep following the target while maintaining the formation.

The target tracking strategies discussed in [4], [10], [7] assumes linear models for each vehicle. These strategies do not take into account the kinematic constraints of the vehicles. In [14], the authors have studied the problem of vision based target

Sangeeta Daingade · Arpita Sinha
Systems and Control Engineering, Indian Institute of Technology Bombay, India
e-mail: {sangeetad,asinha}@sc.iitb.ac.in

M.A. Hsieh and G. Chirikjian (eds.), *Distributed Autonomous Robotic Systems*,
Springer Tracts in Advanced Robotics 104,
DOI: 10.1007/978-3-642-55146-8_2,

tracking among a group of ground robots where it is assumed that the robots can measure the target's position, velocity, and acceleration. They have developed control laws for both single-integrator and double-integrator type robot models.

The unicycle model closely resembles the dynamics of mobile robots and unmanned aerial vehicles (UAVs). The target tracking strategies discussed in [16], [9], [8] , [5], [11], [13], [12] are based on the unicycle model for the vehicles and assumes that all the agents are homogeneous. Switching control law is designed in [16] to track the center of mass of the agents. The agents follow piecewise linear trajectory. In [9], [8] the authors assume all to all communication topology and have presented analysis with only three agents. The strategy discussed in [5] considers a scenario where limited sensing capability of the agents is taken into account. The authors have shown that at equilibrium the agents get distributed around the target in different platoons along the same circle, although the agents are not uniformly distributed. The splay state configuration introduced in [11] enables representation of the equilibrium state of the agents tracking a moving target. The control law proposed by the authors requires knowledge of the derivative of the heading angle which might not always be known or computed with precision.

Our work is based on cyclic pursuit which is a simple strategy derived from the behavior of social insects. In cyclic pursuit, agent i follows agent $i+1$ modulo n. This strategy can be used to obtain various behaviors like rendezvous, motion in formation, target capturing etc. Bruckstein et al. [3] modeled behaviors of ants, crickets and frogs with continuous and discrete cyclic pursuit laws. Stability and convergence of group of ants in linear pursuit are described in [2]. Marshall et al. [15] studied the formations of multivehicle system under linear as well as nonlinear cyclic pursuit. They have analyzed the equilibrium and stability of these formations in case of identical agents. In [18] and [19], Sinha et al. studied generalization of the linear cyclic pursuit and nonlinear cyclic pursuit respectively. The authors have derived a necessary condition for equilibrium formation of heterogeneous agents. Cyclic pursuit based formation control strategies discussed in [15] and [19], deal with formations about a point which cannot be specified a priori. In order to enable enclosure of the target, we should achieve formations about a specific point (target). Rattan et al. [17] proposed a Implicit Leader Cyclic Pursuit (ILCP) law for achieving formations about a given goal point and have used it for rendezvous of multiple vehicles.

We propose a target tracking strategy based on nonlinear cyclic pursuit strategy where each agent needs the position information from one of it's neighbor. This is a type of cyclic pursuit strategy which is suitably modified such that nonholonomic agents can uniformly enclose the target. The agents will move on the concentric circles with the target at the center.The novelty of the work lies in deriving a simple control strategy which can be easily measured and computed. In this paper we study the case when the target is stationary. The extension of this work for moving target is underway and we present a simulation result to show that this strategy is equally applicable for moving targets. The main contributions of the paper are:

- Cyclic pursuit based decentralized target tracking strategy for nonholonomic, heterogeneous agents,

- Necessary conditions for achieving a formation about a stationary target, and
- Stability analysis of the equilibrium formations when the agents are identical.

This paper is organized as follows. Analysis of the proposed strategy begins with the modeling of the system in Section 2 followed by possible equilibrium formations in Section 3. In Section 4, we derive the necessary conditions for equilibrium. Then we discuss a special case in Section 5 where the agents are identical followed by the stability analysis of the proposed control law in Section 6. Simulation results are presented in Section 7 and conclusions and future research directions are discussed in Section 8.

2 Modeling of System

Consider a group of n agents employed to track a target. The kinematics of each agent with a single nonholonomic constraint can be modeled as:

$$\dot{x}_i = V_i \cos h_i; \qquad \dot{y}_i = V_i \sin h_i; \qquad \dot{h}_i = \frac{a_i}{overV_i}, \tag{1}$$

where $P_i \equiv [x_i, y_i]^T$ represents the position of agent i and h_i represents the heading angle of the agent i with respect to a global reference frame. V_i and a_i represent the linear speed and lateral acceleration of agent i respectively. Equation (1) can represent a point mass model of a UAV flying at a fixed altitude or a point mass model of a wheeled robot on a plane. We use a generic term “agent ” to represent the aerial or ground vehicle. We assume that agent i is moving with linear speed V_i which is constant over time. Therefore, the motion of the agent i is controlled using the lateral acceleration, a_i.

Our objective is to enclose the target with n agents. It is assumed that each agent i has the information about the target position and $i+1^{th}$ agent's position. Consider the target to be located at point P as shown in Fig. 1. We modify the classical cyclic pursuit law for enclosing the target such that agent i, positioned at P_i, follows not only $i+1^{th}$ agent at P_{i+1} but also the target at P. Let ρ be a constant which decides the weight agent i gives to the target position P, over the position of the agent $i+1$, P_{i+1}. This weighting scheme is mathematically equivalent to following a virtual leader located at the point P'_{i+1} which is a convex combination of P and P_{i+1}. The point P'_{i+1} is calculated as: $P'_{i+1} = \rho\ P_{i+1} + (1-\rho)\ P$, where $0 < \rho < 1$.

Since we are considering a stationary target, we assume a target centric reference frame and define the following variables (refer to Fig. 1): r_{ig} – Distance between i^{th} agent and the target; r_i – Distance between i^{th} agent and $(i+1)^{th}$ agent; r'_i – Distance between i^{th} agent and virtual leader at P'_{i+1}; f_i – angle made by the vector r_{ig} with respect to the Ref.; ϕ_i – angle between the heading and the line of sight (LOS) $P_iP'_{i+1}$ of agent i. We define the control input to the ith agent, that is, the lateral acceleration a_i, as

$$a_i = k_i \phi_i, \tag{2}$$

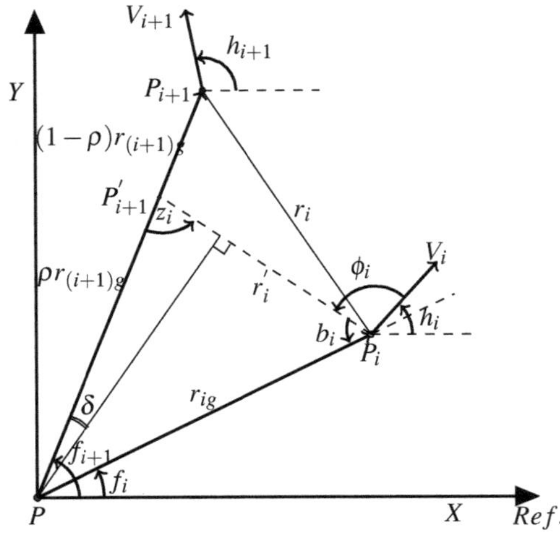

Fig. 1 Positions of the vehicles in a target centric frame

where $k_i > 0$ is the controller gain. We assume $0 \leq \phi_i < 2\pi$ for all time, $t \geq 0$. This condition ensures that the agents always rotate in counter clockwise direction. Let ω_i be angular speed of agent i and $f_i^{i+1} = f_{i+1} - f_i$. Let us define the states of the system as $\mathbf{x}_{i(1)} = r_{ig}$, $\mathbf{x}_{i(2)} = f_{i+1} - f_i$ and $\mathbf{x}_{i(3)} = h_i - f_i$ for $i = 1, 2, \cdots n$. The value of $\dot{r}_{ig} = V_i \cos(h_i - f_i)$ and $\dot{f}_i = \frac{V_i \sin(h_i - f_i)}{r_{ig}}$. Now the kinematics given by (1) can be re-written in the target centric reference frame as,

$$\dot{\mathbf{x}}_{i(1)} = V_i \cos(\mathbf{x}_{i(3)}) \tag{3a}$$

$$\dot{\mathbf{x}}_{i(2)} = \frac{V_{i+1} \sin(\mathbf{x}_{i+1(3)})}{\mathbf{x}_{i+1(1)}} - \frac{V_i \sin(\mathbf{x}_{i(3)})}{\mathbf{x}_{i(1)}} \tag{3b}$$

$$\dot{\mathbf{x}}_{i(3)} = \frac{k_i \phi_i}{V_i} - \frac{V_i \sin(\mathbf{x}_{i(3)})}{\mathbf{x}_{i(1)}}. \tag{3c}$$

Equation (3) gives the kinematics of the ith agent. In the subsequent sections all the analysis are based on this model.

Note 1. During implementation, the agents will have a limit on the maximum lateral acceleration a_{max}, that is, $|a_i| \leq a_{max} \forall i$. We take into account this constraint by putting a bound on the value of k_i as $k_i \leq k_{max}$, where $k_{max} = \frac{a_{max}}{2\pi}$.

3 Formation at Equilibrium

In this section we study the asymptotic behavior of agents under the control law (2).

Theorem 1. *Consider n agents with kinematics (3). At equilibrium the agents move in concentric circles, with (i) the target at the center of concentric circles and (ii) equal angular velocities.*

Proof. At equilibrium $\dot{\mathbf{x}}_{i(j)} = 0$ for $i = 1,...,n$ and $j = 1,2,3$, which implies, $\dot{r}_{ig} = 0$, $\dot{f}_i^{i+1} = 0$ and $\dot{h}_i - \dot{f}_i = 0$. So the distance between the target and agent i (for all i) remains constant at equilibrium. Using (3a) we can write, $h_i - f_i = (2m+1)\frac{\pi}{2}$, where $m = 0, \pm1, \pm2, \cdots$. From (3c) we can write $\frac{k_i\phi_i}{V_i} = \frac{V_i \sin(h_i - f_i)}{r_{ig}} = \pm\frac{V_i}{r_{ig}}$. Since $k_i > 0$, $V_i > 0$, $0 \leq \phi_i < 2\pi$ and $r_{ig} \geq 0$, we get

$$\frac{k_i\phi_i}{V_i} = \frac{V_i}{r_{ig}} \tag{4}$$

and therefore, $m = 0, \pm2, \pm4, \cdots$. Assuming $h_i \in [0, 2\pi)$ and $f_i \in [0, 2\pi)$, we get $(h_i - f_i) \in (-2\pi, 2\pi)$. Therefore $m = 0$ or $m = -2$, which implies the same angle. Therefore

$$h_i - f_i = \frac{\pi}{2}. \tag{5}$$

From (2) and (4), $a_i = \frac{V_i^2}{r_{ig}}$. Since V_i and r_{ig} are constant, a_i is constant for all i. Therefore all the agents move along a circular path with the target at its center and radius r_{ig}. This proves the first part of the theorem. The the angular velocity of agent i can be calculated as, $\omega_i = \frac{a_i}{V_i} = \frac{V_i}{r_{ig}}$. From equation (3b) and (5), we can write

$$\frac{V_i}{r_{ig}} = \frac{V_{i+1}}{r_{(i+1)g}} \tag{6}$$

Using (6), we conclude that for all i, $\omega_i = \omega_{i+1}$. Therefore, all the agents move around the target in concentric circles with equal angular speed. ∎

Corollary 1. *At equilibrium the agents with kinematics (3), form a rigid polygon that rotates about the target.*

Proof. The proof follows directly from the Theorem 1.

4 Conditions for the Existence of Equilibrium

In this section we derive necessary conditions for the existence of equilibrium using geometrical and trigonometric relations. Let the radius of the circle traversed by the first agent at equilibrium be $r_{1g} = R_1$. Using (6), we can write, $r_{ig} = \frac{V_i}{V_1}R_1$ for all i. So from (4) we write,

$$\phi_i = \frac{V_i V_1}{k_i R_1} \tag{7}$$

Consider Fig. 1and let $\angle PP_iP'_{i+1} = b_i$ and $\angle PP'_{i+1}P_i = z_i$. So $\phi_i + b_i + (h_i - f_i) = \pi$ and $b_i + z_i + f_i^{i+1} = \pi$. At equilibrium $\phi_i + b_i = \frac{\pi}{2}$ and $f_i^{i+1} = \frac{\pi}{2} + \phi_i - z_i$. We assume that the angle is positive if measured counterclockwise and negative if clockwise. As the agents are in cyclic pursuit we can write,

$$\sum_{\substack{i=1 \\ (\text{mod } n)}}^{n} (f_i^{i+1}) = 2\pi d, \qquad d = \pm 1, \pm 2, \cdots. \tag{8}$$

Applying the sine rule, from $\triangle PP'_{i+1}P_i$ and (6) we can calculate, $\sin(z_i) = \frac{V_i}{\rho V_{i+1}} \cos(\phi_i)$. Let $\mathscr{X}$ be the set of agents defined as

$$\mathscr{X} = \{i \in \{1, \cdots, n\} | \rho < \frac{V_i}{V_{i+1}}\}.$$

Claim: *The set $\mathscr{X} \neq \emptyset$.*

Proof: We prove it by contradiction. Let us assume that $\mathscr{X} = \emptyset$. Then $\rho V_{i+1} \geq V_i$ for all $i = 1, \cdots, n$, which implies, $\rho^n V_1 \geq V_1$. This can not be true as $0 < \rho < 1$, which leads to a contradiction. So there exits at least one agent in the set $\mathscr{X}$. ■

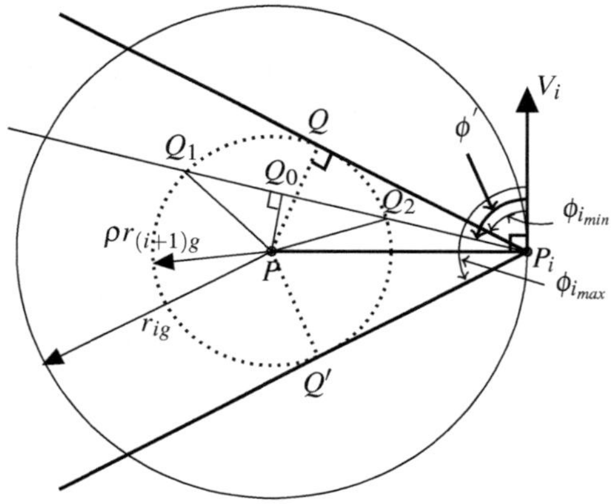

Fig. 2 Formation of agents in set $\mathscr{X}$

For the agents in set $\mathscr{X}$, from (6), $\rho r_{(i+1)g} < r_{ig}$. Now consider an agent $i \in \mathscr{X}$ at equilibrium as shown in Fig. 2. The point P represents the target. At equilibrium the agent i moves along the circle which has radius r_{ig} and center P. Let the agent i be at the point P_i. For a given ρ, the virtual leader of agent $i \in \mathscr{X}$ will lie on the circle of radius $\rho r_{(i+1)g}$ which is less than r_{ig}. Note that $r_{(i+1)g}$ can be greater or less than r_{ig}.

Let P_iQ and P_iQ' be the tangents drawn from point P_i to the circle of radius $\rho r_{(i+1)g}$ and $\phi_{i_{min}}$ and $\phi_{i_{max}}$ are the angles that the velocity vector of ith agent makes with P_iQ and P_iQ'. Then $\phi_{i_{min}} + \phi_{i_{max}} = \pi$. At equilibrium ϕ_i must satisfy

$$0 < \phi_{i_{min}} \leq \phi_i \leq \phi_{i_{max}} < \pi. \tag{9}$$

When $\phi_i = \phi_{i_{min}}$, the virtual leader of i will be at Q and therefore $f_i^{i+1} = \phi_i$. Similarly, when $\phi_i = \phi_{i_{max}}$, $f_i^{i+1} = \pi + \phi_i$. When $\phi_i = \phi'$ which is within the bounds specified by (9), there are two possible locations where the virtual leader can lie (Q_1 and Q_2). So z_i as defined in Fig. 1 can take two values. The two values of z_i are $z_{i_a} = \angle PQ_1P_i$ and $z_{i_b} = \angle PQ_2P_i$, which can be calculated as

$$z_{i_a} = \sin^{-1}\left(\frac{V_i}{\rho V_{i+1}}\cos(\phi_i)\right) \tag{10a}$$

$$z_{i_b} = \pi - \sin^{-1}\left(\frac{V_i}{\rho V_{i+1}}\cos(\phi_i)\right) \tag{10b}$$

Similarly, (f_i^{i+1}) can take two values. At equilibrium RHS of (10) should be real. So the condition for the existence of equilibrium can be stated as:

Theorem 2. *Consider n agent system with kinematics (3). Given ρ the necessary condition for the equilibrium is*

$$\max_{i\in\mathcal{X}} \eta_i \leq \min_{j\in\mathcal{X}} \mu_j, \tag{11}$$

$$\eta_i = \left(\cos^{-1}\left(\frac{\rho V_{i+1}}{V_i}\right)\right)\frac{k_i}{V_iV_1} \quad \text{and} \quad \mu_j = \left(\pi - \cos^{-1}\left(\frac{\rho V_{j+1}}{V_j}\right)\right)\frac{k_j}{V_jV_1}. \tag{12}$$

Proof. Equation (10) holds if, for all i,

$$|\cos(\phi_i)| \leq \frac{\rho V_{i+1}}{V_i} \tag{13}$$

For $i \notin \mathcal{X}$, i.e. if $\rho V_{i+1} \geq V_i$, equation (13) is always true. However, if $i \in \mathcal{X}$, then $p\pi + \cos^{-1}\left(\frac{\rho V_{i+1}}{V_i}\right) \leq \phi_i \leq (p+1)\pi - \cos^{-1}\left(\frac{\rho V_{i+1}}{V_i}\right)$, where $p = 0, \pm 1, \pm 2, \cdots$. From (9), p can take only one value, $p = 0$. Substituting the value of ϕ_i from (7) we get $\frac{k_i}{V_iV_1}\left(\cos^{-1}\left(\frac{\rho V_{i+1}}{V_i}\right)\right) \leq \frac{1}{R_1} \leq \frac{k_i}{V_iV_1}\left(\pi - \cos^{-1}\left(\frac{\rho V_{i+1}}{V_i}\right)\right)$. Let

$$\mathfrak{R}_i = \{R_1 : \eta_i \leq 1/R_1 \leq \mu_i\ ,\ i \in \mathcal{X}\}, \tag{14}$$

where η_i and μ_i are given by (12). For each i, (14) gives the range of values R_1 can take. If $\bigcap_{i\in\mathcal{X}} \mathfrak{R}_i \neq \emptyset$ then, there exists some R_1, such that (13) is satisfied for all i. If $\bigcap_{i\in\mathcal{X}} \mathfrak{R}_i \neq \emptyset$ has to be true, then (11) must hold. Thus the necessary condition for equilibrium is given by (11). ∎

Note 2. For the agents not in the set $\mathcal{X}$, that is, $\rho V_{i+1} \geq V_i$, from (6), $\rho r_{(i+1)g} \geq r_{ig}$. When $\rho r_{(i+1)g} = r_{ig}$, the virtual leader lies on the circle of radius r_{ig} and ϕ_i can take value in $[0, \pi]$. When $\rho r_{(i+1)g} > r_{ig}$, the virtual leader lies outside the circle of radius r_{ig} and $\phi_i \in [0, 2\pi)$. In both the cases, given a ϕ_i, there exists an unique point where virtual leader can lie and hence the position of the next agent $i+1$ is unique. In these cases the value of z_i lies in $[0, \pi/2]$ and is given by (10a).

Note 3. When Theorem 2 is satisfied, (8) may not be satisfied. Therefore, to achieve a formation one requires both Theorem 2 and (8) to be true simultaneously.

5 Formation in Case of Homogeneous Agents

Consider a special case when all the agents are identical, that is, $V_i = V$ and $k_i = k$ for all i. Then at equilibrium, from (6) we have $r_{ig} = R$ for all i. So the agents move on the same circle with the target at its center. From (7),

$$\phi_i = \frac{V^2}{kR} = \phi_{eq} \tag{15}$$

for all i. Also $\phi_{i_{max}} = \phi_{max}$ and $\phi_{i_{min}} = \phi_{min}$ for all i. The value of ϕ_{eq} can be calculated as

$$\cos(\phi_{eq}) = \frac{\rho \sin(f_i^{i+1})}{\sqrt{1+\rho^2 - 2\rho\cos(f_i^{i+1})}}. \tag{16}$$

From $\triangle PQP_i$ (Fig. 2) we can write,

$$\rho = \cos(\phi_{min}) \tag{17}$$

We observe that $\rho V < V$ and therefore all the agents belong to the set $\mathscr{X}$. Then the necessary condition in Theorem 2 can be expressed as $\rho \geq \left|\cos(\phi_{eq})\right|$.

Consider Fig. 2 and assume $\phi_{eq} = \phi'$. Then $\angle Q_0PP_i = \phi_{eq}$. As long as (9) is satisfied, Theorem 2 is trivially satisfied. Let $\angle Q_0PQ_1 = \angle Q_0PQ_2 = \delta$ and

$$f_a = \angle P_iPQ_1 = \phi_{eq} + \delta, \tag{18}$$

$$f_b = \angle P_iPQ_2 = \phi_{eq} - \delta. \tag{19}$$

The value of f_i^{i+1} can be either f_a or f_b. Note that, from (10), $z_{i_a} + \delta = \pi/2$ and $z_{i_b} - \delta = \pi/2$. Also from $\triangle Q_1PP_i$

$$\cos(\phi_{eq}) = \rho\cos(\delta). \tag{20}$$

Let $M_a = \{i : f_i^{i+1} = f_a\}$ and $M_b = \{i : f_i^{i+1} = f_b\}$. Then $|M_a| + |M_b| = n$. We have three possibilities:

Case 1: When M_a is empty, that is, $f_i^{i+1} = f_b\ \forall i$, from (19) , we can write $\cos(f_b) = \cos(\phi_{eq} - \delta)$. Using (16) and (20), $\left(\rho\cos(f_b) - 1\right)\left(\rho - \cos(f_b)\right) = 0$. Since $0 < \rho < 1$, $\rho\cos(f_b) \neq 1$, so $\rho = \cos(f_b)$. From (17), $f_b = \phi_{min} = \pi - \phi_{max}$. Then from Fig. 2, we observe that the agent $i+1$ will be at Q or Q' and thus we have either $\phi_{eq} = \phi_{min}$ or $\phi_{eq} = \phi_{max}$.

Case 2: When M_b is empty, that is, $f_i^{i+1} = f_a,\ \forall i$, following similar procedure as in the previous case, $\cos(f_a) = \cos(\phi_{eq} + \delta)$ needs to be true. Using (16) and (20) and

simplifying, we get $\rho\sin^2(f_a) - (1-\rho\cos(f_a))(\rho-\cos(f_a)) = \cos(f_a)(1+\rho^2 - 2\rho\cos(f_a))$ which is satisfied for all ρ.
Case 3: When both M_a and M_b are non-empty, following similar procedure, it is found that the equilibrium formation is possible if $\cos(\delta) = \rho\cos\left(2\pi\frac{d}{n} - \frac{m}{n}\delta\right)$ is true for the values of δ satisfying (20) and $m = |M_a| - |M_b|$.

For Cases 1 and 2, we have $f_i^{i+1} = \bar{f} \quad \forall i$, where $\bar{f}$ is some constant. However, in Case 3, f_i^{i+1} is not same for all i. A set of 40000 Monte Carlo simulations were run for different values of n and ρ. The value of n was varied from 3 to 11 and ρ from 0.1 to 0.9 in the steps of 0.1. We selected 500 random initial conditions for each pair of n and ρ. It has been observed that, formation at equilibrium was always a regular polygon, which mean case 3 has never occurred. Thus, in this paper, we will concentrate on Cases 1 and 2 only and analyze the conditions for achieving a stable formation uniformly distributed around the target.

In Case 1 and Case 2,

$$\phi_{min} \le \bar{f} \le \pi + \phi_{max} \tag{21}$$

with Case 1 corresponding to equalities in (21). When $f_i^{i+1} = \bar{f} \quad \forall i$, the agents will be uniformly distributed around the target. Then from (8) we get $\bar{f} = 2\pi\frac{d}{n}$. From (15), the distance between an agent and the target is, $R = \frac{V^2}{k\phi_{eq}}$. Then the distance between i^{th} and $(i+1)^{th}$ agent will be $\bar{r} = 2R\sin\left(\frac{\bar{f}}{2}\right) = 2\frac{V^2}{k\phi_{eq}}\sin(\pi\frac{d}{n})$. So the states of the system (3) at equilibrium are,

$$\mathbf{x}_{i_{eq}(1)} = \frac{V^2}{k\phi_{eq}}; \qquad \mathbf{x}_{i_{eq}(2)} = 2\pi\frac{d}{n}; \qquad \mathbf{x}_{i_{eq}(3)} = \frac{\pi}{2}. \tag{22}$$

At equilibrium, the agents arrange themselves in a regular formation around the target. This regular formation of n agents can be described by a regular polygon (n,d), where $d \in \{1,2,...,n-1\}$. This d is reflected in equilibrium state $\mathbf{x}_{i_{eq}(2)}$.

Note 4. When $\phi_{eq} = \phi'$ (as shown in Fig. 2), from (17) and (21), $\rho = \cos(\phi_{min}) \ge \cos(\bar{f}) = \cos(2\pi\frac{d}{n})$, as such, $\rho \ge \cos(2\pi\frac{d}{n})$. Let us replace $\frac{d}{n}$ by a continuous variable $q \in (0,1)$. We can plot $\rho = \cos(2\pi q)$ as shown in Fig. 3. It can be seen that, for a given n, Region I corresponds to Case 2 (when M_b is empty) and the boundary of the Region I corresponds to Case 1 (when M_a is empty).

Since $\rho \in (0,1)$, we ignore the dotted portion of the curve on the left half of Fig. 3.

6 Stability Analysis for Homogeneous Agents

We study the stability of the equilibrium formation of homogeneous agents. The equilibrium points are given by (22). Since d can take $(n-1)$ values, there are $(n-1)$

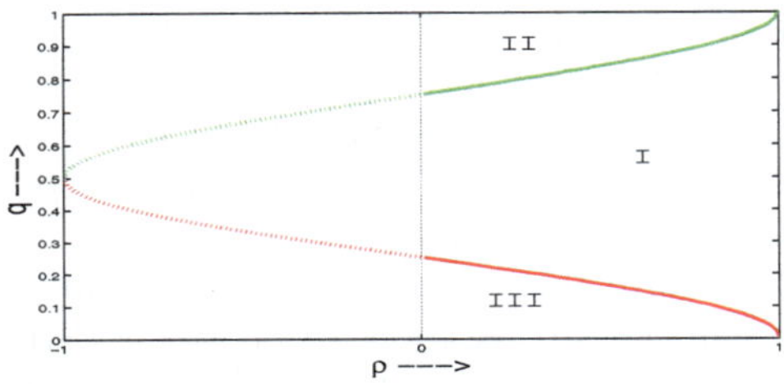

Fig. 3 q versus ρ plot when $\rho = \cos(2\pi q)$

formations possible for a given V, k and ρ. Linearizing (3) about the equilibrium point (22), we get $\dot{\hat{\mathbf{x}}}_i = A\hat{\mathbf{x}}_i + B\hat{\mathbf{x}}_{i+1}$, where

$$A = \begin{bmatrix} 0 & 0 & -V \\ \frac{V}{R^2} & 0 & 0 \\ a_{31} & a_{32} & \frac{-k}{V} \end{bmatrix} \qquad B = \begin{bmatrix} 0 & 0\,0 \\ \frac{-V}{R^2} & 0\,0 \\ \frac{-k\rho\sin(2\pi\frac{d}{n})}{VR\left(1+\rho^2-2\rho\cos(2\pi\frac{d}{n})\right)} & 0\,0 \end{bmatrix},$$

where $a_{31} = \frac{k\rho\sin(2\pi\frac{d}{n})}{VR\left(1+\rho^2-2\rho\cos(2\pi\frac{d}{n})\right)} + \frac{V}{R^2}$ and $a_{32} = \frac{k\rho\left(\rho-\cos(2\pi\frac{d}{n})\right)}{V\left(1+\rho^2-2\rho\cos(2\pi\frac{d}{n})\right)}$. So the system of n vehicles can be written as $\dot{\hat{X}} = \hat{A}\hat{X}$, where $\hat{X} = [\hat{\mathbf{x}}_1, \hat{\mathbf{x}}_2, \cdots, \hat{\mathbf{x}}_n]^T$ and $\hat{A}$ is a circulant matrix given by $\hat{A} = \text{circ}\left(A \; B \; 0_{3\times 3} \; \cdots \; 0_{3\times 3}\right)$. The stability of the formation depends on the eigenvalues of $\hat{A}$.

Theorem 3. *Consider n agents with kinematics (3), moving with unit linear velocity. For a given value of ρ, the equilibrium points given by (22) are locally asymptotically stable if*

$$q_l < \frac{d}{n} < q_u, \quad \textit{where} \tag{23}$$

$$q_l = -0.17\rho + 0.254$$
$$q_u = \begin{cases} 0.17\rho + 0.75 & \textit{for } \rho \leq 0.5305 \\ \frac{2}{\pi}\cos^{-1}(\sqrt{2\rho - 1}) & \textit{for } \rho > 0.5305. \end{cases} \tag{24}$$

Proof. The stability of a formation depends on the eigenvalues of the circulant matrix $\hat{A}$. We can find the eigenvalues of circulant matrix (as given in [6]) using the representer polynomial $P(z)$ of circulant matrix $\hat{A}$. The representer polynomial of a circulant matrix $C = $ circ $(C_1, C_2, \cdots C_n)$ is defined as $P(z) = \sum_{i=1}^{n} C_i(z^{i-1})$. So the representer polynomial of $\hat{A}$ will be, $P(z) = A + zB$. Let $\varsigma = e^{j\frac{2\pi}{n}}$, where $j = \sqrt{-1}$. The block circulant matrix $\hat{A}$ can be diagonalized using a Fourier matrix F_n given by $\hat{A} = (F_n \otimes I_3)D(F_n \otimes I_3)^\star$, where $(\star)$ indicates conjugate transpose. The diagonal matrix D is given by, $D = \text{diag}(P(1), P(\varsigma), \cdots, P(\varsigma^{n-1}))$. So we can write $D_i = A + \varsigma^{i-1}B$, $i \in \{1, 2, \ldots .n\}$.

The eigenvalues of $\hat{A}$ are same as eigenvalues of D_i, $i = 1, \cdots n$. We can comment about the stability of n vehicle system, by observing the eigenvalues of each block D_i. Assuming $V = 1$, each D_i can be factorized as $D_i = \frac{1}{k}T^{-1}\tilde{D}_i T$, where T =diag$[k, 1, 1]$ and

$$\tilde{D}_i = \begin{bmatrix} 0 & 0 & -1 \\ {\phi_{eq}}^2(1-\varsigma^{i-1}) & 0 & 0 \\ \frac{\phi_{eq}\rho\sin(2\pi\frac{d}{n})(1-\varsigma^{i-1})}{1+\rho^2-2\rho\cos(2\pi\frac{d}{n})} + {\phi_{eq}}^2 & \frac{\rho(\rho-\cos(2\pi\frac{d}{n}))}{1+\rho^2-2\rho\cos(2\pi\frac{d}{n})} & -1 \end{bmatrix}.$$

The spectrum $\sigma(\cdot)$ of D_i and $\tilde{D}_i$ are related as $\sigma(D_i) = \frac{1}{k}\sigma(\tilde{D}_i)$. Since $k > 0$, stability of D_i can be determined from the stability of $\tilde{D}_i$. $\tilde{D}_i$ does not have a term containing gain k. So we can conclude that the stability of (n,d) formation is independent of k, as long as $k > 0$. $\tilde{D}_i$ is a complex matrix, since $\varsigma^{i-1} = e^{2\pi j\frac{i-1}{n}}$ is a complex quantity. We can write $\varsigma^{i-1} = \beta_i + j\xi_i \in \mathbb{C}$, where $\beta_i = \cos(2\pi\frac{i-1}{n})$ and $\xi_i = \sin(2\pi\frac{i-1}{n})$. Then the characteristic polynomial of $\tilde{D}_i$ can be written as $P_{\tilde{D}_i}(\lambda) = \lambda^3 + c_1\lambda^2 + c_2\lambda + c_3$, where $c_1 = 1$, $c_2 = a_2 + jb_2$, $c_3 = a_3 + jb_3$. Let H be the Hermitian matrix corresponding to characteristic polynomial $P_{\tilde{D}_i}(\lambda)$. Then the polynomial $P_{\tilde{D}_i}(\lambda)$ is asymptotically stable if and only if the principal minors h_1,h_2 and h_3 of H are positive [1]. In this case $h_1 = 2$, $h_2 = 4(a_2 - a_3 - {b_2}^2)$ and $h_3 = 8({a_2}^2 a_3 - a_2 b_2 b_3 - 2a_2{a_3}^2 - 3a_3 b_2 b_3 - {b_3}^2 - a_3{b_2}^2 a_2 - {b_2}^3 b_3 + {a_3}^3)$. Here,

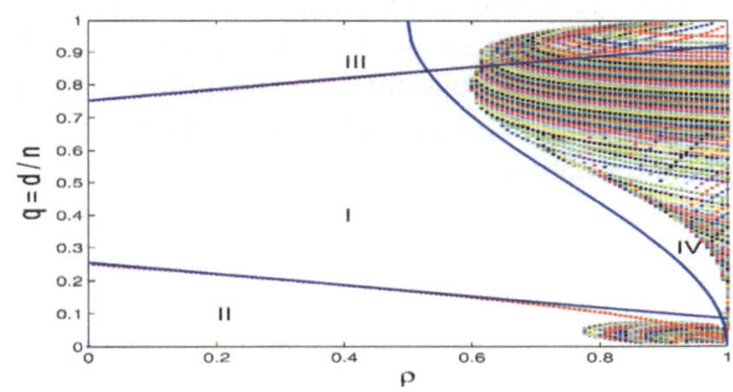

Fig. 4 Stability region for continuous values of q, ρ and $\tilde{\beta}$ (Dotted lines represent locus of points where $h_2 = 0$ or $h_3 = 0$)

h_2 and h_3 depends on $\frac{d}{n}$ and β_i which takes discrete values. To find the range of values of $\frac{d}{n}$ and β_i for which $h_2 > 0$ and $h_3 > 0$, we replace $\frac{d}{n}$ by a continuous variable $q \in (0, 1)$ (as stated in Note 4). Also we replace β_i by a continuous variable $\tilde{\beta} \in [-1, 1]$. For a given $\tilde{\beta}$, we can plot q versus ρ when $h_2 = 0$ and $h_3 = 0$. Figure 4 shows these plots for different values of $\tilde{\beta}$. Let us define

$$S_2 = \{(\rho, q, \tilde{\beta}) : h_2 > 0, \rho \in (0, 1), q \in (0, 1), \tilde{\beta} \in [-1, 1]\}$$
$$S_3 = \{(\rho, q, \tilde{\beta}) : h_3 > 0, \rho \in (0, 1), q \in (0, 1), \tilde{\beta} \in [-1, 1]\}.$$

Then $S = S2 \cap S3$, defines the stability region where both h_2 and h_3 are positive. In Fig. 4, Region I corresponds to S. Region I can be numerically approximated with conservative bounds as

$$q + 0.17\rho - 0.254 = 0 \tag{25}$$

$$q - 0.17\rho - 0.75 = 0 \tag{26}$$

$$\cos^2(\frac{\pi q}{2}) - 2\rho + 1 = 0 \quad \text{for} \quad \rho \geq 0.5. \tag{27}$$

Equations (25) and (26) are the linear approximations of $\rho = \cos(2\pi q)$. The Region I in Fig. 4 is a subset of the Region I of Fig. 3. So for a given ρ, the bounds on q can be defined by (23). Thus we can conclude that for a given ρ and n, if $\frac{d}{n}$ satisfies (23), then the eigenvalues of $\tilde{D}_i$ and hence that of $\hat{A}$ will have negative real part. So the formation will be locally asymptotically stable. ■

Remark: The stability results obtained using linearization technique matches with the analysis done for Case 1 and Case 2 in Section 5.

7 Simulation Results

First we consider five heterogeneous agents which are randomly placed initially and the target is located at the origin $(0,0)$.

Case 1: We consider five agents moving with speed $V = [25,\ 20,\ 15,\ 10,\ 5]$, controller gain $k = 5$and $\rho = 0.8$. In this case $\mathscr{X} = \{1,2,3,4\}$ and $\bigcap \mathfrak{R}_i$ as defined by (14) is an empty set. So the system does not have an equilibrium as the condition in Theorem 2 is violated. Simulation results in Fig. 5-(a) shows that the agents do not settle to rigid polygonal formation.

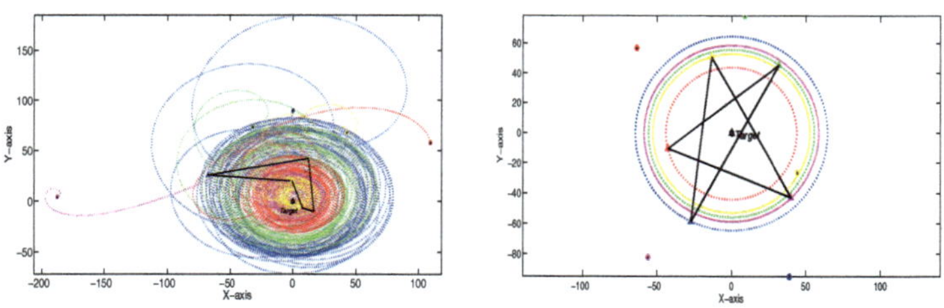

(a) (b)

Fig. 5 Formation of heterogeneous agents (• :- initial position and △ :- final position)

Case 2: Now we modify the speed of the vehicles to $V = [22, 19, 15, 20, 18]$ with controller gain and ρ same as in Case 1. Now we have $\mathscr{X} = \{1,2,4,5\}$. In this case Theorem 2 is satisfied and there exist a range of values of R_1 as $41.48 \leq R_1 \leq 94.23$. However, as stated in Note 2 of Section 4, Equation (8) also needs to be satisfied.

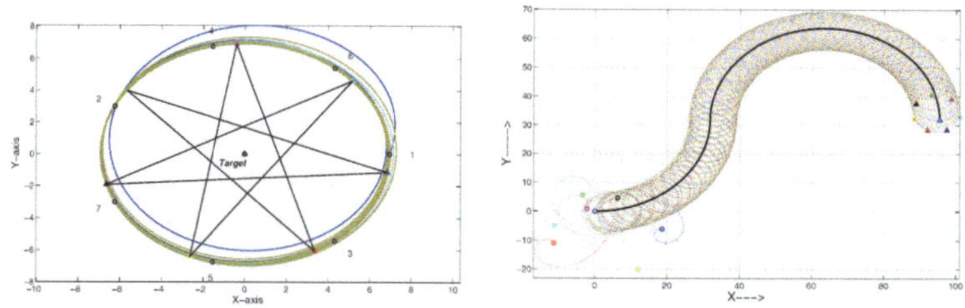

(a) (b)

Fig. 6 Stable formation of seven homogeneous agents (a) Stationary target (b) Moving target ($\bullet$:- initial position and $\triangle$:- final position)

There are two values of $R_1 = \{45.45, 64.52\}$, for which equation (8) is satisfied. Fig. 5-(b) shows that the system of 5 agents settles to $R_1 = 64.52$ with $d = 2$.

Next we consider seven homogeneous agents, all moving with unit speed and having controller gain $k = 0.1$.

Case 3: Here we assume that the target is stationary . The agents are initially in $(7,3)$ formation with $\rho = 0.4$, which corresponds to Region *I* of Fig. 4. At equilibrium the agents stays in $(7,3)$ formation as shown in Fig. 6-(a). The analytical values of inter-agent distance r and target to agent distance R match with the simulation values.

Case 4: In this case we consider a moving target following a sinusoidal path, with a velocity 0.1 unit/sec. The agents start from random initial positions with $\rho = 0.1$. Fig. 6-(b) shows that they get into a $(7,3)$ formation about the target and continue to enclose the target maintaining the formation. This illustrates the potential of the proposed strategy for cooperatively tracking a moving target.

8 Conclusion

The paper has addressed cyclic pursuit based target monitoring using a group of heterogeneous agents modeled as planar kinematic unicycle model moving with constant speed. Mathematical formulation and the analysis are carried out for a stationary target. At equilibrium, the agents move with a polygonal formation around the target with equal angular speed. Necessary condition for the existence of the equilibrium is derived. This leaves us with the future work of deriving sufficient condition for equilibrium formation of heterogeneous agents. In case of homogeneous agents, local stability analysis of the equilibrium formation shows that the parameter ρ plays an important role in the type of formation achieved at the equilibrium. The proposed strategy works well in the case of a moving target. So the next step in this research work would be to analyze moving target tracking scenario. Another direction for future work would be to implement the strategy on a real system or on a Hardware-In-Loop-Simulator (HILS).

References

1. Barnett, S.: Polynomials and linear control systems. Monographs and Textbooks in Pure and Applied Mathematics Series. Marcel Dekker, New York (1983)
2. Bruckstein, A.M.: Why the Ant Trail look so straight and nice. The Mathematical Intelligencer 15(02), 59–62 (1993)
3. Bruckstein, A.M., Cohen, N., Efrat, A.: Ants, crickets and frogs in cyclic pursuit. In: Center for Intelligence Systems. Technical Report 9105, Technion-Israel Institute of Technology, Haifa, Israel (1991)
4. Kobayashi, K., Otsubo, Hosoe, S.: Design of Decentralized Capturing Behavior by Multiple Robots. In: IEEE Workshop on Distributed Intelligent Systems: Collective intelligence and its Applications, Prague, Czech Republic, pp. 463–468 (2006)
5. Ceccarelli, N., Marco, M., Garilli, A., Giannitrapani, A.: Collective circular motion of multi-vehicle systems. Automatica 44, 3025–3035 (2008)
6. Davis, P.J.: Circulant Matrices, 2nd edn., NewYork, Chelsea (1994)
7. Guo, J., Yan, G., Lin, Z.: Local control strategy for moving-target-enclosing under dynamically changing network topology. Systems & Control Letters 59, 654–661 (2010)
8. Klein, D.J., Matlack, C., Morgansen, K.A.: Cooperative target tracking using oscillator models in three dimensions. In: Proceedings of the American Control Conference, New York, NY, USA, pp. 2569–2575 (2007)
9. Klein, D.J., Morgansen, K.A.: Controlled collective motion for mulitvehicle trajectory tracking. In: Proceedings of the American Control Conference, Minneapolis, Minnesota, USA, pp. 5269–5275 (2006)
10. Kim, T.H., Sugie, T.: Cooperative control for target-capturing task based on a cyclic pursuit strategy. Automatica 43, 1426–1431 (2007)
11. Kingston, D., Beard, R.: UAV Splay state configuration for moving targets in wind. In: Pardalos, P.M., Murphey, R., Grundel, D., Hirsch, M.J. (eds.) Advances in Coperative Control and Optimization. LNCIS, vol. 369, pp. 109–128. Springer, Heidelberg (2008)
12. Lan, Y., Lin, Z., Cao, M., Yan, G.: A Distributed Reconfigurable Control Law for Escorting and Patrolling Missions using Teams of Unicycles. In: Proceedings of the 49th IEEE Conference on Decision and Control, Hilton Atlanta Hotel, Atlanta, GA, USA, pp. 5456–5461 (2010)
13. Lan, Y., Yan, G., Lin, Z.: Distributed control of cooperative target enclosing based on reachability and invariance analysis. Systems & Control Letters 59, 381–389 (2010)
14. Ma, L., Hovakimyan, N.: Vision-based cyclic pursuit for cooperative target tracking. In: Proceedings of the American Control Conference, O'Farrell Street, San Francisco, CA, USA, pp. 4616–4621 (2011)
15. Marshall, J.A., Broucke, M.E., Francis, B.A.: Formations of vehicles in cyclic pursuit. IEEE Transaction on Automatic Control 49, 1963–1974 (2004)
16. Paley, D., Leonard, N., Sepulchre, R.: Collective motion: bistability and trajectory tracking. In: Proceedings of the 43rd IEEE Conference on Decision and Control, pp. 1932–1937 (2004)
17. Rattan, G., Ghosh, D.: Nonlinear cyclic pursuit strategies for MAV swarms. In: Technical Report, DRDO-IISc Programme on Advanced Research in Mathematical Engineering, pp. 1–32 (2009)
18. Sinha, A., Ghosh, D.: Generalization of linear cyclic pursuit with application to rendezvous of multiple autonomous agents. IEEE Transactions on Automatic Control 51(11), 1819–1824 (2006)
19. Sinha, A., Ghosh, D.: Generalization of nonlinear cyclic pursuit. Automatica 43, 1954–1960 (2007)

High Resolution Atmospheric Sensing Using UAVs

Bobby Hodgkinson, Doug Lipinski, Liqian Peng, and Kamran Mohseni*

Abstract. A technique to obtain high resolution atmospheric data using small mobile sensors is presented. A fluid based control scheme using smoothed particle hydrodynamics (SPH) is implemented to perform field measurements in a leader-follower arrangement for a team of unmanned aerial vehicles (UAVs) equipped with environmental sensors. A virtual leader is created by using a reduced density SPH particle to guide the unmanned aerial vehicles along a desired path. Simulations using the control scheme demonstrate excellent measurement ability, swarm coherence, and leader following capability for large swarms. A K-means algorithm is used to reduce the measurement error and provide accurate interpolation of the field measurement data. Experimental results are presented which demonstrate the guidance and collision avoidance properties of the control scheme using real UAVs. Readings from the UAV's temperature and humidity sensor suite are used with the K-means algorithm to produce a smooth estimation of the respective distribution fields.

1 Introduction

Several methods of sensing the atmosphere exist and can be classified into two primary groups: *remote sensing* or *in situ*. Well-known systems in the remote sensing group include RADAR, LIDAR, weather satellites, etc. [3, 5, 13, 29]. These systems are typically large (on the order of cubic meters), are generally focused on

Bobby Hodgkinson · Doug Lipinski · Liqian Peng
Department of Mechanical and Aerospace Engineering, Institute for Networked Autonomous Systems, University of Florida, Gainesville, FL
e-mail: {hodgkinson,dmlipinski,liqianpeng}@ufl.edu

Kamran Mohseni
Department of Mechanical and Aerospace Engineering, Department of Electrical and Computer Engineering, Institute for Networked Autonomous Systems, University of Florida, Gainesville, FL
e-mail: mohseni@ufl.edu

* Corresponding author.

M.A. Hsieh and G. Chirikjian (eds.), *Distributed Autonomous Robotic Systems*,
Springer Tracts in Advanced Robotics 104,
DOI: 10.1007/978-3-642-55146-8_3, © Springer-Verlag Berlin Heidelberg 2014

gathering data over very large areas (several kilometers) over a long period time, and are expensive to construct and maintain. The in situ group also contains several well-known systems such as dropsondes, weather balloons, barometers, etc. These systems are typically much smaller than remote sensing systems, and are focused on gathering high accuracy data at specific points in time and space. The data locations can be stationary or time varying but the in situ sensors are categorized by the ability to only provide data at a single location at a given time. Generally speaking, in situ sensors provide more accurate readings at a specific point than remote sensing systems but remote sensing systems are much better suited to give a *snapshot* of a desired measurement in a region of space. The atmospheric sensing community desires a sensor with the accuracy of an in situ device and the ability to construct a large area *pseudo-snapshot* of an environment. By definition, it is not possible for a system of in situ sensors to give a true snapshot, but the ability to gather a large amount of data in a time frame that is shorter than the time scale of the changing environment can be considered a pseudo-snapshot. This is where autonomous robotic aerial systems can find a tremendously beneficial niche. Aircraft with on board sensors are capable of providing high resolution readings at specific locations similar to dropsondes while also adding horizontal and vertical mobility to allow for a wider range pseudo-snapshot over a relatively short period of time.

A large amount of research has been conducted using mobile in situ devices such as dropsondes and radiosondes but the primary drawback of these sensors is that they are typically only able to measure data along a vertical line. Obviously a large number of appropriately spaced sensors are required to obtain a full three dimensional pseudo-snapshot of the environment thus increasing cost of deployment. Furthermore, dropsondes are not able to revisit specific locations and the only way to obtain information on how the environment changes as a function of time is to deploy additional sensors at different times which also increases the cost of deployment. Unmanned aerial vehicles (UAVs) are able to gather information over large areas and have the ability to revisit specific locations at later instances of time thus allowing for information about the time evolution of an environment, making the use of UAVs a viable option in atmospheric sensing. Possibly the most widely known mobile in situ atmospheric sensing system is the Aerosonde UAV [10]. The general concept of the Aerosonde UAV is to equip a small UAV with in situ atmospheric sensors and fly the aircraft in an environment. This concept yields a notable advantage over dropsondes in that the aircraft can maneuver in the horizontal and vertical directions. A sample of the work that has been conducted using similar concepts of the Aerosonde UAV with smaller, lower-cost UAVs can be found in: [6, 7, 31, 32].

The focus of this article is to present a system comprised of multiple small UAVs equipped with a simple atmospheric sensor suite to gather a three dimensional pseudo-snapshot of an environment. A K-means algorithm provides a smooth estimation of the environment using the discrete points obtained by the mobile sensors while also reducing the noise associated with the measurements. Additionally, the objective requires a flexible and easily implementable cooperative control scheme to guide the vehicles and ensure collision avoidance when the vehicles are in close proximity. While many options exist, we have chosen to use a fluid based control implemented with the Smoothed Particle Hydrodynamics (SPH) fluid dynamics scheme due to several desirable characteristics. This scheme treats each vehicle as an individual fluid particle and the control forces are determined by the SPH approximation of the Navier-Stokes equations of fluid motion. This technique ensures collision avoidance and also creates a flexible controller that is computationally reasonable for implementation on resource constrained platforms. While UAVs are used in this demonstration, the SPH control scheme can be applied to ground and underwater robots as well as heterogeneous swarms of robots containing any combination of ground, aerial, and underwater platforms [11].

In the following sections we present an overview of the SPH control scheme, discuss the error reduction technique, and present test results using multiple UAVs demonstrating several highly desirable properties of the SPH control scheme. We also present error reduced results obtained from data gathered by two UAVs flown using the SPH control.

2 SPH Control Scheme

While there are many possible control schemes for use with small UAVs, fluid based control is especially appealing since fluid flows have many properties that a group of vehicles may wish to mimic [11, 15, 17, 18, 26]. Chiefly, fluids exhibit smooth motions and do not penetrate obstacles. In terms of vehicle control, these properties correspond to efficient motion and collision/obstacle avoidance. Additionally, UAVs operate in a fluid environment meaning a fluid based control scheme may allow for easier integration of strong background flows into the vehicle path planning process.

In particular, the smoothed particle hydrodynamics (SPH) discretization has proven to be an effective method of applying the Navier-Stokes equations in a control setting [11]. This Lagrangian technique treats each vehicle as a fluid particle, giving fluid-like motion for vehicle swarms with inherent collision and obstacle avoidance. A more complete discussion of the method is available in a review article by Monaghan [23] or the book by Liu and Liu [19]. Here we present only the aspects of SPH that are used in our cooperative control scheme.

The SPH algorithm is computationally efficient since each vehicle is represented by a single fluid particle. By choosing a compactly supported smoothing kernel for the particles, it is also possible to limit vehicle interactions to a short range. This results in vehicles interacting with only their nearest neighbors (typically no more than six vehicles in 2D). The localized interactions also make long range communication unnecessary. These advantages result in a control algorithm that is simple enough to run in real time using the limited processing capabilities of the robot [30]. Additionally, this is a distributed control scheme that requires no central controller since only local vehicle interactions are used. All the benefits of a distributed control or peer-to-peer control scheme are thereby included as well [12, 25].

The SPH scheme is dependent on choosing a Gaussian-like smoothing kernel which is used to apply fluid properties. We use the cubic spline kernel shown in Figure 1. This kernel is nonzero for $||r|| < 2h$ and defines the interaction range

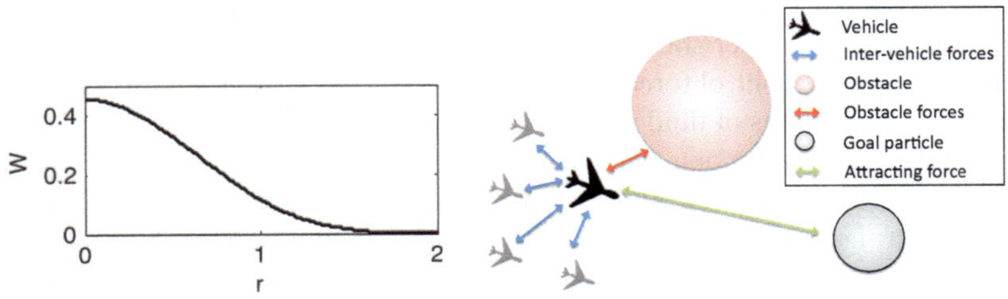

Fig. 1 (Left) The cubic spline smoothing kernel (with $h = 1$) as a function of distance, r, used in this article. (Right) A schematic of the SPH control scheme and the the forces involved.

between particles. Particles interact through pressure and viscous forces that are applied through the smoothing kernel, determining the particle motion. The SPH acceleration is given by

$$\frac{d\mathbf{v}_i}{dt} = -\sum_j m_j \left[\left(\frac{P_i}{\rho_i^2} + \frac{P_j}{\rho_j^2} \right) \nabla_i W(\mathbf{r}_{ij}, h) + \frac{\Pi_{ij}}{||\mathbf{r}_{ij}||} \frac{\partial W(r_{ij}, h)}{\partial ||\mathbf{r}_{ij}||} \right] \tag{1}$$

where P_i is the pressure at particle i, ρ_i is the density, Π_{ij} is a viscous force between particles i and j, and $\mathbf{r}_{ij}$ is the relative position vector $\mathbf{r}_j - \mathbf{r}_i$. In this study, we neglect the viscous force for simplicity. The density is computed by summing over nearby particles

$$\rho_i = \sum_j m_j W(\mathbf{r}_{ij}, h) \tag{2}$$

and the pressure is computed using an equation of state

$$P_i = K\left(\frac{\rho_i}{\rho_0} - 1\right) \tag{3}$$

where K is a positive coefficient.

Previously, SPH control schemes have used external forces for swarm guidance while the SPH forces have mainly provide collision avoidance [11,27,30]. However, we take the approach of using *reduced density virtual particles* for guidance. Just as a reduced density region in a fluid creates pressure gradients that drive fluid to this region, the reduced density particles attract vehicle particles to them.

By defining a particle's mass to be large enough that its density is always at least as large as the reference density, ρ_0, the particle is ensured a non-negative pressure, but if the mass is chosen to be less than this threshold, negative pressures can result. In the SPH control scheme, all particles representing vehicles will be given sufficient mass to ensure positive pressure and a single attracting virtual particle will be used which uses a reduced mass to create a negative pressure region. This negative pressure creates attracting forces and therefore acts as a goal region. In the low mass limit the acceleration terms simplify, slightly reducing computational cost. A schematic showing the interaction of vehicles, obstacles and attracting particles is shown in Figure 1.

In the control scheme, the SPH accelerations for a particle due to all nearby (within $2h$) particles are computed and passed to the vehicle controller, which attempts to enforce the desired motion through a combination of roll, pitch, and thrust commands.

3 UAV Data Sampling and K-means Approximation

Suppose we want to measure the temperature or humidity of a 3-dimensional region. Our UAVs can quickly collect large amounts of data through the custom temperature and humidity sensor suite. However, in experimental flights these sensors may not be highly accurate due to the limitations of sensor response time which is related to the rate of change of temperature and humidity over time and space and vehicle flight speed. For this reason, we use a K-means based error reduction scheme to effectively reduce the noise. K-means algorithms can effectively cluster N data points into K clusters. To this effect, the field function has only K unknown coefficients corresponding to K basis functions. On the other hand, Kriging and Gaussian interpolation are spanned by N basis functions. Therefore, the K-means algorithm can be considered as a dimension reduction technique that approximates the field function without significant loss of information. Additionally, the computational cost is significantly lower for K-means than Kriging and Gaussian process regression if $K \ll N$. One drawback of the K-means algorithm is that it may filter the high frequency components of the original field. However, as long as the ensemble of the data set is large enough and the field is smooth, the K-means algorithm is able to capture the main modes of the field. This method also enables us to approximate data at unsampled locations, and potentially (in future work) suggest a path for the UAVs to follow and collect additional data to minimize the existing uncertainty. This section will discuss the effects of noise on the data and introduce the statistical method for noise mitigation and interpolation.

Let x be any point in the measurement domain and x_i denote a spatial location of a sensor measurement at time step i. Each sensor inevitably introduces some location error, ξ_i, and some measurement error, ε_i. We devise a scheme based on the K-means algorithm to extract information from limited measurements and reduce measurement noise.

Suppose the original field $f(x)$ is a smooth function of position, sampling points x_i near x can be used to approximate the field of x,

$$f(x) = y_i + \varepsilon_i + J(x)\cdot\xi_i + J(x)\cdot(x-x_i) + O(|x-x_i|^2) + O(|x-x_i|\cdot|\xi_i|) + O(|\xi_i|^2)$$

where $J(x)$ is the Jacobian matrix at point x. A better estimator $\hat{f}(x)$ can be obtained through a linear combination of some measurement results y_i. Our goal is to find an optimal weighting function, $\varphi_n(x,x_i)$ so that the error in $\hat{f}$ is minimized.

If many points are measured and the measurement error is potentially large the variance becomes the dominant error. To reduce the noise error, we will use a cluster of points instead of a single point to estimate the unknown field. We use the well know K-means algorithm [21] to cluster the data points into K groups. Each cluster is assigned a value near the mean of its members and interpolation may be performed using a radial basis function network. Let C be the encoder, $C(i) = j$ means the ith points belongs to the jth cluster. To optimize the clustering process, we use the following cost function [8]

$$J(C) = \sum_{j=1}^{K} \sum_{C(i)=j} \left\|x_i - u_j\right\|^2. \tag{4}$$

where u_j is the center of cluster j. To minimize the cost function $J(C)$, we use an iterative descent approach K-means algorithm [20, 22] to find the encoder. The K-means algorithm proceeds with the following two steps iteratively until a convergent encoder has been obtained:

Step 1. Minimize the cluster variance with respect to the cluster means $\{u_j\}_{j=1}^{K}$:

$$\min_{\{u_j\}_{j=1}^{K}} \sum_{j=1}^{K} \sum_{C(i)=j} \left\|x_i - u_j\right\|^2 \quad \text{for a given } C. \tag{5}$$

Step 2. Having computed the optimized cluster means in step 1, we next optimize the encoder as:

$$C(i) = \arg\min_{1 \leq j \leq k} \left\|x_i - u_j\right\|^2. \tag{6}$$

Then, we can build a radial basis function network [4, 24]

$$F(x) = \sum_{j=1}^{K} w_j \varphi(x, u_j), \tag{7}$$

where $\varphi(x, u_j)$ is the Gaussian kernel function. w_j can be seen as an approximation of the temperature at the cluster center. Its value can be trained by a least mean square algorithm.

In the following section, the functionality of this algorithm is verified by a simulation of several aircraft flying through an artificial temperature field. The algorithm is also applied to experimental data of two aircraft equipped with humidity and temperature sensors.

4 Simulation and Experimentation

In the following we describe the experimental platform and use it to demonstrate certain properties of the SPH control scheme. We also verify the K-means error reduction technique using data obtained through simulation and apply the reduction technique to data from an experiment of two vehicles controlled by the SPH scheme. In order to accurately match the simulation to the physical world, the limited flight capabilities of the UAVs are imposed in the simulation through particle velocity and acceleration constrains. Also the inaccuracies of the real world sensor readings are modeled by adding noise to the sensor readings in simulation.

4.1 UAV Hardware and Sensor Suite

A UAV equipped with a simple sensor suite and limited onboard computation capabilities is used for this study (shown in Figure 2). The aircraft has a wingspan of 0.8 m and weighs less than 0.5 kg. The airframe is a single piece of Styrofoam and can be purchased under the name F-27 Styker; additionally the ailerons, flaps, nose cone, propellers, etc. are all mass produced thus making the aircraft inexpensive compared to a custom design. This aircraft has been successfully used in several experiments in our research group and has proven itself robust to experimental mishaps, relatively simple and inexpensive to maintain. Additionally, the small size and simplicity of the aircraft allow for rapid deployment by a single pilot in areas that are not ideal for traditional UAVs. Most importantly, the Delta-Wing UAV is equipped with a custom autopilot to allow for implementation of custom control strategies. The CUPIC is a complete autopilot system developed at the University of Colorado at Boulder [28], and used in a large number of experiments in our (and other) research group(s). Several studies [2, 14] have shown that it is possible to achieve fully autonomous operation of a small UAV by means of this simple autopilot equipped with a limited number of sensors. The work of Floreano et. al [9, 16] has demonstrated the use of a similar fixed wing aircraft in swarm applications by using a different autopilot system. Additionally, Sensefly [1] is a Swiss company that provides a UAV that can be used to gather high resolution imaging using a similar hardware platform.

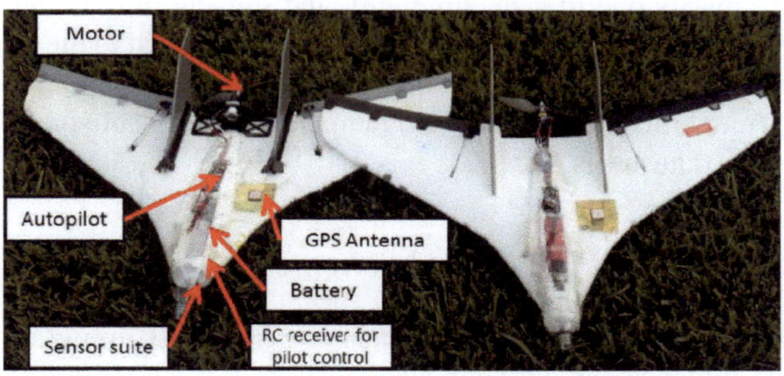

Fig. 2 Resource constrained delta-wing UAVs used in experiments. The 0.8 m wingspan UAV is equipped with a GPS sensor, a roll rate sensor, a communication radio, and autopilot to control the craft during autonomous operation.

Shaw and Mohseni [30] showed that the CUPIC autopilot is also capable of demonstrating fully autonomous, distributed cooperative control of a team of UAVs. The CUPIC, in its most basic design, consists of an on-board processor, a single axis rate gyro to sense roll rates, an absolute pressure sensor for altitude sensing, and a GPS receiver for positioning. The autopilot controls the vertical location of the aircraft through pitch and thrust commands. The pitch and thrust commands are determined from the error between the desired altitude and the current altitude as well as the magnitude of the SPH acceleration from Equation 1. The horizontal location of the aircraft is controlled by varying the roll angle. The desired roll angle is determined by the error between the direction of the SPH acceleration vector (from Equation 1) and the aircraft's current heading vector. Virtual saturation limits are employed in order to avoid commands which would result in aircraft stall or an excessive roll angle. The program has built-in routines to account for short term blackouts in the GPS signals and the noise and drifts in the sensors. The autopilot has been proven to be fully capable of stable autonomous flight on a wide variety of MAVs including Delta-wing aircrafts [2], warping-wing aircrafts, and gust-insensitive aircrafts [14].

The CUPIC autopilot system also includes a complementary ground station which is comprised of a laptop running a MATLAB routine. The autopilot transmits telemetry data to the ground station and the ground station transmits commands and the location of the artificial attracting particle for the SPH control algorithm. The MATLAB routine includes a user interface that provides pilots and observers real time information of the aircraft's GPS position, physical state, sensor suite raw data, and autopilot commands. The graphical interface also includes the ability to alter the artificial particle's location, speed, and path. A detailed communication characterization of the communication scheme used in the autopilot systems is given in [30].

The autopilot was designed with the ability to interface up to 7 additional analog sensors through on board analog to digital converters. For this experiment a custom board was manufactured to house a HIH-5031 humidity sensor and a LM35 temperature sensor. The HIH-5031 humidity sensor and LM35 temperature sensor were chosen primarily due to size, simplicity and sensing range. The sensors raw output voltage is read by the autopilot at 10Hz and transmitted to the ground station along with the aircraft's most recent GPS position. The raw output voltage is then converted to percent relative humidity and temperature using equations found in the sensors respective datasheets.

4.2 Verification of SPH Control Scheme

In applications involving autonomous agents the two most important control aspects are agent guidance and collision avoidance. In this article, agent guidance is accomplished using a reduced density virtual particle which acts to attract agents to a specific region of space. Collision avoidance is accomplished by setting the SPH parameters of the agents such that they are repelled from each when they reach a certain distance. If the vehicle separation is greater than this distance the agents are only attracted to the reduced density particle.

The guidance property is demonstrated by placing a stationary attractor particle at a location in the domain and engaging the autopilot with the UAV at some other point in the domain. Figure 3 shows the GPS position of the UAV demonstrating the guidance property. The marker color indicates the magnitude of the SPH force as calculated by the algorithm. The plane approaches the attracting particle (indicated by a red $\times$), passes almost directly over the virtual particle and then begins to double back as the direction of the SPH force vector points opposite to the plane's heading.

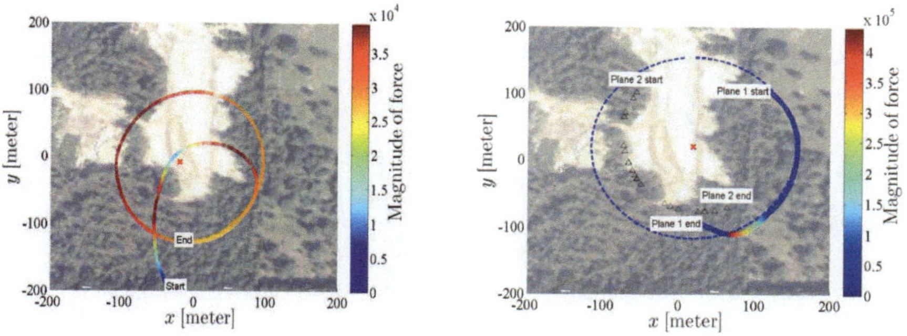

Fig. 3 Experimental demonstration of SPH control properties using small, resource constrained UAVs. (Left) Experimental results of a single vehicle (colored ∘) approaching an attractor particle (red ×) from the south and then beginning a loiter pattern around the attractor particle. Color of the ∘ represents the magnitude of SPH force calculated by the vehicle. Results taken from a small portion of large flight experiment. (Right) Experimental results of a single vehicle (colored ∘) avoiding another aircraft (black △) while loitering an attracting particle (red ×). The blue dashed line indicates the previous loiter circle achieved by plane 1 prior to interaction with plane 2. Color of the ∘ represents the magnitude of SPH force calculated by the vehicle.

The plane then enters into a loiter circle which is a result of a balance between the SPH force magnitude and physical limitations (i.e. turning radius) of the aircraft. For this and all following experiments, aircraft were given an h value of 30 and the attractor particle was given an h value of 200; the average of the two h values were used to determine the SPH force per Equation 1.

The collision avoidance property is demonstrated with two flying aircraft. Figure 3 plots the information received from a single vehicle (plane 1): ∘ represent the vehicle's GPS position colored by the calculated SPH force magnitude, the red × represents the location of the attractor particle, △ represent the GPS position of the other vehicle (plane 2) as known by plane 1. The dashed line represents the loiter circle achieved by plane 1 prior to interaction with plane 2. Plane 1 loitered in a clockwise direction while plane 2 loitered in a counter clockwise direction. As the two planes approach each other, both planes make corrections to their respective courses avoiding potential collision as evident in the course correction of plane 1. The SPH force magnitude range is greater for collision avoidance than attraction due to the fact that the interaction between planes results in a higher repulsive force than the attraction force experienced between a plane and the attracting particle. Simulations have been previously conducted showing the smooth collision avoidance property of the SPH control technique [27] and these experimental results correspond well to the simulations. Although collision avoidance may not be guaranteed in real-world situations where packet loss and location error play a role, incorporating a safety factor into the inter-vehicle spacing provides high confidence that collisions will be avoided.

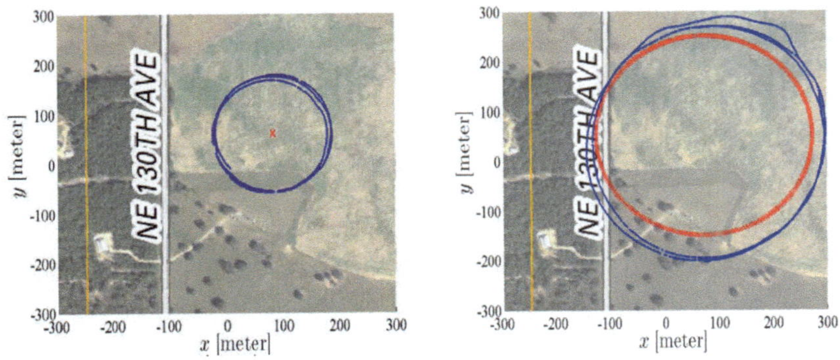

Fig. 4 Two ways to achieve loiter circles using SPH control scheme and UAVs. (Left) GPS coordinates of a single vehicle (blue •) in a series of loiter circles around a stationary attractor particle (red ×). The loiter circles are a result of a balance between the SPH force and the physical limitations of the aircraft. (Right) Experimental results of a single vehicle (blue •) following a moving attracting particle (red ×) demonstrating a different method of achieving a loiter.

Due to the fact that fixed wing aircraft must maintain a forward velocity to stay aloft, a loiter circle is a common technique employed in experiments involving fixed wing aircraft. One way of achieving a loiter circle using SPH control is with a stationary attractor particle and imposed acceleration and velocity constraints on the moving particle. The moving particle will move directly towards the attracting particle until it passes the particle. Due to specified constraints, the moving particle will then bank one way and eventually find and maintain an equilibrium balancing the SPH force and the imposed constraints resulting in a loiter. This property was experimentally verified and the results can be seen in Figure 4 which shows the GPS location for a single aircraft (blue •) loitering around a stationary attractor (red ×).

While a stationary particle allows for a loiter circle, the radius of the circle is a function of several parameters and thus the radius of the loiter is not easy to predict. It is difficult to maintain a uniform loiter since any disturbance to the vehicle's path could result in the plane taking a more direct approach over the attractor particle as evident in Figure 3. A more robust way to achieve a loiter circle is with a moving attractor particle. If the velocity and acceleration of the attracting particle are within the physical limitations of the aircraft, the vehicle will follow closely behind the attractor. This technique is shown in Figure 4 with the GPS location of the aircraft shown as blue • and the attractor location as a red ×. A moving particle allows for a loiter circle of varying radius as well as more complex paths.

4.3 Sensing Using Multiple Vehicles

Next we simulate a more complicated situation involving multiple vehicles taking measurements over a large domain to determine the temperature field. The results are shown in Figure 5. In this simulation, a group of 10 vehicles begins in the lower left corner of the domain and travels back and forth across the domain, finishing in the upper right. Each vehicle records a temperature every 0.2 seconds and sends data back to the base station. The temperature is generated by the function

$$\begin{aligned} T(x,y) =& 75+3[\sin(x/50)+\cos(y/42)\cos(x/100) \\ &+3\tan^{-1}((x+20)/10)+\cos(\sqrt{x^2+y^2}/40)]. \end{aligned} \tag{8}$$

To create a closer approximation of reality and consider the limited accuracy of onboard temperature sensors, white Gaussian noise with a standard deviation of 0.78 is added to the temperature data. Every 5th point of the temperature data is plotted in Figure 5 as a colored circle. The vehicle paths are shown as black curves and the end vehicle positions are shown as black dots. The SPH controller maintains an even vehicle spacing throughout the trajectories. By using this well spaced group of multiple vehicles flying at 15 m/s, a large amount of data is collected over this 250,000 sq. m domain in only 160 seconds.

This procedure produces a large amount of data (8000 data points in 160 s) that can be used to reduce the noise in the resulting interpolations. By using the aforementioned K-means algorithm we are able to interpolate the data over the interior of the domain and reduce the error. The K-means clusters for this example are shown in Figure 5. The error in the final temperature approximation can be computed as

$$E_\infty = ||T - T_{\text{estimate}}||_\infty \text{ or } E_{\text{RMS}} = \sqrt{\frac{\sum_i^n (T(\mathbf{x}_i) - T_{\text{estimate}}(\mathbf{x}_i))^2}{n}}$$

where n is the number of spatial grids, T is the true temperature (given by Equation 8) and T_{estimate} is the temperature estimated by the K-means algorithm. We find that $E_\infty = 2.32$ and $E_{\text{RMS}} = 0.23$ after applying K-means and interpolating. This is an improvement over the raw noisy data that had $E_\infty = 2.97$ and $E_{\text{RMS}} = 0.79$. The largest errors occur at the edge of the domain where interpolations are less likely to be valid due to the limited data in these regions.

In the real world experiments, we measured humidity and temperature using the aforementioned sensor suite placed on the two aircraft used in demonstrating the SPH control scheme. An AcuRite digital humidity and temperature monitor was used to determine the temperature and percent relative humidity at the ground station location as 36 $°C$ and 57% respectively. All the data was collected by two UAVs flying in a series of overlapping loiter circles centered at the location of a stationary attractor particle as seen in Figure 6. The path of UAV 1 is shown as a dashed black line, the path of UAV 2 is shown as a red dash-dot line, and the attractor particle is shown as a blue $\circ$. The loiter circles are a result of a balance between the minimum velocity enforced by the autopilot to keep the plane aloft and the attraction to the artificial particle. After each plane completed at least 3 loiter circles the location of the attractor particle was moved approximately 70 meters to the northeast resulting in the behavior shown in Figure 6. The motion of the aircraft in the horizontal plane was determined by the SPH control law while the vertical location of the aircraft

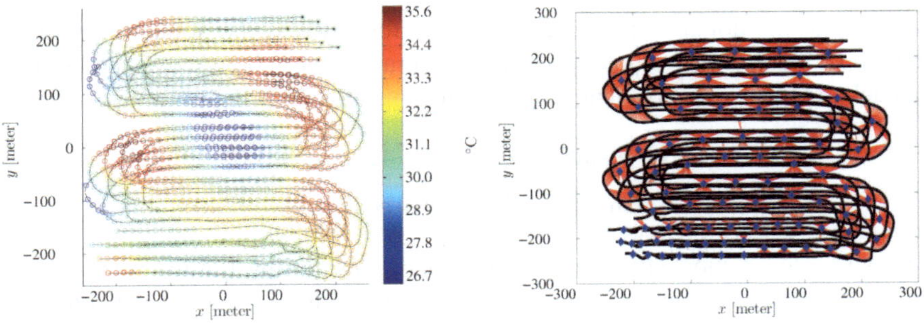

Fig. 5 Simulation of data gathering using UAVs equipped with a temperature sensor. (Left) Simulation of 10 vehicles (black ∘'s) that were guided through a 500×500 domain by a single attracting particle. The black curves show the full vehicle paths and the colored ∘'s denote individual temperature measurements (in °C). (Right) The K-means clusters for the 10 vehicle simulation. Cluster points are shown in blue, data points are shown in black and the data-cluster connections are shown in red.

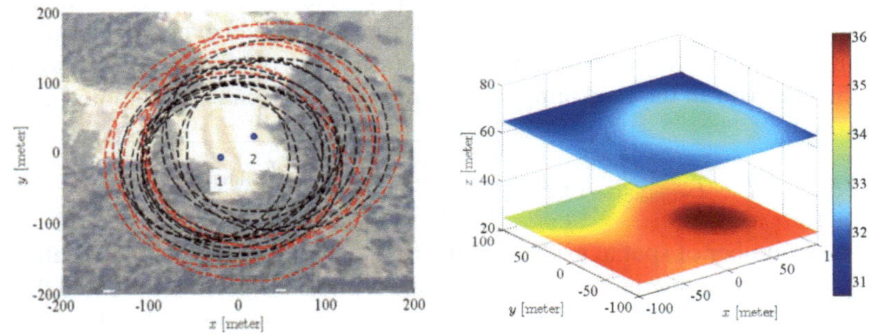

Fig. 6 Flight data of two UAVs demonstrating the SPH control scheme to gather the humidity distribution of an environment. (Left) UAV 1 path in black dashed line, and UAV 2 path in red dash-dot line. An attractor particle (blue ∘) was placed at (0,0) and then moved to the northeast to approximately (50,50). (Right) Humidity field results from two aircraft. The aircraft were flown approximately 40 meters apart (in altitude). The field results show higher relative humidity at a lower altitude and also a region of higher humidity at approximately $x = 50$ m, $y = 0$ m.

was maintained by an altitude controller. Regardless of the altitude difference, the aircraft maintain a safe horizontal separation thus avoiding collision if the UAVs altitude's were the same. The aircraft were flown at an altitude separation of 40 m in order to obtain three dimensional data about the temperature and humidity fields.

The noise of the sensor readings was rather high (approximately 10%), in order to approximate the entire field and minimize error we implemented the K-means method to reduce the noise. The results from the K-means approximations for the two aircraft are then used to find humidity and temperature as a smooth function of space. Figure 6 shows the resulting humidity fields obtained from the two aircraft at their respective average altitude.

The test environment was a dry retention pond surrounded by woods. The results show a higher relative humidity centered at approximately ($x = 50$ m, $y = 0$ m) which is at the edge of the wooded region. While more tests are required to make conclusive remarks, the higher relative humidity over a region of trees is likely indicative of increased evaporation over this region compared to the dry retention pond.

5 Conclusions

An SPH based controller has been successfully implemented that includes a reduced density virtual attracting particle for vehicle guidance of autonomous UAVs with limited processing capabilities. The temperature and humidity data collection capabilities of multiple UAVs were demonstrated. Additionally, these UAVs experimentally verified several desirable properties of the SPH control scheme. Although the examples presented here are two-dimensional, all the techniques used are valid in three dimensions, but have been artificially restricted to constant altitude for simplicity and an added level of safety.

Additional simulations have been implemented to demonstrate the multi-vehicle capabilities of the SPH controller as pertinent to the data collection opportunities made possible by swarms of sensor equipped UAVs that can quickly collect data over a large two or three dimensional region. This is in contrast to the commonly used data collection methods available today (i.e. remote sensing, dropsondes, weather balloons, etc.). Furthermore, uncertainties in the sensor readings are mitigated using a K-means algorithm that is well suited to process and interpolate the data over a desired region to minimize errors. The data processing algorithm was implemented on a set of humidity and temperature data gathered by a pair of aircraft equipped with sensors flying in several loiter patterns as governed by the SPH control.

References

1. Sensefly autonomous ultralight uav for professionals@ONLINE (October 2012)
2. Allred, J., Hasan, A., Pisano, B., Panichsakul, S., Gray, P., Han, R., Lawrence, D., Mohseni, K.: SensorFlock: A mobile system of networked micro-air vehicles. In: The ACM SenSys 2007: The 5th ACM Conference on Embedded Networked Sensor Systems, Sydney, Australia, November 6-9 (2007)
3. Amzajerdian, F., Pierrottet, D., Petway, L., Hines, G., Roback, V.: Lidar systems for precision navigation and safe landing on planetary bodies, p. 7 (2011)
4. Buhmann, M.D.: Radial Basis Functions: Theory and Implementations. Cambridge University (2003)
5. Campbell, J.: Introduction to Remote Sensing. The Guilford Press (1996)

6. Corrigan, C.E., Roberts, G.C., Ramana, M.V., Kim, D., Ramanathan, V.: Capturing vertical profiles of aerosols and black carbon over the Indian Ocean using autonomous unmanned aerial vehicles. Atmospheric Chemistry & Physics Discussions 7, 11429–11463 (2007)
7. Hardin, P., Jensen, R.: Small-scale unmanned aerial vehicles in environmental remote sensing: Challenges and opportunities (2011)
8. Hastie, T., Tibshirani, R., Friedman, J.: The Elements of Statistical Learning: Data Mining, Inference, and Prediction. Springer (2009)
9. Hauert, S., Leven, S., Varga, M., Ruini, F., Cangelosi, A., Zufferey, J., Floreano, D.: Reynolds flocking in reality with fixed-wing robots: communication range vs. maximum turning rate. In: 2011 IEEE/RSJ International Conference on Intelligent Robots and Systems (IROS), pp. 5015–5020. IEEE (2011)
10. Holland, G., Webster, P., Curry, J., Tyrell, G., Gauntlett, D., Brett, G., Becker, J., Hoag, R., Vaglienti, W.: The aerosonde robotic aircraft: A new paradigm for environmental observations. Bulletin of the American Meteorological Society 82(5) (May 2001)
11. Huhn, S., Mohseni, K.: Cooperative control of a team of AUVs using smoothed particle hydrodynamics with restricted communication. In: Proceedings of the ASME 28th International Conference on Ocean, Offshore and Arctic Engineering, Honalulu, HA, May 31-June 5, OMAE 2009-79869 (2009)
12. Jadbabaie, A., Lin, J., Morse, A.S.: Coordination of groups of mobile autonomous agents using nearest neighbor rules. IEEE Transactions on Automatic Control 48(6) (2003)
13. Jensen, J.R.: Remote Sensing of the Environment: An Earth Resource Perspective, 2nd edn. Prentice Hall (May 2006)
14. Lawrence, D., Frew, E., Pisano, W.: Lyapunov vector fields for autonomous unmanned aircraft flight control. Journal of Guidance, Control, and Dynamics 31(5), 1220–1229 (2008)
15. Pimenta, R.M.L.C.A., Mendes, M.L., Pereira, G.: Fluids in electrostatic fields: An analogy for multirobot control. IEEE Transactions on Magnetics 43(4) (2007)
16. Leven, S., Zufferey, J.-C., Floreano, D.: Dealing with Mid-Air Collisions in Dense Collective Aerial Systems. Journal of Field Robotics 28(3), 405–423 (2011)
17. Lipinski, D., Mohseni, K.: Cooperative control of a team of unmanned vehicles using smoothed particle hydrodynamics. AIAA paper 2010-8316, AIAA Guidance, Navigation, and Control Conference, Toronto, Ontario, Canada, August 2-5 (2010)
18. Lipinski, D., Mohseni, K.: A master-slave fluid cooperative control algorithm for optimal trajectory planning. In: 2011 IEEE International Conference on Robotics and Automation (ICRA), pp. 3347–3351. IEEE (2011)
19. Liu, G., Liu, M.: Smoothed particle hydrodynamics: a meshfree particle method. World Scientific (2003)
20. Lloyd, S.P.: Least square quantization in pcm. IEEE Transactions on Information Theory 28(2), 129–137 (1982)
21. MacKay, D.: Information Theory, Inference and Learning Algorithms. Cambridge University Press, New York (2003)
22. MacQueen, J.B.: Some methods for classification and analysis of multivariate observations. In: Proceedings of 5th Berkeley Symposium on Mathematical Statistics and Probability, pp. 281–297 (1967)
23. Monaghan, J.: Smoothed particle hydrodynamics. Annual Review of Astronomy and Astrophysics 30, 543–574 (1992)
24. Moody, J., Darken, C.J.: Fast learning in networks of locally tuned processing units. Neural Computation 1, 281–294 (1989)

25. Olfati-Saber, R.: Flocking for multi-agent dynamic systems: Algorithms and theory. IEEE Transactions on Automatic Control 51(3), 401–420 (2006)
26. Pimenta, L., Michael, N., Mesquita, R., Pereira, G., Kumar, V.: Control of swarms based on hydrodynamic models. In: IEEE International Conference on Robotics and Automation (May 2008)
27. Pimenta, L., Michael, N., Mesquita, R., Pereira, G., Kumar, V.: Control of swarms based on hydrodynamic models. In: IEEE International Conference on Robotics and Automation, pp. 1948–1953 (May 2008)
28. Pisano, W.J., Lawrence, D.A.: Autonomous UAV control using a 3-sensor autopilot. AIAA paper 2007-2756, American Institute of Aeronautics and Astronautics, Infotech Aerospace Conference, Rohnert Park, CA (May 2007)
29. Richards, M., Scheer, J., Holm, W.: Principles of Modern Radar Basic Principles. SciTECH Publishing Inc. (2010)
30. Shaw, A., Mohseni, K.: A fluid dynamic based coordination of a wireless sensor network of unmanned aerial vehicles: 3-d simulation and wireless communication characterization. IEEE Sensors Journal 11(3), 722–736 (2011)
31. Spiess, T., Bange, J., Buschmann, M., Vörsmann, P.: First application of the meteorological Mini UAV M2AV. Meteorologische Zeitschrift 16(2), 159–169 (2007)
32. van den Kroonenberg, A., Martin, S., Beyrich, F., Bange, J.: Spatially-Averaged Temperature Structure Parameter Over a Heterogeneous Surface Measured by an Unmanned Aerial Vehicle. Boundary-Layer Meteorology 142, 55–77 (2011)

A Physics Inspired Finite State Machine Controller for Mobile Acoustic Arrays*

Thomas Apker and Eric Martinson

Abstract. Applying coherent array processing to sound source localization when individual sensors are attached to heterogeneous platforms is a multi-faceted challenge for both perception and mobility. Recent technical advances in robot localization have made such mobile acoustic arrays possible, but the multi-robot coordination problem remains incomplete. How can a team of robots coordinate in cluttered environments, both with each other and static mounted sensors to effectively localize sound sources? This work proposes and implements a physicomimetics based robot control system with solid, liquid, and gas phase finite states. Applying these different phases appropriately enables efficient navigation through clutter and localization of both exposed and buried or hidden sound sources by teams of mobile robots.

1 Introduction

Many sensing modalities require multiple, dispersed transducers to measure their target signal and produce a meaningful output. Modern phased-array and bistatic radars depend on networked receivers to compare the arrival time of radio frequency (RF) signals to localize sources and even image whole regions [5]. Chemical and biological agent detection requires sensors spread across likely plume paths to determine the extent and concentration of the hazard[17]. Acoustic surveillance for 3D sound source localization can be accomplished by teams of mobile sensors combining simple clustering and repulsion behaviors [11]. In this work, we focus on designing and implementing a distributed controller to allow autonomous ground

Thomas Apker · Eric Martinson
NRC/NRL Postdoctoral Fellow, Naval Research Laboratory,
4555 Overlook Ave SW, Washington, DC 20375, USA
e-mail: {thomas.apker.ctr,eric.martinson.ctr}@nrl.navy.mil

* This work was performed under Office of Naval Research Work Orders N0001409 WX30013 and 55-9019-B2-5.

M.A. Hsieh and G. Chirikjian (eds.), *Distributed Autonomous Robotic Systems*, Springer Tracts in Advanced Robotics 104,
DOI: 10.1007/978-3-642-55146-8_4, © Springer-Verlag Berlin Heidelberg 2014

robots to move through cluttered environments and form a coherent microphone array for sound source localization.

Maneuvering mobile robots into position to form effective array geometries is an ongoing challenge. Motion constraints, uncertain navigation and cluttered environments pose challenges for single robot path planning that Shah *et al.* [19] addressed with a two-stage approach consisting of a deterministic global path planner and local stochastic controller designed to bound a robot's position error relative to the reference path. For multi agent systems, Krontiris *et al.* [10] found that operating in cluttered environments required scheduled changes in formation, and thus interactions between the agents, to guarantee movement without collisions. This mode-switching behavior appears in biological systems as well. Balch *et al* [4] found that ants are directed by an internal finite state machine (FSM) with behaviors that are appropriate for each step of the foraging process. They implemented this control scheme on a pair of robots programmed to accomplish a cooperative foraging task.

We found that a robot team that can cluster around peers or waypoints, transit a cluttered environment and recover from getting lost required fundamentally different behaviors in each case. Early work by Gage [7] suggested that these clustering, transiting and wandering behaviors resembled the solid, liquid and gas behaviors of matter. We approach this problem by designing a finite state machine controller based on the physicomimetics framework developed by Spears [21] with three modes. Near their objective, the agents entered the "solid," or clustering, mode, designed to cause them to settle into a good array shape autonomously. When the objective was farther away, the agents were biased towards movement in a "liquid" mode designed to allow them to flow around obstacles. Rather than treat a lost robot as a failure, we implemented a wandering, or "gas," mode to give the robot a chance to return to the network. Collectively, this approach proved very effective at allowing four mobile robots to augment a static array of four microphones to localize sound sources in a simulated disaster area.

2 Background

Few robotic auditory systems are capable of 3D sound source localization (SSL). Time-delay on arrival based SSL cannot generally estimate range to a target because inter-microphone distances on a robot are too small. When microphones are too close together, bearing and azimuth are straightforward, but more than time delays are required to determine range to a sound source. Previous robotic efforts have addressed this limitation for nearby sources using biologically inspired software and hardware [18]. Alternatively, moving the robot to multiple locations and combining all data together in an evidence grid [13] enables localization over larger areas, but fails for short duration sounds. To handle 3D localization over distances and transient noise, microphones must be spread out over greater distances [12].

Given fixed-position microphones distributed about the environment, a robot can contribute to auditory scene analysis by providing different vantage points. There are, however, at least two ways in which their data can be integrated with a robotic auditory system. In the first method, the on-robot and room-mounted arrays are treated as separate arrays. Their information is fused probabilistically to estimate location and/or directivity of a sound source [16]. Alternatively, if the on-robot microphones can be localized accurately enough with respect to the room-mounted sensors, the two systems could be treated as a single, large-baseline, coherent array. Martinson and Fransen [12] demonstrated such a coherent system using visual feedback for inter-microphone localization. Thrun [22] and Nakadai [16] also localized distributed microphones using aural feedback.

Given the technical capabilities for building mixed mobility coherent arrays, a significant remaining challenge is building effective array geometries. The positioning of microphones in factory built array systems is optimized for the mathematics of sound source localization, enhancing both computational speed and location accuracy. This is not feasible when microphones are distributed across multiple devices. Theoretically, given large enough numbers of sensors in a network, a random distribution is a good alternative choice for monitoring a large area [3]. In practice limitations on communication, power, processing and localization accuracy limit the number of streaming audio sensors that can be included in a coherent array. With smaller numbers of sensors, random array shapes have a random localization performance over a known region. Simple behavioral rules for clustering about the region of interest and avoiding other robots ameliorate this drawback, enabling localization of sound sources over small regions [11]. Larger, cluttered environments, however, require enhanced multi-robot coordination to reach the area of interest and form an effective array.

Artificial potential fields are a well established tool for robot path planning and coordination [2]. Spears [21] developed a physics-based swarm intelligence framework in which the agents determine their trajectory based fields generated by their neighbors. Most applications use rules designed to drive the agents into a desired formation about a point in space [9]. In addition to static formations, physicomimetics interactions can allow agents to follow a plume to its source by computing the flux across the swarm [20].

Several authors have studied finite state machine approaches to robotic team coordination on tasks with discrete phases. Worcester *et al.* [23] developed a finite state representation of a construction task, and used a three step algorithm to partition its components among a team of homogeneous robots. Mather and Hsieh [14] applied a similar approach to task scheduling for robots engaged in surveillance tasks, and found that probabilistic transitions between observation and navigation states would ensure statistically even coverage of numerous targets. We applied this concept of breaking tasks into discrete parts with appropriate behaviors for each to the physicomimetics framework to allow agents to autonomous form a recursively self-optimizing microphone array.

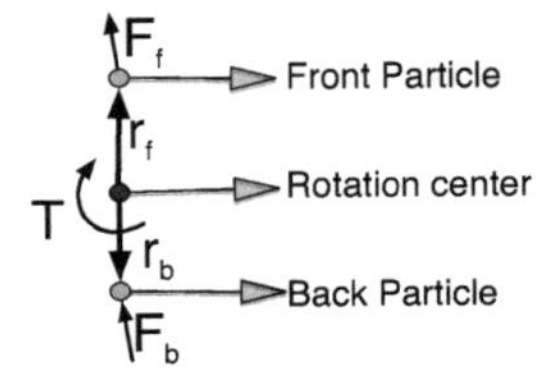

Fig. 1 Schematic model of the dumbbell agent that allows particle forces and virtual torques to direct the agent model

3 Physicomimetics FSM Controller Design

We designed the physicomimetics guidance system with three basic modes inspired by the behavior of a molecule within the three main physical states of matter. The details of the behaviors of each mode were defined to allow us to localize a sound source in a cluttered environment by reposition the array based on knoweldge of the source source location.

3.1 Agent Model Design

The agent model converted physicomimetics forces into motion commands to guide the mobile agents. We accounted for nonholomic motion constraints by adding extend to the point-mass model in [21] as shown in figure 1. It consisted of a pair of particles located fore and aft of the robot's rotation center at positions $\mathbf{r_f}$ and $\mathbf{r}_b$, respectively. At each time-step k, the the agent set a forward speed u and turn rate Ω using equations 1-2. The agent mass m and moment of inertia I_{zz} were defined based on the physical robot's acceleration capabilities. The friction terms μ_u and μ_Ω proved useful a means implmementing mode switching.

$$
\begin{aligned}
F_t &= (\mathbf{F}_f + \mathbf{F}_b) \cdot \begin{bmatrix} \cos\Psi \\ \sin\Psi \end{bmatrix} \\
\mathbf{M} &= \mathbf{F_f} \times \mathbf{r_r} + \mathbf{F_b} \times \mathbf{r_b} \\
u_{j,k+1} &= (u_{j,k} + \Delta t F_t/m)(1 - \mu_u); && (1) \\
\Omega_{j,k+1} &= (\Omega_{j,k} + \Delta t \Sigma (M_z + T_z)/I_{zz})(1 - \mu_\Omega) && (2)
\end{aligned}
$$

3.2 Basic Interaction Types

We used two basic types of interaction fields in this study, radial and vortex. The radial force laws define the lowest potential energy locations for each member of the swarm, within 1.5m of the sound source candidate and 1.5m away from other microphones. Vortex forces were used to navigate around obstacles.

The radial interaction used in this study is shown in equations 3-5, and is a function of the distance, r_{ij}, between agents i and j; a mass power, p_m, distance power, p_d; attract-repel (AR) distance, d_{AR}; maximum distance; r_{max}; and scaling term, G. The version proposed by Spears in [21] included a discontinuous jump that led to undesired "jumping" of the agents near $r_{ij} = d_{AR}$, and so they introduced a damping term η to enforce continuity.

$$F_{ij} = \begin{cases} G\frac{(m_i m_j)^{pm}}{r_{ij}^{p_d}} & \text{if } r < d_{AR} \\ -G\frac{(m_i m_j)^{pm}}{r_{ij}^{p_d}} & \text{if } d_{AR} \leq r \leq d_{max} \\ 0 & \text{if } r > d_{max} \end{cases} \tag{3}$$

$$\eta = \left(\frac{\|(d_{AR} - r)\|}{d_{AR}}\right)^{d_p} \quad \text{if } r < 2d_{AR} \tag{4}$$

$$F_{ij} = F_{ij}\eta \tag{5}$$

Vortex forces were used to guide the mobile agents around obstacles. Our implementation included a radial element out to a specified boundary, σ, and a tangential component at all points within d_{max}. The strength of the components is a function of the vortex magnitude Γ and r_{ij} as shown in equations 6 and 7, and the sign of the tangential component was set to augment the agent's tangential speed around the obstacle.

$$F_r = \Gamma / r_{ij} \text{ if } r_{ij} < \sigma \text{ else } 0 \tag{6}$$

$$F_t = \Gamma / r_{ij} \text{ if } (\hat{t} \cdot \mathbf{v}) \geq 0 \text{ else } -\Gamma / r_{ij} \tag{7}$$

3.3 Solid Mode

In stop the agents in potential wells, we defined conditionally increasing friction terms μ_u and μ_Ω and a static friction check when the mobile agents were within $1.5d_{AR}$ of an attractive particle, *e.g.* a waypoint. The friction terms increased by 80% each time the agent's commanded direction of motion changed in solid mode, indicating passage through a potential well. The agents were commanded to stop once their applied force was less than the friction times their virtual weight. These parameters were manually tuned for this study.

Motion in the solid mode was characterized by slowing and settling into formation. As such, it required navigation that is precise but can be slow, such as periodic chirps to allow the robots to compute their peers' locations. While the force laws presented will lead to even lattices of holonomic agents, motion constraints force asymmetry into the final arrangement that our approach exploits to improve SSL performance.

3.4 Liquid Mode

In the liquid mode, the goal was guide the agents to within $1.5d_{AR}$ of an attractive particle. Both friction terms were set to .1, allowing the relatively weak long-range attraction of the next sound source candidate to guide the agents in the right general direction, while the vortex and repulsive forces guided the agents around obstacles and each other, respectively. Since obstacles and teammates could prevent an agent from reaching the solid mode range, we added a rule that liquid-mode robots would freeze in place after a majority of the swarm had settled or three minutes after the attractive particle changed location.

Since liquid mode motion was characterized by constant motion towards an objective, navigation could be less precise than in the solid mode but needs to be fast enough to keep the estimate bounded as the vehicle moves. The key to operating in unmapped, cluttered environments was to use the energy-preserving vortex force instead of potential-well creating repulsion to avoid obstacles.

3.5 Gas Mode

To implement the wandering behavior, we commanded the mobile agent to change its particle type so that it would not continue to interact with anything except locally detected obstacles while engaging in a random walk biased towards moving backwards. Since moving robots generate sounds that could confuse the array's primary task, we added a condition that the agent only move in the gas mode if there was an attractive particle on the map. In this study, the attractive particle was cleared approximately one minute after the liquid-mode robots were frozen to give the "lost" robots a chance to rejoin the team.

4 Sound Source Localization

The SSL algorithm used for this study was generalized cross correlation (GCC). This estimates the energy associated with different possible time delays between a detected source and each pair of microphones. The results are organized by spatial likelihood, graphing relative energy vs 3D position estimates for possible sources. The resulting cross-correlation value, adjusted for the predicted time difference on arrival, was highest for those position/time differences corresponding most closely with the true value. The GCC value was determined separately for each microphone pair, and then summed across all microphone pairs a and b for every position:

$$F_l = \sum_{a=1}^{N} \sum_{b=1}^{N} \int_{\omega} W(\omega) M_a(\omega) \overline{M_b(\omega)} \exp^{-j\omega(T(l,a)-T(l,b))} d\omega \tag{8}$$

where (M_a) was the Fourier transform of a's signal, $\overline{M_b}$ was the complex conjugate of M_b, ω was the frequency in radians per second, and W was a frequency dependent weighting function based on an average noise sample, if known, or using the PHAT transform if not.

To localize a sound source, the spatial likelihood was estimated for every 0.1sec of recorded audio signal. These are linearly scaled to between $[0.1, 0.99]$ and each cell in a global auditory evidence grid [13] was updated using log-odds notation to reflect this new measurement. From the resulting evidence grid, cells whose value were less than 90% of maximum were discarded, and the remaining cells are clustered together. For each cluster c, the combined log-likelihood L_c and the weighted centroid μ_c are identified. The centroid of the cluster with the greatest L_c is the most likely sound source position. Sound source location is an important factor for sound amplification using beamforming [6] and the more general problem of auditory scene analysis.

Although a properly spread array can theoretically determine 3D positioning from a single sample location, or snapshot, obstacles can interfere with location estimates. As demonstrated in [12], multiple snapshots can be fused together to improve localization using the same evidence grid representation. Therefore, we will be examining both the snapshot and fused localization estimates at the end of a search.

5 Results

Multi-agent acoustic arrays were evaluated as part of a simulated disaster response. In this scenario, a remote location had been surveyed from the air, and small sensor packages distributed across the region. The role of these static sensors was to identify regions of greatest interest to initial responders. Mobile robotic agents were dispatched to augment the static array and localize the source.

The testing arena for this work is an 8m$\times$12m area of the Laboratory for Autonomous Systems Research. This area had been instrumented prior to robot deployment with a microphone in each corner to listen for unexpected acoustic events (e.g. explosions, shifting rubble, speech, etc). In this environment with a 60 dB average background noise, initial microphone placements were close enough together to detect initial sounds of $>$ 70 dB. After detecting the presence of such an event, 4 mobile robots were sent into the arena to augment the static array, creating a very large baseline coherent array. An initial acoustic measurement was taken when all robots have entered the testing arena. Then they proceeded towards the location of the sensor that detected the greatest change in ambient noise level during the acoustic event. Building on the authors' previous work in [11], we defined mobile sound source localization as a three step iterative process:

1. Define initial best location candidate
2. Move microphones to surround that candidate location
3. Sample the sound source again and produce a new best candidate

In our experiment, we repeated steps 2 and 3 four times or until the best candidate location moved by less than 1m. We used the three state physicomimetics controller described above for the mobile agents to navigate the cluttered field, settle into good array shapes and, if necessary, stop on command without explicit teleoperation of any individual robot.

5.1 Hardware Setup

We used Pioneer3-AT robots with SICK laser scanners. These robots are connected to each other via a shared wireless network. Each robot also has an attached wireless microphone. All microphone signals were fed into an RME OctaMic II preamp at a central location to provide the time synchronization necessary for a coherent array. Up to 24 microphones can be synchronized in this fashion. For larger arrays, work by Girod *et al.* [8], as well as [15], have demonstrated how such synchronization can potentially be achieved across distributed computing platforms without a centralized data acquisition system.The location of all mobile and static microphones was determined using an overhead camera system, simulating GPS or an aerial photography.

5.2 Sound Source Localization

A total of 11 trials were completed, varying by sound source type and location. In all cases, the source was a computer speaker was placed inside the arena. First a recording of an impulse noise was used to trigger the robot investigation. The static sensor with the highest SNR was used as the focal point for the initial investigation. After the initial sound finished, either a gas leak recording (63 dB at 1m) or a speech recording (65 dB max) was looped while the robots localized the source. 6 gas leaks and 5 speech source trials were completed. Of these 11 total trials, 6 speaker locations were exposed, meaning only the back and/or base of the speaker was blocked by an obstacle. The remaining locations were buried. In 4 instances, there was only a single unobstructed direction, resulting in a highly directional sound source. In the last instance, the speaker was fully enclosed by a box.

In the unobstructed trials, the sound source was localized rapidly and accurately. As such, each trial consisted of at least three measurements (1 initial measurement, 1 measurement near the microphone with the highest recorded SNR, and 1 measurement at the suspected SSL location). Only 1 trial failed to converge after 3 measurements. Table 1 describes the results in detail.

Table 1 SSL performance for exposed sources

Trial	Num. Samples	Initial Error (m)	Final Error (m)	Fused Error (m)
1	3	7.0	0.2	0.2
2	3	7.0	0.9	0.2
3	3	0.5	0.4	0.2
4	3[a]	0.5	0.4	0.4
5	3	0.8	0.5	0.5
6	3	3.4	0.5	0.3
	Mean	3.2	0.5	0.3

[a] Post analysis of the trial showed the robots should have continued one additional run.

In all cases, moving the array elements robotically substantially improved the localization accuracy of the source, averaging a 0.5 m final error in 3D position the final configuration. Fusing the data from all three sample locations in an evidence grid improved accuracy another 0.2m.

Table 2 SSL performance for buried sound sources

Trial	Num. Samples	Initial Error (m)	Final Error (m)	Fused Error (m)
1	3	0.7	0.6	0.4
2	5	3.4	5.7	0.2
3	5	1.6	0.2	0.5
4	5	1.4	4.3	0.4
5	3	0.7	0.2	0.8
	Mean:	1.6	2.2	0.5

When the sound source was buried, however, the results were not as clean. In 3 out of 5 trials, the team did not stop after 3 samples, but instead completed 5 samples. Furthermore, the snapshot performance was extremely variable. The average SSL error after the robots localize a sound source and relocated themselves was 2.8m with $\sigma = 2.0$m. This is in contrast to the exposed sources, which had $\mu = 0.5$m and $\sigma = 0.2$m. This variability was suppressed in the fused evidence grid, which had significantly lower error than the final snapshot, only 0.5m. However, the fused data was not always more accurate than the last snapshot.

This variability was due to our use of the time-delay on arrival algorithms for SSL. When the sound source was exposed, there were direct paths between the sound source and most locations in the environment. A direct path was important because reflections make the signal path longer, meaning greater measured delay and, therefore, greater localization error. With an exposed source, a majority of the robots ended their motion with a direct path between source and receiver. With the buried sources, however, especially where there is only a narrow window in which a direct path can occur, only a very few robots end up with a direct path to the source, while most end up around the edges listening to reflections. Furthermore, even if the robots do localize the buried source correctly, when they moved to surround the source, they often lost the source again as more robots had indirect paths between the source and their microphones. This direct vs indirect path problem is illustrated in Figure 2 . Because very few microphones had a direct path to the source, the snapshot evidence (top) is dominated by a line (as if from a single microphone pair). A good snapshot, however, should have looked more like the fused evidence (bottom), as multiple pairs of microphones strongly contribute to localizing the sound source.

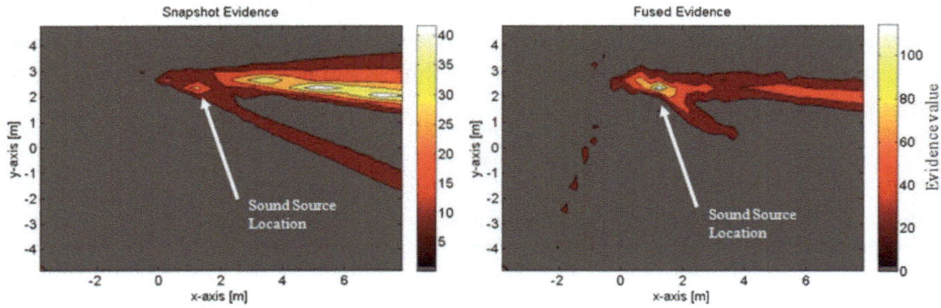

Fig. 2 SSL likelihood plot of a snapshot (left) and fused (right) localization

Fig. 3 Plot of the paths followed by the robotic agents (solid lines) from where they began in the staging area (pentagons) to their first (squares) and second (diamonds) sampling point, followed by their egress path back to the staging area. The sound source location (large x) and static microphones (small triangles) are shown for reference.

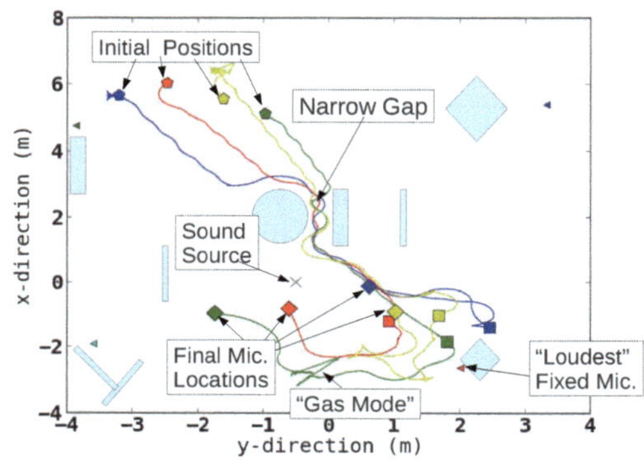

5.3 *Mobility*

Between the vortex obstacle force and low-friction liquid mode, the agents were able to transit between very quickly between sampling locations despite having to navigate around each other and numerous obstacles. Physicomimetic parameters were set based on the authors' experience with this platform in a similar environment [11]. The microphone locations, obstacles and mobile agent paths from exposed sound source trial 3 are shown in figure 3 . Note that all four agents autonomous chose and were able to follow a path through a 1.2m gap as they moved from the staging area to the first sample point, suggesting that this approach allowed reactive navigation through doorways.

6 Conclusion and Future Work

We successfully demonstrated the design of a distributed, reactive control system to move microphones to recursively improve an array. Since different behaviors were required for different phases of this process, we implemented a finite state machine controller with a "solid" mode for clustering near a desired point, a "liquid" mode for transiting cluttered fields and a "gas" mode to allow lost robots to wander until properly localized. This system based on a very simple reactive controller worked very well to form arrays of microphones for GCC sound localization of exposed sound sources, and was substantially better than a static array at locating buried sound sources.

As we are now confident in the basic GCC and physicomimetics FSM approach, we expect to explore this problem in three ways in the future. First of all, we intend to implement alternative localization metrics discussed in Section 2 for achieving coherent microphone arrays. Secondly, we intend to address this problem by examining the SNR of each recording location. As sound travels, the pressure wave amplitude decays resulting in a lower SNR. Therefore, the SNR for locations with a direct path should be higher, even if it occurs in only a narrow band. If the robots could recognize that they were in a good location and use that to adapt physicomimetics parameters then more robots could be encouraged to stay within the optimal localization region for the source, and localization accuracy should improve. Finally, the apparent need for exploration, transit and clustering behaviors suggests that there is much richer design space for a physicomimetics FSM than the one presented here.

Beamforming and multiple sound source localization represent straightforward extensions of this work. The key challenges are identifying which microphones are receiving enough of a signal from a given location or sound source to make a meaningful contribution, and repositioning microphones to address ambiguities.

References

1. Apker, T.B., Lennon, J., Potter, M.A., Sofge, D.: Past point models: Physicomimetics on nonholonomic vehicles. In: AIAA INFOTECH@Aerospace, St. Louis, Missouri, March 29-31 (March 2011)
2. Arkin, R.: Behavior-based Robotics. MIT Press, Boston (1998)
3. Azimi-Sadjadi, M.R., Jiang, Y., Wichern, G.: Properties of randomly distributed sparse acoustic sensors for ground vehicle tracking and localization. In: Proc. of 2006 SPIE Conference on Defense and Security, Orlando, FL, April 2006, vol. 6201 (April 2006)
4. Balch, T., Dellaert, F., Feldman, A., Guillory, A., Isbell, C.L., Khan, Z., Pratt, S.C., Stein, A.N., Wilde, H.: How multirobot systems research will accelerate our understanding of social animal behavior. Proceedings of the IEEE 94(7), 1445–1463 (2006)
5. Coker, J.D., Tewfik, A.H.: Performance synthesis of uav trajectories in multistatic sar. IEEE Transactions on Aerospace and Electronic Systems 47(2), 848–863 (2011)
6. Frechette, M., Létourneau, D., Valin, J.-M., Michaud, F.: Integration of sound source localization and separation to improve dialogue management on a robot. In: Proceedings of IEEE/RSJ International Conference on Intelligent Robots and Systems, IROS (2012)
7. Gage, D.W.: Command and control for many-robot systems. Unmanned Systems 10(4), 28–34 (1992)
8. Girod, L., Estrin, D.: Robust range estimation using acoustic and multimodal sensing. In: Proceedings of the 2001 IEEE/RSJ International Conference on Intelligent Robots and Systems 2001, vol. 3, pp. 1312–1320 (2001)

9. Kira, Z., Potter, M.: Exerting human control over decentralized robot swarms. In: Proceedings of the 4th International Conference on Autonomous Robots and Agents, Wellington, NZ, February 10-12, pp. 566–571 (2009)
10. Krontiris, A., Louis, S., Bekris, K.E.: Multi-level formation roadmaps for collision-free dynamic shape changes with non-holonomic teams. In: 2012 IEEE International Conference on Robotics and Automation (ICRA), pp. 1570–1575 (May 2012)
11. Martinson, E., Apker, T., Bugajska, M.: Optimizing a reconfigurable robotic microphone array. In: 2011 IEEE/RSJ International Conference on Intelligent Robots and Systems (IROS), pp. 125–130 (September 2011)
12. Martinson, E., Fransen, B.: Dynamically reconfigurable microphone arrays. In: 2011 IEEE International Conference on Robotics and Automation (ICRA), pp. 5636–5641 (May 2011)
13. Martinson, E., Schultz, A.: Discovery of sound sources by an autonomous mobile robot. Autonomous Robots 27, 221–237 (2009), doi:10.1007/s10514-009-9123-1
14. William Mather, T., Ani Hsieh, M.: Ensemble synthesis of distributed control and communication strategies. In: 2012 IEEE International Conference on Robotics and Automation (ICRA), pp. 4248–4253 (May 2012)
15. Miura, H., Yoshida, T., Nakamura, K., Nakadai, K.: Slam-based online calibration of asynchronous microphone array for robot audition. In: 2011 IEEE/RSJ International Conference on Intelligent Robots and Systems (IROS), pp. 524–529 (September 2011)
16. Nakadai, K., Nakajima, H., Murase, M., Kaijiri, S., Yamada, K., Nakamura, T., Hasegawa, Y., Okuno, H.G., Tsujino, H.: Robust tracking of multiple sound sources by spatial integration of room and robot microphone arrays. In: Proceedings of the 2006 IEEE International Conference on Acoustics, Speech and Signal Processing, ICASSP 2006, vol. 4, p. IV (May 2006)
17. Nofsinger, G., Cybenko, G.: Distributed chemical plume process detection: Milcom 2005 #1644. In: IEEE Military Communications Conference, MILCOM 2005, vol. 2, pp. 1076–1082 (October 2005)
18. Rodemann, T.: A study on distance estimation in binaural sound localization. In: 2010 IEEE/RSJ International Conference on Intelligent Robots and Systems (IROS), pp. 425–430 (October 2010)
19. Shah, S.K., Pahlajani, C.D., Lacock, N.A., Tanner, H.G.: Stochastic receding horizon control for robots with probabilistic state constraints. In: 2012 IEEE International Conference on Robotics and Automation (ICRA), pp. 2893–2898 (May 2012)
20. Spears, D.F., Thayer, D.R., Zarzhitsky, D.V.: Foundations of swarm robotic chemical plume tracing from a fluid dynamics perspective. International Journal of Intelligent Computing and Cybernetics 2(4), 745–785 (2009)
21. Spears, W.M., Spears, D.F., Hamann, J.C., Heil, R.: Distributed, physics-based control of swarms of vehicles. Autonomous Robots 17(2-3), 137–162 (2004)
22. Thrun, S.: Affine structure from sound. In: Proceedings of Conference on Neural Information Processing Systems (NIPS). MIT Press, Cambridge (2005)
23. Worcester, J., Rogoff, J., Hsieh, M.A.: Constrained task partitioning for distributed assembly. In: 2011 IEEE/RSJ International Conference on Intelligent Robots and Systems (IROS), pp. 4790–4796 (September 2011)

Vision Based Mobile Target Geo-localization and Target Discrimination Using Bayes Detection Theory

Rajnikant Sharma, Josiah Yoder, Hyukseong Kwon, and Daniel Pack

Abstract. In this paper, we develop a technique to discriminate ground moving targets when viewed from cameras mounted on different fixed wing unmanned aerial vehicles (UAVs). First, we develop a extended kalman filter (EKF) technique to estimate position and velocity of ground moving targets using images taken from cameras mounted on UAVs. Next, we use Bayesian detection theory to derive a log likelihood ratio test to determine if the estimates of moving targets computed at two different UAVs belong to a same target or to two different targets. We show the efficacy of the log likelihood ratio test using several simulation results.

1 Introduction

Over the past decade, there has been an increase in the use of Unmanned Aerial Vehicles (UAVs) in several military and civil application that are considered dangerous for human pilots. These applications include surveillance [1], reconnaissance [2], search [3], and fire monitoring [4, 5]. Among the suite of possible sensors, a video camera is inexpensive, lightweight, fits the physical requirements of small fixed wing UAVs, and has a high information to weight ratio. One of the important applications of camera equipped fixed wing UAVs is determining the location of a ground target when imaged from the UAV. The target is geo-localized using the pixel location of the target in the image plane, the position and attitude of the air vehicles, the camera's pose angles, and knowledge of the terrain elevation. Previous target localization work using a camera equipped UAV is reported in [6, 7, 8, 9] and references therein. Barber *et al.* [7] used a camera, mounted on a fixed-wing UAV, to geo-localize a stationary target. They discussed recursive least square (RLS) filtering, bias estimation, flight path selection, and wind estimation to reduce the localization

Rajnikant Sharma · Josiah Yoder · Hyukseong Kwon · Daniel Pack
Academy Center for UAS Research, US Air Force Academy, Colorado
e-mail: {rajnikant.sharma.ctr.in, josiah.yoder.ctr, hyukseong.kwon, daniel.pack}@usafa.edu

M.A. Hsieh and G. Chirikjian (eds.), *Distributed Autonomous Robotic Systems*, Springer Tracts in Advanced Robotics 104,
DOI: 10.1007/978-3-642-55146-8_5, © Springer-Verlag Berlin Heidelberg 2014

errors. Pachter *at el.* [6] developed a vision-based target geo-location technique that uses camera equipped unmanned air vehicles. They jointly estimate the target's position and the vehicles's attitude errors using linear regression resulting in improved target geo-localization. A salient feature of target geo-localization using bearing and range based sensors is the dependence of the measurement uncertainty on the position of the sensor relative to the target. Therefore, the influence of input parameters on nonlinear estimation problems, can be exploited to derive the optimal geometric configurations of a team of sensing platforms. However, maintenance of optimal configurations is not feasible given constraints on the kinematics of typical fixed wing aircraft. Frew [8] evaluated the sensitivity of the target geo-localization to orbit coordination, which enables the design of cooperative line of sight controllers that are robust to variations in the sensor measurement uncertainty and the dynamics of the target tracked. The existing geo-localization techniques are developed for stationary targets, in this paper, we detail a geo-localization technique for mobile ground targets using a camera mounted on a small fixed wing UAV.

Several researchers have used Multiple UAVs for cooperative geo-localization [10, 11, 12], because multi-agent platform provides several advantages including robustness, scalability, and access to more target localization information. The localization information computed at different UAVs can be fused to improve the target geo-location accuracy. Information fusion can be performed either in a centralized or distributed manner. However, irrespective of the information fusion method, centralized or distributed, it is important to determine if geo-location estimates computed by two different UAVs belong to the same target or to two different targets. The first approach is known as a distance based approach where if the Mahalonobis distance [13] between two estimates is less than a threshold then the state estimates belong to same target otherwise to two different targets. Some of the distance based data association methods include nearest neighbors (NN) method [14], probabilistic data association (PDA) [13], and joint probabilistic data association (JPDA) [15]. The second approach is known as appearance or view based data association, which is described in [16] and references therein. The existing data association techniques had been successfully demonstrated either for stationary targets or moving targets sensed either from stationary sensors or sensors mounted on ground robots or small UAVs used for indoor navigation (e.g., quad-rotors). However, the existing data association techniques cannot provide desired accuracy for small fixed wing UAVs. Small fixed wing UAVs imagery, attitude estimates, and UAVs position estimates from GPS are highly noisy which results in target geo-location estimates with high uncertainty. Therefore the distance based data association techniques will lead to high rates of false alarms and miss detections. On the contrary, appearance based data association methods are not affected by uncertainties in target location estimates. However, these methods are computationally expensive, and since small fixed UAVs have limited computation resources, they cannot be implemented onboard small UAVs . Furthermore, the low resolution imagery and altitude of UAVs result in small number target pixels in image, therefore, there are not enough features to perform view based data association.

In this paper, we discuss geo-localization of moving targets using pixel measurements from a camera onborad small fixed wing UAVs. Next, we formulate a hypothesis based on Bayes detection theory and develop a log likelihood ratio test to find if target location estimates from two different UAVs using electro-optical (EO) sensors belong to the same target or to two different targets. We show that the log likelihood ratio test has high probability of correct data association and low probability of false alarm under high errors in position and attitude of the UAVs. Also the test is computationally inexpensive and can be implemented in real-time onboard small fixed wing UAVs.

The rest of the paper is organized as follows. In Section 2, we detail the vison based geo-location of a moving target using a gimballed electro-optical/infrared (EO/IR) camera on board a fixed-wing UAV. In Section 3, we develop a log likelihood ratio for determining the correspondence among the target state estimates. In Section 4, we present simulation results and probability of correct association and probability of false alarm. Finally, in Section 5, we give our conclusions.

2 Geo-location

In this section, we detail a technique of geo-localizing a mobile target in inertial coordinate using gimballed EO/IR camera on board a small fixed-wing UAV. We assume that the targets are detected with probability one. The technique presented here is an extension of geo-localization technique for a stationary target [17] to a mobile target.

In this paper, we use the camera model detailed in [17], which is shown in Figure 1, where f is the focal length in pixel units and P converts pixels to meters. The location of the projection of the target is expressed in the camera frame as $(P\varepsilon_x, P\varepsilon_y, Pf)$, where ε_x and ε_y are the pixel location, in pixels, of the target. The distance from the origin of the camera frame to the pixel location $(\varepsilon_x, \varepsilon_y)$, as shown in Figure 1, is PF where

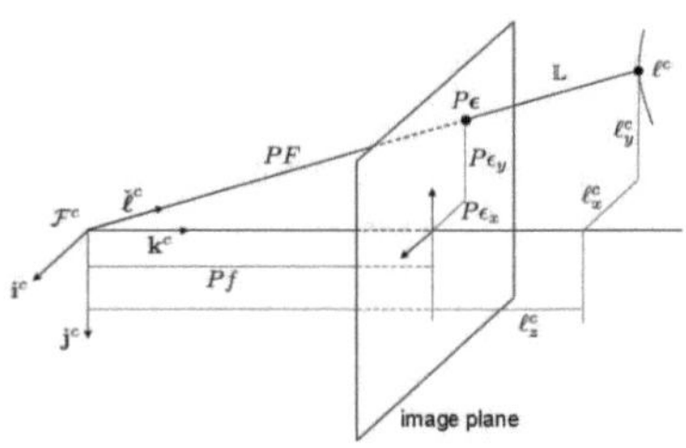

Fig. 1 [17] The camera frame. The target in the camera frame is represented by l^c. The projection of the target onto the image plane is represented by ε. The pixel location $(0,0)$ corresponds to the center of the image, which is assumed to be aligned with the optical axis. The distance to the target is given by $\mathbb{L}$. ε and f are in units of pixels. l is in units of meters.

$$F = \sqrt{f^2 + \varepsilon_x^2 + \varepsilon_y^2}. \tag{1}$$

Finally using Figure 1, we can write the expression for the unit direction vector as

$$\check{l}^c = \frac{l^c}{\mathbb{L}} = \frac{1}{\sqrt{\varepsilon_x^2 + \varepsilon_y^2 + f^2}} \begin{pmatrix} \varepsilon_x \\ \varepsilon_x \\ f \end{pmatrix}. \tag{2}$$

Let $l = p_t^i - p_{uav}^i$ be the relative position vector between the moving target and the UAV, where $p_{uav}^i = (p_n, p_e, p_d)^\top$ is the UAV's (north, east, down) position in inertial frame measured by an onboard GPS receiver and $p_t^i = (t_n, t_e, 0)^\top$ is the target's (north, east, down) position in the inertial frame. We define $\mathbb{L} = ||l||$ and $\check{l} = \frac{l}{\mathbb{L}}$. From the described geometry, we obtain the following relationship

$$\begin{aligned} p_t^i &= p_{uav}^i + \mathscr{R}_b^i \mathscr{R}_g^b \mathscr{R}_c^g l^c, \\ &= p_{uav}^i + \mathbb{L}(\mathscr{R}_b^i \mathscr{R}_g^b \mathscr{R}_c^g \check{l}^c), \end{aligned} \tag{3}$$

where $\mathscr{R}_b^i = \mathscr{R}_b^i(\phi, \theta, \psi)$ is the rotation matrix from the body to the inertial frame, $\mathscr{R}_g^b = \mathscr{R}_g^b(\alpha_{az}, \alpha_{el})$ is the rotation matrix from the gimbal to the body frame, and $\mathscr{R}_c^g$ is the rotation matrix from the camera to the gimbal frame. We assume that the UAV's attitude $(\phi,\ \theta, \psi)^\top$ (roll, pitch, yaw) is available for geo-localization. We also assume that the gimbal azimuth and elevation angles $(\alpha_{az}, \alpha_{el})$ are available and use the controller detailed in [17] to point the camera in the direction of the target.

The objective of the geo-localization problem is to estimate the range to the target, $\mathbb{L}$, which can be estimated using the flat earth model as shown in Figure 2. If the UAV's height above the ground, $h = -p_d$, is known then the range estimate can be computed as

$$\mathbb{L} = \frac{h}{\mathbf{k}^i \cdot \mathscr{R}_b^i \mathscr{R}_g^b \mathscr{R}_c^g \check{l}^c}, \tag{4}$$

where $\mathbf{k}^i$ is the unit vector in inertial frame pointing towards the center of the earth.

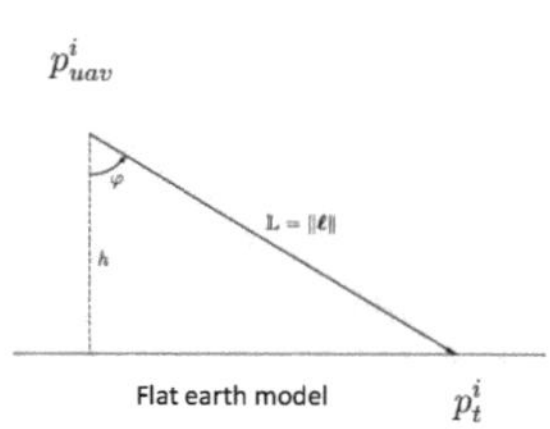

Fig. 2 Range estimation using flat-earth assumption

Using the flat earth model, we can write the expression for the geo-location estimate as

$$p_t^i = p_{uav}^i + h \frac{\mathscr{R}_b^i \mathscr{R}_g^b \mathscr{R}_c^g \check{l}^c}{\mathbf{k}^i \cdot \mathscr{R}_b^i \mathscr{R}_g^b \mathscr{R}_c^g \check{l}^c}. \tag{5}$$

2.1 Geo-location Using Extended Kalman Filter

The geo-location estimate in Equation (5) provides a one-shot estimate of the target location. Unfortunately, this equation is highly sensitive to measurement errors, especially attitude estimation errors of the airframe. Also, the velocity and the heading of the target also need to be estimated. In this section we will describe the use of the Extended Kalman Filter (EKF) to estimate the location, velocity, and heading of a mobile ground target. Rearranging (5), we get

$$p_{uav}^i = h(x) = p_t^i - \mathbb{L} \mathscr{R}_b^i \mathscr{R}_g^b \mathscr{R}_c^g \check{l}^c. \tag{6}$$

Since p_{uav}^i is measured by the GPS, it will be used as the measurement equation, assuming that the GPS noise is Gaussian with zero-mean.

We assume that the ground mobile target moves with constant velocity but can change its heading instantaneously. The target position $p_t^i = [t_n \; t_e \; 0]$ consist of north and east position coordinates and the target motion model can be written as

$$\dot{p}_t^i = \begin{pmatrix} v_n \\ v_e \\ 0 \end{pmatrix}, \tag{7}$$

where v_n is the velocity of UAV in north direction and v_e is the velocity of UAV in east direction.

Since $\mathbb{L} = ||p_t^i - p_{uav}^i||$, we have

$$\dot{\mathbb{L}} = \frac{(p_t^i - p_{uav}^i)\top(\dot{p}_t^i - \dot{p}_{uav}^i)}{\mathbb{L}}, \tag{8}$$

where for constant altitude flight, $\dot{p}_{uav}^i$ can be approximated as

$$\dot{p}_{uav}^i = \begin{pmatrix} \hat{V}_g \cos\hat{\chi} \\ \hat{V}_g \cos\hat{\chi} \\ 0 \end{pmatrix}. \tag{9}$$

In the above equation, $\hat{V}_g$ and $\hat{\chi}$ are the estimated ground speed and the course angle obtained using the EKF as presented in [17]. The input to the geo-localization algorithm is the position and the ground speed of the MAV in the inertial frame as obtained from GPS described in [17], the estimate of the line-of-sight vector as given in (2), and the attitude as estimated by the EKF described in [17].

The geo-localization algorithm is an EKF with state

$$\hat{x}_t = [\hat{t}_n, \hat{t}_e, \hat{\mathbb{L}}, \hat{v}_n, \hat{v}_e]^\top, \tag{10}$$

where $\hat{t}_n$ is the estimated target north position, $\hat{t}_e$ is the estimated target east position, $\hat{\mathbb{L}}$ is the estimated distance between target and the UAV, $\hat{v}_n$ is the estimated target velocity in north direction, and $\hat{v}_e$ is the estimated target velocity in east direction. The prediction equation can be written as

$$\dot{\hat{x}}_t = f(x) = \begin{pmatrix} \hat{v}_n \\ \hat{v}_e \\ \frac{(\hat{p}_t^i - \hat{p}_{uav}^i)^\top (\dot{\hat{p}}_t^i - \dot{\hat{p}}_{uav}^i)}{\hat{\mathbb{L}}} \\ 0 \\ 0 \end{pmatrix}. \tag{11}$$

3 Target Discrimination Using Bayes Detection Theory

In this section, we discuss the problem of target discrimination required for information fusion. Before fusing any information, it is important to find if the state estimates from two different platforms belongs to the same target or to two different targets. To solve this problem we use Bayes Detection theory [18] to develop a log likelihood ratio test.

Let us assume that, under hypothesis H_1, the estimates from two different UAV platforms are of same target, and under H_0 the estimates from two UAV platforms belong to two different targets.

Let $X_1 \in N[m_1(t), P_1(t)]$ and $X_2 \in N[m_2(t), P_2(t)]$ be the target state estimate of two UAVs (excluding range between an UAV and a target), where m and P are the mean and covariance respectively. We define a new random variable $Y = X_1 - X_2$, drawn from a Gaussian probability density function (p.d.f.), $Y \in N[m^y, P^y]$, where $m^y = m_1 - m_2$ and $P^y = P_1 + P_2$. If X_1 and X_2 are estimates of the same target then $m^y = 0$, otherwise $m^y \neq 0$. We assume the following initial distribution

$$P[X_1 = X_2] = \tau,$$
$$P[X_1 \neq X_2] = 1 - \tau,$$

where $0 \leq \tau \leq 1$ is the probability of two targets being same. We sample the output each sampling period and obtain n samples. In other words

$$H_1 : Y_i \in N[0, P_i^y],$$
$$H_0 : Y_i \in N[m_i^y, P_i^y].$$

The probability density of Y_i under each hypothesis is

$$f_{(Y_i|X_1=X_2)} = \frac{1}{(2\pi)^{\frac{n}{2}}|P_i^y|^{\frac{1}{2}}} \exp\left[-\frac{1}{2}(y_i^\top (P_i^y)^{-1} y_i)\right],$$
$$f_{(Y_i|X_1\neq X_2)} = \frac{1}{(2\pi)^{\frac{n}{2}}|P_i^y|^{\frac{1}{2}}} \exp\left[-\frac{1}{2}(y_i - m_i^y)^\top (P_i^y)^{-1}(y_i - m_i^y)\right].$$

Because the Y_i's are not independent, we cannot write the joint probability density of $Y_1,\cdots,Y_n$ simply as the product of the individual probability. However, we can write the joint probability using Bayes chain rule and conditional probability.

$$\begin{aligned} f_{(Y_1,\cdots,Y_n|X_1=X_2)} &= f_{(Y_n|Y_n-1)} f_{(Y_n-1|Y_n-2)} \cdots f_{(Y_2|Y_1)} f_{(Y_1)}, \\ &= \prod_{i=1}^{n} \frac{1}{(2\pi)^{\frac{n}{2}}|P_i^y|^{\frac{1}{2}}} \exp\left[-\frac{1}{2}(y_i^\top (P_i^y)^{-1} y_i)\right], \\ f_{(Y_1,\cdots,Y_n|X_1\neq X_2)} &= f_{(Y_n|Y_n-1)} f_{(Y_n-1|Y_n-2)} \cdots f_{(Y_2|Y_1)} f_{(Y_1)}, \\ &= \prod_{i=1}^{n} \frac{1}{(2\pi)^{\frac{n}{2}}|P_i^y|^{\frac{1}{2}}} \exp\left[-\frac{1}{2}(y_i - m_i^y)^\top (P_i^y)^{-1}(y_i - m_i^y)\right]. \end{aligned}$$

The likelihood ratio can be written as

$$\begin{aligned} l(y_1,\cdots,y_n) &= \frac{f_{(Y_1,\cdots,Y_n|X_1=X_2)}}{f_{(Y_1,\cdots,Y_n|X_1\neq X_2)}}, \\ &= \frac{\prod_{i=1}^{n} \frac{1}{(2\pi)^{\frac{n}{2}}|P_i^y|^{\frac{1}{2}}} \exp\left[-\frac{1}{2}(y_i^\top (P_i^y)^{-1} y_i)\right]}{\prod_{i=1}^{n} \frac{1}{(2\pi)^{\frac{n}{2}}|P_i^y|^{\frac{1}{2}}} \exp\left[-\frac{1}{2}(y_i - m_i^y)^\top (P_i^y)^{-1}(y_i - m_i^y)\right]}. \end{aligned}$$

After canceling common terms and taking the logarithm, we have

$$\log l(y_1,\cdots,y_n) = \frac{1}{2}\sum_{i=1}^{n}(m_i^y)^\top (P_i^y)^{-1} m_i^y - \sum_{i=1}^{n}(m_i^y)^\top (P_i^y)^{-1} y_i,$$

from which the likelihood ratio test can be written as

$$\phi_\tau(y_1,\cdots,y_n) = \begin{cases} 1 \; if \; \log l(y_1,\cdots,y_n) > \log\frac{1-\tau}{\tau} \\ 0 \; if \qquad otherwise \end{cases}. \tag{12}$$

In other words, $\phi_\tau(y_1,\cdots,y_n) = 1$ means that target state estimates computed at different UAVs belong to the same target, and $\phi_\tau(y_1,\cdots,y_n) = 0$ means that the target state estimates belong to two different targets. We can compute probability of false alarm or incorrect data association as

$$\begin{aligned} P_{FA} &= P[\phi_\tau(y_1,\cdots,y_n) = 1|Y_1 \neq Y_2], \\ &= E_{Y_1\neq Y_2}\phi_\tau(y_1,\cdots,y_n). \end{aligned}$$

Similarly, we can compute probability of correct data association

$$P_D = P[\phi_\tau(y_1,\cdots,y_n) = 1|Y_1 = Y_2],$$
$$= E_{Y_1=Y_2}\phi_\tau(y_1,\cdots,y_n).$$

Clearly, P_{FA} and P_D depends on the assumed a priori distribution parameter τ, the number of samples, m^y, and P^y. Since, we are dealing with multi-dimensional normal p.d.f.s, the integration in above expressions cannot be computed directly. However in the next section, we will compute P_{FA} and P_D using simulations.

4 Results

In this section, we develop a simulation environment in MATLAB/Simulink to geo-localize mobile targets using two fixed wing UAVs and analyze the log likelihood ratio test developed in the previous section. Figure 3(a) and Figure 3(b) show the simulation snap shots of two different scenarios considered in this paper. In Figure 3(a) two different UAVs are geo-localizing the same moving ground target, which is in field-of-view of both the UAVs. On the other hand, in Figure 3(b) two different UAVs are geo-localizing two different ground targets moving in same direction with same velocity. The objective is to use the log likelihood test (12) to determine if the target state estimates computed at two different UAVs are of a same target or two different targets.

First, we show results of geo-location of a target using a single UAV. Some of the important parameters used in the simulations are as follows

- GPS error variance: $[\sigma_n = 15m,\ \sigma_e = 15m,\ \sigma_d = 20m]$,
- UAV attitude error variance: $\sigma_{att} = 0.1\ rad$,
- Target speed $V_t = 5\ m/s$,
- UAV air speed $V_a = 20\ m/s$.

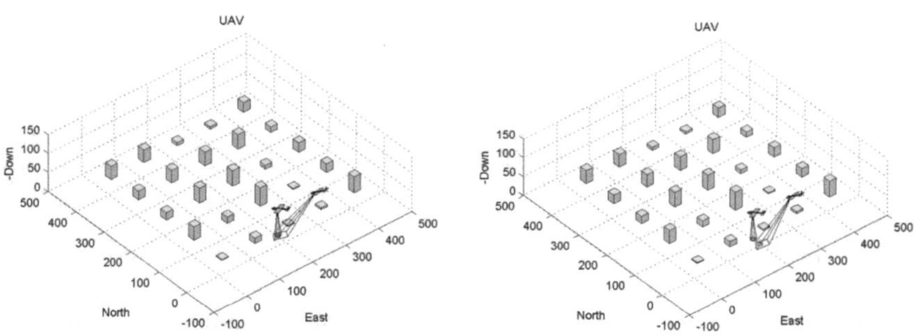

(a) (b)

Fig. 3 (a) UAV based geo-location: Two UAVs geo-localizing a single ground moving target. (b) UAV based geo-location: Two UAVs geo-localizing two different ground moving target.

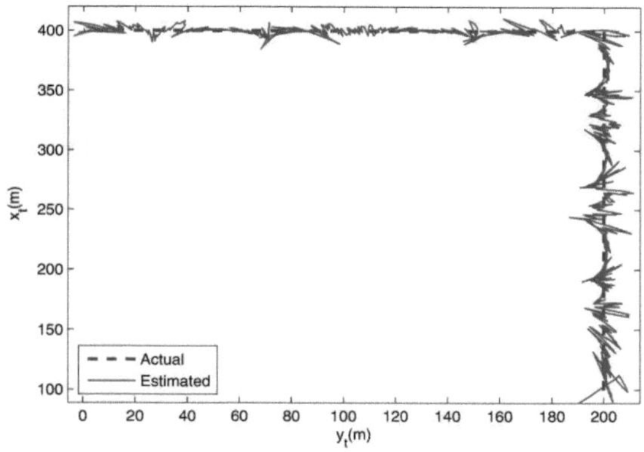

Fig. 4 UAV based geo-location: Actual and estimated trajectories of a ground moving target. The blue dashed curve and solid red curve represents the actual and estimated trajectories respectively

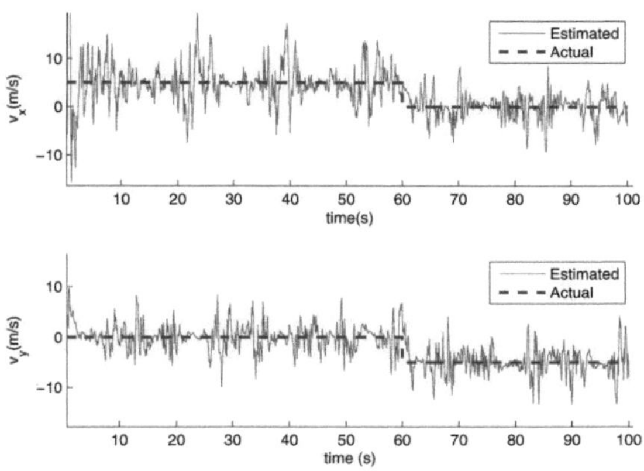

Fig. 5 UAV based geo-location: Velocity estimates of a ground moving target in x and y direction. The blue dashed curve and solid red curve represents the actual and estimated velocities respectively

Figure 4 shows the actual and estimated trajectory of a ground moving target. It can be seen that the estimated trajectory has the same behavior but has significant amount of uncertainty. The location estimates are very sensitive to the UAV attitude errors, the more errors in the UAV attitude the more the uncertainty in the location estimates of the target. Figure 5 shows the actual and estimated velocities of the target in x and y direction. It can be seen that the velocity estimates are quite noisy but unbiased because there is no measurement of absolute velocity of the target, rather it is obtained from the target position estimates.

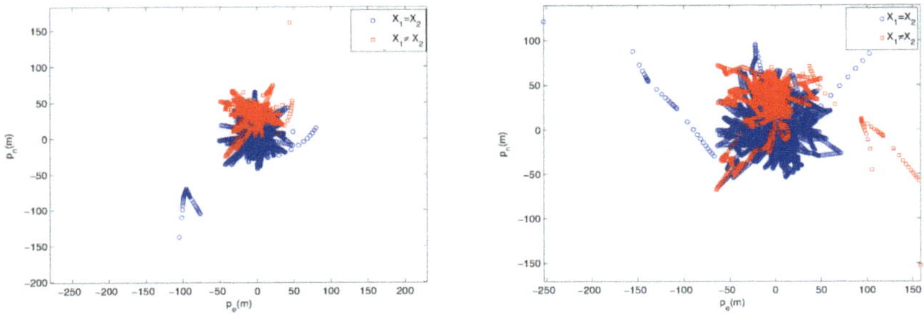

(a) (b)

Fig. 6 (a) This figure shows the distribution of random variable $Y = X_1 - X_2$ for $X_1 = X_2$ and $X_1 \neq X_2$ at attitude errors $\sigma_{att} = 0.1$ *rad* and $35m$ of distance between two targets. The blue circles represent distribution of $X_1 = X_2$ and red squares represent $X_1 \neq X_2$. (b) This figure shows the distribution of random variable $Y = X_1 - X_2$ for $X_1 = X_2$ and $X_1 \neq X_2$ at attitude errors $\sigma_{att} = 0.6$ *rad* and $35m$ of distance between two targets. The blue circles represent distribution of $X_1 = X_2$ and red squares represent $X_1 \neq X_2$.

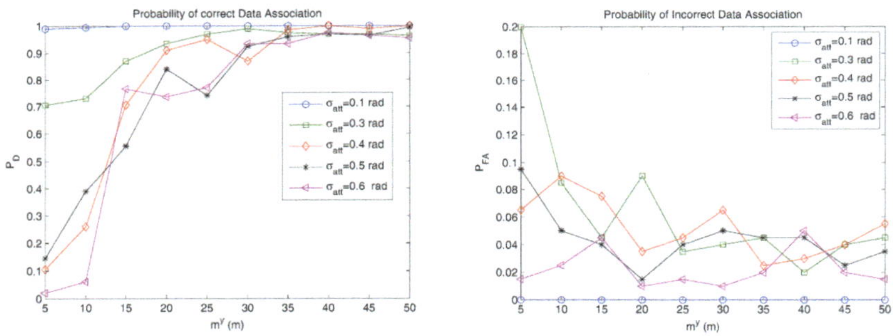

(a) (b)

Fig. 7 (a) This figure shows the probability of correct data association obtained using the likelihood test for different attitude errors and m^y. Each probability number is generated over 200 runs over $n = 20$ sample for each run. (b) This figure shows the probability of false alarm obtained using the likelihood ratio test for different attitude errors and m^y. Each probability number is generated over 200 runs over $n = 20$ sample for each run. Also for to generate these probabilities we assume that actual distance between two targets is equal to m^y.

Next we compute the distribution of $Y = X_1 - X_2$ for $X_1 = X_2$ and $X_1 \neq X_2$ for the attitude errors $\sigma_{att} = 0.1$ *rad* and $\sigma_{att} = 0.6$ *rad* as shown in Figure 6(a) and Figure 6(b) respectively. It can be seen that the distributions of $X_1 = X_2$ and $X_1 \neq X_2$ overlap and the overlap region increases with larger attitude errors. These distributions show the challenge in discriminating the two distributions using limited amount of samples.

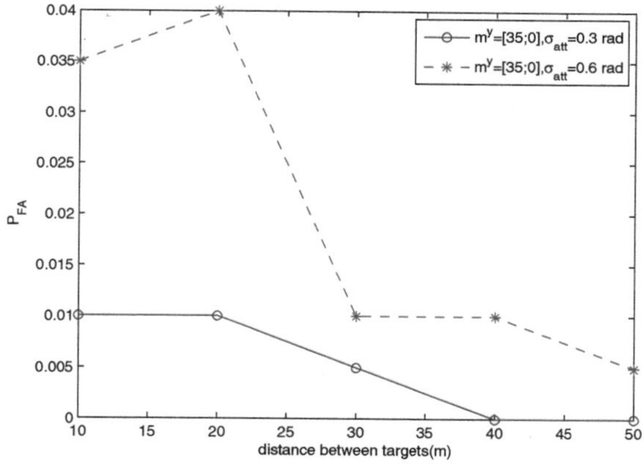

Fig. 8 This figure shows the probability of false alarm obtained using the likelihood ratio test for different attitude errors and distance between targets at a constant $m^y = [35;0]$

Figure 7(a) and Figure 7(b) show the probability of correct association and probability of false alarm with respect to the parameter m^y at different levels of UAV attitude errors. In Figure 7(b), we kept the distance between two the UAVs equal to m^y to compute the P_{FA}. It can be seen, from Figure 7(a), that the test provides accuracy greater then 90% for $m^y \geq 35$ m with low probability of false alarm. In Figure 8, we keep $m^y = 35m$ constant and plot P_{FA} with respect to the actual distance between two targets for different UAV attitude errors. It can be seen that P_{FA} is smaller than 0.01 after the distance between two targets is greater than $m^y = 35m$.

Simulation results presented in this section show that the Bayes log likelihood ratio test developed in this paper can discriminate targets that are very close and are moving in the same direction with the same velocity most of the time without requiring the appearance features. The test is computationally inexpensive and can be implemented in real-time onboard UAVs to fuse information to accurately geo-localize ground moving targets.

5 Conclusions

In this paper, we have detailed a technique for geo-localization of moving targets using pixel measurements from a camera mounted on a fixed wing UAV. We have formulated a hypothesis based on Bayes's detection theory and developed a log likelihood ratio to find if the target state estimates computed by two different UAVs using electro-optical sensors are of the same target or of two different targets. We have shown that the test for finding correspondence has high probability of correct data association and low probability of false alarm under high errors in UAV position and attitude errors. Also the Bayes log likelihood ratio test is computationally inexpensive and can be computed in real-time onboard on smaller UAVs.

Currently, we have only the addressed data association problem using simple binary hypothesis. Our future work will be to extend this test to discriminate multiple moving targets geo-localized by multiple UAVs. Furthermore, we will explore the option of incorporating UAV path planning to improve the probability of correct association while minimizing the probability of false alarm. Lastly, we are also interested in determining how the altitude of the UAV affects the accuracy of the algorithm.

References

1. Quigley, M., Goodrich, M.A., Griffiths, S., Eldredge, A., Beard, R.W.: Target acquisition, localization, and surveillance using a fixed-wing mini-UAV and gimbaled camera. In: Proc. IEEE Int. Conf. Robotics and Automation ICRA 2005, pp. 2600–2605 (2005)
2. Ayyagari, A., Harrang, J.P., Ray, S.: Airborne information and reconnaissance network. In: Proc. IEEE Military Communications Conf. MILCOM 1996, vol. 1, pp. 230–234 (1996)
3. Beard, R.W., McLain, T.W.: Multiple UAV cooperative search under collision avoidance and limited range communication constraints. In: Proc. 42nd IEEE Conference on Decision and Control, December 9-12, vol. 1, pp. 25–30 (2003)
4. Casbeer, D.W., Beard, R.W., McLain, T.W., Li, S.-M., Mehra, R.K.: Forest fire monitoring with multiple small UAVs. In: Proc. American Control Conf the 2005, pp. 3530–3535 (2005)
5. Sujit, P.B., Kingston, D., Beard, R.: Cooperative forest fire monitoring using multiple UAVs. In: Proc. 46th IEEE Conf. Decision and Control, pp. 4875–4880 (2007)
6. Pachter, M., Ceccarelli, N., Chandler, P.R.: Vision-based target geo-location using camera equipped MAVs. In: Proc. 46th IEEE Conf. Decision and Control, pp. 2333–2338 (2007)
7. Barber, D.B., Redding, J.D., McLain, T.W., Beard, R.W., Taylor, C.N.: Vision-based target geo-location using a fixed-wing miniature air vehicle. Journal of Intelligent and Robotic Systems 47(4) (December 2006)
8. Frew, E.W.: Sensitivity of cooperative target geolocalization to orbit coordination. Journal of Guidance, Control, and Dynamics 31(4) (August 2008)
9. Ross, J.A., Geiger, B.R., Sinsley, G.L., Horn, J.F., Long, L.N., Niessner, A.F.: Vision-based target geolocation and optimal surveillance on an unmanned aerial vehicle. In: AIAA, Guidance, Navigation, and Control Confrence, Honolulu, Hawaii (August 2008)
10. Olfati-Saber, R.: Distributed kalman filter with embedded consensus filters. In: 44th IEEE Conference on Proc. and 2005 European Control Conference Decision and Control CDC-ECC 2005, December 12-15, pp. 8179–8184 (2005)
11. Niehsen, W.: Information fusion based on fast covariance intersection filtering. In: Proc. Fifth International Conference on Information Fusion, July 8-11, vol. 2, pp. 901–904 (2002)
12. Casbeer, D.W., Beard, R.: Distributed information filtering using consensus filters. In: American Control Conference on Proc. ACC 2009, June 10-12 (2009)
13. Kirubarajan, T., Yakov, B.-S.: Probabilistic Data Association Techniques for Target Tracking in Clutter. Proceedings of the IEEE 92(3), 536–537 (2004)
14. Leonard, J.J., Durrant-Whyte, H.F., Cox, I.J.: Dynamic map building for an autonomous mobile robot. The International Journal of Robotics Research 11(4), 286–298 (1992), `http://ijr.sagepub.com/content/11/4/286.abstract`

15. Yaakov, B.-S., Thomas, E.F.: Tracking and Data Association. Academic, NewYork (1988)
16. Gil, A., Reinoso, O., Mozos, O., Stachnissi, C., Burgard, W.: Improving data association in vision-based SLAM. In: 2006 IEEE/RSJ International Conference on Intelligent Robots and Systems (2006)
17. Beard, R., McLain, T.W.: Small Unmanned Aircraft: Theory and practice. Princeton University Press (2011)
18. Moon, T.K., Stirling, W.C.: Mathematical Methods and Algorithms in Signal Processing. Prentice-Hall, Upper Saddle River (2000)

Part II
Task Allocation and Coordination Strategies

A Real World Coordination Framework for Connected Heterogeneous Robotic Systems

Nicola Bezzo, Mike S. Anderson, Rafael Fierro, and John Wood

Abstract. In this paper we consider the problem of coordinating robotic systems with different kinematics, sensing and vision capabilities, to achieve certain mission goals. An approach that makes use of a heterogeneous team of agents has several advantages when cost, integration of capabilities, or possible large search areas need to be considered. A heterogeneous team allows for the robots to become "specialized", accomplish sub-goals more effectively, thus increasing the overall mission efficiency. We consider connectivity constraints and realistic communication, exploiting mobility to implement a power control algorithm that increases the *Signal to Interference plus Noise Ratio* (SINR) among certain members of the network. We also create realistic sensing fields and manipulation by using the geometric properties of the sensor field-of-view and the manipulability metric, respectively. The control strategy for each agent of the heterogeneous system is governed by an artificial physics law that considers the different kinematics of the agents and the environment, in a decentralized fashion. We show that the network is able to stay connected at all times and covers the environment well. We demonstrate the applicability of the proposed strategy through simulation results implementing a pursuit-evasion game in a cluttered environment.

1 Introduction

In recent years we have witnessed an increase in the use of mobile robots for different applications spanning from military to civilian operations. Search and rescue

Nicola Bezzo · Mike S. Anderson · Rafael Fierro
Department of Electrical & Computer Engineering, University of New Mexico, Albuquerque, NM 87131, USA
e-mail: {nicbezzo,msander,rfierro}@ece.unm.edu

John Wood
Department of Mechanical Engineering, University of New Mexico, Albuquerque, NM 87131, USA
e-mail: jw@unm.edu

M.A. Hsieh and G. Chirikjian (eds.), *Distributed Autonomous Robotic Systems*, Springer Tracts in Advanced Robotics 104,
DOI: 10.1007/978-3-642-55146-8_6, © Springer-Verlag Berlin Heidelberg 2014

missions, disaster relief operations, and surveillance are just few examples of scenarios where the use of autonomous and intelligent robotic systems is preferred over the use of human first responders. In such operations wireless communication needs to be reliable over the robotic network to maneuver the unmanned vehicles and transmit information. We are interested in heterogeneous robotic systems with agents having different kinematics, sensing behaviors, and functionalities. For instance we will consider quadrotor aerial vehicles, that can be approximated as holonomic agents, interacting with ground robots (*e.g.*, non-holonomic, car-like, etc.), both with different communication ranges and sensing and manipulation patterns. Fig. 1 shows an example of heterogeneous systems with quadrotors cooperating with ground vehicles and crawling agents, all systems acting as communication and sensing relays.

Within this paper the contribution to the current research on distributed robotic systems is fourfold: *i)* we consider heterogeneous robotic systems with different dynamics and realistic communication analysis, sensing geometries, and manipulation constraints, in a decentralized fashion; *ii)* we build a power control algorithm for communication purposes to improve the SINR among certain members of the network; *iii)* we extend our previous work [1] considering a more general and realistic scenario while showing that the heterogeneous system stays connected all the time; and *iv)* in the simulation result we implement a game theoretical case with an adversarial opponent that defends a certain goal target from the heterogeneous system. Throughout this work we integrate together several tools for coordination and control of distributed heterogeneous robotic systems.

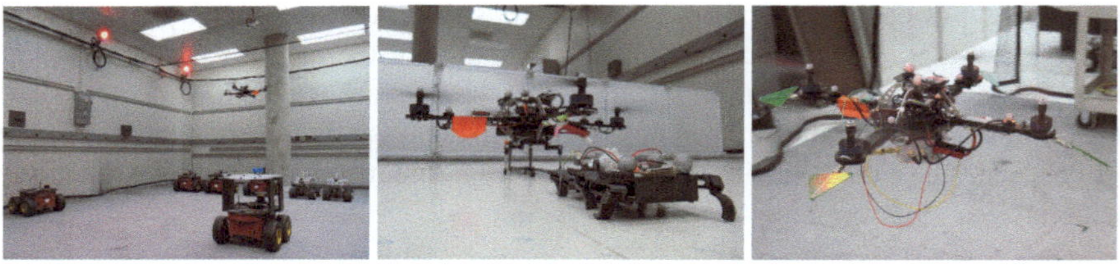

(a) (b) (c)

Fig. 1 (a) A group of Pioneer P3-AT vehicles cooperating with a quadrotor [2] part of the heterogeneous test bed at the MARHES Laboratory; (b) Deployment of the OctoRoACH crawling robot [3] using a quadrotor; and (c) An example of aerial mobile relay with four directional antennas

1.1 Related Work

The use of heterogeneity in robotic applications is a recent topic of research that is attracting several researchers because of the challenges created when dealing with multi-agent systems having different kinematics, sensing, and manipulation capabilities. Authors in [4] consider formally a heterogeneous system and analyze its properties based on graph coloring techniques to assign colors to different types of agents. Similarly to the work presented in this paper, authors in [5] use agents with different dynamics and capabilities to execute multiple missions in a decentralized fashion considering task sequencing and a consensus-based technique.

In missions involving multi-agent systems, it is necessary to consider wireless communication to maintain network connectivity at all times. The robotics and control community are very active in investigating the integration of communication in robotics applications, because the uncertainties found in wireless channels can compromise the performance of the entire multi-agent system. For instance, authors in [6] propose a modified Traveling Salesperson Problem to navigate an underwater vehicle in a sensor field, using a realistic model that considers acoustic communication fading effects. In [7] a Rician fading model for the communication channel is utilized in a pursuit-evasion game with two mobile agents moving in a cluttered environment. Similarly to the work presented in this paper, the authors in [8] present a multi-agent system with interaction between aerial and ground vehicles based on task assignment for complex missions. From a graph-theoretical point of view and more recently in [9], a survey about graph connectivity is provided in mobile robot swarms, discussing different approaches and algorithms to maintain and optimize connectivity among mobile robot networks.

The communication community has been recently investigating cognitive radio antennas to improve the SINR in cellular networks, [10]. The main idea around this type of device is to change the transmission and reception parameters to improve the overall communication quality. One of the most common ways to improve the SINR is to use *Power Control* (PC) algorithms in which all wireless devices adjust their power level to reach a desired SINR threshold [11, 12]. In the work presented in this paper we consider a similar PC approach, but we exploit the mobility of the mobile agents to change the received power at a certain location and reach a desired SINR.

Finally from a sensing point of view, authors in [13] present an optimization framework to maneuver aerial vehicles equipped with cameras to perceive a certain area based on field of view properties.

The remainder of this paper is organized as follows. In Section 2, we define the heterogeneous system and formulate the connectivity problem considering relay, sensor, and manipulator agents. Then, in Section 3, we present the connectivity constraints and a consensus algorithm to equilibrate the network distribution, followed by the motion constraints in Section 4. In Section 5, we analyze a power control method to improve the SINR over the network and in Section 6, we present the sensing and manipulation constraints. Finally in Section 7, we show simulation results and draw conclusions in Section 8.

2 Heterogeneous Connected Robotic System

In this section we give a formal definition of a heterogeneous robotic network followed by the problem formulation and connectivity constraints used to create interactions among the hybrid network.

Definition 1. (Heterogeneous System): A network of $\mathcal{N}$ robots is called *heterogeneous* if the members of the network are interconnected, act together toward a common objective and if the following conditions hold:

- one or more agents in the network have different motion dynamics with respect to other agents in the system;
- one or more agents in the network have different sensing/manipulation constraints or improved wireless communications abilities with respect to other agents in the systems, but all agents have at least some wireless communication capabilities.

2.1 Heterogeneous Network Topology

While the theoretical analysis presented here can be generalized for any type of network, we decide to focus on heterogeneous groups made of three types of mobile agents:

- $\mathcal{N}_c$ communication relays with communication range $R_c > 0$ and holonomic kinematics (*i.e.*, aerial vehicles like quadrotors). The set of relays is denoted by $\mathcal{A}_c$.
- $\mathcal{N}_s$ mobile sensors with communication range $0 < R_s < R_c$ and non-holonomic kinematics given by the bicycle model

$$\mathbf{u}_k = \begin{bmatrix} \dot{x} \\ \dot{y} \\ \dot{\theta} \\ \dot{\gamma} \end{bmatrix} = \begin{bmatrix} \cos(\theta) & 0 \\ \sin(\theta) & 0 \\ \frac{1}{L}\tan(\gamma) & 0 \\ 0 & 1 \end{bmatrix} \begin{bmatrix} v \\ w \end{bmatrix}, \qquad w = \lambda_s(\gamma_d - \gamma) \tag{1}$$

  where L is the distance between the front and rear axles, v is the velocity, w is the steering command described by a 1st order linear servo model, λ_s is the servo gain, and γ_d is the desired steering angle. The set of all mobile sensors is denoted by $\mathcal{A}_s$.
- $\mathcal{N}_m$ manipulator agents with communication range $0 < R_m (= R_s) < R_c$ and non-holonomic kinematics (1).The set of all manipulators is denoted by $\mathcal{A}_m$.

The specific problem we are interested in this paper is the following:

Problem 1. Deployment of Heterogeneous Robotic Networks: Given a heterogeneous robotic network of $\mathcal{N}$ agents partitioned by $\mathcal{N}_c$, $\mathcal{N}_s$, and $\mathcal{N}_m$, find a set of feasible policies $\mathbf{u}_i \in \mathcal{U}$ for each agent such that the workspace of interest $\mathcal{W}$ is well covered, the network is always connected to a fixed base station b, and it is possible to reach and manipulate a target $\mathcal{D}$ protected by an adversarial opponent $\mathcal{T}$ having unknown dynamics $\mathbf{u}_{\mathcal{T}}$.

The adversarial opponent attempts to capture the manipulator agents while the mobile sensors try to pursue and capture the adversary, if detected.

Each agent in the group has some sensing capabilities that are explored in detail in the following sections. For now we will focus on the connectivity problem and formulate an algorithm to expand the network and cover a specific environment.

3 Connectivity Constraints

Following our previous work [1] we build a connectivity algorithm by taking advantage of the communication properties of the heterogeneous network. Specifically we formulate connectivity constraints to expand the input set (accelerations, velocities, and in turn positions) the agents can choose from, while still guaranteeing connectivity at all times.

We define that a relay agent i can communicate with another relay agent j if and only if $j \in \mathcal{B}_c^i$ with $\mathcal{B}_c^i = \mathcal{B}(\mathbf{x}_i, R_c)$ the ball centered in i of radius R_c. A mobile sensor k (or equivalently a mobile manipulator q) can communicate with a relay i if and only if $k(\text{or}\, q) \in \mathcal{B}_s^i = \mathcal{B}(\mathbf{x}_i, R_s)$. However a relay agent i can communicate with a mobile sensor k (or equivalently a mobile manipulator q) if $k(\text{or}\, q) \in \mathcal{B}_c^i = \mathcal{B}(\mathbf{x}_i, R_c)$. Therefore, by exploiting this last constraint we can expand the sensor and manipulator agents in the environment relaxing continuous bidirectional communication constraints and thus explore a larger area of the workspace. These agents return within range of bidirectional communication with the relay when they have information relevant to the entire network.

At the beginning of a mission we consider a connected graph having the nodes placed in random positions. Also we create the following initial conditions

$$\begin{aligned} &\forall i \in \mathcal{A}_c, \exists j \in \mathcal{A}_c, i \neq j \text{ s.t. } i \in \mathcal{B}(\mathbf{x}_j, R_c) \\ &\forall k \in \{\mathcal{A}_s, \mathcal{A}_m\}, \exists j \in \mathcal{A}_c \text{ s.t. } k \in \mathcal{B}(\mathbf{x}_j, R_c) \end{aligned} \tag{2}$$

In order to have a uniform graph and maximize the coverage of a space, while maintaining connectivity, the connections between the agents of the heterogeneous system are biased based on the geometry of the communication radii. Since the sensor and manipulator agents have limited communication capabilities, the main idea is to have the communication relays connect to each other and expand the entire network in the environment. We consider that each communication relay is equipped with a high performance rf device that offers a large range and bandwidth to handle the communication with multiple nodes. Hence, each sensor and manipulator will be connected directly to a specific communication relay based on the minimum euclidean distance to the closest relay. The *sensor/relay* and *manipulator/relay* assignments are built based on a local consensus algorithm described in Algorithm 1.

Specifically in Algorithm 1, C_c^i and C_s^i are the set of neighbor relays and mobile sensors connected to the ith communication relay, respectively. n_i is the number of sensors connected to the ith relay and $\tilde{n}_i$ the updated number of sensors after running the algorithm. Finally $\hat{\mathcal{A}}_j^i$ is the set containing $\hat{\mathcal{N}}_j^i$ communication relays connected to i with $n_j \leq n_i$. Note that Algorithm 1 applies also to the manipulator agents in which we will have to consider C_m^i mobile manipulators' neighbors connected to

Algorithm 1. Heterogeneous Local Consensus Algorithm

while $t < t_{\text{final}}$ **do**
  for $i = 1, \ldots, \mathcal{N}_c$ **do**
    Calculate the round down average number of sensor agents in the neighborhood of i
    $\underline{\tau}_i = \frac{n_i + \sum_{\{j \in C_c^i | n_j \leq n_i\}} n_j}{\hat{\mathcal{N}}_j^i + 1}$
    for $j = 1, \ldots, \hat{\mathcal{N}}_j^i$ **do**
      if $\exists k \in \hat{\mathcal{A}}_j^i$ s.t. $n_i = n_j \forall j \in \{\hat{\mathcal{A}}_j^i \setminus k\}$ and $n_k \leq (n_i - 2)$ **then**
        $\tilde{n}_k = n_k + \frac{(n_k + n_i)}{2}$
        $\tilde{n}_i = n_i - \frac{(n_k + n_i)}{2}$
      else
        $\tau_j^i = \underline{\tau}_i - n_j$
        $\tilde{n}_j = n_j + \tau_j^i$
        $\tilde{n}_i = n_i - \tau_j^i$
        for $l = 1, \ldots, \tau_j^i$ **do**
          if $\exists p \in (C_s^i \cap \mathcal{B}_c^j)$ s.t. $||\mathbf{x}_p - \mathbf{x}_j|| = \min ||\mathbf{x}_q - \mathbf{x}_j|| \forall q \in C_s^i$ **then**
            $p \in C_s^j$ and $p \notin C_s^i$
          end if
        end for
      end if
    end for
  end for
  return $n_i = \tilde{n}_i$
end while

i. If the graph is connected, then we can guarantee the network will reach at least a local consensus that is given by the average number of sensors and manipulators connected to the relays in the neighborhood of the ith relay [14].

4 Motion Constraints

• *Relay agent*: Given (2), $\forall$ sensors $k \in C_s^i$ (or manipulators $q \in C_m^i$), if $||\mathbf{x}_i - \mathbf{x}_{k(q)}|| \leq R_\varepsilon$, with $R_s(R_m) < R_\varepsilon < R_c$, the motion of the ith communication relay follows the spring-mass interaction

$$\ddot{\mathbf{x}}_i = \mathbf{u}_i,$$

$$\mathbf{u}_i = \left[\sum_{j \in C_c^i} \kappa_{ij} (l_{ij} - R_\varepsilon) \hat{\mathbf{d}}_{ij} \right] - \delta_i \dot{\mathbf{x}}_i - \nabla_{\mathbf{x}_i} \varsigma(\mathbf{x}_i), \tag{3}$$

where $\mathbf{u}_i \in \mathcal{U}$ ($\mathcal{U} \in \Re^3$) is the control input, $\mathbf{x}_i = (x_i, y_i, z_i)^T$ is the position vector of the ith relay relative to a fixed Euclidean frame, and $\dot{\mathbf{x}}_i$, $\ddot{\mathbf{x}}_i$ denote the velocity and acceleration (control input), respectively. C_c^i is the set of neighbor relays connected to the ith relay. C_c^i is built using the Gabriel Graph rule [15] in which between any two nodes i and j, we form a virtual spring if and only if there is no robot k inside the circle of diameter $\overline{ij}$, [15]. l_{ij} and $\hat{\mathbf{d}}_{ij}$ are the length and direction of force of the virtual spring between robot i and j while κ_{ij} and δ_i are the spring constant and damping coefficient, respectively. Here, we assume $\kappa_{ij} = \kappa_{ji}$ and $\delta_i > 0$. $\nabla_{\mathbf{x}_i}\varsigma(\mathbf{x}_i) = A_\varsigma(\mathbf{x}_i - c_\varsigma)$ is the gradient of $\frac{A_\varsigma}{2}\|\mathbf{x}_i - c_\varsigma\|^2$, a quadratic attractive potential function where c_ς is the center of the region where the target is located and it is known *a priori*.

Theorem 1. *A network of communication relays having switching dynamics depicted in* (3) *is guaranteed to eventually reach stability in which all agents converge to a null state.*

Proof. The proof for this theorem can be formulated using Lyapunov theory and can be found in our previous work [15].

If $\exists k \in C_s^i$ such that $R_\varepsilon < ||\mathbf{x}_i - \mathbf{x}_k|| < R_c$ then the relay node believes that the specific sensor agent k is in pursuit mode; therefore it switches into a follower mode with dynamics

$$\mathbf{u}_i = \alpha(\mathbf{x}_k - \mathbf{x}_i) \tag{4}$$

where $\alpha \in \Re^+$.

- *Sensor agent*: We consider the following interaction

$$\mathbf{u}_k = \begin{cases} \kappa_{ki}(l_{ki} - R_s)\,\hat{\mathbf{d}}_{ki} - \delta_k\dot{\mathbf{x}}_k & \text{if } k \in \mathcal{B}(\mathbf{x}_i, R_s) \\ \mathbf{u}_{\text{search}} & \text{if } k \in \{\mathcal{B}(\mathbf{x}_i, R_\varepsilon) \setminus \mathcal{B}(\mathbf{x}_i, R_s)\} \\ \mathbf{u}_{\text{pursuit}} & \text{if } k \text{ is in pursuit mode} \end{cases}, \tag{5}$$

in which with $\mathbf{u}_{\text{search}}$ we intend a random motion within the toroid centered in the ith relay controlling the kth mobile sensor. $\setminus$ is the set-minus operator.

- *Manipulator agent*: We similarly consider the following logic

$$\mathbf{u}_q = \begin{cases} \kappa_{qi}(l_{qi} - R_m)\,\hat{\mathbf{d}}_{qi} - \delta_i\dot{\mathbf{x}}_q & \text{if } q \in \mathcal{B}(\mathbf{x}_i, R_m) \\ \mathbf{u}_{\text{search}} & \text{if } q \in \{\mathcal{B}(\mathbf{x}_i, R_\varepsilon) \setminus \mathcal{B}(\mathbf{x}_i, R_m)\} \\ \mathbf{u}_{\text{manip}} & \text{if } q \text{ is in manipulation mode} \end{cases}, \tag{6}$$

Specifically for the search modes in both (5) and (6) we create the following connectivity constraints: at every Δt time the communication relay transmits its location to its neighbors. Given $\mathcal{V}^c_{\max}$ the maximum velocity of each relay, the maximum distance the relay can travel in Δt time is $D^c_{\max} = \mathcal{V}^c_{\max}\Delta t$. Thus when in search mode, agents are guaranteed to stay in search mode if and only if in an interval of time Δt they don't enter inside the region $\{\mathcal{B}(\mathbf{x}_i, (R_\varepsilon - D^c_{\max})) \setminus \mathcal{B}(\mathbf{x}_i, (R_{s(m)} + D^c_{\max}))\}$. It is important to note that when in pursuit mode, if $k \in \{\mathcal{B}(\mathbf{x}_i, R_c) \setminus \mathcal{B}(\mathbf{x}_i, R_\varepsilon)\}$ the communication relay i is not attracted anymore toward the target region but switches into following mode with dynamics (4) to maintain connectivity to pursuer k.

We can formulate the following theorem that guarantees connectivity among the heterogeneous network at all times:

Theorem 2. *Given an initially connected heterogeneous robotic system made of* $\mathcal{N}_c$ *relays,* $\mathcal{N}_s$ *mobile sensors, and* $\mathcal{N}_m$ *manipulator agents with switching topologies expressed by* (3), (4), (5), (6), *the network is guaranteed to maintain connectivity if for an interval of time* $\Delta t > 0$, *each* $j \in C_c^i$, $k \in C_s^i$, *and* $q \in C_m^i$ *take goal points* $g^i_{j(k)(q)} \in \mathcal{B}(\mathbf{x}_i(t), (R_c - \varepsilon(\Delta t)))$, *with* $\varepsilon(\Delta t) \geq \mathcal{V}^i_{max}\Delta t$

Proof. Assuming that the dynamics of each agent are stable or at least stabilizable, if $g^i_j \in \mathcal{B}(\mathbf{x}_i(t), (R_c - \varepsilon(\Delta t)))$, then $\mathbf{x}_i(t+\Delta t) \in \mathcal{B}(\mathbf{x}_i(t), \varepsilon(\Delta t))$ shifting the communication region of $\varepsilon(\Delta t)$, thus leaving $\mathbf{x}_j(t+\Delta t) \in \mathcal{B}(\mathbf{x}_j(t), \varepsilon(\Delta t)) \subset \mathcal{B}(\mathbf{x}_j(t), (R_c - \varepsilon(\Delta t)))$. Hence, j is always connected to i. Note that $\varepsilon(\Delta t)$ is an upper (and safe) bound for the sensor k and manipulator q agents since $\mathcal{V}^i_{\max} > \mathcal{V}^k_{\max} = \mathcal{V}^q_{\max}$.

Corollary 1. *A sensor agent k connected to the* i^{th} *relay with dynamical topology* (5) *is guaranteed to be in* search mode *at all times if* $||\mathbf{x}_k - \mathbf{x}_i|| > R_s$ *and for an interval of time* $\Delta t > 0$, $g^i_k \in ((\mathcal{B}(\mathbf{x}_i(t), (R_c - \varepsilon(\Delta t))) \setminus \mathcal{B}(\mathbf{x}_i(t), (R_s + \varepsilon(\Delta t))))$ *with* $\varepsilon(\Delta t)$ *as of Theorem 2.*

Proof. The proof for this corollary extends from the proof of Theorem 2 and it is based on the geometrical properties of the connectivity constraints imposed in (5). Corollary 1 applies to the manipulators agents, as well.

Within Corollary 1 we allow the communication relays to move freely and expand in the environment while the sensor and manipulator agents are in search mode without entering in other regions. Thus k can take a goal $\in (\mathcal{B}(\mathbf{x}_i, (R_c)) \setminus \mathcal{B}(\mathbf{x}_i, (R_c - \varepsilon(\Delta t))))$ only if it detects an intruder. In the same way, q can take a goal $\in (\mathcal{B}(\mathbf{x}_i, (R_c))) \setminus \mathcal{B}(\mathbf{x}_i, (R_c - \varepsilon(\Delta t))))$ if and only if it detects the target $\mathcal{T}$.

5 Communication Power Control

In this section we introduce a power control algorithm to maintain a certain SINR level between the agents of the heterogeneous robotic network.

Depending on the distance between the robots as well as path loss, fading, and shadowing, the power received at a certain mobile sensor (or manipulator) k that is transmitted by a relay i is attenuated by a gain $g_{ik} = K\left(\frac{l_0}{l_{ik}}\right)^\beta + \frac{\psi_{ik}}{P_i}$ where K is a constant that depends on antenna properties and channel attenuation, l_0 is a reference distance, and l_{ik} is the separation distance between robots i and k. β is the path-loss exponent and finally ψ_{ik} is an additional attenuation due to shadowing and that is usually a log-normal distributed random variable [12]. i can communicate with k provided its (SINR) is above a certain threshold T. Thus the goal is to find whether there exists an assignment of power levels and distance between robots so that each robot's SINR is acceptable. Since g_{ik} depends on the distance separation l_{ik} between node i and j, instead of regulating the power, we can think of changing l_{ik} to maintain the SINR above a desired value. Therefore, we can implement the following algorithm

$$\text{find} \sum_{k \in C_c^i} g_{ik} \ \ \forall i = 1, \ldots, \mathcal{N}_c, \tag{7}$$

subject to:

$$\begin{aligned}
&\left\{\frac{P_i g_{ik}}{\sum_{j\neq i, j\in\mathcal{B}(\mathbf{x}_k,R_s)} P_j g_{jk}+\nu_k}\right\} \geq T, \\
&0 < P_i \leq P_{\max}, \\
&l_{ik} \geq l_{\min} \qquad\qquad \forall i = 1,\ldots,\mathcal{N}_c, \forall k \in \mathcal{C}_c^i.
\end{aligned} \tag{8}$$

where $l_{\min}$ is the minimum distance separation between any agent in order to avoid collision. $P_{\max}$ is an upper bound for the maximum power each agent can have, P_i is the power level of agent i, and ν_k the receiver noise at the kth sensor. For simplicity's sake we assume that the ν_k can be neglected. In other words, by implementing this algorithm we guarantee a certain quality of service (QoS) among the robotic team and adjust the received power at k through mobility.

6 Sensing and Manipulation

In this work we consider two distinct sensing behaviors: a *vision capability* and a *obstacle avoidance characteristic*, as depicted in Fig.2(a-b). The former implies the use of camera like systems in which we are able to perceive different features in the environment and for instance recognize friends and foes, targets and other characteristics of the environment. Within the obstacle avoidance, we consider laser range finder type sensors that are capable of measuring distances with high precision and thus can be employed for navigation. Finally the manipulator agents are equipped with a planar arm to lift or move objects (2(c)) .

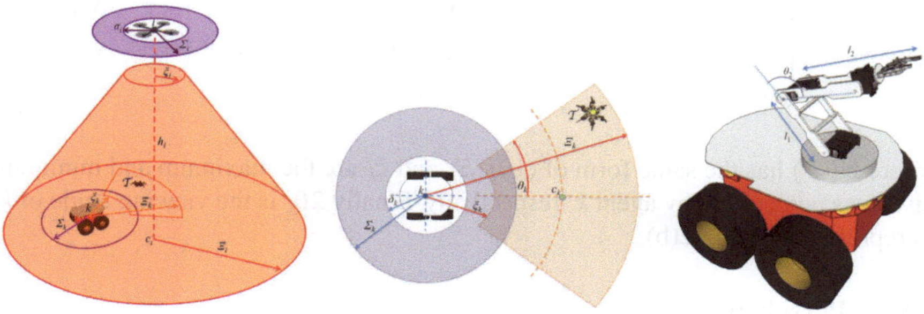

(a) (b) (c)

Fig. 2 Representation of the sensing capabilities for the aerial relay and mobile sensor. (a) The aerial relay i hovers at a certain height h_i and has a toroidal sensing for obstacle avoidance and a conical field of view over the ground. (b) The sensor agent k (and also the manipulator q) has a toroidal sensing for obstacle avoidance and a limited field of view in front of it. $\mathcal{T}$ is a target of interest. (c) Representation of the manipulator configuration used in this work. For ease, here we consider a two link planar arm with end effector, installed on the top, frontal position of the robot.

6.1 Vision Detection

In the scenario envisioned in this paper, each robot has some degree of vision capability. The aerial relay can see a large area but with low resolution, while the sensor and the manipulator agents on the contrary can perceive a smaller area but with higher resolution. Following Fig.2, we use the following probability of detection field for the aerial relays

$$\mathcal{S}_i(\mathbf{x}_{\mathcal{T}}) = \begin{cases} \mathrm{N}\left(c_i, \varphi_i^2\right) & \text{if } \|\mathbf{x}_{\mathcal{T}} - c_i\| \leq \Xi_i \text{ and } h_i = const. \\ 0 & \text{otherwise} \end{cases} \tag{9}$$

where $\mathbf{x}_{\mathcal{T}}$ is the state (*i.e.*, the position) of a target $\mathcal{T}$ located in a 3D workspace. $\mathrm{N}(c_i, \varphi_i^2)$ is the field of view normal distribution centered in c_i with variance φ_i^2. $\|\cdot\|$ is the euclidean distance norm and Ξ_i is the maximum vision range for the ith aerial relay. Model (9) holds if the quadrotor is hovering at a constant altitude (hence $h_i = const.$). By using this model, the probability of detection is higher moving toward the centroid of the field of view of i. Thus if a target $\mathcal{T}$ in position $\mathbf{x}_{\mathcal{T}}$ is such that $\|\mathbf{x}_{\mathcal{T}} - c_i\| \leq \Xi_i$, the probability of detection (pr) is given by

$$\mathrm{pr}(\mathbf{x}_{\mathcal{T}}) = \mathrm{N}(c_i, \varphi_i^2) = \frac{1}{\varphi_i\sqrt{2\pi}} e^{-\frac{(\mathbf{x}_{\mathcal{T}} - c_i)^2}{2\varphi_i^2}} \tag{10}$$

A centroid motion scheme, centering the target in this field of view is discussed in the following section.

For the sensor and manipulator agents, similarly we use the following constraint

$$\mathcal{S}_k(\mathbf{x}_{\mathcal{T}}) = \begin{cases} \mathrm{N}\left(\left(\frac{(\Xi_k - \mathbf{x}_k) - (\xi_k - \mathbf{x}_k)}{2}\right), \varphi_k^2\right) & \text{if } \mathbf{x}_{\mathcal{T}} \in \theta_k(\|\Xi_k - \mathbf{x}_k\|^2 - \|\frac{(\Xi_k + \xi_k)}{2} - \mathbf{x}_k\|^2). \\ \mathrm{N}\left(\left(\frac{(\Xi_k - \mathbf{c}_k) - (\xi_k - \mathbf{c}_k)}{2}\right), \varphi_k^2\right) & \text{if } \mathbf{x}_{\mathcal{T}} \in \theta_k(\|\frac{(\Xi_k + \xi_k)}{2} - \mathbf{x}_k\|^2 - \|\xi_k - \mathbf{x}_k\|^2). \\ 0 & \text{otherwise} \end{cases}, \tag{11}$$

where $\mathrm{N}(\cdot,\cdot)$ has the same form of (10). Ξ_k and ξ_k are the maximum and minimum distances perceptible by agent k, respectively. Finally $2\theta_k$ is the viewing angle of k, as represented in Fig.2(b).

6.1.1 Pursuit and Evasion

If a mobile sensor k detects an adversarial opponent $\mathcal{T}$ inside its field of view, it switches from *search mode* into *pursuit mode* (5) with the following dynamics

$$\mathbf{u}_{\text{pursuit}} = \begin{cases} \alpha(\mathbf{T}_k^{\mathcal{W}}(\mathbf{x}_{\mathcal{T}}^k - c_k^k)) & \text{if } \mathbf{x}_{\mathcal{T}} \in \theta_k(\|\Xi_k - \mathbf{x}_k\|^2 - \|\frac{(\Xi_k + \xi_k)}{2} - \mathbf{x}_k\|^2) \\ \alpha(\mathbf{x}_{\mathcal{T}} - \mathbf{x}_k) & \text{if } \mathbf{x}_{\mathcal{T}} \in \theta_k(\|\frac{(\Xi_k + \xi_k)}{2} - \mathbf{x}_k\|^2 - \|\xi_k - \mathbf{x}_k\|^2) \end{cases}, \tag{12}$$

in which $\mathbf{T}_k^{\mathcal{W}}$ is the transformation matrix that converts the kth robot frame into the world $\mathcal{W}$ frame. $\mathbf{T}_k^{\mathcal{W}} = \mathbf{R}_k^{\mathcal{W}}\mathbf{D}_k^{\mathcal{W}}$ where $\mathbf{R}_k^{\mathcal{W}}$ is the rotation matrix and $\mathbf{D}_k^{\mathcal{W}}$ is the translation matrix, [16]. $\mathbf{x}_{\mathcal{T}}^k$ and c_k^k are the position of the target and field of view centroid in the kth robot frame, respectively.

Thus, within the first equation in (12) we navigate the centroid c_k of the field of view of k toward $\mathbf{x}_{\mathcal{T}}$. Once $\mathcal{T}$ is within the region of the constraint in the second equation of (12), k is guided toward the evader through an attractive potential force. Here we assume that based on the velocities of both k and $\mathcal{T}$, $\mathcal{T}$ is capturable if $\mathbf{x}_{\mathcal{T}} \in \theta_k(||\frac{(\Xi_k+\xi_k)}{2} - \mathbf{x}_k||^2 - ||\xi_k - \mathbf{x}_k||^2)$ (that is the constraint in the second equation of (12)).

6.2 Obstacle Avoidance

For the obstacle avoidance effect the reader is referred to the toroidal shapes in Fig.2. The workspace, $\mathcal{W}$, is populated with $\mathcal{N}_o$ fixed polygonal obstacles $\{O_1, \ldots, O_{\mathcal{N}_o}\}$, whose geometries and positions are assumed unknown. In order to avoid obstacles we model a ray field of view around the agents, similarly to a laser range finder footprint, and we create a repulsive potential whose value approaches infinity as the robot approaches the obstacle, and goes to zero if the robot is at a distance greater than Φ_i or smaller than ϕ_i from the obstacle.

$$W_{O,i} = \begin{cases} \frac{1}{2}\eta_i \left(\frac{1}{\rho(\mathbf{x}_i)} - \frac{1}{\rho_0}\right)^2 & \text{if } \phi_i \leq \rho(\mathbf{x}_i) \leq \Phi_i \\ 0 & \text{if } \rho(\mathbf{x}_i) > \Phi_i \text{ or } \rho(\mathbf{x}_i) < \phi_i \end{cases}, \tag{13}$$

where $\rho(\mathbf{x}_i)$ is the shortest distance between the agent and any detected obstacle in the workspace and η_i is a constant.

The repulsive force is then equal to the negative gradient of $W_{O,i}$ For simplicity sake, here we assume that the aerial relay, sensor and manipulator agents have all the same obstacle avoidance constraint, as depicted in Fig.2(a-b).

6.3 Manipulability

For the manipulation behavior we assume that $\mathcal{N}_m$ agents are equipped with an articulated arm having $\mathcal{N}_l$ links. From a classical robotics book, [16], it is well known that the *manipulability metric* offers a quantitative measure of the relationship between differential change in the end-effector pose relative to differential change in the joint configuration. In this work, we use this concept to define the best configuration of the manipulator agent such that when a certain target object needs to be handled, the manipulability measure is maximized.

Let us define the Jacobian relationship

$$\zeta = J\dot{a} \tag{14}$$

that specifies the end-effector velocity that will result when the joint move with velocity $\dot{a}$.

If we consider the set of joint velocities $\dot{a}$ such that $||\dot{a}||^2 = \dot{a_1}^2 + \dot{a_2}^2 + \dots \dot{a_{\mathcal{N}_i}}^2 \leq 1$, then we obtain

$$||\dot{a}||^2 = \zeta^T (JJ^T)^{-1} \zeta = (U^T \zeta)^T \Sigma_m^{-2} (U^T \zeta) \tag{15}$$

in which we have used the singular value decomposition SVD $J = U \Sigma V^T$ [16].

If the Jacobian is full rank (rank $J = m$), (15) defines the manipulability ellipsoid. It is easy to show that the manipulability measure is given by

$$\mu = \sigma_1 \sigma_2 \dots \sigma_m \tag{16}$$

with σ_i the diagonal elements of Σ.

For convenience in this work we consider that each manipulator robot is equipped with a two-link planar arm (Fig.2(c)) in which the manipulability measure is given by

$$\mu = l_1 l_2 |\sin \theta_2| \tag{17}$$

where l_1 and l_2 are the length of the two links of the manipulator and θ_2 is the angle between the two links. Therefore, the highest manipulability measure is obtained when $\theta_2 = \pi/2$.

6.3.1 Target Manipulation

If a manipulator agent q detects a fixed target $\mathcal{D}$ that needs manipulation (*e.g.*, lifting, moving, grabbing, etc.) then we apply a similar law as in (12) with the only difference that once the target is detected, we compute the configuration of the manipulator arm such that we obtain the highest manipulability measure μ. μ will translate into a certain spatial configuration and position of the end effector $\mathbf{x}_\zeta$. Finally, $\mathbf{x}_\zeta$ becomes the input for the controller. Thus the control law for q in *manipulation mode* becomes

$$\mathbf{u}_{\text{manip}} = \alpha(\mathbf{T}_q^{\mathcal{W}}(\mathbf{x}_{\mathcal{D}}^q - \mathbf{x}_\zeta(\mu))) \text{ if } \mathbf{x}_{\mathcal{D}} \in \theta_q(||\Xi_q - \mathbf{x}_q||^2 - ||\mathbf{x}_\zeta(\mu) - \mathbf{x}_q||^2) \tag{18}$$

7 Simulation Results

In the simulation of Fig. 3 we assemble together all the pieces descried in the previous sections and consider a search and rescue/pursuit-evasion scenario. A heterogeneous system made of the same number and type of agents as in the first simulation, explores an environment in search of a target $\mathcal{D}$ that needs to be manipulated, while maintaining connectivity with a fixed base station. The target is protected by an opposing player $\mathcal{T}$ that circles around its perimeter. $\mathcal{T}$ tries to capture the manipulator agent q, while avoiding any mobile sensor k. If a mobile sensor k detects $\mathcal{T}$, it switches into pursuit mode to capture the opponent. Here we assume that $\mathcal{V}_{\max}^k > \mathcal{V}_{\max}^{\mathcal{T}} > \mathcal{V}_{\max}^q$.

Specifically, during the simulation, while the consensus algorithm 1 is run to equilibrate the network, the agents are attracted to a region where the target is

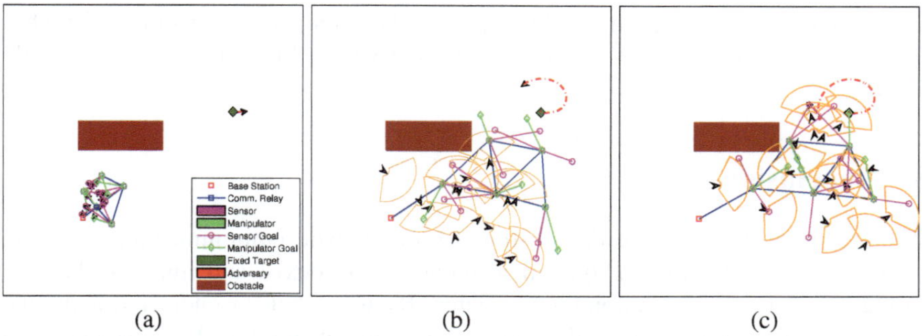

Fig. 3 Pursuit-evasion simulation. (a) The heterogeneous system is in spring-mass mode. (b) The sensor and the manipulator agents are in search mode while the communication relays expand the network in the environment. (c) A sensor detects and pursue an adversarial player (top right of the figure making a circular trajectory), while a manipulator moves toward the fixed target.

located. The relays maintain network connectivity and enforce the power control algorithm (7) keeping the SINR above a certain threshold (Fig. 4(a)). In Fig. 4(b) it is plotted the case in which the agents don't follow the PC algorithm. Specifically in this case the SINR can take values below the threshold, obtaining a lower quality of communication. In Fig. 3(b) the sensor and manipulator robots switch into search mode and explore the area surrounding their assigned communication relay. Finally, in Fig. 3(c) a mobile sensor detects the opposing player and switches into pursuit mode while a manipulator moves toward the fixed target.

Additionally, through equivalence comparisons, we found that the average coverage for a heterogeneous system is 63% while for the homogeneous scenario is reduced to 22% of the workspace. Therefore, by decoupling the tasks of sensing

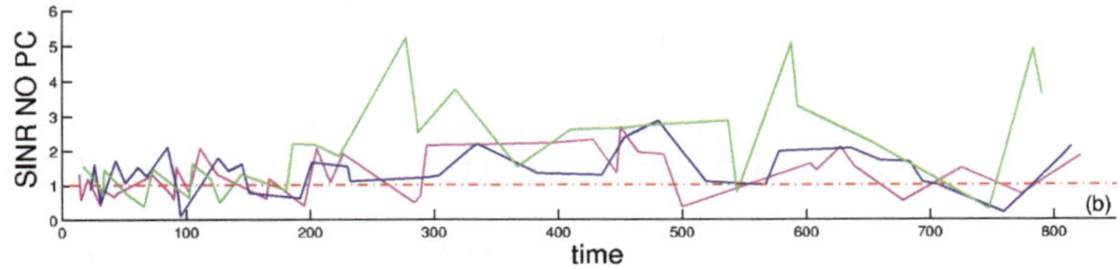

Fig. 4 Comparison between SINR (dB) with Power Control (top) and without Power Control (bottom) for two mobile sensors and one robotic manipulator assigned to one of the five relays in Fig. 3

and relaying communication, and by imposing the connectivity constraints for a heterogeneous system we can cover a larger area with a limited number of mobile agents.

8 Conclusion

Within this work, we have demonstrated how realistic communication, sensing, manipulation, and different dynamical models can coexist and improve the performance of a group of autonomous agents. By using heterogeneous systems we can manage different agents and improve the coverage and thus the sensed areas in a workspace. Besides using realistic sensing and vision capabilities, we have analyzed: (i) a communication connectivity algorithm to equilibrate and maintain the network connectivity while exploring the environment and (ii) a power control algorithm to guarantee a certain SINR level between the relay and the sensor and manipulator agents. We presented a simulation scenario in which we implemented a search/pursuit-evasion scenario with a heterogenous network. The robots expand in the environment attracted by potential functions, avoid a large obstacle and reach a partially known target that is manipulated by a specific agent of the system.

Future work will consist in expanding the theoretical results of this paper, presenting more realistic environments with non-convex obstacles and more opponents and considering failures in the system (*e.g.*, some agents stop working). Also we intend to implement the proposed framework through hardware experiments using our heterogeneous test bed of quadrotors, car-like and differential-drive ground vehicles, and crawling agents (Fig. 1).

Acknowledgements. This work was supported by NSF grants ECCS #1027775 and IIS #0812338, by DOE URPR grant #DE-FG52-04NA25590, and by SNL-1 PO #1071006.

References

1. Cortez, R., Fierro, R., Wood, J.: Connectivity maintenance of a heterogeneous sensor network. In: Proc. of the 10th International Symposium on Distributed Autonomous Robotic Systems (DARS), Lausanne, Switzerland, November 01-03, pp. 1–12 (2010)
2. Ascending Technologies GmbH, `http://www.asctec.de/`
3. Pullin, A., Kohut, N., Zarrouk, D., Fearing, R.: Dynamic turning of 13 cm robot comparing tail and differential drive. To appear in IEEE International Conference of Robotics and Automation, ICRA (2012)
4. Abbas, W., Egerstedt, M.: Distribution of agents in heterogeneous multi agent systems. In: IEEE International Conference on Robotics and Automation, pp. 976–981 (December 2011)
5. Di Paola, D., Gasparri, A., Naso, D., Ulivi, G., Lewis, F.: Decentralized task sequencing and multiple mission control for heterogeneous robotic networks. In: 2011 IEEE International Conference on Robotics and Automation (ICRA), pp. 4467–4473. IEEE (2011)

6. Hollinger, G., Mitra, U., Sukhatme, G.: Autonomous data collection from underwater sensor networks using acoustic communication. In: 2011 IEEE/RSJ International Conference on Intelligent Robots and Systems (IROS), pp. 3564–3570. IEEE (2011)
7. Bhattacharya, S., Basar, T., Hovakimyan, N.: Singular surfaces in multi-agent connectivity maintenance games. In: Proc. of IEEE Conference on Decision and Control and European Control Conference (CDC-ECC), pp. 261–266 (December 2011)
8. Whitten, A., Choi, H., Johnson, L., How, J.: Decentralized task allocation with coupled constraints in complex missions. In: Proc. of American Control Conference (ACC), San Francisco, CA (June 2011)
9. Zavlanos, M., Egerstedt, M., Pappas, G.: Graph-theoretic connectivity control of mobile robot networks. Proceedings of the IEEE 99, 1–16 (2011)
10. Ao, W., Cheng, S., Chen, K.: Connectivity of multiple cooperative cognitive radio ad hoc networks. IEEE Journal on Selected Areas in Communications 30(2), 263–270 (2012)
11. Zhou, P., Chang, Y., Copeland, J.: Reinforcement learning for repeated power control game in cognitive radio networks. IEEE Journal on Selected Areas in Communications 30(1), 54–69 (2012)
12. Goldsmith, A.: Wireless communications. Cambridge Univ. Pr. (2005)
13. Schwager, M., Julian, B., Angermann, M., Rus, D.: Eyes in the sky: Decentralized control for the deployment of robotic camera networks. Proceedings of the IEEE 99, 1–21 (2011)
14. Olfati-Saber, R., Fax, J., Murray, R.: Consensus and cooperation in networked multi-agent systems. Proceedings of the IEEE 95(1), 215–233 (2007)
15. Bezzo, N., Yuan, Y., Fierro, R., Mostofi, Y.: A decentralized connectivity strategy for mobile router swarms. In: Proc. of The 18th World Congress of the International Federation of Automatic Control (IFAC), Milan, Italy (August 30, 2011)
16. Spong, M., Hutchinson, S., Vidyasagar, M.: Robot modeling and control. Wiley, New Jersey (2006)

Optimizing for Transfers in a Multi-vehicle Collection and Delivery Problem

Brian Coltin and Manuela Veloso

Abstract. We address the Collection and Delivery Problem (CDP) with multiple vehicles, such that each collects a set of items at different locations and delivers them to a dropoff point. The goal is to minimize either delivery time or the total distance traveled. We introduce an extension to the CDP: what if a vehicle can transfer items to another vehicle before making the final delivery? By dividing the labor among multiple vehicles, the delivery time and cost may be reduced. However, introducing transfers increases the number of feasible schedules exponentially. In this paper, we investigate this Collection and Delivery Problem with Transfers (CDP-T), discuss its theoretical underpinnings, and introduce a two-approximate polynomial time algorithm to minimize total distance travelled. Furthermore, we show that allowing transfers to take place at any location for the CDP-T results in at most a factor of two improvement. We demonstrate our approximation algorithms on large simulated problem instances. Finally, we deploy our algorithms on robots that transfer and deliver items autonomously in an office building.

1 Introduction

Previously, we deployed a team of robots in an office building. The robots navigate autonomously on multiple floors to complete tasks requested by users at a centralized website. These tasks include retrieving and delivering objects. For example,

Brian Coltin
The Robotics Institute, Carnegie Mellon University, 5000 Forbes Ave., Pittsburgh, PA 15213
e-mail: bcoltin@cs.cmu.edu

Manuela Veloso
Computer Science Department, Carnegie Mellon University, 5000 Forbes Ave., Pittsburgh, PA 15213
e-mail: veloso@cs.cmu.edu

M.A. Hsieh and G. Chirikjian (eds.), *Distributed Autonomous Robotic Systems*, Springer Tracts in Advanced Robotics 104,
DOI: 10.1007/978-3-642-55146-8_7,

users can ask the robots to bring them printouts from the printer or coffee from the kitchen, or to transfer written messages, USB sticks, or other items between offices.

Forming a schedule for the robots to fulfill all requests is an instance of the Vehicle Routing Problem (VRP). In this work, we propose and examine an extension of the VRP: what if the robots can *transfer* items between each other? By having one robot pick up an object and transfer it to a different robot (or a series of robots) for delivery, we can conserve both time and the battery life of the robots. Transferring items makes the problem significantly harder because this creates exponentially more possible schedules. Scheduling with transfers has potential energy-saving and productivity-increasing applications in both transportation domains (e.g., taxis that exchange occupants heading to nearby locations) and in warehouse automation.

In particular, we examine the role of transfers in the *collection and delivery problem* (CDP), a subproblem of the VRP in which the robots retrieve a set of items and deliver them to a single central location. In our building, popular requests for the robots include retrieving mail from the secretaries' offices and delivering it to the central office, as well as delivering business cards, fliers or candy to many offices and returning the leftovers. Both problems are instances of the CDP. We call the CDP with transfers the CDP-T. The CDP-T is NP-hard, so we introduce algorithms both to solve it optimally and approximately.

In this paper, we first discuss selected related work, introduce the CDP-T, and formulate the problem as a delivery tree. We consider CDP with transfers anywhere (CDP-TA), where transfers are not limited to pickup locations, and show that the optimal solution to the CDP-T costs at most twice that of the CDP-TA in a metric space. We propose a two-approximate polynomial time algorithm for the CDP-T, and a metaheuristic to improve on this solution. We show the effectiveness of these algorithms in simulation and on physical robots, and demonstrate an approach for two robots to autonomously transfer objects. To our knowledge, this is the first time multiple robots have created and executed a schedule with transfers.

2 Related Work

Extensive research has been done on the Vehicle Routing Problem (VRP) in which a set of vehicles visit a set of locations to service customer requests. Variants of the VRP include the Pickup and Delivery Problem (PDP), in which a set of goods must be picked up from one set of locations and delivered to another set; and the Capacitated PDP (CPDP). The VRP and its variants are generally NP-hard, and are often solved optimally with branch and bound algorithms such as mixed integer programming. See [19] for an extensive discussion of the VRP.

A plethora of approaches have been proposed to solve the PDP, including branch and bound methods, heuristics, and metaheuristics [3, 16]. The Collection and Delivery Problem (CDP) is a subproblem of the PDP in which there is only a single delivery point. Approximation algorithms with guarantees have been developed for the VRP with release times at which jobs can first be performed but without deadlines [4], for the VRP with time windows [2], and for the CPDP [6, 11].

Less work has focused on robots that transfer items. Mail couriers and the transportation industries use fixed "transshipment" points, hubs such as post offices and airports, where objects are deposited for another agent to retrieve. Algorithms have been developed to find the optimal solution [8] and heuristics [14] for the PDP with fixed transshipment points, and for the CPDP with time windows and a single transshipment point [15]. Solving the PDP with online requests to minimize delivery delay has been considered in [20] with a single relay between the pickup and delivery vehicles. In [18], transfers at any pickup or delivery location are considered in the construction of heuristics for the PDP with time windows. In [10], an approximation is given for the preemptive CPDP, where objects are dropped off at pickup points and retrieved by other vehicles later.

Robotics researchers have extensively studied the task allocation problem, in which robots are assigned to (independent) tasks to maximize utility. In the VRP and CDP, tasks are not independent [9]. Other researchers have devised algorithms to find the optimal rendezvous locations given a schedule of meetings, which could be used in conjunction with the schedules we create [1].

In this work, we consider a version of the PDP with transfers and only a single dropoff point, and provide a constant factor approximation algorithm. We prove that this is also a constant factor approximation to the CDP-TA, where transfers take place at any location (see Section 3.4). We develop further heuristics to improve these solutions and demonstrate their effectiveness in real-world scenarios. To our knowledge, we are the first to deploy robots which plan and execute a schedule with autonomous transfers.

3 The Collection and Delivery Problem with Transfers

We are given a set of robots, located at positions $R \subseteq M$, where M is a metric space with distance metric d. The robots must retrieve items from pickup locations $L \subseteq M$, then deliver all the objects to a final drop-off location $f \in M$. Robots may meet up to transfer objects, but only at the pickup locations. We assume that the robots have infinite capacity.

We construct the complete graph $G = (V, E)$ where $V = R \cup L \cup \{f\}$ and the edge weights are defined by the metric d. The goal is to find a path P_i on G for each robot such that all the objects are delivered. In a valid solution, the endpoint of every path P_i is either f or, in the case of a transfer, an intermediate node in another robot's path. Every vertex $l \in L$ is part of at least one path P_i since every item is delivered.

Many such sets of paths exist, and we consider two objectives to distinguish between them: a) minimize the total movement energy of the robots, corresponding to the sum of edge weights, and b) minimize the completion time, corresponding to the length of the longest path from any robot to the destination.

Now, consider the directed graph T which is the union of all the paths of used robots P_i, where the paths are chosen to either minimize the total distance traveled or the maximum distance traveled to deliver any object.

Claim. There exists an optimal directed graph T that is a tree.

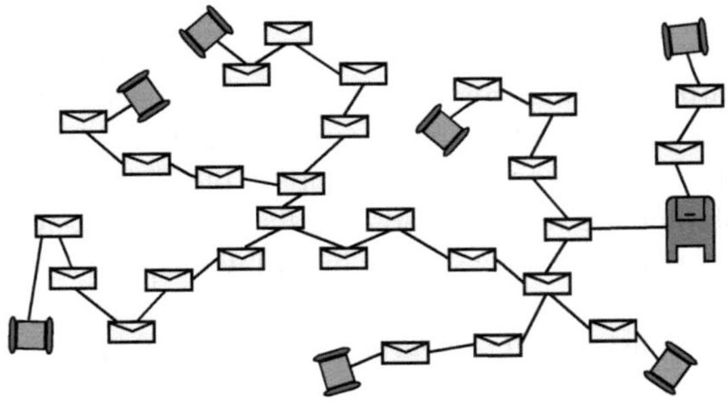

Fig. 1 The optimal solution to the CDP-T can be formulated as a delivery tree. Edges represent the motion of a single robot. Where the tree branches, all the robots at that branch point transfer their entire load to a single robot which continues alone.

Proof. T is a tree if and only if it is connected and has $|E| = |V| - 1$ edges. T is connected, since every pickup location and used robot is part of a path to f. Every robot route P_i either ends at f or at a transfer point. If not, traveling to the final location on the path is either needless or retrieves an item that is not delivered to f. Furthermore, every route P_i contains f or a transfer point where items are given to a different robot nowhere else in the path. If these points do exist elsewhere, a solution of less than or equal cost can be constructed by removing the points due to the triangle inequality. Any items transferred earlier can still be delivered by transferring them at the route's final point instead. Hence, there is an optimal set of paths where each robot transfers or delivers items exactly once at the end of its path. Furthermore, there is an optimal solution where each node $l \in L$ is not the endpoint of exactly one path P_i, as otherwise a point could be omitted from one of the paths to construct a solution with no worse cost. So there exists a set of paths T which has $|E| = |L| + |R \cap T|$ edges, as each vertex aside from f is not the endpoint of exactly one path P_i. T has $|V| = |L| + |R \cap T| + 1$ vertices, and is hence a tree.

An equivalent formulation for the CDP-T is to construct a *delivery tree* D with the following properties (see Figure 1):

1. The interior nodes are the pickup locations $l \in L$ and the final delivery point f.
2. The leaf nodes are a subset of R, the starting locations of the used robots.
3. Branch points represent transfers of one robot's load to another. All but one robot at a transfer point remain behind and are not used again.

Our goal is to either find the delivery tree of minimum weight $w(D)$ (minimum movement cost), of minimum weighted depth $\texttt{depth}(d)$ (minimum delivery time), or to minimize a linear combination $\alpha w(D) + (1-\alpha)\texttt{depth}(D)$ where $0 \leq \alpha \leq 1$.

We show that the CDP-T problem is NP-hard by reducing the TSP to the CDP-T. In the TSP, a salesman aims to find the minimum distance Hamiltonian tour that visits every city in a set V exactly once, and return to the initial city. The TSP can be reduced to the CDP-T problem by setting the cities V as the pickup locations L. We have one robot r, which begins at the same location as the traveling salesman. We set the dropoff location f to be the same location, r. So if the CDP-T can be solved in polynomial time, so can the TSP. Hence the CDP-T is NP-hard.

3.1 Optimal Approach

The CDP-T can be formulated as a mixed integer program. We solve for binary variables $x_{a,b}$, which indicate whether the directed edge from node a to node b is in the solution. The final destination f has zero outgoing edges ($\sum_{v \in V} x_{f,v} = 0$), each location $l \in L$ has exactly one outgoing edge ($\forall l \in L \ \sum_{v \in V} x_{l,v} = 1$), and each robot $r \in R$ has zero incoming edges and at most one outgoing edge, but may have zero ($\forall r \in R \ \sum_{v \in V} x_{r,v} \leq 1, \sum_{v \in V} x_{v,r} = 0$). The vertex f and each point $l \in L$ have at least one incoming edge ($\forall n \in (L \cup \{f\}) \ \sum_{v \in V} x_{v,n} \geq 1$).

The formulation as it stands still allows subtrees, in which the delivery tree is not connected. To address this, we introduce constraints similar to the subtour elimination constraints that are used to formulate the TSP as an Mixed-Integer Program (MIP): $\forall U \subset V, \ U \neq \emptyset \ \sum_{e \in \delta(U)} x_e \geq 1$, where $\delta(U)$ is the set of edges which link U and $V \setminus U$.

This MIP formulation gives paths which solve the CDP-T. The graph is connected (aside from unused robots), so every item is part of a robot's path to f. The graph has $|R| + |L| = |V| - 1$ edges since each used robot and item has exactly one outgoing edge, and is hence a tree. The robots are the only leaf nodes in the tree, since they have one edge while the items have at least two. We minimize the total distance $D = \sum_{e \in E} d(e)$ traveled by all the robots.

Alterntiavely, to minimize the time taken to deliver all items (the length of the longest path from a robot to the goal) we define binary variables $p_{e,r}$ which indicate whether the directed edge e is part of the path from r to f. Edges from robots are part of that robot's path if the edge exists. No edges exist *to* robots, and f has no outgoing edges. We add the constraints $\forall r \in R, \ l \in L \quad p_{e_{r,l},r} = x_{r,l}, \ p_{e_{l,r},r} = 0, \ p_{e_{f,l},r} = 0$. Edges with robots as nodes are not part of another robot's extended path: $\forall r_1, r_2 \in R, l \in L \cup \{f\} \ p_{e_{r_2,l},r_1} = 0$. Each edge between retrieval and delivery points that is in the solution is part of the path for at least one robot (or more, with transfers). $\forall l_1, l_2 \in L \cup \{f\}, l_1 \neq l_2 \ \sum_{r \in R} p_{e_{l_1,l_2},r} \geq x_{l_1,l_2}$. Finally, edges between two retrieval or delivery locations are only on a robot's path to f if an incoming node was also on the extended path. $\forall l_1 \in L, l_2 \in L \cup \{f\}, l_1 \neq l_2, r \in R, \ p_{e_{l_1,l_2},r} \geq x_{l_1,l_2} - 1 + \sum_{l_3 \in L \setminus l_1, l_2 \cup \{f\}} p_{e_{l_3,l_1},r}$.

We define a variable G to be the length of the longest extended path. Then we add constraints so that G is greater than the length of every robot's extended path: $\forall r \in R \ G \geq \sum_{e \in E} p_{e,r} d(e)$. Our objective function is then $\min G$, or a weighted sum of the two objectives, $\min \alpha D + (1 - \alpha) G$.

3.2 Minimum Length Approximation

Computing the exact solution to the CDP-T is difficult since the problem is NP-hard. Hence we are interested in approximation algorithms, which find a solution in polynomial time that is not optimal, but provably close to optimal.

The approximation algorithm we introduce generates two-approximate solutions in terms of total distance to the CDP-T using only a single robot. In fact, a two-approximate solution to the CDP-T is the best guarantee we can possibly make

Algorithm 1. $\texttt{cdp\_t}(L,f,R)$: Construct a delivery tree given the set of pickup points L, the dropoff point f, and robots R. $N_T(v)$ gives the set of neighbors of v in T.

$G \leftarrow \texttt{complete\_graph}(L \cup \{f\}, d)$
$T \leftarrow \texttt{mst}(G)$
$s, v \leftarrow \text{argmin}_{s \in R, v \in V(G)} d(s,v)$
$T' \leftarrow \texttt{deliv\_tree}(R, L, f, T \cup \texttt{edge}(s,v))$
for $r \in R, r \notin T'$ **do**
  $v \leftarrow \text{argmin}_{v \in V(G), N_{T'}(v) \cap R = \emptyset} d(r,v)$
  $T'' \leftarrow \texttt{deliv\_tree}(R, L$
    $f, T' \cup \texttt{edge}(r,v))$
  if $\texttt{cost}(T'') < \texttt{cost}(T')$ **then**
    $T' \leftarrow T''$
  end if
end for
return T'

Algorithm 2. $\texttt{deliv\_tree}(R,L,f,T)$: Construct a delivery tree given the set of pickup points L, the dropoff point f, robots R, and an intermediate delivery tree T.

$A = \{l \in L : \exists r \in R \text{ s.t. } l \in \texttt{path(r, f)}\}$
$T' =$ Null Graph
for $r \in R$ **do**
  $n = r, P = []$
  while $n \notin T'$ **do**
    Append n to P
    If $n = f$, break
    Choose $n' = v \in N_T(n)$ s.t. $v \notin P$ and $v \notin A$

    If $\nexists n'$, $n' = v \in N_T(n)$ s.t. $v \notin P$ and $v \in A$

    $n = n'$
  end while
  $P' = P$ with duplicate vertices removed
  $T' = T' \cup P$
end for
return T'

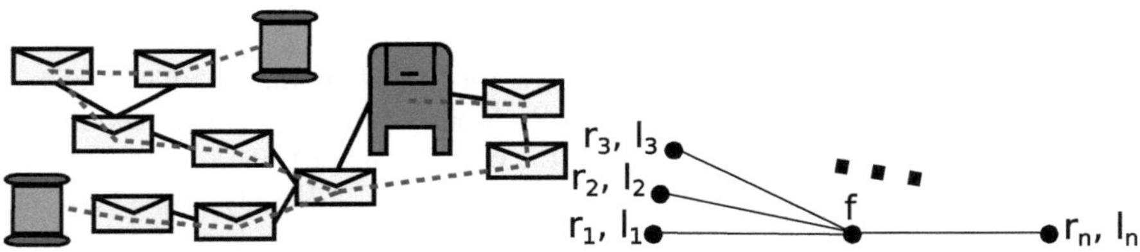

Fig. 2 (a) An example of the approximation algorithm. Solid black lines indicate edges on the minimum spanning tree, and dashed lines show the generated delivery tree. (b) Robots and items are situated at the end of n hallways of length one emanating from the delivery point f. $O_n = n$, where every robots travels to f. $O_1 = 2n - 1$, where one robot retrieves every item.

with only a *single* robot. If O_n is the optimal solution with n robots, and O_1 the optimal solution with any one of those robots, Figure 2b shows a case where $w(O_1)$ approaches arbitrarily close to $2w(O_n)$. With multiple robots, the heuristic further reduces both the distance traveled and the delivery time.

Our approximation is based on the minimum spanning tree two-approximate heuristic for the TSP, but is extended for multiple robots that transfer items. The algorithm is shown in Algorithm 1. First, we construct the complete graph G with

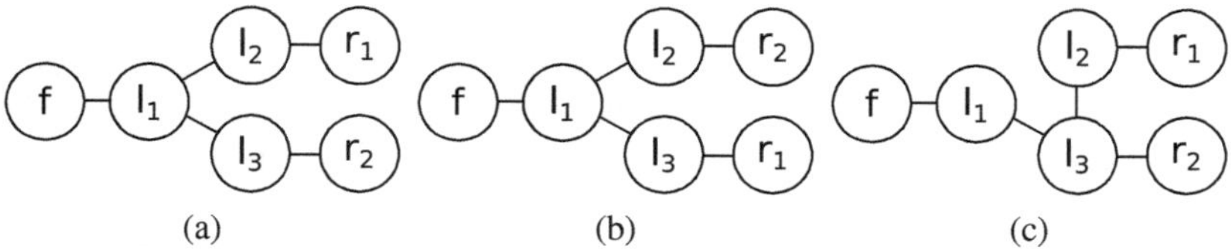

Fig. 3 (a) The initial state. (b) A neighbor found by swapping l_1 and l_2 (the edges linking them to other nodes are different). (c) A neighbor with l_2 grafted from l_1 to l_3.

pickup locations L and drop-off location f with edge lengths determined by the distance metric d, and its minimum spanning tree T. Next, choose the edge e of lowest weight from a node in T to a node in $r \in R$, and add e and r to T to construct T'.

If O is the delivery tree for the optimal solution, then $w(T) \leq w(O \setminus R)$, since T is the minimum spanning tree over the same nodes. Since a valid delivery tree must have at least one edge connected to a robot, and we chose the minimum one, $w(T') \leq w(O)$. We call T' an *intermediate delivery tree*, since it can be used to construct a delivery tree but not all leaf nodes are robots.

We then construct a tour P, starting at r and ending at f, which visits each vertex in T' at least once with the procedure `deliv_tree` (see Alg. 2). When T' contains only a single robot, this algorithm is equivalent to the two-approximate TSP approximation, which traverses each edge at most twice. The `deliv_tree`(T) algorithm extends this heuristic to multiple robots which can transfer items, and creates a valid delivery tree with weight at most $2w(T)$. Thus, $w(P) \leq 2w(T) + w(e) \leq 2w(O \setminus R) + w(e) \leq 2w(O \setminus R)$. So our single robot algorithm is two-approximate to the optimal multi-robot solution in terms of total distance.

We can lower the cost further with multiple robots, although no better guarantees on the approximation bound are obtained. We begin with the constructed intermediate delivery tree for a single robot, T', and for each robot r, greedily add the shortest edge from r to a node in T to the tree T'. We construct a new delivery tree from T' with `deliv_tree`, and keep the edge and robot in the intermediate delivery tree T' if and only if the delivery tree's cost decreases according to our objective function. We then attempt to add the next robot to the updated T', iterating through every robot. This procedure still gives a two-approximation to the CDP, and additional robots will sometimes decrease both the total distance traveled and the time to completion, depending on the problem instance. See Figure 2a for an example of the algorithm's results.

We have constructed a two-approximate delivery tree in polynomial time that uses multiple robots, since the cost of the multi-robot heuristic is at most the cost of the single-robot heuristic which is two-approximate.

3.3 *Improvement with Local Search*

Next, we introduce a metaheuristic to improve upon the two-approximate solution with local search techniques. Specifically, we make use of simulated annealing [13].

Simulated annealing is a metaheuristic that begins at some state, and chooses a random "neighbor" of that state. With probability $accept(e, e', t)$ the new state is accepted as the current state, where e is the "energy" (in our case, the cost) of the current state, e' is the energy of the new state, and t is the temperature, or the fraction of iterations of the algorithm currently completed. If the new state is rejected we remain at the current state and repeat with a new neighbor. The algorithm continues either for a fixed number of iterations or until the energy crosses some threshold, when the best solution that has been encountered thus far is returned.

To apply simulated annealing to the CDP, we must define a starting point, an energy function, an acceptance probability, and a function to return random neighbors of a state. We search over the underconstrained intermediate delivery trees rather than strict delivery trees, as this allows us to develop a broader concept of neighboring solutions that is more closely tied to the approximation heuristic. We use as a starting point the tree generated by our fast multi-robot heuristic.

The energy function is the cost function of the delivery tree constructed with `deliv_tree`, and it incorporates both the total weight and depth of the tree as a function of α. The acceptance probability is 1 if $e' < e$, and $e^{\frac{e-e'}{t}}$ otherwise, a standard acceptance function frequently used in the literature.

Neighboring states are found either by *swapping* two non-robot, non-destination nodes on the intermediate delivery tree (that are not robots or f), swapping their neighbor sets, or by *grafting* one branch of the tree at a transfer point onto a neighboring node, replacing a single edge. See Figure 3 for examples.

We run the simulated annealing algorithm for a thousand iterations. Every hundred iterations we restart from the best solution found thus far to explore the most promising regions more thoroughly. In practice, simulated annealing improves upon the solutions of the two-approximate heuristic.

3.4 Transfers at Any Location

Next, we consider the general case of CDP-TA, where items can be transferred anywhere, rather than only at vertices in L as in the CDP-T. We show that the optimal solution for the metric CDP-T, O_v, is within a factor of two of the cost of the optimal solution to the CDP-TA, O_a, when minimizing total distance traveled.

The CDP-TA is closely related to the Steiner tree problem, in which a set of points must be connected by the shortest possible set of edges. "Helper" points may be added (transfer points) to construct a solution. The CDP-TA is different in that all the leaf nodes of a valid delivery tree must be robots or the final destination.

Claim. $O_v \leq 2O_a$

Proof. Given a delivery tree T with optimal cost O_a with transfers at any point (the transfer points are separate nodes in the tree), we construct a delivery tree T' with transfers only at vertices and cost $O_v \leq 2O_a$. Let X be the set of transfer points in T.

First, construct the forest F with vertices $X \cup N(X)$, where $N(X)$ is the set of neighboring vertices to X in T. Add all edges between the vertices of F in T to

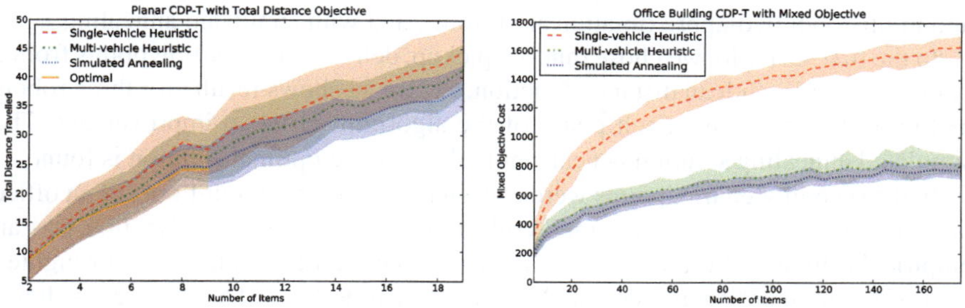

(a) (b)

Fig. 4 Results for the CDP-T algorithms with (a) three robots and up to twenty items to retrieve under the total distance objective function in the planar domain, and (b) 25 robots and up to 175 items to retrieve under the mixed objective function in the office building domain. Lines indicate the mean objective function cost, and the filled area shows the standard deviation across trials. Darker areas indicate overlap.

F as well. For each connected component F_i of F (which may include multiple transfer points), define S_i to be the complete graph with vertices $F_i \setminus X$. Then $w(F_i) = w(\texttt{steiner}(S_i))$, where `steiner` gives the minimum Steiner tree. If this were not the case, then either the minimum Steiner tree could be used to construct a delivery tree of lower cost, or we do not have the minimum Steiner tree.

Let $S = \cup S_i$. Construct $T' = (T \setminus X) \cup \texttt{msf}(S)$, where $\texttt{msf(S)} = \cup_i \texttt{mst}(S_i)$. T' is a valid delivery tree where the items are transferred at retrieval or dropoff points.

Then $O_a = w(T) = w(T \setminus X) + w(F)$, since $T = (T \setminus X) \cup F$. Furthermore, $w(F) = \sum w(F_i) = \sum w(\texttt{steiner}(F_i \setminus X))$. Similarly, $O_v \leq w(T') = w(T \setminus X) + w(\texttt{msf}(S)) = w(T \setminus X) + \sum_i w(\texttt{mst}(F_i \setminus X))$. The weight of a minimum spanning tree is bounded by the Steiner ratio κ such that $w(\texttt{mst}(G)) \leq \kappa w(\texttt{steiner}(G))$. Hence, $O_v \leq w(T \setminus E) + \kappa \sum_i R_i \texttt{steiner}(R_i \setminus E) \leq \kappa O_a$.

If our distance function is a metric, than $\kappa = 2$ [17], and the optimal solution to the CDP-T is two-approximate to the CDP-TA. If the distance function is Euclidean, κ is conjectured to be $\frac{2}{\sqrt{3}} \approx 1.1547$. This conjecture is believed to be true, but is still open and unproven [12]. If it holds, then for the Euclidean case CDP-T is a $\frac{2}{\sqrt{3}}$-approximation to CDP-TA. Hence, our two-approximate heuristic for the CDP-T provides a four-approximate heuristic for metric CDP-TA and a $\frac{4}{\sqrt{3}} \approx 2.31$-approximate for the Euclidean CDP-TA, assuming the conjecture regarding the Euclidean Steiner ratio holds.

4 Simulation Results

To validate our approach, we tested the algorithms in two scenarios. We varied the number of items to retrieve, and created 50 random problem instances for each set of parameters. We solved each problem optimally (when feasible), with the single and multi-robot two-approximation algorithms, and with simulated annealing.

In the first scenario, we chose random pickup, delivery, and starting points from a plane, using a Euclidean distance function. Figure 4a shows results for three robots and up to twenty items. Each phase of the algorithm gives an improvement. The simulated annealing solution is near-optimal, when the optimal solution is found.

In the second scenario, points were chosen randomly from four floors of an office building in a simulation of the mail collection task. The distance function was an empirical estimate of robot travel times. The objective was to minimize a weighted sum of the moment energy cost and the delivery time ($\alpha = 0.5$). Figure 4b shows the results with 25 robots and up to 175 items. Each level of the algorithm offers a significant improvement, particularly the multi-vehicle heuristic, since it achieves much better depths than the single robot approximation. We are unable to find the optimal solutions for problems of this size.

On an Intel 2.83 GHz Core2 Quad CPU, the two heuristics ran in under a second for all instances of up to 500 vertices (these instances are not shown). Simulated annealing ran in under fifteen seconds for every instance. The optimal MIP formulations were solved with `lpsolve`, with each attempt aborted after thirty seconds. These results show that the approximation algorithms can solve large problems quickly, and provide near-optimal solutions when comparison is possible.

5 Illustrative Deployments

We have deployed our CDP-T heuristic on two robot platforms: the CoBot and CreBot robots (see Figure 5). The CoBots autonomously navigate over large distances indoors, and with them we show that transferring items can reduce energy consumption and delivery time. The CoBots cannot physically transfer items, and so they ask humans to help. With the CreBots, we demonstrate that autonomous transfers are feasible by presenting a method to transfer items with a tilting tray.

Both the CoBots and CreBots autonomously localize and navigate in the building using software developed for the CoBots [5]. A centralized web server accepts user requests and sends schedules to the robots [7]. The robots collect and deliver items by arriving at an office and asking a human to place or remove the items.

5.1 Transfers with CoBots

We examine three scenarios where two CoBots were deployed to collect and deliver objects in an office building. The problem setups and the planned paths, both with transfers and without, are shown in Figure 6. For each scenario, we recorded the time it took the robot(s) to complete the task and the total combined distance traveled by all of the robots. To reduce the variance in our experiments, we assumed that humans were always immediately available to place items in the CoBots' baskets and to help transfer items. However, the times recorded with transfers do include the time to ask and thank a human for their help, and are shown in Table 1.

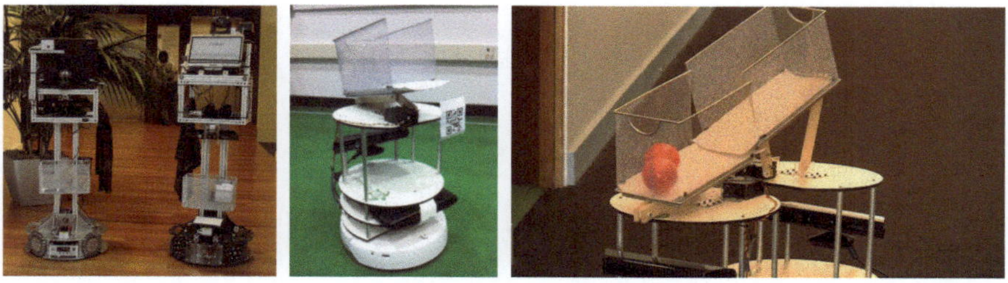

(a) (b) (c)

Fig. 5 (a) Two CoBot robots. (b) A CreBot robot, with a Create base, laptop computer, Kinect RGB-D camera, tilting tray, and QR code for alignment. (c) One CreBot transfers an item to another during a collection and delivery task.

Table 1 Deployment Results for Selected Scenarios

	With Transfers		Single Robot	
Scenario	Time (min.)	Dist.(m)	Time (min.)	Dist. (m)
1	4:22	229.35	8:18	287.12
2	5:52	220.68	7:55	238.66
3	7:00	230.56	9:36	272.37

In all three scenarios, using multiple robots with transfers reduced both the time to complete the task and the total distance traveled. This may not be the case in all scenarios, depending on the problem setup. However, we have demonstrated that in many scenarios, transferring items between robots can and does reduce the energy consumed and the time taken to complete the task.

5.2 Autonomous Transfers

We have designed the CreBots to transfer items autonomously. They consist of an iRobot Create as a base, with Willow Garage Turtlebot shelves, a laptop, Kinect RGB-D camera, and a custom-built tilting tray on top to transfer items (see Figure 5b).

Unlike the CoBots, the CreBots transfer items autonomously. To do so, two CreBots head towards the same location based on their localization information. The robots stop either when they reach the destination, or are blocked within two meters of the destination (presumably by the other robot). Next, the robots send each other their localization positions wirelessly and turn to face each other.

Localization is accurate and robust enough for a rough alignment, but not precise enough to transfer objects based solely on localization. For fine alignment, the robots are each equipped with a QR code which is detected by the Kinect and used for precise docking. The transferrer advances slowly until it comes within the range it can detect the QR code using an off the shelf library, which measures the QR code's bounding box. The transferrer computes the distance and angle to the QR

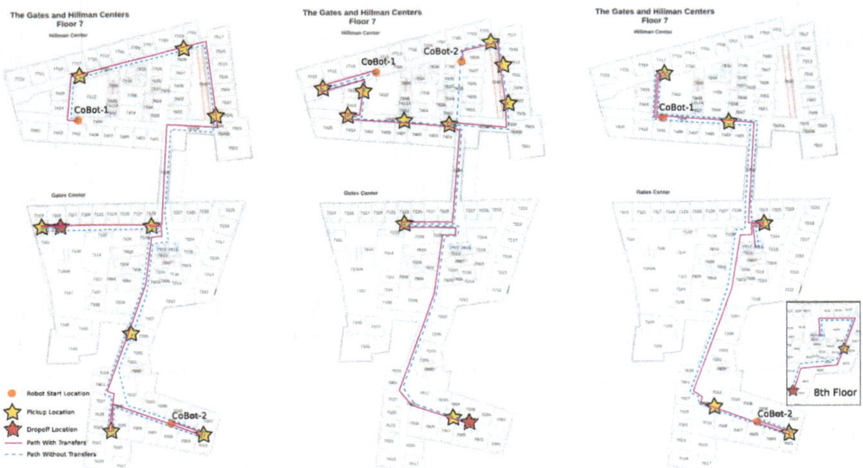

(a) (b) (c)

Fig. 6 The paths planned by the approximation algorithm and taken by the robots, with and without transfers, for (a) Scenario 1, (b) Scenario 2, and (c) Scenario 3

code from its known size and camera parameters, then rotates in place to align with the QR code. The transferrer continues to move forwards until its bump sensor is triggered, and dumps its load. When dumping, the tilting tray shakes back and forth to ensure that all the objects are dislodged. The CreBots have successfully executed collection and delivery tasks with transfers.[1]

6 Conclusion

We have introduced the Collection and Delivery Problem with Transfers, and shown that the solution comes in the form of a delivery tree. We solved the CDP-T optimally with an MIP, and bounded the cost with transfers anywhere in terms of the cost when transfers occur only at vertices. We introduced a two-approximate algorithm under the minimum length objective, and proposed a heuristic and metaheuristic to find solutions of lower depth while maintaining the bound on total length. Furthermore, we demonstrated the effectiveness of these heuristics on large real-world problem instances, and showed the feasibility of transfers between physical robots.

Acknowledgements. This research was partially sponsored by the Office of Naval Research under grant number N00014-09-1-1031, and by the National Science Foundation under award number NSF IIS-1012733 . The views and conclusions expressed are those of the authors only.

Special thanks to Thomas Charley for designing and building the CreBots' tilting trays.

[1] Video available at `http://youtu.be/pzXv7p_aZhE`

References

1. Alton, K., Mitchell, I.: Efficient dynamic programming for optimal multi-location robot rendezvous. In: IEEE Conference on Decision and Control, pp. 2794–2799. IEEE (2008)
2. Bansal, N., Blum, A., Chawla, S., Meyerson, A.: Approximation algorithms for deadline-tsp and vehicle routing with time-windows. In: Proceedings of the ACM Symposium on Theory of computing, pp. 166–174. ACM (2004)
3. Berbeglia, G., Cordeau, J., Gribkovskaia, I., Laporte, G.: Static pickup and delivery problems: a classification scheme and survey. Top 15(1), 1–31 (2007)
4. Bhattacharya, B., Hu, Y.: Approximation algorithms for the multi-vehicle scheduling problem. In: Cheong, O., Chwa, K.-Y., Park, K. (eds.) ISAAC 2010, Part II. LNCS, vol. 6507, pp. 192–205. Springer, Heidelberg (2010)
5. Biswas, J., Coltin, B., Veloso, M.: Corrective gradient refinement for mobile robot localization. In: Proc. of IEEE Conf. on Intelligent Robots and Systems (IROS), pp. 73–78. IEEE (2011)
6. Charikar, M., Raghavachari, B.: The finite capacity dial-a-ride problem. In: Proc. of 39th Annual Symposium on Foundations of Computer Science 1998, pp. 458–467. IEEE (1998)
7. Coltin, B., Veloso, M., Ventura, R.: Dynamic user task scheduling for mobile robots. In: Proc. of the Work. on Automated Action Planning for Autonomous Mobile Robots. AAAI (2011)
8. Cortés, C., Matamala, M., Contardo, C.: The pickup and delivery problem with transfers: Formulation and a branch-and-cut solution method. European Journal of Operational Research 200(3), 711–724 (2010)
9. Gerkey, B., Matarić, M.: A formal analysis and taxonomy of task allocation in multi-robot systems. The International Journal of Robotics Research 23(9), 939–954 (2004)
10. Gørtz, I., Nagarajan, V., Ravi, R.: Minimum makespan multi-vehicle dial-a-ride. In: Algorithms-ESA 2009, pp. 540–552 (2009)
11. Gupta, A., Hajiaghayi, M., Nagarajan, V., Ravi, R.: Dial a ride from k-forest. ACM Transactions on Algorithms (TALG) 6(2), 41 (2010)
12. Ivanov, A., Tuzhilin, A.: The steiner ratio gilbert–pollak conjecture is still open. Algorithmica, 1–3 (2011)
13. Kirkpatrick, S., Gelatt, C., Vecchi, M.: Optimization by simulated annealing. Science 220(4598), 671 (1983)
14. Mitrovic-Minic, S., Laporte, G.: The pickup and delivery problem with time windows and transshipment. Information Systems and Operational Research 44(3), 217–228 (2006)
15. Nakao, Y., Nagamochi, H.: Worst case analysis for a pickup and delivery problem with single transfer. In: Numerical Optimization Methods, Theory and Applications, vol. 1584, pp. 142–148 (2008)
16. Parragh, S., Doerner, K., Hartl, R.: A survey on pickup and delivery problems. Journal für Betriebswirtschaft 58(2), 81–117 (2008)
17. Takahashi, H., Matsuyama, A.: An approximate solution for the steiner problem in graphs. Math. Japonica 24(6), 573–577 (1980)
18. Thangiah, S., Fergany, A., Awan, S.: Real-time split-delivery pickup and delivery time window problems with transfers. Central European Journal of Op. Research 15(4), 329–349 (2007)
19. Toth, P., Vigo, D.: The vehicle routing problem, vol. 9. Soc. for Industrial Mathematics (2002)
20. Waisanen, H., Shah, D., Dahleh, M.: Fundamental performance limits for multi-stage vehicle routing problems. Operations Research (2007)

Distributed Amorphous Ramp Construction in Unstructured Environments*

Nils Napp and Radhika Nagpal

Abstract. We present a model of construction using iterative amorphous depositions and give a distributed algorithm to reliably build ramps in unstructured environments. The relatively simple local strategy for interacting with irregularly shaped, partially built structures gives rise robust adaptive global properties. We illustrate the algorithm in both the single robot and multi-robot case via simulation and describe how to solve key technical challenges to implementing this algorithm via a robotic prototype.

1 Introduction

Robots are best suited for dirty, dull, and dangerous tasks. This paper focuses on algorithms for the dirty and dangerous task of construction in unstructured terrain. Applications range from rapid disaster response, like building levees and support structures, to remote construction in mines or space. The requirement of working in unstructured terrain frequently coincides with a lack of infrastructure, such as global positioning or a consistent shared global state estimate, that simplify coordination of multiple robots and deliberative planing. Distributed algorithms that use limited local information and coordinate through stigmergy solve this problem and provide scalable solutions. Robustness to poor sensing and irregular terrain can further be improved by using *amorphous* construction materials that comply to obstacles. Such construction is locally reactive, both on an algorithmic level, i.e. where robots

Nils Napp · Radhika Nagpal
Harvard University
33 Oxford St
Cambridge MA 02138, USA
e-mail: nnapp@wyss.harvard.edu, rad@eecs.harvard.edu

* This work was supported by the Wyss Institute for Bioinspired Engineering.

M.A. Hsieh and G. Chirikjian (eds.), *Distributed Autonomous Robotic Systems*, Springer Tracts in Advanced Robotics 104,
DOI: 10.1007/978-3-642-55146-8_8,

Fig. 1 Examples of amorphous construction. (a) Amorphous construction in biology. A termite preparing an amorphous dollop of mud for deposition. Inset shows a mound built around a tree. (b) Prototype of a construction robot. The robot was remote controlled to build a ramp using amorphous foam depositions. Inset shows a cone-shaped deposition.

deposit based on local cues, and a physical level, i.e. amorphous construction materials react by changing shape to conform to their environment.

Our approach is inspired by biological systems, such as mound building termites [18], that are very adept at building in unstructured terrain, Fig. 1(a). Their skill combines scalable coordination through stigmergy and the use of amorphous building materials that interface with an irregular environment. We would like to endow scalable robot teams with similar skill, however an algorithmic foundation for doing so is lacking. Current models for autonomous robotic construction focus on assembling pre-fabricated building materials and cannot accommodate the continuous nature of amorphous building materials. The contribution of this paper is twofold:(A) A mathematical framework for reasoning about robots that construct with amorphous materials, and (B) a distributed, locally reactive algorithm for adaptive ramp building. This work is a step away from robots assembling discrete pre-fabricated components and instead embracing the messy continuous world, Fig. 1(b).

Section 2 presents mathematical models for amorphous construction and adaptive ramp building. Section 3 gives a local strategy for creating structures that robots can climb; Sec. 4 extends those results to include physical constraints for single and multiple robots. Section 5 discusses future work.

1.1 Related Work

Currently, there is much interest in the topic of robotic construction with mobile robots [3, 4, 6, 10, 11, 15], as well as decentralized algorithms by which multiple robots can coordinate construction [1, 9, 12, 17]. These systems are mainly focused on building pre-specified structures using lattice-based building materials [5, 20]. Lattice-based building blocks have good structural properties—being strong, stiff, and light—but place assumptions on the initial environment being level and devoid of obstacles. These methods are difficult to extend to unstructured environments with irregularly shaped obstacles. Furthermore, alignment and attachment restrictions affect all other aspects of design, for example adding complex assembly order constraints.

In contrast, amorphous building materials—e.g. foam, mud, sandbags or compliant blocks—sidestep these limitations [13]. They can help compensate for uncertainty and measurement errors without requiring complex sensing or reasoning. For example, compliant and amorphous materials are used for rapidly building flood protection in disaster zones [7, 19] or pouring foundations over irregular terrain. Similarly, the requirement of fixed attachment orientations can be relaxed by using adhesive in the autonomous robotic construction of curved walls [2, 3]. The closely related work in [16] uses foam to rapidly adapt robot parts to a unknown tasks instead of adapting structures to unknown terrain. Digital manufacturing via CAD/CAM, and some large-scale robotic construction systems, such as [8], also use amorphous materials to build continuous shapes. While these systems are not specifically focused on construction in unstructured environments, we can exploit the materials and design principles to design robots that utilize amorphous materials.

2 Problem Formulation and Questions

We present a solution to the *adaptive ramp building* problem as a particular example of a distributed construction task in unstructured terrain. The problem is to design a deposition and motion strategy that allows reach an arbitrary goal position, despite irregularly shaped obstacles. Robots can sense the goal direction, move on partially built structures, and deposit amorphous materials to make non-climbable structures climbable. Adaptive ramp building is an example of how amorphous construction materials can be used to create robust behavior and also provides a primitive behavior for solving more complex tasks. The remainder of this section presents mathematical models for continuous structures, amorphous depositions, and climbable structures.

2.1 Mathematical Model for Continuous Structures

We model construction in two or three dimensions. Gravity constrains robots to move along surfaces on which they can incrementally deposit construction material. We assume that the *construction area* Q is a connected, compact, and finite subset of $\mathbb{R}^1$ (or $\mathbb{R}^2$) and the domain of a bounded, non-negative height function $h : Q \rightarrow \mathbb{R}^+$. The graph of h, $(x, h(x))$ $x \in Q$, describes a *structure*. Robots move on structures and modify them.

If structures are modeled as functions, depositions are operators on functions. To distinguish the two, function spaces are denoted by scripted letters. For example, let $\mathcal{Q}$ be the space of real-valued, bounded functions on Q, and $\mathcal{Q}^+ \subset \mathcal{Q}$ the subset of non-negative ones. Function application to points is denoted by parentheses $(\cdot)$ and operator application to functions by brackets $[\cdot]$. For example, applying function $h \in \mathcal{Q}^+$ to a point $x \in Q$ is written as

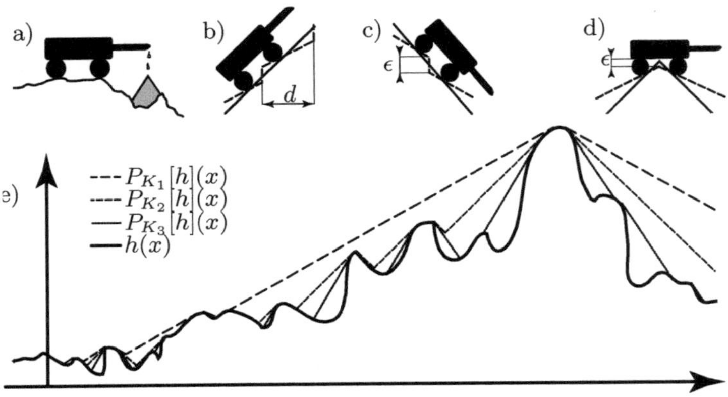

Fig. 2 Parameter Geometry. (a) Robot making an amorphous deposition. (b,c) Relation of K to the maximal steepness a robot can climb and descend, (solid) without discontinuity (dashed) with discontinuity. (d) Relation of steepness K to the required ground clearance to drive over the apex of a cone. (e) A height function on $h \in \mathcal{Q}^+$ and its projections onto Lipschitz functions with different parameters $K_3 > K_2 > K_1$.

$h(x)$, and applying an operator $D : \mathcal{Q}^+ \rightarrow \mathcal{Q}^+$ to h is denoted by $D[h]$. In the case of functions, all relational symbols should be interpreted pointwise, e.g. given $h, g \in \mathcal{Q}^+$, $h \leq g \equiv h(x) \leq g(x)\ \forall x \in Q$.

One limitation of modeling structures as functions is that many physical terrains have overhangs and are not functions. However, the benefit of this restrictive model is that it comes with analysis tools, such as continuity and integration, that can be used to reason about construction algorithms.

2.2 Model for Amorphous Deposition

Robots can deposit amorphous construction material and control its volume and position, Fig. 1(b). The free surface of each deposition is modeled by a *shape function* $f \in \mathcal{Q}$ while the bottom conforms to the structure, Fig. 2(a). As a simple, yet sufficiently general, family of shape functions we use cones. Given an apex position $(\phi, \sigma) \in Q \times \mathbb{R}^+$ and steepness $K_D \in \mathbb{R}^+$ let

$$f_{(\phi,\sigma)}(x) = \sigma - K_D|\phi - x|. \tag{1}$$

The deposition operator $D : \mathcal{Q} \times \mathcal{Q}^+ \rightarrow \mathcal{Q}^+$ models interactions of depositions with the environment, here simply covering it. Given a structure $h \in \mathcal{Q}^+$ with $h(\phi) < \sigma$, the new structure after deposition $f_{(\phi,\sigma)}$ is given by

$$D[f_{(\phi,\sigma)}, h](x) = \max_{x \in Q}(f(x), h(x)). \tag{2}$$

Given an initial structure $h_0 \in \mathcal{Q}^+$ a structure is built by a sequence of depositions characterized by their shape parameters $(\phi_1, \sigma_1), (\phi_2, \sigma_2), (\phi_3, \sigma_3), \ldots$. The height function h_n after n depositions is defined recursively by

$$h_n(x) = D[f_{(\phi_n, \sigma_n)}, h_{n-1}](x). \tag{3}$$

After the n-th deposition, the local reactive rules of each robot direct it to move on h_n and possibly make a deposition resulting in a new structure h_{n+1}.

This deposition model preserves continuity, independent of the particular parameter choices (ϕ_n, σ_n). In this and the following proofs, let $B_\varepsilon(x)$ denote the *open ball* of radius ε around x, i.e. $y \in B_\varepsilon(x)$ if and only if $|y - x| < \varepsilon$.

Lemma 1. *Given a continuous $h_0 \in \mathcal{Q}^+$ and $\varepsilon \in \mathbb{R}^+$ then $\exists \delta$ s.t. $\forall x \in Q$, $\forall y \in B_\delta(x)$ and any h_n created according to* (3)*, $h_n(y) \in B_\varepsilon(h_n(x)))$.*

Proof. By continuity of h_0 and compactness of Q, for any given $\varepsilon \in \mathbb{R}$ $\exists \delta'$ s.t. $\forall y \in B_{\delta'}(x)$, $h_0(y) \in B_\varepsilon(h(x))$. By construction of h_n, $\delta = \min\{\delta', \varepsilon / K_D\}$ has the above property.

Our proposed solution to the ramp building problem can accommodate uncertainty in both the deposition location and size, see Sec. 4.1 end. However, for clarity we assume an exact shape function f in the following proofs.

2.3 Navigable Structures

Building a ramp means turning a structure that robots cannot climb into one they can climb. As such, any algorithm to adaptively build ramps needs a tractable description of climbable structures. This section defines the notion of *navigable* functions on Q, which represent climbable physical structures.

We use three parameters to describe robot specific motion constraints: $K \in \mathbb{R}^+$, to model the maximum steepness robots can drive up or down, $\epsilon \in \mathbb{R}^+$, to model the largest discontinuity robots can freely move past, and $d \in \mathbb{R}^+$, to limit the amount of discontinuity in a small area (i.e. robot length), Fig. 2(b)–2(d). Formally, navigable structures are locally (parameter d) close (parameter ϵ) to K–Lipschitz continuous [14, p. 594], i.e

$$|h(x) - h(y)| \le K|x - y| \ \forall x, y \in Q. \tag{4}$$

Specifically, a function $h \in \mathcal{Q}$ is called navigable if and only if

$$|h(x) - h(y)| \le \epsilon + K|x - y| \ \forall x, y \in Q \text{ and } |x - y| \le d. \tag{5}$$

To reason about global guarantees of our local algorithms, we construct the operator P_K, defined by (7). It maps any structure to the *closest* K-Lipschitz function that can be built by only adding material, Fig. 2(e). At a given point $x \in Q$, P_K takes the maximum value of any cones that need to be added so all other points fulfill condition (4). There are two important properties of P_K. Firstly, by construction

$$P_K[h](x) \geq h(x) \; \forall h \in \mathcal{Q}. \tag{6}$$

Since depositions are additive, it is important $P_K[h]$ can be reached by only adding to h. Secondly, $P_K[h]$ returns the smallest function in $\mathcal{L}_K$, the space of K-Lipschitz functions on Q, in the following sense, see Sec. 6 for proof.

Theorem 2. *Given any two functions $h \in \mathcal{Q}$ and $g \in \mathcal{L}_K$ with $g \geq h$, the operator*

$$P_K[h](x) = \max_{y \in Q}\{h(y) - K|y - x|\} \tag{7}$$

with $K \in \mathbb{R}^+$, has the following properties:

1. $P_K[h]$ is K-Lipschitz,
2. $g \geq P_K[h]$.

The following theorem shows that if steeper features are allowed, less material needs to be added, Fig. 2(d).

Theorem 3. *Given an arbitrary function $h \in \mathcal{Q}$ and $K_1, K_2 \in \mathbb{R}^+$ with $K_1 \leq K_2$ the projections onto $\mathcal{L}_{K_1}$ and $\mathcal{L}_{K_2}$ follow $P_{K_2}[h] \leq P_{K_1}[h]$.*

Proof. For a given point $y \in Q$ in (7), $h(y) - K_2|y-x| \leq h(y) - K_1|y-x|$ since the $|y-x|$ is non-negative. □

Given an initial function h_0, the next section gives a locally reactive deposition strategy such that after N depositions h_N is navigable, i.e. fulfills (5), and is bounded above by $P_K[h_0]$.

3 Local Reactive Deposition Algorithm

In a *local* deposition strategy, robots with limited sensing range $r \in \mathbb{R}^+$ (with $r > d$) move on top of the structure and *react* to features in their sensing range. Algorithm 1 relates local checks and depositions to global properties. It checks for points that imply a non-navigable feature and deposits in such a way as to decrease the distance from the current structure to closest K-Lipschitz structure. Specifically, Alg.1 searches for points $|y - x| \leq d$ s.t.

$$|y - x|K + \epsilon < |h(y) - h(x)|. \tag{8}$$

3.1 Correctness of Local Deposition Strategy

The correct behavior of Alg. 1 is that after a finite number of depositions the resulting structure h_N is navigable. The proof proceeds in two steps. (A) Thm. 4 shows progress, i.e. every deposition has a strictly positive volume.

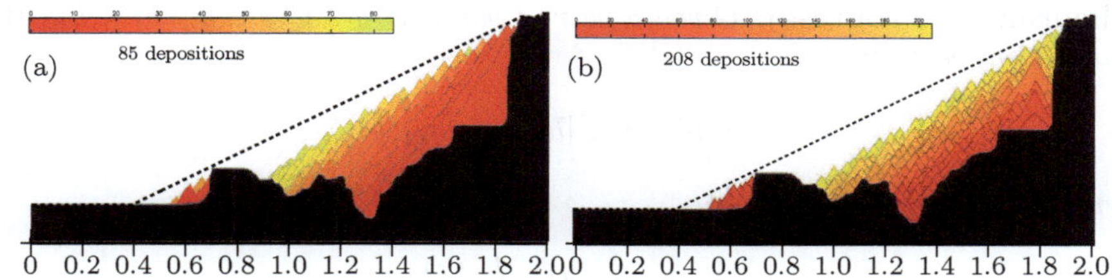

Fig. 3 Simulations of deposition algorithms. The initial structure h_0 is shown as solid black and the upper bound $P_K[h_0]$ as a dashed black line. The simulation parameters are: $Q = [0, 2]$, $K = 0.5$, $K_D = 1.5$, $\epsilon = 0.05$, and $d = 0.2$. Depositions progressively change color, see color-bar. (a) Deposition locations are picked randomly and the height according to Alg. 1. (b) Deposition locations and heights are picked according to Alg. 2, with $x_0 = 0.2$ and $x_* = 1.9$. As the colors shows, in both cases information about the cliff on the right propagated backward through stigmergy. Additional motion and deposition height constraints in Alg. 2 result in a layered structure and smaller depositions. The simulations incorporate additive noise to the deposition shape function, see Sec. 4.1.

Algorithm 1. Local Deposition Strategy. Pick point pairs that imply a local non-navigable feature and deposit on the lower one.

Given $h \in \mathcal{Q}^+$.
$h_0 \leftarrow h$
while $\exists\, x, y \in Q$ s.t. $|x - y| \leq d$, $K|y - x| + \epsilon < |h_n(y) - h_n(x)|$ **do**
  if $h_n(x) < h_n(y)$ **then**
    $x' \leftarrow x$
    $y' \leftarrow y$
  else
    $x' \leftarrow y$
    $y' \leftarrow x$
  end if
  Pick any $\omega \in [\epsilon, h_n(y') - h_n(x') - K|x' - y'|]$
  Deposit at x' with height ω, i.e. $h_{n+1} = D[f_{(x', \omega + h_n(x'))}, h_n]$
end while

(B) Thm. 5 shows depositions obey the invariant upper bound $P_K[h_0]$. By combining them, Thm. 6 shows correct behavior. Note that since $P_K[h_0]$ is the smallest dominating K-Lipschitz function, Alg. 1 is also efficient in the sense that it avoids unnecessary depositions, i.e. construction beyond the conservatively navigable $P_K[h_0]$, see Fig. 3(a).

The *volume* of the difference between two structures $g, h \in \mathcal{Q}^+$ is given by

$$V(g, h) = ||g - h||_1 \equiv \int_Q |g(x) - h(x)|\; dx. \tag{9}$$

Similarly, the volume of a particular deposition is given by $V(D[f_{(\phi,\sigma)}, h], h)$.

Theorem 4 (Progress). *Given a pair of points $x, y \in Q$ s.t. $h_n(x) < h_n(y)$ and the property that*

$$|x-y|K + \epsilon < |h_n(x) - h_n(y)|,$$

depositing on x with a height

$$\omega \in [\epsilon, \frac{h_n(y) - h_n(x)}{K|x-y|}]$$

results in a deposition volume $V(D[f_{(x,\omega)}, h_n], h_n) > \varepsilon$ that is bounded below by a strictly positive number.

Proof. Note that the deposition height is at least ϵ. By Lem. 1 there exists some δ s.t. h_n maps every $B_\delta(x) \subset Q$ into $B_{\epsilon/3}(h_n(x))$. As a result, $\forall p \in B_\delta(x)$, $h(p) < h(x) + \frac{\epsilon}{3}$ and $h(x) + \frac{2\epsilon}{3} < D[f_{(x,\omega)}, h_n](p)$. Therefore, $V(D[f_{(x,\omega)}, h_n], h_n) > \int_{B_\delta(x)} \frac{\epsilon}{3} = \varepsilon > 0$. □

Theorem 5 (Invariant). *Assuming that $K_D > K$, depositions made with Alg. 1 leave the mapping onto $\mathcal{L}_K$invariant, i.e. $P_K[h_n] = P_K[h_0]$.*

See Sec. 6 for proof.

Theorem 6. *Given an initial structure $h_0 \in \mathcal{Q}^+$, following Alg. 1 terminates after a finite number of steps, N; and for no points in Q does h_N fulfill non-navigability condition (8), i.e. $\forall z \in Q$ and $x, y \in B_{\frac{d}{2}}(z)$,*

$$|x-y|K + \epsilon \geq |h_N(x) - h_N(y)|.$$

Proof. The expression for the remaining volume $V(P[h_0], h_n) = ||P[h_0] - h_n||_1 = \int_Q |P[h_0](x) - h_n(x)|dx$ can be rewritten as

$$\int_Q |P[h_0](x) - h_{n+1}(x) + h_{n+1}(x) - h_n(x)|dx.$$

By Thm. 5 and (6), $P[h_0](x) - h_{n+1}(x) \geq 0$ and $h_{n+1}(x) - h_n(x) \geq 0$, therefore

$$\begin{aligned} V(P[h_0], h_n) &= \int_Q |P[h_0](x) - h_{n+1}(x)|dx + \int_Q |h_{n+1}(x) - h_n(x)|dx \\ &= V(P[h_0], h_{n+1}) + V(h_{n+1}, h_n). \end{aligned}$$

By Thm. 4 the second term is bounded below by a positive number ε, thus

$$V(P[h_0], h_{n+1}) < V(P[h_0], h_n) - \varepsilon.$$

Since volume is always non-negative, condition (8) for making depositions must be violated after a finite number of steps N. □

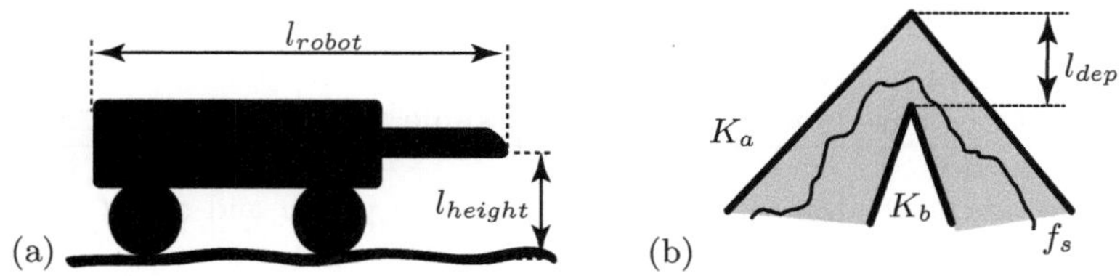

Fig. 4 Physical parameters. (a) Relevant robot dimension based on the prototype shown in Fig. 1(b). (b) Parameters for bounds of an arbitrary deposition shape function.

4 Adaptive Ramp Building

The local deposition algorithm Alg. 1 does not specify which points to pick if the non-navigability condition (8) is true for multiple pairs, neither does it consider physical robot size or whether robots can reach deposition locations. The benefit of this vagueness is generality. Algorithm 1 works in arbitrary dimensions with an arbitrary number of robots making depositions in any order. It forms the theoretical underpinning for Alg. 2, Fig. 3, which takes such physical considerations into account. It gives a local deposition and motion strategy that allows robot from an arbitrary starting position $x_0 \in Q$ to reach a goal position $x_* \in Q$. By using a more or less conservative ϵ the built structures can be made more or less smooth.

4.1 Adaptive Ramp Building with a Single Robot

To solve the adaptive ramp building problem via Alg. 1, robots need to identify point pairs that imply non-navigable features and make depositions. The strategy in Alg. 2 is to move toward the goal x_* unless a robot encounters a non-navigable feature that impedes its progress. In that case, a robot deposits according to Alg. 1 and backs up to check that the new deposition does not itself preset a non-navigable feature.

Since deposition and motion constraints depend on the robot's physical dimensions, Fig. 4(a), additional parameter constraints are necessary to prove correctness of Alg. 2. First, to guarantee that robots have enough room to back up we assume they start at a point $x_0 \in Q$ on the initial structure h_0 and can move freely within a radius $r_0 \in \mathbb{R}^+$ without making any depositions,

$$P_K[h](y) = h(y), \quad \forall y \in B_{r_0}(x_0) \subset Q. \tag{10}$$

Second, key dimensions of the robot as well as the deposition parameter K_D need to obey the following constraints, Fig. 4(a):

Algorithm 2. Adaptive ramp building. Given a structure h_0, an initial position x_0, and a goal position x_*, the following algorithm builds a ramp over irregular structures based on local sensing. Assume, w.l.o.g. that $x_0 < x_*$.

```
while x ≠ x* do
    Move toward goal until ∃y ∈ [x, x + r] that the pair y and x + d violate
    condition (8) , or x = x*
    if x ≠ x* then
        Move to the lower the point. (Note that all points in [x0, x + r) are
        climbable.)
        Pick height according to Alg.1 and condition (12).
        x ← x − 2d
    end if
end while
```

$$K_D \geq K + \frac{\epsilon + l_{height}}{d}, \tag{11}$$

$$l_{height} > \epsilon, \tag{12}$$

$$r_0 > 2d + l_{robot}. \tag{13}$$

Condition (11) limits how far backward new depositions can extend into previously navigable terrain. It ensures that the motion and deposition strategy will not direct robots to deposit directly underneath themselves. Condition (12) ensures that the deposition mechanism has enough clearance to make depositions that conform with the assumptions in Alg.1. Condition (13), conservatively, ensures that a physical robot has enough space to back up.

Theorem 7. *Given a robot that fulfills parameter conditions (11)-(13) with starting position x_0 that fulfills (10) following Alg. 2 will reach a goal point x_* after a finite number of steps.*

Proof. Denote the interval $[x_0 - r_0, x + d]$ in which no point pairs fulfill (8) by A (*accessible region*). Robots stay inside the accessible region at all times while finding points to deposit on. First, condition (12) guarantees a robot can make a deposition of height ϵ, as required by Alg. 1. Second, condition (11) guarantees that depositions with a maximum height of l_{height} made in the interval $[x, x+d]$ will not extend into $[x_0 - r_0, x - d]$. As a result, moving to $x - 2d$ after a deposition guarantees that no point pairs in A fulfill (8). By (10) and the deposition strategy there are always accessible points, i.e. $[x_0 - r_0, x_0] \subset A$. By Alg. 1 this algorithm terminates after a finite number of depositions with $x = x_*$. □

Figure 3(b) shows a series of depositions made via Alg. 2. This strategy also guarantees that robots can always reach x_0 without requiring additional depositions, which could allow robots to replenish supplies. Conversely, the accessible region provides cooperating robots access the deposition site, Sec. 4.2.

Physical depositions are not perfect cones, Fig. 1(b). Algorithm 2 explicitly allows for uncertainty in the target structure (via ϵ), but not for deposition

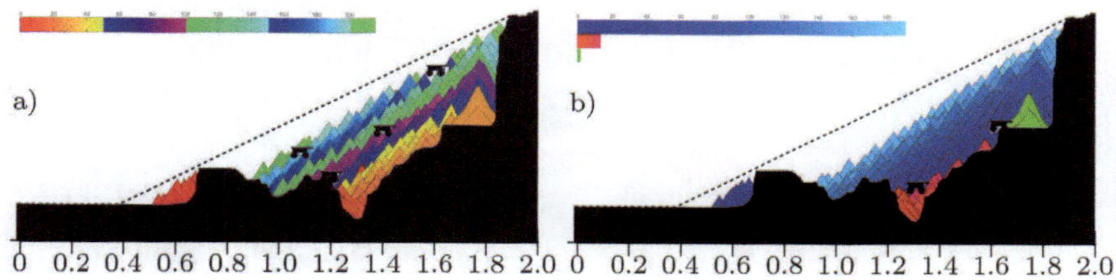

Fig. 5 Simulations of adaptive ramp building. Parameters are $x_0 = 0.2$, $x_* = 1.9$, $d = 0.1$ and otherwise the same as in Fig. 3. (a) Example of cooperative ramp building. Each robot is limited to making 25 depositions (indicated by a different color gradients), after which the active robot signals it is out of material and a new robot begins. (b) Multiple robots start simultaneously. If a robot becomes stuck, it is treated as a obstacle by other robots.

uncertainty. In fact, the upper bound for target structures requires that no depositions accidentally make intermediate structures larger than $P_K[h_0]$. Following is a short description on how to address this problem and allow depositions with arbitrary continuous shape functions f (and bounded derivative f'_{max}), as long as f can be sandwiched between two cones, Fig. 4(b). As long as $l_{dep} < \epsilon$. Alg. 1 (and as a result Alg. 2) still work with the following substitutions: In Lem. 1 f'_{max} takes the place of K_D. In Thm. 4 the minimum height is $\epsilon - l_{dep}$ instead of ϵ. In Thm. 5 and condition (11)K_D is replaced with K_a. In addition to uncertainty in shape, this approach of bounding cones also allows for uncertainty in the exact deposition location and volume.

4.2 Adaptive Ramp Building with Multiple Robots

The locally reactive nature of Alg. 2 makes extension to multiple robots easy. Without giving detailed motion and communication strategies, this section outlines two approaches. First, imagine multiple robots with limited deposition capacity cooperatively building a ramp. One robot starts executing Alg. 2 while the others follow. Once a robot runs out of building material, it signals for another robot to execute Alg. 2 and returns to a base station at x_0, or it can stop and be treated as an obstacle by other robots, Fig. 5(a). This coordination strategy works due to the distributed nature of Alg. 2. Information about deposition locations is communicated through stigmergy.

Second, imagine multiple robots can start at different locations and execute Alg. 2 concurrently. For example, to build a large ramp toward a beacon multiple robots could be dropped along the construction path. Each robot starts building a ramp. However, without initially fulfilling starting condition (10) robots might become stuck, i.e. cannot move to an appropriate place to make a deposition, Fig. 5(b) right. Further, without coordination one robot might deposit on another, Fig. 5(b) middle. Despite these failures,

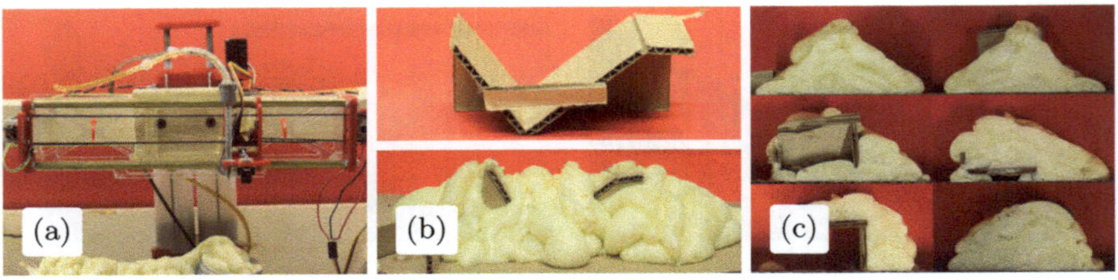

Fig. 6 Scanning foam deposition mechanism. (a) A scanning carriage holds a downward facing IR-distance sensor and mixing nozzle. Pressurized foam precursors are delivered to the nozzle by flexible tubing. (b) Top, Initial obstacle before leveling deposition. Bottom, final structure after deposition episode. (c) Cross sections of final structure. Each leveling deposition episode represents one cone-like deposition in Alg. 2.

if one robot initially fulfills (10) the process with successfully complete. Other robots can provide speed up through parallelism until they become stuck.

4.3 Physical Implementation and Experimental Results

We built a remote controlled prototype robot, Fig. 1(b), and a scanning foam deposition mechanism, Fig. 6(a), for testing solutions to the key technical challenges presented by Alg. 2. The prototype shows that robots can, in principle, build and navigate relatively large foam structures. The scanning deposition mechanism demonstrates autonomous leveling behavior that can be used to turn the physical three dimensional construction problem into the simplified two dimensional problem solved by Alg. 2.

One major challenge is designing a deposition mechanism and selecting an appropriate material [13]. The prototype robot and scanning deposition mechanism both use two compartment syringes with mixing nozzles (McMaster-Carr PN: 74695A11 with 74695A63, 7451A22 with 7816A32) and high expansion poly-urethane casting foam (US-Composites 2 lb foam) to make amorphous depositions.

The scanning deposition mechanism consists of a mixing nozzle and distance sensor mounted on moving carriage, Fig 6(a). By running a Alg. 1 along the direction of carriage travel (with $K = 0, \epsilon = 2\,\text{cm}$ and d covering all of Q) this mechanism autonomously creates a level structure from amorphous depositions. Mounting this mechanism on the front of a robot and treating each leveling deposition episode as a single deposition in Alg. 2, turns the physical construction problem into the simplified model. Viewed from the side, each leveled line under the carriage represents the apex of a conical deposition. Algorithm 2 simply picks the next point to level.

5 Conclusion

We developed a continuous model for amorphous depositions, and used it to prove correctness of a distributed algorithm that solves the adaptive ramp building problem. This example application illustrates how locally reactive behavior and amorphous building material together can create reliable building behavior in unstructured terrain.

Adaptive ramp building can also serve as a base behavior for composing more complicated behaviors. For example, it could guarantee accessibility to locations where support structures need to be built. With the ability to consistently encode virtual points in a group of robots, adaptive ramp building could be used to build arbitrary (K-Lipschitz) structures by building ramps to a carefully chosen set of virtual points: an approach we plan to explore.

There are a number of ways the presented algorithms could be improved. Our presentation focused on correctness, not optimality. Robots could be much smarter about coordination between robots and selecting deposition points to maximize the volume of each deposition, especially if their sensing radius was much larger than d.

6 Proofs

Proof (Thm. 2). 2.1) Assume to the contrary that $\exists x, y \in Q$ s.t.

$$|P_K[h](x) - P[h](y)| > K|x - y|. \tag{14}$$

Assume w.l.o.g. that $P_K[h](y) \leq P_K[h](x)$ and since $P_K[h]$ is a positive scalar function $|P_K[h](x) - P[h]_K(y)| = P_K[h](x) - P_K[h](y)$. Rearranging the terms in (14) leads to the contradiction $P_K[h](x) - K|x - y| > P_K[h](y)$, since the max in $P_K[h](y)$, see (7), is taken over the entire domain, including x. Therefore points violating the Lipschitz condition cannot exist in $P[h]$. □

2.2) Assume to the contrary that there exists a point $x \in Q$ s.t. $P_K[h](x) > g(x) \geq h(x)$. Since there cannot be equality between $P_K[h](x)$ and $g(x)$ the maximization in (7) must take its maximum value at some other point $y \in Q$. Rearranging $P_K[h](x) = h(y) - k|x - y| > g(x)$ results in $h(y) - g(x) > k|x - y|$, and since $g > h$ $g(y) - g(x) > k|x - y|$ which is a contradiction, as it would violate the Lipschitz continuity of g. □

Proof (Thm. 5). First, note that P can be applied to non-continuous functions, specifically continuous structures with a single discontinuous point. Let $\widetilde{h}_{n,(\phi,\sigma)}(x) = h_n(x) + (\sigma - h_n(\phi))\delta_\phi x$ where δ denotes the Kronecker delta.

Next, since ϕ is in the search set of max for point $P_K[h_n](x)$ in (7) $h_n(\phi) \leq \sigma = h_n(\phi) + \omega \leq P_K[h_n](\phi)$, consequently

$$\widetilde{h}_{n,(\phi,\sigma)} \leq P_K[h_n]. \tag{15}$$

Finally, since restricting $y \in \{x, \phi\} \subset Q$ in (7) results in the same expression as (2) $D[f_{(\phi,\sigma)}, h_n] = h_{n+1} \leq P_{K_D}[\widetilde{h}_{n,(\phi,\sigma)}]$. Thus, $h_{n+1} \leq P_{K_D}[\widetilde{h}_{n,(\phi,\sigma)}]$.

By Thm. 3 and assuming that $K_D > K$, $P_{K_D}[\widetilde{h}_{n,(\phi,\sigma)}] \leq P_K[\widetilde{h}_{n,(\phi,\sigma)}]$. Together Thm. 2.2 and (15) imply that $P_K[\widetilde{h}_{n,(\phi,\sigma)}] \leq P_K[h_n]$, which results in the series of relations $h_{n+1} \leq P_K[\widetilde{h}_{n,(\phi,\sigma)}] \leq P_K \leq P_K[h_n]$. And again, by Thm. 2.2 $P_K[h_{n+1}] \leq P_K[h_n]$. However, $h_{n+1} \geq h_n$ implies $P_K[h_{n+1}] \geq P_K[h_n]$, thus $P_K[h_{n+1}] = P_K[h_n]$. By induction, $P_K[h_n] = P_K[h_0]$. □

References

1. Berman, S., Halasz, A., Hsieh, M.A., Kumar, V.: Optimized stochastic policies for task allocation in swarms of robots. IEEE Transactions on Robotics 25(4), 927–937 (2009)
2. Bonwetsch, T., Gramazio, F., Kohler, M.: Digitally fabricating non-standardised brick walls. In: ManuBuild, Rotterdam, Netherlands, pp. 191–196 (2007)
3. D'Andrea, R.: Flying machine enabled construction, `http://www.idsc.ethz.ch/Research_DAndrea/fmec` (ongoing)
4. Galloway, K.C., Jois, R., Yim, M.: Factory floor: A robotically reconfigurable construction platform. In: IEEE International Conference on Robotics and Automation (ICRA), pp. 2467–2472 (May 2010)
5. Grushin, A., Reggia, J.A.: Automated design of distributed control rules for the self-assembly of prespecified artificial structures. Robotics and Autonomous Systems 56(4), 334–359 (2008)
6. Hjelle, D., Lipson, H.: A robotically reconfigurable truss. In: Proceedings of ASME/IFToMM International Conference on Reconfigurable Mechanisms and Robots (June 2009)
7. Khalili, N.: Emergency Sandbag Shelter: How to Build Your Own. CalEarth Press (2008)
8. Khoshnevis, B.: Automated construction by contour crafting related robotics and information technologies. Journal of Automation in Construction Special Issue: The Best of ISARC 2002 13, 5–19 (2004)
9. Ladley, D., Bullock, S.: Logistic constraints on 3d termite construction. In: Dorigo, M., Birattari, M., Blum, C., Gambardella, L.M., Mondada, F., Stützle, T. (eds.) ANTS 2004. LNCS, vol. 3172, pp. 178–189. Springer, Heidelberg (2004)
10. Lindsey, Q., Mellinger, D., Kumar, V.: Construction of cubic structures with quadrotor teams. In: Proceedings of Robotics: Science and Systems, Los Angeles, CA, USA (June 2011)
11. Magnenat, S., Philippsen, R., Mondada, F.: Autonomous construction using scarce resources in unknown environments. Autonomous Robots 33, 467–485 (2012)
12. Napp, N., Klavins, E.: A compositional framework for programming stochastically interacting robots. The International Journal of Robotics Research 30(6), 713–729 (2011)
13. Napp, N., Rappoli, O., Wu, J., Nagpal, R.: Materials and mechanisms for amorphous robotic construction. In: Proceedings of IEEE/RSJ International Conference on Intelligent Robots and Systems, IROS (2012)
14. Naylor, A.W., Sell, G.R.: Linear Opearator Theory in Engineering and Science. In: Applied Mathematical Sicences. Springer (1982)

15. Petersen, K., Nagpal, R., Werfel, J.: Termes: An autonomous robotic system for three-dimensional collective construction. In: Proceedings of Robotics: Science and Systems, Los Angeles, CA, USA (June 2011)
16. Revzen, S., Bhoite, M., Macasieb, A., Yim, M.: Structure synthesis on-the-fly in a modular robot. In: Proceedings of IEEE/RSJ Conference on Intelligent Robots and Systems, IROS (2011)
17. Theraulaz, G., Bonabeau, E.: Coordination in distributed building. Science 269 (1995)
18. Turner, J.S.: The Externded Organism. The Physiology of Animal-Built Structures. Harvard University Press (2004)
19. Northwestern Division US Army Corps of Engineers. Sandbagging Techniques (2004)
20. Werfel, J., Nagpal, R.: Three-dimensional construction with mobile robots and modular blocks. Int. J. Rob. Res. 27, 463–479 (2008)

On Segregative Behaviors Using Flocking and Velocity Obstacles

Vinicius Graciano Santos, Mario F.M. Campos, and Luiz Chaimowicz

Abstract. This paper presents a novel approach to swarm navigation that combines hierarchical abstractions, flocking behaviors, and an efficient collision avoidance mechanism. Our main objective is to keep large groups of robots segregated while safely navigating in a shared environment. For this, we propose the *Virtual Group Velocity Obstacle*, which is an extension of the Velocity Obstacle concept for groups of robots. By augmenting velocity obstacles with flocking behaviors and hierarchical abstractions, we are able to navigate robotic swarms in a cohesive and smooth fashion. A series of simulations and real experiments were performed and the results show the effectiveness of the proposed approach.

1 Introduction

The use of large groups of robots in the execution of complex tasks has received much attention in recent years. Generally called *Robotic Swarms*, these systems employ a large number of simpler agents to perform different types of tasks, oftentimes inspired by their biological counterparts.

A basic requirement for most robotic swarms is the ability for safe navigation in shared environments, *i.e.*, the ability of moving to specific goals while avoiding collisions with obstacles, teammates, and other groups. A desired property in this case is to keep robots close to their kins and avoid merging with other groups. This is called segregation and, as discussed previously [17], is a phenomenon that is seen in several biological systems.

This paper presents a novel approach to swarm navigation that keeps large groups of robots segregated while safely navigating in a shared environment. Our approach consists of extending the concept of Velocity Obstacles [10] with flocking behaviors

Vinicius Graciano Santos · Mario F.M. Campos · Luiz Chaimowicz
Vision and Robotics Laboratory (VeRLab), Computer Science Department,
Universidade Federal de Minas Gerais, Brazil
e-mail: {vgs,mario,chaimo}@dcc.ufmg.br

M.A. Hsieh and G. Chirikjian (eds.), *Distributed Autonomous Robotic Systems*, Springer Tracts in Advanced Robotics 104,
DOI: 10.1007/978-3-642-55146-8_9,

and hierarchical abstractions. Basically, Velocity Obstacles define the set of robot velocities that would result in a collision between the robot and an obstacle moving at a given velocity. We first augment this concept with flocking behaviors [23], which enables different robot groups to avoid collisions and navigate in a cohesive fashion using global perception. But when we restrict the robot's perception to a smaller area, this approach is unable to keep groups segregated. To solve this problem, we employ hierarchical abstractions [12]: we consider a group of robots as a single entity and make individual robots avoid velocities that would result in collisions with this entity, *i.e.*, we prevent robots from mingling with other groups.

We call this approach *Virtual Group Velocity Obstacles* and we present a series of experiments that show its capability of attaining our goal of safely navigating large robotic groups while keeping them segregated.

This paper is organized as follows: Section 2 discusses related work in the field of collision avoidance, swarm navigation, and hierarchical abstractions. The Velocity Obstacle concept is presented in Section 3. Section 4 introduces our methodology used to develop the Virtual Group Velocity Obstacle. Experimental results in simulated and real environments are presented in Section 5, while Section 6 brings the conclusion and directions for future work.

2 Related Work

One of the earliest works that considered the problem of controlling a swarm of agents was presented by Reynolds [23] with the aim of realistically simulating a flock of birds, known as *boids*. In that work, local interactions among agents within a neighboring area define an emergent behavior for the whole flock. Such interactions can be modeled as a special case of the social potential field method [22], an extension of the classical artificial potential field technique [16] that specifically deals with multi-agent systems.

The artificial potential field approach has been widely employed for controlling multi-robot and swarm systems [24]. Moreover, many works have focused on its use together with flocking principles in order to obtain specific behaviors, such as area coverage [15], moving in formation [3], converging into shapes [8], shepherding [18], segregation [17], and so on. However, it is known that the method is not oscillation-free and suffers from local minima, which is an intrinsic property that can arise from the combination of potentials, specially in unknown environments.

Recent work on robot collision avoidance has provided methods that are guaranteed to be collision-free and oscillation-free [5, 14, 25], even under nonholonomic constraints [2]. These methods rely on the concept of Velocity Obstacles [10], which is an extension of the Configuration Space Obstacle [19] for a time-varying system. A Velocity Obstacle defines the set of robot velocities that would result in a collision between the robot and an obstacle moving at a given velocity. Thus, the robot performs an avoidance maneuver at a specific time by selecting velocities that do not belong to that set. This approach has been widely used and extended for multi-agent navigation [1, 5, 6, 14, 25], even when considering uncertainties in position, shape, and velocity of the obstacles [11, 25]. An important extension was the development of the Reciprocal Velocity Obstacle [6], which acknowledged that most works on collision avoidance have not taken into account that obstacles' motion may be affected by the presence of the agent. That is, they have overlooked the reciprocity that arises when those obstacles are in fact other agents that can also react according to the robot's behavior, which could lead to oscillations in the system.

A different hierarchical paradigm considers the whole group as a single entity, sometimes called virtual structure [26], which effectively reduces the dimensionality of the control problem. The desired behavior is assigned to this structure, which implicitly controls the robotic swarm. Based on a mapping of the swarm's configuration space to a lower dimensional manifold, whose dimension is independent of the number of robots, a formal hierarchical abstraction that allows decoupled control of the pose and shape of a team of robots, was developed in [4]. This work was extended in [21], to account for three dimensional swarms. Based on the former, a hierarchical cooperation mechanism between multiple unmanned aerial and ground vehicles was developed [7], where UAVs are responsible for estimating the configuration of the ground robots and for sending control messages to the groups.

In our previous work [12], we have shown that hierarchical abstraction paradigms can be used in conjunction with simple collision avoidance techniques in order to achieve complex behaviors for swarm systems. We proposed a mechanism that allows large groups to deviate from each other during navigation and we have applied it to the problem of traffic control. In the present paper, we propose a different approach. We take advantage of the robust collision avoidance techniques that have been developed [6] and use them in conjunction with simple flocking rules and hierarchical abstractions for swarm navigation. Specifically, our goal is to ensure greater cohesion between robotic groups that navigate in the same workspace. That is, robots in the same group must stay together while avoiding merging with distinct groups. Henceforth, we denote this desired property as a *segregative behavior*.

3 Velocity Obstacles

Let A and B be two robots moving on the plane. Each robot i is fully actuated with kinematic model given by $\dot{\mathbf{p}}_i = \mathbf{v}_i$, where $\mathbf{p}_i = [x_i, y_i]^T$ is its pose and $\mathbf{v}_i$ its velocity. The velocity obstacle $VO_B^A(\mathbf{v}_B)$ of B to A is defined as the set of all velocities $\mathbf{v}_A$ that will result in a collision among robots A and B at some instant in time [10]. More precisely, we define $\lambda(\mathbf{p}, \mathbf{v})$ as a ray starting at $\mathbf{p}$ heading in the direction of $\mathbf{v}$ and $B \oplus -A$ as the Minkowski sum of B and $-A$, where $-A$ represents robot A reflected about its reference point.

$$\lambda(\mathbf{p}, \mathbf{v}) = \{\mathbf{p} + t\mathbf{v} \,|\, t \geq 0\} \tag{1}$$

Given these definitions, we can say that a velocity $\mathbf{v}_A \in VO_B^A(\mathbf{v}_B)$ if and only if the ray starting at $\mathbf{p}_A$ heading in the direction $\mathbf{v}_A - \mathbf{v}_B$ intersects $B \oplus -A$. Therefore, the full set of velocities that specifies a Velocity Obstacle can be denoted as:

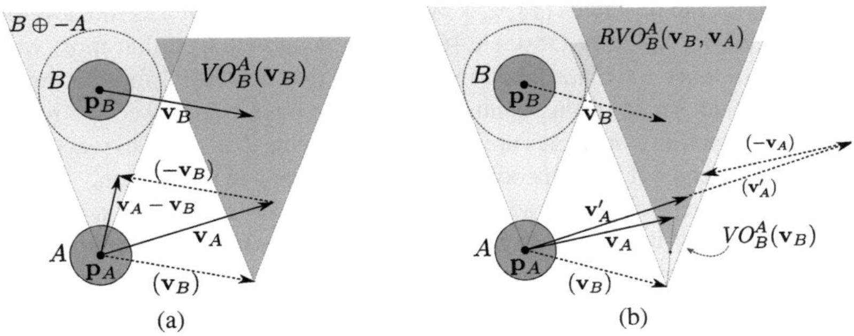

Fig. 1 (a) The Velocity Obstacle $VO_B^A(\mathbf{v}_B)$. (b) The Reciprocal Velocity Obstacle $RVO_B^A(\mathbf{v}_B, \mathbf{v}_a)$.

$$VO_B^A(\mathbf{v_B}) = \{\mathbf{v}_A \,|\, \lambda(\mathbf{p}_A, \mathbf{v}_A - \mathbf{v}_B) \cap (B \oplus -A) \neq \emptyset\}. \tag{2}$$

This set has an interesting property: if A selects a velocity that is outside its Velocity Obstacle induced by B, and if B maintains its current velocity, it is guaranteed that a collision between them will not occur [10]. Figure 1(a) shows an example of a Velocity Obstacle in a system with two circular robots. As it can be seen, $VO_B^A(\mathbf{v}_B)$ is a cone with its apex at $(\mathbf{v}_B)$.

When dealing with multi-robot systems, navigation based on the original Velocity Obstacle suffers from oscillation issues. In order to overcome this problem, a new velocity is chosen such that it is the average of the robot's current velocity and a velocity that lies outside the Velocity Obstacle [6]. Formally, the Reciprocal Velocity Obstacle is defined as:

$$RVO_B^A(\mathbf{v}_B, \mathbf{v}_A) = \{\mathbf{v}'_A \,|\, 2\mathbf{v}'_A - \mathbf{v}_A \in VO_B^A(\mathbf{v}_B)\}. \tag{3}$$

This new set contains all velocities that are the average of $\mathbf{v}_A$ and a velocity within $VO_B^A(\mathbf{v}_B)$, which can be seen as the cone $VO_B^A(\mathbf{v}_B)$ translated such that its apex lies at the mean of $\mathbf{v}_A$ and $\mathbf{v}_B$, as shown in Figure 1(b). Selecting the velocity which is closest to the robot's prior velocity and that also lies outside the set $RVO_B^A(\mathbf{v}_B, \mathbf{v}_A)$, guarantees a collision-free and oscillation-free navigation between a pair of robots [6].

Finally, in order to select inputs when dealing with Velocity Obstacles, an optimization problem must be solved. Several different approaches have been proposed in [5, 6, 10, 14, 25]. In the following section, we discuss how to couple flocking behaviors with sampling-based velocity selection [6].

4 Methodology

As mentioned, our main objective is to safely navigate large groups of robots in a shared environment while maintaining segregation among groups. Our approach consists of extending the concept of Reciprocal Velocity Obstacles with flocking behaviors and hierarchical abstractions.

We consider that robots are assembled together into a set of groups $\Gamma = \Gamma_1 \cup \Gamma_2 \cup \ldots \cup \Gamma_N$, where $\forall j,k : j \neq k \rightarrow \Gamma_j \cap \Gamma_k = \emptyset$. Let $\Phi_k \subseteq \Gamma_k$ be the set of robots belonging to group Γ_k that are within the neighborhood N_i, thus being perceived by robot i; and $\mathbf{p}(\Phi_k)$, $\mathbf{v}(\Phi_k)$ be the average position and velocity of group Φ_k, respectively.

Flocking behaviors can be achieved by a set of simple rules defined among neighboring robots [23]. Usually, robot controllers are derived by giving feedback on errors, such as relative velocities and relative positions. In our approach, we extend the velocity selection process presented in [6], which fast samples the set of admissible velocities and selects the best one according to an utility function.

Let $\mathbf{v}_i^{\text{pref}}$ be the preferred velocity of robot i, such as the vector pointing at the goal's direction with magnitude that is equal to the maximum allowed speed. In each iteration, velocities are sampled using a uniform distribution from the set of admissible velocities $AV^i(\mathbf{v}_i)$, that comprises the possible new velocities given the kinematic and dynamic constraints:

$$AV^i(\mathbf{v}_i) = \{\mathbf{v}'_i \mid \|\mathbf{v}'_i\| < v_i^{max} \wedge \|\mathbf{v}'_i - \mathbf{v}_i\| < a_i^{max}\Delta t\}, \tag{4}$$

where v_i^{max} and a_i^{max} are the maximum speed and maximum acceleration of robot i, respectively, and Δt is the time step of the system.

Ideally, the selected velocity $\mathbf{v}_i^{\text{new}}$ among the sampled set should lie outside the union of all RVOs generated by other robots and VOs generated by dynamic and static obstacles. However, as the environment may become crowded to the point that no admissible velocities will exist, the algorithm is allowed to selected a velocity that belongs to the generated set of Velocity Obstacles, but it is penalized by this choice according to the following function:

$$\mathbf{v}_i^{\text{flock}} = \mathbf{v}_i^{\text{pref}} + \alpha(\mathbf{v}(\Phi_k) - \mathbf{v}_i) + \beta(\mathbf{p}(\Phi_k) - \mathbf{p}_i) \tag{5}$$

$$P_i(\mathbf{v}'_i) = \frac{w}{c_i(\mathbf{v}'_i)} + \|\mathbf{v}_i^{\text{flock}} - \mathbf{v}'_i\|, \tag{6}$$

with $i \in \Gamma_k$, where α rules the convergence of the robot's current velocity to its neighbors' average velocity, β is the weight that governs the behavior which makes robots move toward the centroid of their neighboring agents, and w regulates the avoidance behavior between sluggish and aggressive. Function $c_i(\mathbf{v}'_i)$ is the expected time to collision, which is computed by solving the set of ray intersection equations induced by (2), (3) and taking their minimum. Note that, by setting $\alpha = \beta = 0$, the selection approach is reduced to the original RVO method [6]. Thus, we select a new velocity $\mathbf{v}_i^{\text{new}}$ that minimizes the penalty function P_i over the sampled set of velocities $S \subseteq AV^i(\mathbf{v}_i)$. Figure 2 exemplifies this sampling-based selection procedure.

$$\mathbf{v}_i^{\text{new}} = \underset{\mathbf{v}'_i \in S}{\text{argmin}} P_i(\mathbf{v}'_i) \tag{7}$$

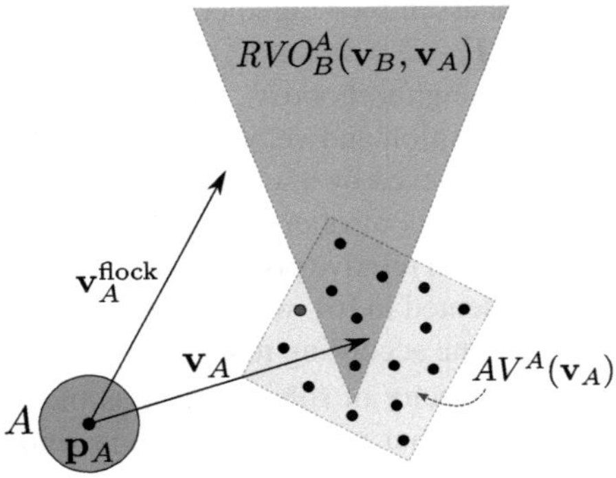

Fig. 2 Sampling-based velocity selection. Admissible velocities which were sampled are represented by small circles. The red sample is chosen as it minimizes the penalty function.

Equation (5) is responsible for the flocking behavior, which can be achieved by fine tuning its constants. For instance, there are triples of (α, β, w) that can lead robots into tightly aggregated groups. In these scenarios, groups moving in opposite directions show an emerging segregative behavior. That is, they tend to smoothly avoid each other without merging, as seen in Figure 3.

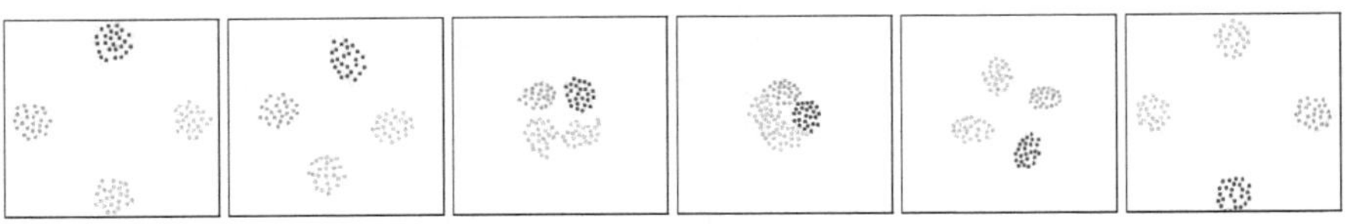

(a) Four groups.

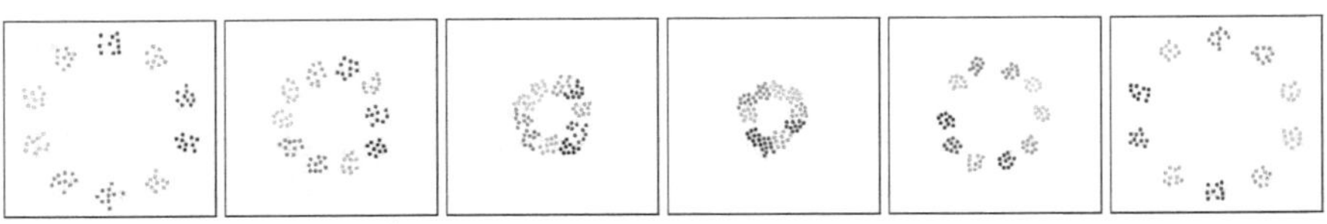

(b) Ten groups.

Fig. 3 Simulated execution of the flocking algorithm with distinct groups moving in opposite directions using global sensing

As groups move toward each other, robots will try to select velocities that lie outside of the combined RVOs induced by the other group. Since this group is tightly packed, robots will select velocities that may lead them to avoid the group as a whole. Given the reciprocity, the other group will select its velocities accordingly.

Nevertheless, the same behavior cannot be guaranteed when dealing with a small sensing neighborhood, since robots will not be able to predict incoming groups. In this case, merging can occur if the robot is not able to correctly maneuver given its current velocity. To account for this problem, we introduce a virtual Velocity Obstacle that is responsible for blocking velocities which may lead to merging among groups. Specifically, we denote this virtual obstacle as the *Virtual Group Velocity Obstacle* (VGVO), which is shown in Figure 4.

The concept of VGVO is simple: each robot i senses the presence of every robot j within a neighborhood N_i and builds the shape of each group of robots, but of its own group. These shapes are considered as virtual obstacles in the workspace of robot i that move with the average velocity of the respective group that has been used in the building process. Hence, a virtual Velocity Obstacle can be built for each shape in order to define the set of velocities that would lead the robot to merge with a different group if the latter maintains its current average velocity.

Let $R(\mathbf{p}_i)$ denote all the points that represent robot i in its workspace. Consequently, the Virtual Group Velocity Obstacle of robot i induced by group Φ_k can be defined as:

$$VGVO^i_{\Phi_k}(\mathbf{v}(\Phi_k)) = \{\mathbf{v}'_i \,|\, \lambda(\mathbf{p}_i, \mathbf{v}'_i - \mathbf{v}(\Phi_k)) \cap C(\mathbf{p}_i, \Phi_k) \neq \emptyset\}, \tag{8}$$

$$C(\mathbf{p}_i, \Phi_k) = \text{Shape}(\bigcup_{j \in \Phi_k} R(\mathbf{p}_j)) \oplus -R(\mathbf{p}_i), \tag{9}$$

where Shape(Q) is the shape of the set of points Q, which could be represented as the smallest enclosing disc, the convex hull or the more general class of α-shapes [9].

Equation (9) refers to the idea of the hierarchical abstraction paradigm presented in Section 2, in which the whole group is considered as a single entity. In every iteration, groups are abstracted into single entities that move according to their average velocities. In this way, single robots navigate using the RVO in conjunction with the VGVO, which guarantees a collision-free navigation among groups while maintaining segregation. Furthermore, we can assume that groups will deviate from single robots given the reciprocity of the RVO. Therefore, we can also specify a Virtual Group Reciprocal Velocity Obstacle with the following definition:

$$VGRVO^i_{\Phi_k}(\mathbf{v}(\Phi_k), \mathbf{v}_i) = \{\mathbf{v}''_i \,|\, 2\mathbf{v}''_i - \mathbf{v}_i \in VGVO^i_{\Phi_k}(\mathbf{v}(\Phi_k))\}. \tag{10}$$

5 Experiments

To study the feasibility of the proposed approach, we executed a series of simulations and real experiments[1]. For the simulations, we used Player/Stage [13], a well known framework for robot programming and simulation. Real experiments were performed with a dozen *e-puck* robots, a small-sized differential robot equipped with a ring of 8 IR sensors for proximity sensing and a DS-PIC processor. A bluetooth wireless interface allows local communication among robots along with a remote computer.

[1] A video showing the experiments is available at:
`http://www.youtube.com/watch?v=ylYMbwOduqg`

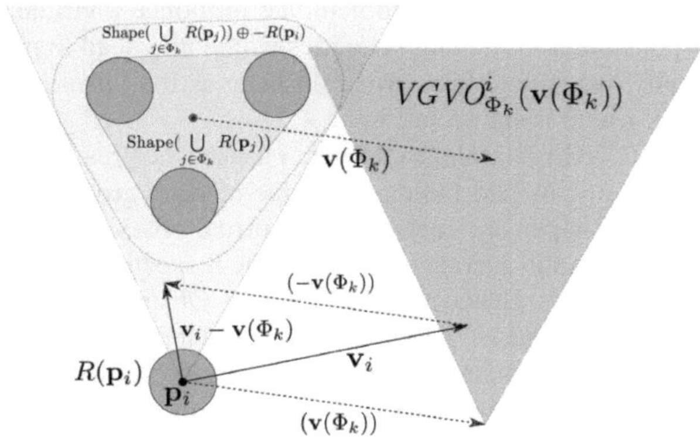

Fig. 4 The Virtual Group Velocity Obstacle $VGVO^i_{\Phi_k}(\mathbf{v}(\Phi_k))$

In all experiments, constants α, β, and w were set manually. When comparing the behavior of different algorithms, these constants remain the same between executions. Additionally, the neighborhood area is defined as five times the robot's radius in experiments with restricted sensing. Furthermore, we assume that each robot has the ability to map every other robot to its respective group.

5.1 Simulations

Figure 3 presents the execution of the flocking algorithm in a simulated environment consisting of 100 virtual robots evenly distributed into four and ten groups that move in opposite directions. Both simulations rely on global sensing, thus robots can perceive each other regardless of their relative distances. Through visual inspection we can see that, robots tend to remain segregated into homogeneous groups using our approach.

When we restrict robot perception to a local sensing, the flocking approach alone does not guarantee segregation, as shown in Figure 5(a). On the other hand, as depicted in Figure 5(b), the VGRVO is able to keep robots segregated even within a local sensing scenario. In this case, the virtual obstacle was built upon the groups' α-shape [9]. Thus, the addition of VGRVOs has led to a successful navigation while maintaining the desired segregation property.

In a recent work [17], a formal way of measuring segregation among groups of agents has been proposed. Two groups Γ_A and Γ_B are said to be segregated if the average distance between robots in the same group is less than the average distance between robots in distinct groups. Thus, the following restriction must hold

$$(d^{AA}_{\text{avg}} < d^{AB}_{\text{avg}}) \wedge (d^{BB}_{\text{avg}} < d^{AB}_{\text{avg}}), \tag{11}$$

where d^{AB}_{avg} is the average distance between robots in group Γ_A and Γ_B.

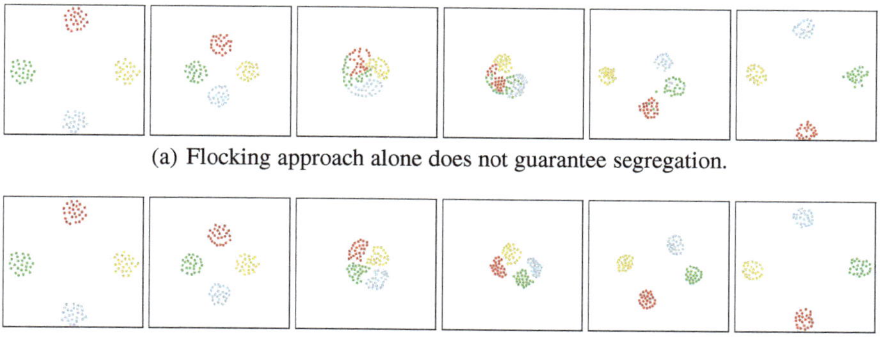

(a) Flocking approach alone does not guarantee segregation.

(b) Virtual Group Reciprocal Velocity Obstacle.

Fig. 5 Simulated execution of the segregative behavior algorithm with distinct groups moving in opposite directions using local sensing

Figure 6 depicts a comparative analysis about the emerging segregative behaviors using the proposed algorithms. For these simulations, a scenario consisting of 200 virtual robots evenly divided into two distinct groups was used. Initially, agents were randomly positioned into a circular area according to a normal distribution. After that, groups were ordered to swap their positions. Figures 6(a) and 6(b) show the results for the original RVO approach and our flocking extension, respectively, while relying on global sensing capabilities for all robots. As it can be seen, constraint (11) holds true for all time slices of the second simulation, which indicates that a segregative behavior has emerged. On the other hand, since the RVO algorithm tends to form merged lanes of moving robots during navigation, (11) will not hold, as expected. This can be confirmed visually in Figure 6(a), as there exists intersection points between the curve d_{avg}^{AB} and one of the others. Figures 6(c) and 6(d) compare our flocking algorithm without and with the use of VGRVO, respectively. In these simulations, robots had their sensors capabilities constrained into a small local neighborhood when compared to the size of their own group. Due to this restriction, groups were not able to remain segregated during navigation by means of flocking only, since (11) did not hold for all time steps in Figure 6(c). However, the use of VGRVOs under these conditions has shown to be effective for maintaining a segregative behavior throughout the simulation, as shown in Figure 6(d).

Although our flocking mechanism can improve group navigation in terms of cohesion and symmetry, specially when dealing with multiple groups, such as in Figure 3(b), the system's performance in terms of elapsed time may decrease. For instance, in Figure 6, when curve d_{avg}^{AB} reaches again its initial value, it means that both groups have swapped their average positions. Hence, it is easy to see that robots using the original RVO approach have reached their goals faster than when applying flocking. Intuitively, the loss in performance happens because the second and third terms of equation (5) may play a damping role. Furthermore, as neighbors get closer

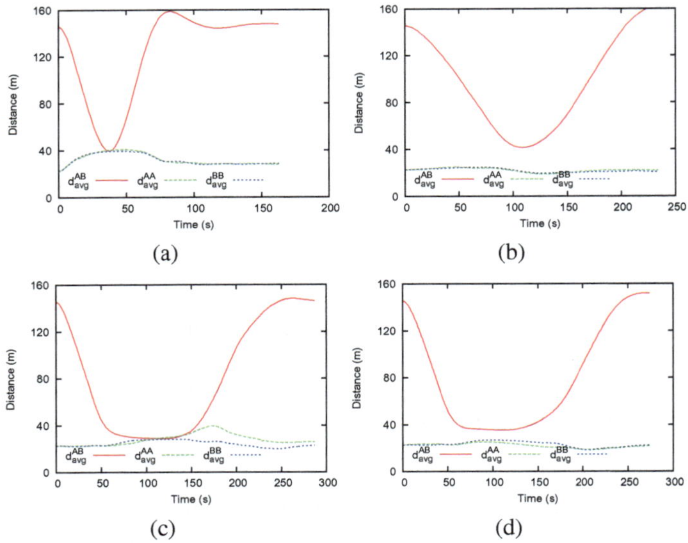

Fig. 6 Segregative behavior analysis for 200 robots evenly distributed into two groups that swap positions. (a) RVO with global sensing. (b) Flocking with global sensing. (c) Flocking with local sensing. (d) VGRVO with local sensing.

to each other, robots will select safer velocities in order to avoid collisions, while trying to maintain the flock. Thus, slower speeds will have higher priority during the selection process.

5.2 Real Robots

Real experiments were conducted indoors using twelve *e-puck* robots. These experiments are important in order to show the feasibility of the algorithm in real scenarios, where all uncertainties caused by sensing and actuation errors may have an important role on the results.

In these experiments, we used a swarm localization framework based on an overhead camera and fiduciary markers for estimating robot's pose and orientation. Also, as the *e-puck*'s IR sensors have a very small range, we implemented a virtual sensor based on the localization system to detect neighboring agents. In order to control the differential drive robots, a damping term based on the current velocity was added to (4) in order to achieve better stability. To account for nonholonomic constraints, input velocities were transformed following the approach presented in [20].

Figure 7 shows snapshots from an execution of VGRVO with limited sensing with two and four groups. We can visually inspect that the behaviors obtained with the real robots is pretty similar to the simulation results. In terms of segregation, the average distances for two groups follow the trend showed in Figure 6(d) for the simulated experiments: the average distance between robots in the same group

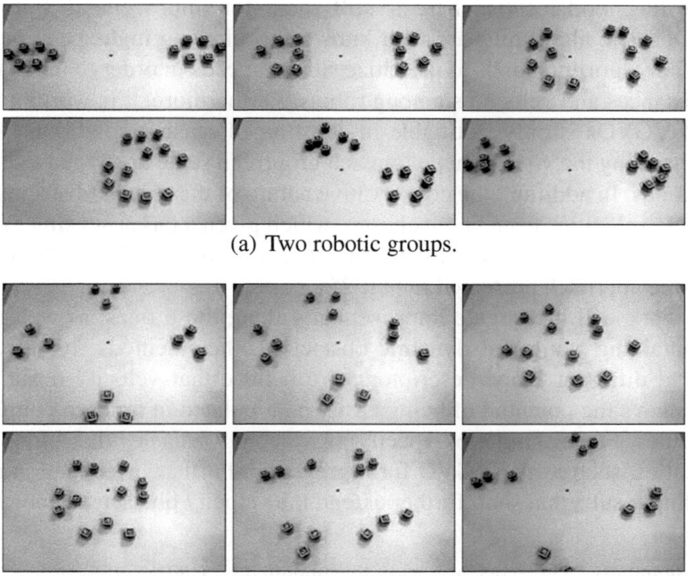

(a) Two robotic groups.

(b) Four robotic groups.

Fig. 7 Real execution of the VGRVO algorithm with different group sizes

is always less than the average distance between robots in distinct groups. These proof of concept experiments indicate that the algorithm can work well to coordinate groups of real robots, allowing them to navigate while maintaining a segregative behavior in an efficient way.

6 Conclusion

This paper presented an approach that ensures greater cohesion between robotic groups that navigate in shared environments. That is, robots flock together with their own group while remaining segregated from others. Our method does not rely on any communication to achieve coordination and works in a distributed fashion.

We achieve flocking behaviors for robotic swarms using Velocity Obstacles by choosing a proper utility function for selecting velocities. Results have shown that segregative behaviors have emerged during our experiments. We realized that this was not the case when dealing with restricted sensing capabilities. Therefore, we introduced a novel concept: the Virtual Group Velocity Obstacle, which is a virtual obstacle that is created in order to prevent aggregation between distinct robotic groups. The VGVO resembles ideas from the hierarchical abstraction paradigm, where groups are abstracted into single entities with the aim of reducing the dimensionality of the control problem. Several experiments were performed in simulated and real scenarios, which demonstrated the effectiveness of the proposed approach.

Despite the good results, there is still room for improvement. For instance, Velocity Obstacle algorithms are well known for allowing high-speed navigation. However, our algorithm tends to prioritize slower speeds in order to maintain stable relative distances and velocities among robots. Furthermore, it is worth noting that the use of VGVOs is only justifiable in constrained sensing conditions, since the cost of computing the virtual shape for each group can scale up given larger sensing neighborhoods. In addition, the conservative nature of the virtual obstacle may prevent feasible velocities from being chosen, which given a larger sensing radius may turn to be overly conservative. Finally, uncertainty models could be easily included following the approach presented in [11, 25].

Future work will investigate improvements along these lines, as well as experiments containing static and dynamic obstacles, other methods for selecting velocities, and different shape descriptors. We believe that velocity obstacle based algorithms have the potential of being easily manipulated in order to couple simple behavior rules. For example, by specifying velocity obstacles that block undesirable velocities, such as the VGVO for segregation. Further investigation may lead to interesting results that shall further extend the velocity obstacle framework.

Acknowledgements. This work was partially supported by CNPq and Fapemig. We would like to thank Prof. Vijay Kumar for some insights on segregative behaviors.

References

1. Abe, Y., Yoshiki, M.: Collision avoidance method for multiple autonomous mobile agents by implicit cooperation. In: Proceedings of the IEEE/RSJ International Conference on Intelligent Robots and Systems, vol. 3, pp. 1207–1212 (2001)
2. Alonso-Mora, J., Breitenmoser, A., Rufli, M., Beardsley, P., Siegwart, R.: Optimal reciprocal collision avoidance for multiple non-holonomic robots. In: Martinoli, A., Mondada, F., Correll, N., Mermoud, G., Egerstedt, M., Hsieh, M.A., Parker, L.E., Støy, K. (eds.) Distributed Autonomous Robotic Systems. STAR, vol. 83, pp. 203–216. Springer, Heidelberg (2013)
3. Balch, T., Hybinette, M.: Social potentials for scalable multirobot formations. In: Proceedings of the IEEE International Conference on Robotics and Automation, pp. 73–80 (2000)
4. Belta, C., Kumar, V.: Abstraction and control for groups of robots. IEEE Transactions on Robotics 20(5), 865–875 (2004)
5. van den Berg, J., Guy, S.J., Lin, M., Manocha, D.: Reciprocal n-body collision avoidance. In: Pradalier, C., Siegwart, R., Hirzinger, G. (eds.) Robotics Research. STAR, vol. 70, pp. 3–19. Springer, Heidelberg (2011)
6. van den Berg, J., Lin, M., Manocha, D.: Reciprocal velocity obstacles for real-time multiagent navigation. In: IEEE International Conference on Robotics and Automation, pp. 1928–1935 (2008)
7. Chaimowicz, L., Kumar, V.: Aerial shepherds: Coordination among uavs and swarms of robots. In: 7th International Symposium on Distributed Autonomous Robotic Systems (2004)

8. Chaimowicz, L., Michael, N., Kumar, V.: Controlling swarms of robots using interpolated implicit functions. In: Proceedings of the IEEE International Conference on Robotics and Automation, pp. 2498–2503 (2005)
9. Edelsbrunner, H., Kirkpatrick, D., Seidel, R.: On the shape of a set of points in the plane. IEEE Transactions on Information Theory 29(4), 551–559 (1983)
10. Fiorini, P., Shillert, Z.: Motion planning in dynamic environments using velocity obstacles. International Journal of Robotics Research 17, 760–772 (1998)
11. Fulgenzi, C., Spalanzani, A., Laugier, C.: Dynamic obstacle avoidance in uncertain environment combining pvos and occupancy grid. In: IEEE International Conference on Robotics and Automation, pp. 1610–1616 (2007)
12. Santos, G., Chaimowicz, V., Hierarchical, L.: congestion control for robotic swarms. In: Proceedings of the IEEE/RSJ International Conference on Intelligent Robots and Systems, pp. 4372–4377 (2011)
13. Gerkey, B.P., Vaughan, R.T., Howard, A.: The player/stage project: Tools for multi-robot and distributed sensor systems. In: Proceedings of the 11th International Conference on Advanced Robotics, pp. 317–323 (2003)
14. Guy, S.J., Chhugani, J., Kim, C., Satish, N., Lin, M., Manocha, D., Dubey, P.: Clearpath: highly parallel collision avoidance for multi-agent simulation. In: Proceedings of the ACM SIGGRAPH/Eurographics Symposium on Computer Animation, pp. 177–187. ACM, New York (2009)
15. Howard, A., Mataric, M., Sukhatme, G.: Mobile sensor network deployment using potential fields: A distributd, scalable solution to the area coverage problem. In: Proceedings of the 6th International Symposium on Distributed Autonomous Robotic Systems (2002)
16. Khatib, O.: Real-time obstacle avoidance for manipulators and mobile robots. In: Proceedings of the IEEE International Conference on Robotics and Automation, vol. 2, pp. 500–505 (1985), doi:10.1109/ROBOT.1985.1087247
17. Kumar, M., Garg, D., Kumar, V.: Segregation of heterogeneous units in a swarm of robotic agents. IEEE Transactions on Automatic Control 55(3), 743–748 (2010)
18. Lien, J.M., Bayazit, O., Sowell, R., Rodriguez, S., Amato, N.: Shepherding behaviors. In: Proceedings of the IEEE International Conference on Robotics and Automation, pp. 4159–4164 (2004)
19. Lozano-Perez, T.: Spatial planning: A configuration space approach. IEEE Transactions on Computers C-32(2), 108–120 (1983)
20. Luca, A.D., Oriolo, G., Vendittelli, M.: Stabilization of the unicycle via dynamic feedback linearization. In: 6th IFAC Symposium on Robot Control, pp. 397–402 (2000)
21. Michael, N., Belta, C., Kumar, V.: Controlling three dimensional swarms of robots. In: Proceedings of the IEEE International Conference on Robotics and Automation, pp. 964–969 (2006)
22. Reif, J.H., Wang, H.: Social potential fields: A distributed behavioral control for autonomous robots. Robotics and Autonomous Systems 27(3), 171–194 (1999)
23. Reynolds, C.W.: Flocks, herds, and schools: A distributed behavioral model. In: Computer Graphics, pp. 25–34 (1987)
24. Siegwart, R., Nourbakhsh, I.R.: Introduction to Autonomous Mobile Robots. Bradford Company, Scituate (2004)
25. Snape, J., van den Berg, J., Guy, S., Manocha, D.: The hybrid reciprocal velocity obstacle. IEEE Transactions on Robotics 27(4), 696–706 (2011)
26. Tan, K.H., Lewis, M.: Virtual structures for high-precision cooperative mobile robotic control. In: Proceedings of the IEEE/RSJ International Conference on Intelligent Robots and Systems, vol. 1, pp. 132–139 (1996)

Object Transportation by Granular Convection Using Swarm Robots

Ken Sugawara, Nikolaus Correll, and Dustin Reishus

Abstract. We propose a novel method for object transport using granular convection, in which the granular material is a robot swarm consisting of small robots with minimal sensors. Granular convection is commonly observed in the "Brazil Nut Effect". In this work, we consider the transported object to be passive, however, and not actuated like the surrounding granular material. We show that the passive object can be transported to a given destination in spite of the fact that each robot does not know the location of the object being transported nor the location of the destination. Each robot moves based solely on a weak repulsive force from the destination and stochastic perturbations. We first show fundamental characteristics of a system with no communication between robots. We observe that very high or very low robot densities are detrimental to object transport. We then show that heterogeneous swarms increase performance. We propose two types of heterogeneous swarm systems: a swarm in which robots switch states probabilistically, and a swarm in which state propagates using local communication. The signal propagation system shows the best performance in terms of success rate and accuracy in a wide range of densities.

1 Introduction

Object transportation is a fundamental task in robotics and frequently arises in various situations from micro to macro scale. Our main interest focuses on micro swarm robotic systems in which the agents have minimal sensing requirements,

Ken Sugawara
Tohoku Gakuin Univ., Japan
e-mail: sugawara@cs.tohoku-gakuin.ac.jp

Nikolaus Correll · Dustin Reishus
Univ. of Colorado Boulder, CO, USA
e-mail: {nikolaus.correll,reishus}@colorado.edu

M.A. Hsieh and G. Chirikjian (eds.), *Distributed Autonomous Robotic Systems*,
Springer Tracts in Advanced Robotics 104,
DOI: 10.1007/978-3-642-55146-8_10,

have minimal communication abilities, and can manipulate objects only by pushing them. In this paper, we propose a series of distributed algorithms that are based on granular convection [1], also known as the "Brazil Nut Effect," and a constant external field.

Granular convection is commonly observed in granola boxes in which the larger items tend to aggregate at the surface after the box has been shaken. This is interesting, as these objects seemingly defy gravity when traveling to the top of the box instead of sinking to the bottom, and has potential applications in manipulating objects at the nano- and micro-scales.

This paper studies the Brazil Nut Effect and its potential applications in robotic object transportation. Instead of gravity and external actuation (by shaking), our system is driven by self-propelled robots, which can be steered by the application of an external field. Depending on the capabilities of the robots, this external field can range from gravitational or magnetic fields to infrared or chemical gradients.

Such robots could be manufactured at very low cost and/or miniaturized and have applications from garbage collection at sea to transporting micro-scale objects in a self-assembly scenario. Currently, object manipulation at this scale is accomplished by application of external fields, e.g., magnetic forces [2]; specially designed environmental templates [3]; manipulation via micro-tweezers [4]; or by enclosing a swarm of agents, e.g., magneto-tactic bacteria [5], inside the object.

1.1 Related Work

Object manipulation using a swarm of robots is a canonical problem also known as "box-pushing task" [6, 7, 8, 9, 10]. In these works, robots rely on sensors to detect when the box is reached and on communication to coordinate pushing direction. Some robotic researchers have also tried to apply protozoa for object manipulation [11, 12].

The work described in this paper addresses a subset of the box-pushing domain as it allows to push cylindrical objects, but no robot explicitly recognizes its behavior as object handling. No robot knows where the destination of the object is, nor where the object itself is. This allows the robots to be very simple, low-power, and inexpensive.

The key mechanism here is granular convection, which has been leveraged in a robot context for segregation and pattern formation [13, 14]. There is also a larger body of work on the physics underlying granular convection on transport and segregation processes [15, 16, 17]. The object undergoes a biased random walk, similar to bacterial chemotaxis [18, 19].

1.2 Contribution of This Paper

We propose a novel method for object transportation in small-size, large scale distributed swarm robotic systems. The object, which is completely passive, has a given starting point and is required to reach a specific destination. We study this problem using a simple kinematic model that takes into account the robots' limited sensors and functions. Each robot does not know the location of the destination explicitly, but instead moves randomly in a closed environment.

We first discuss fundamental properties of the homogeneous system, i.e., all robots are identical, with no explicit communication. We then propose a heterogeneous system composed of robots that change their direction with two different probabilities. Finally, we introduce a heterogeneous system that uses local communication to adjust the fraction of robots that change their direction with higher probability. We show that heterogeneous robotic swarms can improve the success rate of object transportation, and that local communication can further improve performance.

2 Dynamical System Model

Let each robot move using the following dynamics:

$$m d\vec{v}_i / dt = \alpha \vec{F}_1(p) + \beta \vec{F}_2 - \gamma \vec{v}_i \tag{1}$$

where m, $\vec{v}_i$ denote robot's mass and the velocity of the i-th robot, respectively. $\vec{F}_1(p)$ denotes a unit vector and its orientation changes randomly in the range of $(-\pi, \pi)$ according to the probability p. $\vec{F}_2$ denotes repulsive unit vector from the destination. It is calculated as

$$\vec{F}_2 = (\vec{r}_i - \vec{r}_d) / |\vec{r}_i - \vec{r}_d| \tag{2}$$

where $\vec{r}_i$, $\vec{r}_d$ denote the position of the i-th robot and the position of the destination, respectively. γ is a resistive force. For simplicity, we simulate the friction of the robots by this term. Simulation conditions are shown in Table 1.

Table 1 Simulation conditions

Field size:	4.0 x 2.0
Boundary:	with walls
Robot's diameter:	0.2
Object's diameter:	0.6
Robot density:	ρ (fraction of environment occupied by robots)
Resistive force:	γ (fixed at 5.0)
Initial position of robots:	random
Initial position of object:	(1.0, 0.0)
Destination:	(-1.0, 0.0)
Simulation time:	equivalent to 10800 sec (= 3 hours) per a trial
Simulation frequency:	100 trials in each condition

Fig. 1(a) and (c) show snapshots of the simulation. Green, Red, and Blue circles indicate the robots, object, and destination, respectively. Fig. 1(b) and (d) show a typical example of trajectories of the object. The fluctuation becomes small as the density increases (lower row).

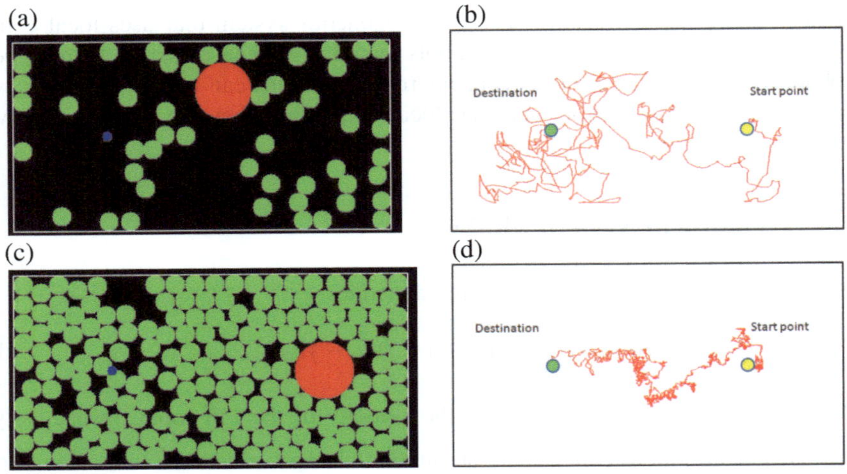

Fig. 1 Snapshots of simulation and the object trajectory. (a) and (b) are the case of N=50 (ρ~0.21), and (c) and (d) are the case of N=180 (ρ~0.75).

3 Robots with Static Characteristics

In this section, we describe the behavior of the system in the case where all robots behave identically. This corresponds to the Brazil Nut Effect, except for the fact that the object itself is not subject to actuation.

As defined in eq. (1), all robots move according to a "random force" of strength α and a "repulsive force from the destination" of strength β. Fig. 2(a) indicates the relation between α, β and the success rate in case where the robots occupy 42% of the area in the environment, i.e., $\rho = 0.42$. The success rate is defined as the fraction of trials in which the object is carried to within 20 pixels of the destination by the end of the trial. We observe that when $\alpha < 0.2$, transportation consistently fails and the success rate increases as α becomes large.

When the repulsive force is dominant, i.e., β is high, no transportation occurs because the robots drive away from the destination without performing random side movements that eventually propel the object toward the destination. When the random force is dominant, i.e., high values of α, transportation to the destination occurs only by chance because their behavior is equivalent to Brownian motion of the object. A green line in Fig. 2(a) indicates β =0, i.e., the success rate by

Brownian motion. Fig. 2(b) shows the success rate after discounting the success rate due to pure Brownian motion, i.e., after subtracting the green line from Fig. 2(a).

In order to clarify the effectiveness of the proposed method, we focus on the region where the success rate exceeds the result of Brownian motion. The effectiveness depends on the density of robots ρ, and Fig. 2(c) indicates the region where the success rate by the proposed method exceeds that of Brownian motion on the α-β plane. The effective region in this plane is relatively small under the condition of low density such as $\rho < 0.2$ or high density such as $\rho > 0.7$, but it works well in a range of α={10, 50} and β={0.2, 5}.

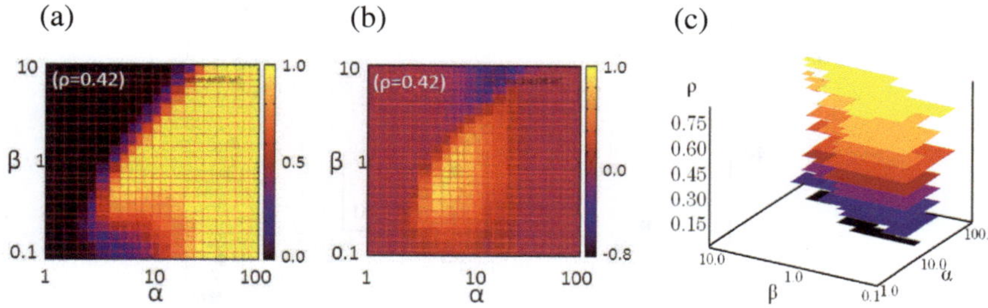

Fig. 2 (a) Success rate with α (random force) and β (repulsive force from the destination) in case of ρ = 0.42. (b) Success rate after discounting success due to pure Brownian motion. (c) Region where the success rate by the proposed method exceeds the success rate of Brownian motion on the α-β plane for various robot densities ρ.

Fig. 3 shows the relationship between the success rate and the density of the robots. As shown in Fig. 2(b), the system shows better performance in a wide range of densities around β =1.0. Fig. 3 shows the success rate is high for a wide range of ρ in the region α>10.

From these results, we can empirically conclude that the system shows reasonable performance around α=10~50 and β~1.0. For the remainder of this paper, we fix α=20.0 and β= 1.0 as typical values for the homogeneous system.

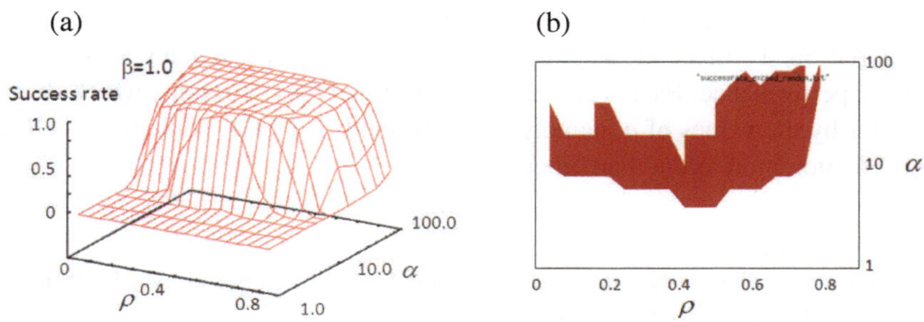

Fig. 3 (a) Success rate on the α–ρ plane. (b) Region where the success rate by the proposed motion exceeds that of Brownian motion on the α–ρ plane. Here, β is fixed as 1.0.

Next, we explore to the relation between p, the probability that the robots change direction, and the system performance. The smaller p is, the longer a robot maintains the direction imposed by the repulsive force. Simulation results in all previous experiments have been obtained with p=0.01. Fig. 4 shows the success rate as a function of p. Empirically, p=0.1 is better than p=0.01 for high densities but worse in low density. Low values for p, such as p=0.001, never perform well. (Low values for p induce a similar effect as low values for α).

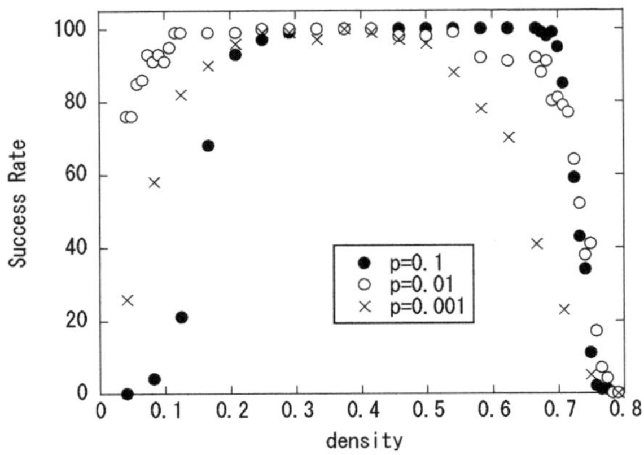

Fig. 4 Success rate as a function of density at different p

4 Extension to Heterogeneous System

When all robots are identical (homogeneous system) and the environment is finite, the density of robots gradually increases with growing distance from the destination as they are driven away from the destination due to the repulsive force. We have shown that the performance of object transportation decreases as density increases above 0.5. We are therefore interested in decreasing robot density in the vicinity of the object. We hypothesize that a heterogeneous swarm, in which some robots exhibit density-lowering behavior, can increase performance of object transportation. To this end, we propose two types of simple dynamics, and evaluate their performance. Each system consists of two types of robots, which dynamics differ by the values of p, namely p = 0.01 and 0.1, that is, some robots change their direction more often than others.

1) *Probabilistic model*

Each robot changes its type asynchronously. The change obeys the transition probabilities, p_{t12} and p_{t21} (Fig. 5(a)).

2) *Signal propagation model*

In order to decrease the density around the object, we attempt to propagate "waves" of different behaviors through the swarm. To this end, we assume the robots have a simple communication protocol in which they can transmit and detect simple signals. The destination is continually emitting signals to all robots within a fixed radius. The robots use the following algorithm, depicted in Fig. 5(b):

- Initially, all robots are in state 1.
- If a robot receives a signal, it changes to state 2 and remains in state 2 for time T_1.
- After time T_1, the robot transmits a signal to all other robots within radius *R*, changes to state 3 for time T_2. In state 3, the robot is "blind" to additional signals.
- After time T_2, the robot reverts to state 1

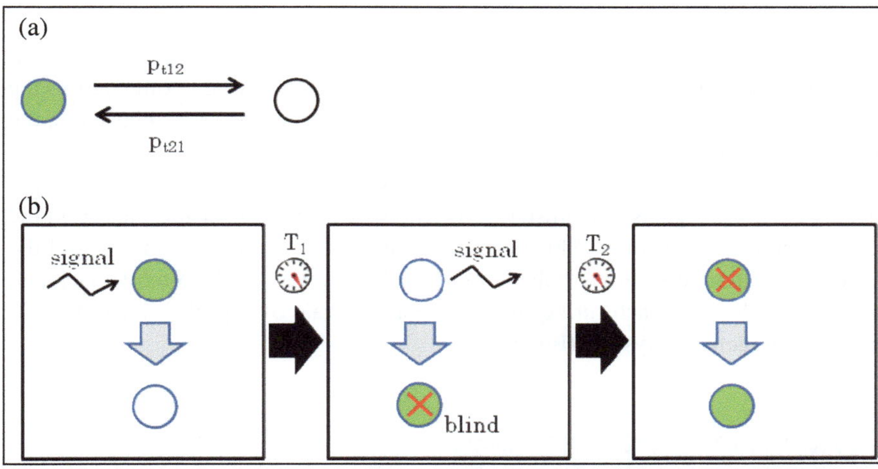

Fig. 5 Schematics of dynamic heterogeneous system. Robots in state 1 are shown as green circles, robots in state 2 are shown as blue circles; and robots in state 3 are shown with a red X. (a) Robots in state 1 change to state 2 with probability p_{t12}; robots in state 2 change to state 1 with probability p_{t21} (b) Robots receive a signal from either the destination or from a neighboring robot inducing it to change state, emit a signal, and revert to the original state after a refractory period.

4.1 Results

Using preliminary experimental results, we chose for p_{t12} and p_{t21} in the probabilistic model to be 1.0×10^{-4} and 2.0×10^{-4}, and $T_1$and T_2 in signal propagation model as 100 sec and 500 sec, respectively.

The success rates of the two heterogeneous systems with respect to the homogeneous system for various densities are plotted in Fig. 6. Data is only reported for high density scenarios because there is little difference in the case of low density.

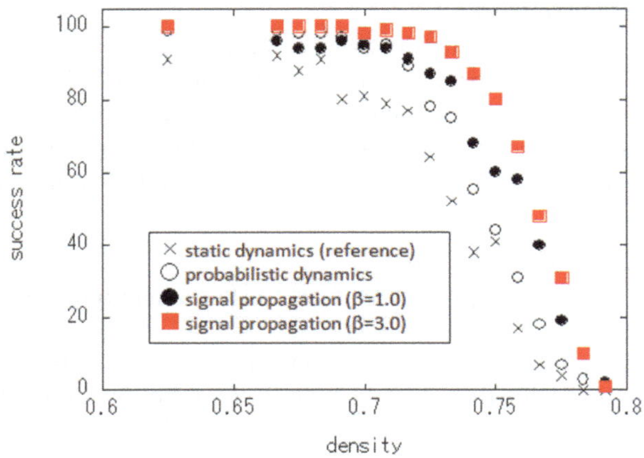

Fig. 6 Success rate of each model

We observe that the heterogeneous system in which robots switch types probabilistically slightly, but consistently, improves success rate when compared to the baseline performance of the homogenous systems that has been tuned based on the empirical results from Section 3.

Fig. 7 shows a snapshot of an experiment with the signal propagation model. As the signal propagates, the cluster of the robots with p=0.01 propagates from the destination to the boundary. This behavior is similar to peristaltic motion, and the object is transported showing pulsing motion (Fig. 8), i.e., tends to move back and forth. Using this approach further increases success rate over probabilistic dynamics and homogeneous system (Fig. 6).

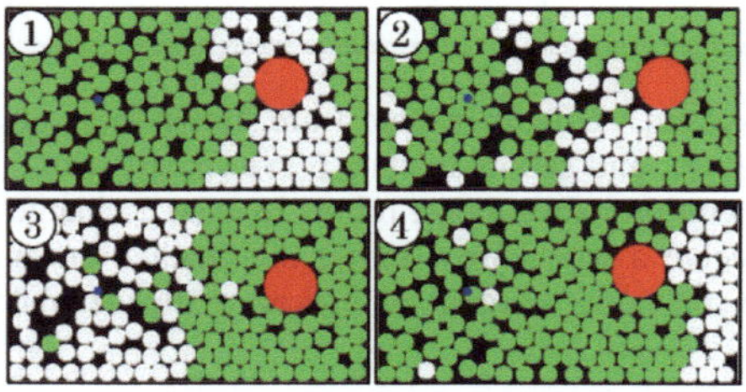

Fig. 7 Snapshot of signal propagation dynamics

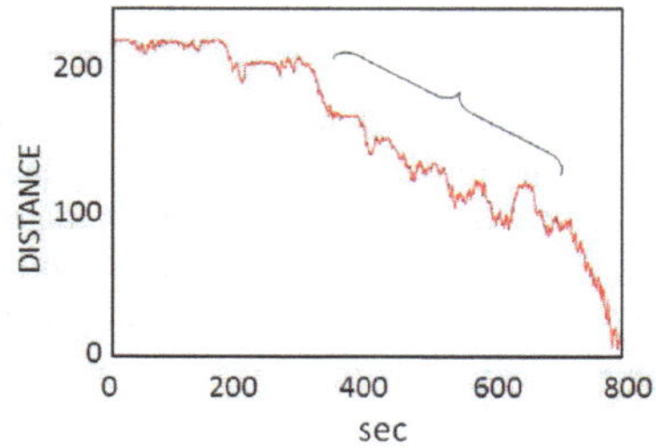

Fig. 8 Distance from the destination by signal propagation dynamics in case of ρ=0.75. Pulsing motion appears in this behavior in the area indicated.

For simplicity, we choose β=1.0 for both states, but if we introduce β in state 1 as 3.0 instead of 1.0, the success rate in this region is improved more as shown by the red rectangles in Fig. 8.

Next, we study the actual completion time, which corresponds to the speed or the efficiency of object transportation. Fig. 9 shows the average completion time of each of the dynamics for different densities. Here the completion time is defined as the first time the object reaches the destination. There is a tendency that the time becomes longer in low and high density region and is minimum around ρ = 0.4~0.5.

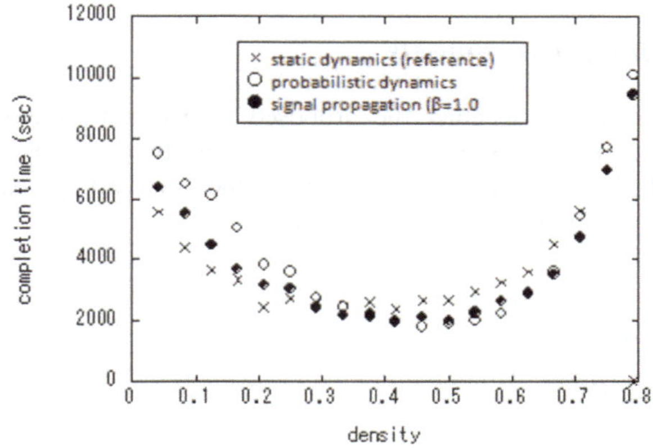

Fig. 9 Completion time of each dynamics

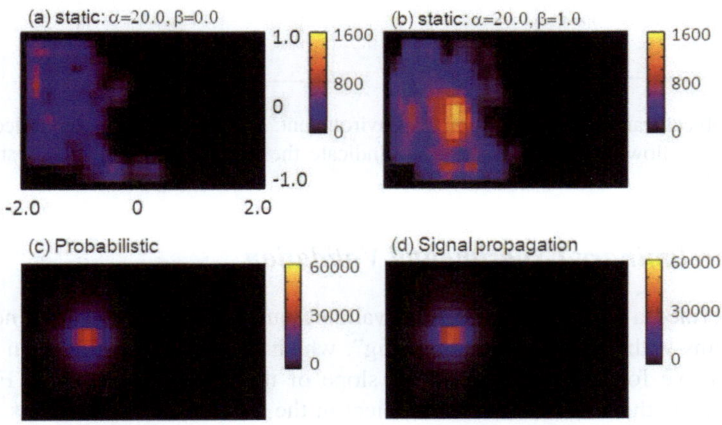

Fig. 10 Sojourn time of object position in case of ρ =0.63. (a) static system with β=0. (b) static system with β=1.0. (c) probabilistic dynamics (d) signal propagation dynamics.

It is also important to evaluate how long the system keeps the object in the vicinity of the destination once the object has reached the destination. A simple way of evaluation is to measure sojourn time around the destination. The experimental field is divided into 40 x 20 cells, and we recorded the total sojourn time in each region (Fig. 10). The count starts when the object once reached destination. Fig. 10 shows the accumulation of the result of 100 trials. The unit of color bar is "sec." In the homogenous system, the distribution of the object position over multiple experiments is wide. In the heterogeneous systems, however, the distribution becomes narrow, and the distribution of the signal propagation dynamics is narrower than that of the probabilistic dynamics.

4.2 Complex Environments

We are also interested whether the proposed object transportation mechanism works in more complex environments. Fig. 11 shows two sample environments of many we tried. These experiments confirm that the proposed approach generalizes to environments with more complex geometries.

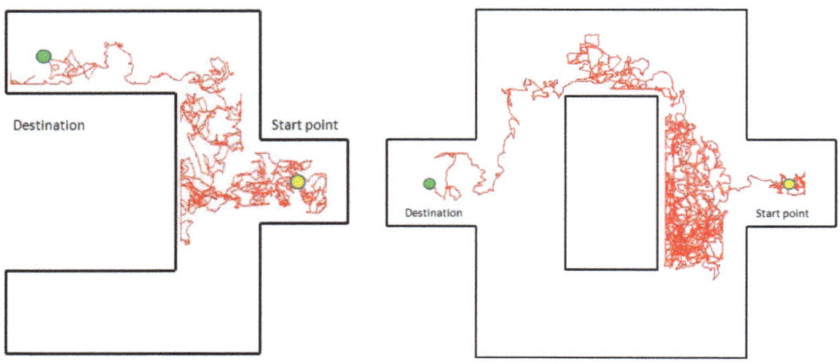

Fig. 11 Object transportation in complex environment. Red line indicates the trajectory of the object. Yellow circle and green circle indicate the starting point and the destination, respectively.

4.3 Preliminary Experimental Validation

We performed a simple experiment to validate our approach for homogenous robot systems with the toy robot "Hex-bug", which is driven by a vibration motor. The repulsive force is realized by the slope of the field. As shown in Fig. 12, which depicts the distribution of the object in the environment, the robots are capable of moving the objects consistently against the slope. The difference between Fig. 12(e) and Fig. 12(f) indicates that the object was transported to the right, or upper, side of the environment.

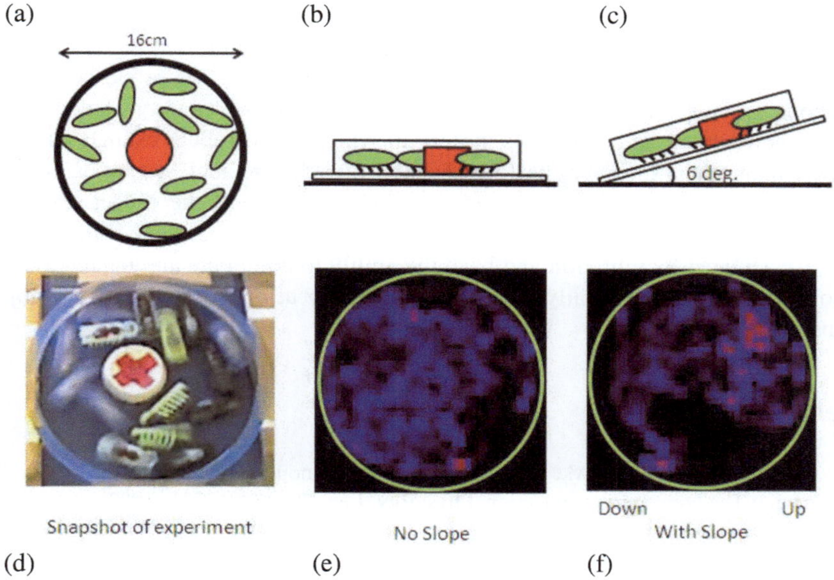

Fig. 12 Experiment by simple toy robots. (a) Cartoon of overhead view of the experiment. The red circle represents the object to be transported and the green ovals represent the Hex-bugs. (b) Cartoon of experimental control, in which the environment was horizontal. (c) Cartoon of experiment, in which the environment was tilted at 6° from horizontal, imparting a force on the Hex-bugs. (d) Photograph of the experiment. The object to be transported is marked with a red cross. (e) Heat map of the position of the object for the control. (f) Heat map of the position of the object for the experiment.

5 Discussion

Results suggest that granular convection has the potential to allow for controlled object transportation. Whereas all our experiments rely on a constant external force, finer control can potentially be achieved by manipulating the external field and robots with appropriate sensor for magnetic, electric, or chemical fields.

Parameters, particularly the ratio between randomness and directed motion, have been chosen ad-hoc in this paper and are functions of the specific simulation environment including the simulated friction, and sizes of robots and object. While this ad-hoc method is potentially applicable to real systems, in future work we are interested in grounding these parameters in first principles.

6 Conclusion

We propose a novel method for object transportation inspired by the Brazil Nut Effect, in which the robots have minimal sensors. Regardless of the simple dynamics of the robots based on random force and weak repulsive force from the destination, the object can be transported to the destination. Based on the observation that high density is detrimental to object transportation, we propose a heterogeneous system that is geared to lower the density in the vicinity of the object. Whereas a simple heterogeneous system in which some robots change direction slower than others improves performance, best results are achieved by using local communication to direct density-reducing behavior toward the object. In future work, we wish to confirm these results with a large number of very simple robots that are actuated by vibration and have the ability to communicate locally. We are also interested in analytically showing how density affects performance of object transportation.

References

1. Knight, J.B., Jaeger, H.M., Nagel, S.R.: Vibration-induced size separation in granular media: The convection connection. Phys. Rev. Lett. 70, 3728–3731 (1993)
2. Sacconi, L., Romano, G., Capitanio, M., De Pas, M., Giuntini, M., Dunlap, D., Rinzi, L., Pavone, F.S.: Three-dimensional magneto-optic trap for micro-object manipulation. Optics Lett. 26, 1359–1361 (2001)
3. Shin, J.S., Pierce, N.: A synthetic DNA walker for molecular transport. J. Am. Chem. Soc. 126, 10834–10835 (2004)
4. Fuchiwaki, O., Ito, A., Misaki, D., Aoyama, H.: Multi-axial micromanipulation organized by versatile micro robots and micro tweezers. In: IEEE Int. Conf. on Robotics and Automation, pp. 893–898 (2008)
5. Martel, S., Tremblay, C., Ngakeng, S., Langlois, G.: Controlled manipulation and actuation of micro-objects with magnetotactic bacteria. Applied Phys. Lett. 89, 233904 (2006)
6. Mataric, M.J., Nilsson, M., Simsarian, K.T.: Cooperative multi-robot box-pushing. In: Proc. of Int. Conf. on Intelligent Robots and Systems, pp. 556–561 (1995)
7. Kube, C.R., Zhang, H.: The use of perceptual cues in multi-robot box-pushing. In: Proc. of Int. Conf. on Robotics and Automation, pp. 2085–2090 (1996)
8. Donald, B.R., Jennings, J., Rus, D.: Information invariants for distributed manipulation. The International Journal of Robotics Research 16, 673–702 (1997)
9. Spletzer, J., Das, A.K., Fierro, R., Taylor, C.J., Kumar, V., Ostrowski, J.P.: Cooperative localization and control for multi-robot manipulation. In: Proc. of the IEEE/RSJ Intl. Conf. on Intelligent Robots and Systems, pp. 631–636 (2001)
10. Kriesel, D.M.M., Cheung, E., Sitti, M., Lipson, H.: Beanbag Robotics: Robotic Swarms with 1-DoF Units. In: Dorigo, M., Birattari, M., Blum, C., Clerc, M., Stützle, T., Winfield, A.F.T. (eds.) ANTS 2008. LNCS, vol. 5217, pp. 267–274. Springer, Heidelberg (2008)
11. Takahashi, K., Ogawa, N., Oku, H., Hashimoto, K.: Organized Motion Control of a lot of Microorganisms Using Visual Feedback. In: Proc. of the IEEE Intl. Conf. on Robotics and Automation, pp. 1408–1413 (2006)
12. Itoh, A.: Motion control of protozoa for bio-MEMS. IEEE-ASME Trans. Mechatron. 5, 181–188 (2000)

13. Groß, R., Magnenat, S., Mondada, F.: Segregation in swarms of mobile robots based on the brazil nut effect. In: Proc. of Int. Conf. on Intelligent Robots and Systems, pp. 4349–4356 (2009)
14. Chen, J., Gauci, M., Price, M.J., Groß, R.: Segregation in swarms of e-puck robots based on the brazil nut effect. In: Proc. of the 11th Int. Conf. on Autonomous Agents and Multiagent Systems (AAMAS 2012), pp. 163–170. ACM Press (2012)
15. Williams, J., Khan, M.: The mixing and segregation of particulate solids of different particle size. Chem. Eng. 269, 19–25 (1973)
16. Jaeger, H.M., Nagel, S.R.: Granular solids, liquids, and gases. Reviews of Modern Physics 68, 1259–1273 (1996)
17. Nowak, E.R., Knight, J.B., Ben-Naim, E., Jaeger, H.M., Nagel, S.R.: Density fluctuations in vibrated granular materials. Phys. Rev. E 57, 1971–1982 (1998)
18. Alt, W.: Biased random walk models for chemotaxis and related diffusion approximations. J. Math. Biology 9, 147–177 (1980)
19. Dhariwal, A., Sukhatme, G., Requicha, A.: Bacterium-inspired robots for environmental monitoring. In: Proc. of the IEEE Intl. Conf. on Robotics and Automation, pp. 1436–1443 (2004)

Multi-Robot Control for Circumnavigation of Particle Distributions

Sarah Tang, Dylan Shinzaki, Christopher G. Lowe, and Christopher M. Clark

Abstract. In this work, we present a decentralized controller for the tracking and following of mobile targets, specifically addressing considerations of: 1) not altering target behavior, 2) target states represented by multiple hypotheses, and 3) limited information from bearing-only sensors. The proposed controller drives a team of n robots to circumnavigate an arbitrary distribution of target points at a desired radius from the targets. The controller also dictates robot spacing around their circular trajectory by tracking a desired relative phase angle between neighbors. Simulation results show the functionality of the controller for arbitrary-sized teams and arbitrary stationary and moving particle distributions. Additionally, the controller was implemented on OceanServer Iver2 AUVs. Tracking results demonstrate the controller's capability to track a desired radius as well as maintain phase with respect to a second AUV.

1 Introduction

In recent years, multi-robot systems have been developed for a variety of applications. Compared to single robot systems, they have better spatial-temporal coverage,

Sarah Tang
Princeton University, Princeton, NJ, USA
e-mail: sytang@princeton.edu

Dylan Shinzaki
Stanford University, Stanford, CA, USA
e-mail: shinzaki@stanford.edu

Christopher G. Lowe
CSU Long Beach, Long Beach, CA, USA
e-mail: clowe@csulb.edu

Christopher M. Clark
Harvey Mudd College, Claremont, CA, USA
e-mail: clark@hmc.edu

M.A. Hsieh and G. Chirikjian (eds.), *Distributed Autonomous Robotic Systems*,
Springer Tracts in Advanced Robotics 104,
DOI: 10.1007/978-3-642-55146-8_11,

greater manipulation force, and are more robust to mission failure. In particular, they can be applied to tracking problems, where a team of robots attempt to gather information about or follow one or more targets. Yet while these teams can be more effective, they also come with a number of challenges. The robots themselves may have nonholonomic kinematics. Sensor information might be limited, noise and external disturbances could make available sensor information hard to process, and maintaining multiple hypotheses as well as fusing sensor information are nontrivial design challenges. In addition, the team often has to maintain a certain distance from their targets so as to not influence natural behavior.

This work is motivated by the scientific goal of autonomous tracking and following of long migratory fish species (e.g. sharks) using Autonomous Underwater Vehicles (AUVs). Previous work [6] has shown the ability of a single AUV equipped with bearing-only sensors to track and follow tagged leopard sharks using a Particle Filter (PF) for estimating target states. In moving towards tracking species whose movement behaviors would be altered by the presence of tracking unit, a multi-AUV controller is required. This controller needs to enable active state estimation using multiple sensor vantage points to reduce the effects of bearing-only sensors, while simultaneously ensuring that all AUVs maintain some predetermined distance from the target.

The remainder of this section will present related work. Section 2 will define the problem at hand in terms of four subproblems and detail our proposed solutions. Section 3 will present experimental results from a controller implementation, and Section 4 will conclude the paper.

1.1 Background

Numerous approaches have been proposed to the general problem of multi-robot control. Control methods are either centralized, where a central processor controls the entire team, or decentralized, where each robot maintains its own controller using sensor information about other robots and its environment. A number of techniques have been proposed for both methods.

A behavior-based controller maintains different modes of action that, depending on the goal of the robot, are weighed differently to give rise to different group dynamics. For example, Balch and Arkin [2] appropriately weight motor schemas for moving to destination, obstacle avoidance, and waypoint navigation. Similarly, Bougherty et al. [5] implement the basic behaviors of move-to-goal, avoid-obstacle, maintain-relative-distance, maintain-relative-angle, and stop, to form the control laws for their robotic team. Generally, such behavior-based methods are successful at maintaining desired group dynamics, however, are hard to analyze for stability guarantees.

The leader-following approach designates robots in the team as leaders or followers. Leader robots simply travel the desired trajectory of the team while follower robots are responsible for maintaining group formation by tracking desired bearing angles and distances to designated leaders. For example, Desai et al. [4] describe decentralized control laws that drive a follower to either maintain distance and bearing to one leader or two bearings to two leaders. Such methods tend to be simpler in implementation, but they are heavily dependent on effective communication between robots. The system is also prone to single-point failure, and the leader robot typically has no means of detecting or compensating for lost followers.

With a virtual structure approach, a group of virtual positions are arranged in a rigid desired formation and moved along a desired trajectory. Each robot follows the specific path traced out by one of the virtual positions. This method is used by Young et al. [17], who drive robots to their desired virtual positions with a Proportional-Derivative (PD) control law, and Tan and Lewis [9], who apply this technique to a vision-based tracking system. One big advantage of such methods is that it is relatively straightforward to specify desired behavior and analytically guarantee stability.

These control methods have proven effective in the domain of multi-robot tracking. Liu et al. [10] propose behavioral-based control that uses reinforcement learning to adjust behavior weights for a target-tracking controller, Chung et al. [3] implement a gradient-based decentralized control law for dynamic target tracking, and Mazo et al. [11] use the virtual vehicle approach. In addition to these main approaches, Lee et al. [8] explore target tracking control for unicycle mobile robots using a nonlinear state feedback controller, while Papanikolopoulos et al. [15] propose a method where output of a visual tracking system serve as input to a Proportional-Integral (PI) controller, pole assignment controller, or an Linear-Quadratic-Gaussian (LQG) controller. However, these methods track a only single estimate of either one or more moving targets.

A few methods do combine control with state estimation. Mottaghi and Vaughan [12] use a Particle Filter to create a potential field for a virtual structure based robotic controller, and Wang et al. [16] propose a flocking control method for multiple robots moving to an estimated position from a distributed Kalman filter. However, these methods drive robots directly to estimated destinations rather than circumnavigation. To accommodate for limited sensing, Zhou et al. [18] propose an iterative Gauss-Seidel Relaxation (GSR) algorithm to drive robots to the best sensing location for a team of heterogeneous robots tracking a moving target, accounting for limited measurements and motion constraints (e.g. maintaining a minimum distance from the target). Paley [14] propose control methods for circular motion of robots moving in external flow fields, however, only allows for circling of one center point. Lan et al. [7] propose a hybrid control method to capture targets with circular motion around the target, but does not combine this method with state estimation.

In our work, we use a stability-guarantee based approach to develop a distributed control method that will simultaneously accommodate all three of the previously mentioned issues, namely, accounting for multiple hypotheses, limited sensing, and maintaining distance from the target. To this end, we use a proportional control law to drive arbitrary-sized teams to circumnavigate target distributions at a specified radius from all targets and angle offset from each other.

2 Multi-Robot Circumnavigation

The aim of this work is to design decentralized control laws to drive a team of n robots to circumnavigate any distribution of p target points. There are two main objectives: First, each robot should circumnavigate all targets at a constant desired radius. Second, each robot should maintain a constant phase difference between itself and the robots in front and behind it. Spacing robots apart in this manner helps avoid collisions and enables information gain from different sensor vantage points with minimal overlap. Fig. 1 illustrates this ideal configuration with four robots circling clockwise around a single target point.

To accomplish this design task, the problem is broken down into four subproblems: 1) design a single robot controller that drives the robot to circumnavigate a single target, 2) design a multi-robot controller that extends the previous controller to enable robots to track a desired phase difference as they circumnavigate a single target, 3) design a target allocation system that further extends the controller to multiple targets, and 4) design an ordering algorithm that initializes each robot's relative position with respect to the circumnavigation loop. Preliminaries are first presented followed by a proposed solution to each of the four subproblems.

2.1 Preliminaries

In this work, a team of n robots operate within a 2D obstacle-free workspace. Each robot i is represented at time t by a state vector $\mathbf{X}_{i,t}^{robot}$ comprised of its position, ($x_{i,t}$, $y_{i,t}$), and yaw orientation, $\theta_{i,t}$, with respect to an inertial cartesian coordinate frame:

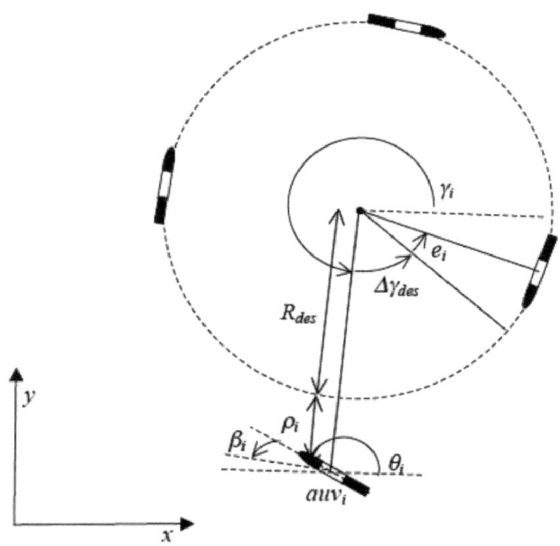

Fig. 1 This picture exemplifies the desired circling behavior, with a team of four AUVs circling one target point. The AUVs circle the target at a desired R_{des} and are spaced around the circle with a desired phase difference $\Delta\gamma_{des}$. The error variables β, ρ, and e, presented in Eq. 8 - 10, are also illustrated for AUV i.

$$\mathbf{X}_{i,t}^{robot} = [x_{i,t}\; y_{i,t}\; \theta_{i,t}]^T. \tag{1}$$

The kinematics of the system are assumed to follow the first order discrete time equations in Eqs. 2 - 4. In this case, $v_{i,t}$ and $\omega_{i,t}$ are the forward and rotational velocities of the ith robot, respectively, and serve as the robot control inputs.

$$x_{i,t+1} = x_{i,t} + v_{i,t}\cos(\theta_{i,t})\Delta t \tag{2}$$
$$y_{i,t+1} = y_{i,t} + v_{i,t}\sin(\theta_{i,t})\Delta t \tag{3}$$
$$\theta_{i,t+1} = \theta_{i,t} + \omega_{i,t}\Delta t \tag{4}$$

To facilitate control with respect to a target at position ($x_{target,t}$, $y_{target,t}$), three additional states are defined. First, let $r_{i,t}$ be the distance between the target and robot i at time t. Second, let $\gamma_{i,t}$ be the relative bearing angle of robot i relative to the target. Third, let $\theta_{des_i,t}$ be the desired yaw angle for robot i, which is in a direction tangent to a circle centered on the target.

$$r_{i,t} = \sqrt{(x_{i,t} - x_{target,t})^2 + (y_{i,t} - y_{target,t})^2} \tag{5}$$

$$\gamma_{i,t} = \tan^{-1}\left((y_{i,t} - y_{target,t})/(x_{i,t} - x_{target,t})\right) \tag{6}$$

$$\theta_{des_i,t} = \gamma_{i,t} - \frac{\pi}{2} \tag{7}$$

Given a desired radial distance from all targets, R_{des}, and a desired relative phase offset, $\Delta\gamma_{des}$, the system can now be described in terms of error variables $\rho_{i,t}$, $\beta_{i,t}$, $e_{i,t}$.

$$\rho_{i,t} = \mathrm{R}_{des} - r_{i,t} \tag{8}$$

$$\beta_{i,t} = \theta_{des_i,t} - \theta_{i,t} \tag{9}$$

$$e_{i,t} = \Delta\gamma_{des} - (\gamma_{i,t} - \gamma_{i-1,t}) \tag{10}$$

The term $e_{i,t}$ represents the error in phase between two vehicles i and $i-1$. Although $e_{i,t}$ is indexed with a robot index i, it actually represents the angle difference between a pair of robots, and so, there is no $e_{0,t}$ term. These errors form the state vector associated with each robot:

$$\chi_{i,t} = [\rho_{i,t}\; \beta_{i,t}\; e_{i,t}]^T. \tag{11}$$

Fig. 1 illustrates these variables for a sample tracking scenario. The kinematics of these controlled variables can be derived from substituting the derivatives of Eqs. 5 - 7 into the derivatives of Eqs. 8 - 10:

$$\rho_{i,t+1} = \rho_{i,t} + v_{i,t}\sin(\beta_{i,t})\Delta t, \tag{12}$$

$$\beta_{i,t+1} = \beta_{i,t} + \left(-\frac{v_{i,t}\cos(\beta_{i,t})}{R_{des} - \rho_{i,t}} - \omega_{i,t}\right)\Delta t, \tag{13}$$

$$e_{i,t+1} = e_{i,t} + \left(\frac{v_{i,t}\cos(\beta_{i,t})}{R_{des} - \rho_{i,t}} - \frac{v_{i-1,t}\cos(\beta_{i-1,t})}{R_{des} - \rho_{i-1,t}}\right)\Delta t. \tag{14}$$

2.2 Single Robot Circumnavigation of a Single Target

With the preliminaries presented above, the first design problem can be stated as follows: Given a robot i that behaves according to Eqs. 2 - 4, determine the discrete time control update laws for the control vector:

$$\mathbf{U_{i,t}} = [v_{i,t}\ \omega_{i,t}]^T \tag{15}$$

that drive errors $\rho_{i,t}$, $\gamma_{i,t}$, and $e_{i,t}$ to 0 as time t approaches infinity.

In the single robot case, the robot is assumed to travel at some nominal velocity $v_{i,t} > 0$. The rotational velocity, $\omega_{i,t}$, is used to drive each robot towards its desired path around the target. The proposed control law is shown in Eq. 16, where R_{des} is the desired radius of the circle centered on the target being tracked, K_β and K_ρ are proportional control gains, and Δt is the time step in seconds between control signal updates.

$$\omega_{i,t} = -\frac{v_{i,t}\cos(\beta_{i,t})}{R_{des} - \rho_{i,t}} + \frac{K_\beta}{\Delta t}\beta_{i,t} + \frac{K_\rho}{\Delta t}\rho_{i,t} \tag{16}$$

Note that the first term in Eq. 16 is a feedback linearization term to accommodate the nonlinear dynamics in Eq. 13. The singularity in Eq. 16 is avoided by initializing robots such that they do not invoke the controller to begin circumnavigation until they are within some minimum ρ_o of all targets. Thus, for all $t \geq 0$, $\rho_{i,t} \leq \rho_o < R_{des}$. This initialization scheme is described in detail in Sec. 2.5.

To derive the system's stability conditions, the control law in Eq. 16 is substituted into Eqs. 12 and 13 to arrive at the closed-loop error dynamic equations. These equations can further be simplified using a small angle approximation $\sin(\beta_{i,t}) \approx \beta_{i,t}$. The error dynamics and corresponding eigenvalues are described in matrix formulation in Eq. 17 and Eq. 18, respectively.

$$\begin{bmatrix}\rho_i \\ \beta_i\end{bmatrix}_{t+1} = \begin{bmatrix}1 & v_{i,t}\Delta t \\ -K_\rho & 1-K_\beta\end{bmatrix}\begin{bmatrix}\rho_i \\ \beta_i\end{bmatrix}_t \tag{17}$$

$$\lambda = \frac{(K_\beta - 2) \pm \sqrt{K_\beta^2 - 4K_\rho v_{i,t}\Delta t}}{2} \tag{18}$$

The discrete system is asymptotically stable if and only if all eigenvalues have magnitude less than 1. The problem is not fully constrained and has many solutions, but for $\Delta t > 0$ and positive gains, gain and velocity bounds are given in Eqs. 19 - 21.

$$K_\rho > 0 \tag{19}$$

$$0 < K_\beta < 4 \tag{20}$$

$$\frac{2(K_\beta - 2)}{K_\rho \Delta t} < v_{i,t} \leq \frac{K_\beta^2}{4K_\rho \Delta t} \tag{21}$$

2.3 Multi-Robot Circumnavigation of a Single Target

In the multi-robot case, the controller in Eq. 16 is used to set $\omega_{i,t}$ for each robot, while $v_{i,t}$ is modulated to drive phase errors $e_{i,t}$ to 0 while still ensuring it lies within the bounds defined by Eq. 21. The following controller for $v_{i,t}$ is proposed:

$$v_{i,t} = \frac{R_{des} - \rho_{i,t}}{R_{des}\cos(\beta_{i,t})}\left(v_{nom} + \frac{R_{des}K_\gamma}{\Delta t}(e_{i+1,t} - e_{i,t})\right). \tag{22}$$

Substituting Eq. 22 into Eq. 14 yields the closed loop kinematic equation for $e_{i,t+1}$. As $e_{i,t}$ represents the error between two vehicles, the $e_{i,t}$ term is dropped for the 0th vehicle and the $e_{i+1,t}$ term is dropped for the $n-1^{th}$ vehicle. Since $e_{i,t+1}$ is dependent on $e_{i-1,t+1}$ and $e_{i,t+1}$, the matrix formulation is required:

$$\begin{bmatrix} e_1 \\ e_2 \\ e_3 \\ \vdots \\ e_n \end{bmatrix}_{t+1} = \begin{bmatrix} 1-2K_\gamma & K_\gamma & 0 & 0 & \cdots & 0 & 0 \\ K_\gamma & 1-2K_\gamma & K_\gamma & 0 & \cdots & 0 & 0 \\ 0 & K_\gamma & 1-2K_\gamma & K_\gamma & \cdots & 0 & 0 \\ & & \ddots & & & & \\ 0 & 0 & 0 & 0 & \cdots & K_\gamma & 1-2K_\gamma \end{bmatrix} \begin{bmatrix} e_1 \\ e_2 \\ e_3 \\ \vdots \\ e_n \end{bmatrix}_t . \tag{23}$$

Again, all eigenvalues must have magnitude less than 1 for the system to be asymptotically stable. The eigenvalues vary with n; the case of three robots is presented in Eqs. 24 - 26 to exemplify how stability criteria can be used to determine appropriate gain values.

$$\begin{bmatrix} e_1 \\ e_2 \end{bmatrix}_{t+1} = \begin{bmatrix} 1-2K_\gamma & K_\gamma \\ K_\gamma & 1-2K_\gamma \end{bmatrix} \begin{bmatrix} e_1 \\ e_2 \end{bmatrix}_t \tag{24}$$

$$\lambda = 1-3K_\gamma, 1-K_\gamma \tag{25}$$

$$0 < K_\gamma < \frac{2}{3} \tag{26}$$

2.4 Multi-Robot Circumnavigation of Multiple Targets

The control laws can be further extended to allow circumnavigation of a set P of p different targets, denoted:

$$P = [x_j\ y_j\ \theta_j\ v_j\ \omega_j], j \in [1:p]. \tag{27}$$

Each robot invokes the control laws from Eqs. 16 and 22 to track the closest target in P at the current time step t. In other words, control variables $\rho_{i,t}$, $\beta_{i,t}$, and $e_{i,t}$ are calculated with respect to only one target j. In this way, robots can, for example, track a set of particles from a particle filter (PF) state estimator [6] that represent one or many objects in the workspace.

To make the transitions between targets smoother, the tracked target at time step t is that which minimizes the Euclidian distance between candidate targets and the predicted robot state at $t+\tau$. Here, τ is some positive integer, and the state is predicted assuming $v_{i,k} = v_{i,t}$ and $\omega_{i,k} = \omega_{i,t}$ for $k \in [1, \tau]$. A minimal τ based on a robot's minimum turning radius can be used to ensure $\rho_{i,t+1} > 0$ when tracking a new target location (*i.e.*, the robot will not come within R_{des} of the next target).

2.5 *Multi-Robot Ordering during Circumnavigation*

The control laws presented in Eq. 16 and 22 ensure that a team of robots will eventually converge to circumnavigation of a collection of targets with relative phase tracking for collision-free motion and multiple sensor vantage points. To prevent collisions from occurring before system convergence, Alg. 1 is used to assign a static order to all robots based on their initial positions with respect to targets.

In Alg. 1, Line 1, each target from P introduced in Eq. 27 is grouped into one of C *clusters*. The cth *cluster* is a subset of the p targets in P that contains p_c targets, where $p_c > 0$ and the Euclidean distance of each target to at least one other target in the group is $< 2R_{des}$. In the case where $C \leq n$, the number of robots assigned to each *cluster*, denoted n_c, is proportional to the number of targets in the *cluster*. If $C > n$, only the largest *clusters* will be assigned robots.

The following procedure is carried out within each *cluster*; for simplicity, Alg. 1 assumes only one *cluster*, so $n_c = n$ and $p_c = p$. Robots identify a circle O centered at the average position of all targets, defined here as $\mathbf{X}_o = (x_o, y_o)$. The radius r_{max} of O is the minimum radius that encapsulates all target circumnavigation loops and robots. A set of n boundary points are evenly distributed on O, to be used as initial destinations for the n robots, as seen in Lines 4 - 6. The robots are matched to *boundary points* according to the relative bearing angles κ from center (x_o, y_o), as calculated on Lines 7-9. That is, the robot with the kth greatest bearing angle among robots will be matched to the *boundary point* with the kth greatest bearing angle among *boundary points*, as in Lines 10 - 15.

Upon initialization, each robot i drives directly towards its individual *boundary point* $\mathbf{X}_{i,t}^{desBP}$ and pauses motion when within a predetermined distance error ρ_o of O, where $\rho_o < R_{des}$ as noted in Sec. 2.2. Only when all robots have distance to $O < (R_{des} + \rho_o)$ does the group start circumnavigating its designated cluster of targets. Note this algorithm is run only once, after which robot order remains constant.

Algorithm 1 findOrder($\mathbf{X}_{i,t}^{robot}\ \forall i = 1..n$, $\mathbf{X}_{j,t}^{target}\ \forall j = 1..p$, R_{des}, t)

$clusters \leftarrow$ clustering($\mathbf{X}_{i,t}^{robot}\ \forall i = 1..n$, $\mathbf{X}_{j,t}^{target}\ \forall j = 1..p$)
$\mathbf{X}_o \leftarrow$ average over $\mathbf{X}_{j,t}^{target}\ \forall j = 1..p$
$r_{max} \leftarrow \max_{\forall j=1..p} (|\mathbf{X}_{j,t}^{target} - \mathbf{X}_o| + R_{des})$
for k=1..n robots **do**
    $\mathbf{X}_{k,t}^{boundarypoint} \leftarrow [x_o + r_{max}\cos(2k\pi/n) \quad y_o + r_{max}\sin(2k\pi/n)]^T$
end for
for i=1..n robots **do**
    $\kappa_i \leftarrow \tan^{-1}((y_{i,t} - y_o)/(x_{i,t} - x_o))$
end for
$order \leftarrow$ sorted list of indices of κ_i in ascending order from $\theta = 0$
for i = 1..n robots **do**
    $k \leftarrow i^{th}$ element of $order$
    $\mathbf{X}_{i,t}^{desBP} \leftarrow \mathbf{X}_{k,t}^{boundarypoint}$
end for
return $order, \mathbf{X}_{i,t}^{desBP}\ \forall i = 1..n$

3 Results

The proposed control laws were simulated in Matlab to verify tracking performance. Controller gains and velocity constraints were selected to ensure controller stability as dictated by the inequalities given in Eqs. 19 - 21 and exemplified by Eqs. 24 - 26. Table 1 lists the chosen constants.

Table 1 Controller constants

Simulation					
K_ρ	0.3	K_β	1.5	K_γ	0.5
v_{nom}	2.5m/s	v_{min}	1m/s	v_{max}	3m/s
R_{des}	8m	$\Delta\gamma_{des}$	$\frac{\pi}{2}$ if $R = 2$, otherwise $\frac{2\pi}{n}$		
Field tests					
K_ρ	0.1	K_β	0.4	K_γ	0.5
v_{nom}	0.56m/s	v_{min}	0.3m/s	v_{max}	0.8m/s

Fig. 2a illustrates a team of 10 robots circumnavigating a single target, and Fig. 2b shows the corresponding error terms converging to zero over time. To demonstrate the versatility of the controller, Fig. 3 illustrates a variety of different scenarios involving multiple targets (e.g. particle distributions), both stationary and moving.

To validate the control system on a real test platform and progress towards autonomous fish tracking with AUVs, the proposed controller was implemented on

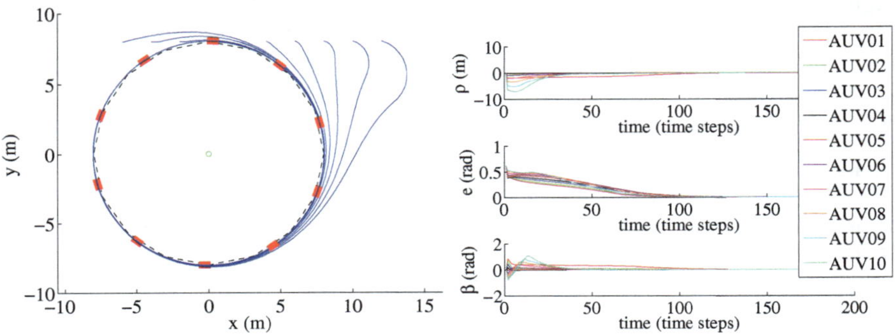

(a) Trajectory (b) Control variables

Fig. 2 Circumnavigation of one target by 10 AUVs

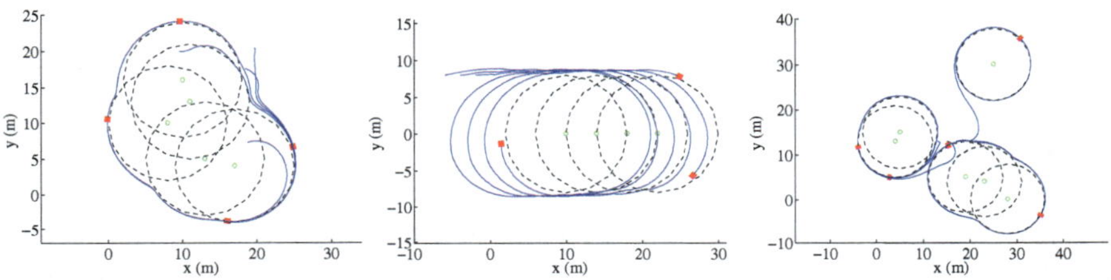

(a) 4 robots, 5 targets (b) 3 robots, 4 moving targets (c) 5 robots, 6 targets, 3 clusters

Fig. 3 Circumnavigation of various target distributions

an OceanServer Iver2 AUV [13], pictured in Fig. 4a. The vehicle is approximately 1.27m in length, 0.15m in diameter, and 20kg in weight, with two fins to control pitch, two rear fins to control yaw, and a three-blade propeller to provide forward velocity. A 24V servo-controlled DC motor gives the AUV a speed range of approximately 0.5m/s to 2.0m/s. A built-in GPS receiver provides longitude and latitude measurements and a three-degree-of-freedom compass provides heading measurements. The AUV runs two processors, a main Intel 1.6 GHz ATOM processor with Windows XP and an additional low power Intel 1.6GHz ATOM processor; the first is responsible for fundamental control of actuators and sensors, while the second is designated for external programs, such as the controller. The AUV communicates with laptop via wireless ethernet. Deployments were conducted in Carnegie Lake, Princeton, NJ and Fisherman's Cove, Santa Catalina Island, CA.

As the following experimental procedure applies to each individual AUV independently, the subscript i will be dropped from future notation. To appropriately actuate the AUV, the relationships from controller outputs, v_t and ω_t, to actuator values for the motor and yaw fin, denoted $u_{v,t}$ and $u_{\omega,t}$, respectively, were determined. Specifically, two functions - one relating v_t to $u_{v,t}$, the other relating $u_{v,t}$ and

ω_t to $u_{\omega,t}$ (as the motor speed of the AUV also dictates the maximum achievable angular velocity) were sought. Position data of the AUV running at various constant $u_{\omega,t}$ and $u_{v,t}$ values were gathered. Offline, actual forward and angular velocities of the AUV were estimated as the secant line between position measurements at $t-1$ and $t-T$ for $t \in [T : t_{end}]$ and averaged to approximate the v_t or ω_t corresponding to a setting of $u_{v,t}$ or $u_{\omega,t}$. The desired functions were then derived from linear regressions, where T was chosen between 1 and 8 to yield the highest r^2 value.

Additionally, there was a clear time delay between the actuation of the AUV and actual achievement of the actuated angular velocity. In this light, the AUV's speed and angular velocity are estimated in real time, denoted $v_{est,t}$ and $\omega_{est,t}$, respectively, using a decaying weighing scheme. The real time estimate of v_t provides more accurate control when used within Eq. 16, and ω_t is estimated for for offline analysis. In future work, a more sophisticated real time estimation method, such as Kalman Filtering or Extended Kalman Filtering, can be implemented.

Results from using one AUV to track various (known) target distributions are summarized in Table 2. For these experiments, $R_{des} = 8$m. Note that average and standard deviations do not increase significantly when tracking moving distributions, demonstrating the controller's stability even for tracking moving targets.

Table 2 Results for single AUV tests

Test	ρ_{avg} (m)	σ_ρ (m)	β_{avg} (rad)	σ_β (rad)
single target	2.7070	1.5325	0.4678	0.5598
3 targets	2.6239	0.4830	0.4472	0.1386
moving target	2.6508	0.8504	0.4474	0.2679

Fig. 4 plots the AUV motion for moving target tracking; Fig. 4b plots the actual AUV trajectory in black with AUV heading every ten time steps in red while Fig. 4c plots controller variables over time. Fig. 4b shows that the AUV is able to follow and circle the target consistently, and Fig. 4c additionally shows that ρ_t and β_t decay towards steady-state values over time, through there seems to be a steady-state error present in both variables. From Table 3, average β_t is 0.45 radians, or approximately 26 degrees, and average ρ_t is 2.65 meters. An approximate three second time lag in actuation effect is a potential cause of this error, and methods to compensate for the time delay could be explored. In addition, though K_β and K_ρ were chosen for practical performance, they might not have been optimal gains. There are several techniques for theoretically optimal gain selection [1] that could be investigated.

Note that looking at the trajectory in Fig. 4b, there seems to be a delay in GPS measurements from corresponding compass readings. The AUV's heading is not tangent to its path at its current position, but rather, to the curvature a small distance forward, suggesting that GPS coordinates are received slower than compass headings.

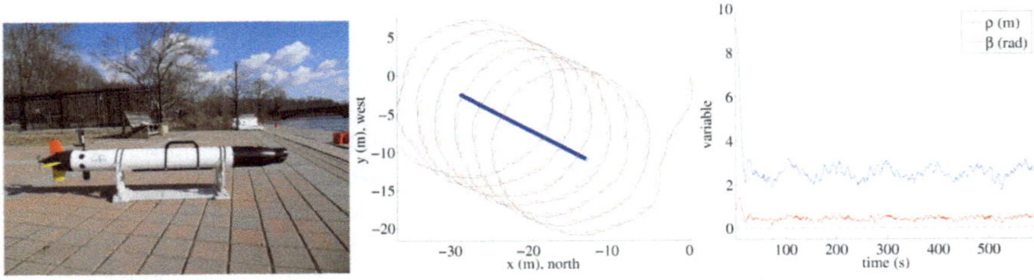

(a) Iver2 AUV (b) Tracking trajectory (c) Error variables

Fig. 4 Single AUV experiment

Next, the phase tracking capabilities of the controller are tested. First, a second AUV was simulated in code. At each time step, the simulated AUV was propagated forward using control law outputs and derived kinematics, and its resulting position was passed to the real AUV, assuming perfect and instantaneous communication between vehicles. For these experiments, $R_{des} = 8$m and $\Delta\gamma_{des} = \frac{\pi}{2}$. Tracking statistics are presented in Table 3. In addition to minimizing error in ρ_t and β_t, the AUV is able to approach the desired γ_t as well.

Table 3 Results for simulated multi-AUV tests

Test	ρ_{avg} (m)	σ_ρ (m)	β_{avg} (rad)	σ_β (rad)	$\gamma_{avg}(rad)$	σ_γ (rad)
stationary target	2.1773	0.5199	0.4670	0.1484	-1.5796	0.0303
moving target	2.0083	0.3342	0.4215	0.1043	-1.5505	0.0277

Additionally, experiments were done with two AUVs, with $R_{des} = 10\text{m}, 12\text{m}$ and $\Delta\gamma_{des} = \pi$. R_{des} of the two AUVs were set differently to avoid collision due to malfunctions during test (e.g. an AUV running out of battery). The two AUVs tracked a single target that, rather than moving continuously, jumped to a new location after a given amount of time. This behavior simulates the possibility of sudden changes in state estimate due to new location measurements after losing measurements for a span of time. Fig. 5 illustrates the tracking trajectories. Again, it can be seen that both AUVs successfully track their respective desired radius while driving phase error e_t to 0. Note that the sudden increase in e_t around 210s results from the movement of the target to a new location; the AUVs, however, are able to minimize e_t again after some time.

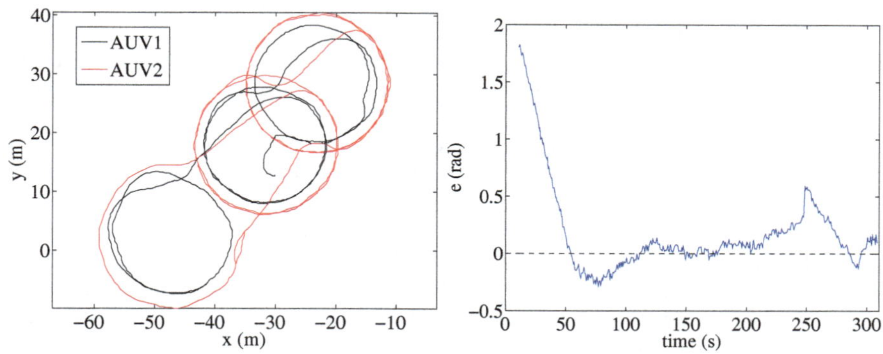

(a) Tracking trajectory (b) e_t over time

Fig. 5 Two-AUV experiment

4 Conclusions

In this work, we present a new multi-robot control strategy that enables multiple nonholonomic robots to circumnavigate an arbitrary distribution of targets while maintaining a constant standoff distance between robots and targets and a desired spacing between robots. A benefit of this strategy is that the control is decentralized and robots only need information regarding the positions of their neighbors. Bounds on controller gains to ensure system stability and convergence to the desired circumnavigation behavior were derived. Simulations were presented with various particle distributions and velocities. To demonstrate the applicability of the strategy to real systems, it was implemented on an AUV subject to real world disturbances (e.g. currents). In the future, this controller will be applied to a multi-AUV system and use target sensors and estimators to determine target states in real-time.

Acknowledgements. This material is based upon work supported by the National Science Foundation under Grant No. 1245813.

References

1. Åström, K.J., Murray, R.M.: Feedback Systems: An Introduction for Scientists and Engineers. Princeton University Press (2008), `http://press.princeton.edu/titles/8701.html`
2. Balch, T., Arkin, R.C.: Motor schema-based formation control for multiagent robot teams. In: Proceedings of the First International Conference on Multi-Agent Systems, pp. 10–16. AAAI Press (1995)
3. Chung, T.H., Burdick, J.W., Murray, R.M.: A decentralized motion coordination strategy for dynamic target tracking. In: Proceedings of the 2006 IEEE International Conference on Robotics and Automation, ICRA 2006, Orlando, Florida, USA, May 15-19, pp. 2416–2422. IEEE (2006)

4. Desai, J., Ostrowski, J., Kumar, V.: Modeling and control of formations of nonholonomic mobile robots. IEEE Transactions on Robotics and Automation 17(6), 905–908 (2001)
5. Dougherty, R., Ochoa, V., Randles, Z., Kitts, C.: A behavioral control approach to formation-keeping through an obstacle field. In: Proceedings of the IEEE Aerospace Conference 2004, vol. 1, p. 175 (2004)
6. Forney, C., Manii, E., Farris, M., Moline, M.A., Lowe, C.G., Clark, C.M.: Tracking of a Tagged Leopard Shark with an AUV: Sensor Calibration and State Estimation (2011)
7. Lan, Y., Yan, G., Lin, Z.: A hybrid control approach to cooperative target tracking with multiple mobile robots. In: Proceedings of the 2009 conference on American Control Conference, ACC 2009, pp. 2624–2629. IEEE Press, Piscataway (2009)
8. Lee, S.O., Cho, Y.J., Hwang-Bo, M., You, B.J., Oh, S.R.: A stable target-tracking control for unicycle mobile robots. In: Proceedings of the 2000 IEEE/RSJ International Conference on Intelligent Robots and Systems, vol. 3, pp. 1822–1827 (2000)
9. Lewis, M.A., Tan, K.H.: High precision formation control of mobile robots using virtual structures. Auton. Robots 4, 387–403 (1997)
10. Liu, Z., Ang, M.H.J., Seah, W.K.G.: A stable target-tracking control for unicycle mobile robots. In: Proceedings of the 2005 IEEE/RSJ International Conference on Intelligent Robots and Systems (2005)
11. Mazo Jr., M., Speranzon, A., Johansson, K.H., Hu, X.: Multi-robot tracking of a moving object using directional sensors. In: Proceedings of the International Conference on Robotics and Automation, ICRA, pp. 1103–1108 (2004)
12. Mottaghi, R., Vaughan, R.: An integrated particle filter and potential field method applied to cooperative multi-robot target tracking. Auton. Robots 23(1), 19–35 (2007), doi:10.1007/s10514-007-9028-9
13. OceanServer Technology Inc.: Iver2 Autonomous Underwater Vehicle (2012), `http://www.iver-auv.com`
14. Paley, D.A.: Cooperative control of an autonomous sampling network in an external flow field. In: CDC, pp. 3095–3100. IEEE (2008)
15. Papanikolopoulos, N., Khosla, P., Kanade, T.: Visual tracking of a moving target by a camera mounted on a robot: A combination of control and vision. IEEE Transactions on Robotics and Automation 9(1), 14–35 (1993)
16. Wang, Z., Gu, D.: Cooperative target tracking control of multiple robots. IEEE Transactions on Industrial Electronics 59(8), 3232–3240 (2012)
17. Young, B., Beard, R., Kelsey, J.: A control scheme for improving multi-vehicle formation maneuvers. In: Proceedings of the 2001 American Control Conference, vol. 2, pp. 704–709 (2001), doi:10.1109/ACC.2001.945797
18. Zhou, K., Roumeliotis, S.: Multirobot active target tracking with combinations of relative observations. IEEE Transactions on Robotics 27(4), 678–695 (2011)

Part III
Modular Robots and Novel Mechanisms and Sensors

A One-Hour Curriculum to Engage Middle School Students in Robotics and Computer Science Using Cubelets

Nikolaus Correll, Chris Wailes, and Scott Slaby

Abstract. Robotics has become a standard K-12 outreach tool and for attracting students to the STEM disciplines. Performing these activities in the class room usually requires substantial time commitment by the teacher and integration into the curriculum requires major effort, which makes spontaneous and short-term engagements difficult. This paper studies using "Cubelets", a modular robotic construction kit, which requires virtually no setup time and allows substantial engagement and change of perception of STEM in as little as a 1-hour session. This paper describes the constructivist curriculum and provides qualitative and quantitative results on perception changes with respect to STEM and computer science in particular as a field of study.

1 Introduction

Robots are computers that are extended by sensing, actuation, and communication capabilities. As such, they provide students access to a variety of engineering concepts [11]. Robots also stimulate people's imagination, a development heavily supported by arts, media and toys. Robotics is therefore considered as an ideal tool to provide young children with first exposure to science, technology engineering and math at the K–12 levels [2, 3, 7, 10].

Prominent examples that have found their place in middle- and high school curricula are LEGO Mindstorms [4, 5, 15], Vex robotic kits [1], the FIRST robotics

Nikolaus Correll · Chris Wailes
Department of Computer Science,
University of Colorado at Boulder, 80309 USA
e-mail: `firstname.lastname@colorado.edu`

Scott Slaby
Angevine Middleschool in Lafayette, CO
e-mail: `scott.slaby@gmail.com`

M.A. Hsieh and G. Chirikjian (eds.), *Distributed Autonomous Robotic Systems*,
Springer Tracts in Advanced Robotics 104,
DOI: 10.1007/978-3-642-55146-8_12,

competition [16], or platforms specifically developed for robotics education [6, 8]. An often cited drawback of these systems and tools is that they tend to attract student populations that are already interested in STEM. Consequently, others have proposed to design robotic activities specifically targeted at girls by adopting robotic building kits that are thematically more associated with girls' interest such as puppetry and arts and crafts [9].

An additional drawback of "standard" robotic tool kits is that they require a certain level of engagement for the students to become productive. For example, LEGO Mindstorms requires assembling a large number of parts, attaching the robot to a computer, and learning a graphical programming system before students can obtain results. In other words, curricula designed around these tools put certain minimum time requirements on the participants that make spontaneous activities such as substituting a lecture in middle or high school impossible.

Fig. 1 "Cubelets" manufactured by Modrobotics are a modular robotic construction kit consisting of various cubes with specific actuation (drive, rotation), communication (light, sound), sensing (distance, temperature, knob, brightness), and computation (min, max, inverse) capabilities, as well as structural parts (blocker, passive, battery). Cubelets exchange sensor information and allow the construction of simple autonomous robots.

This paper studies the use of "Cubelets" (Figure 1) to convey basic concepts of computer science using a constructivist educational model. Cubelets are a modular robotic construction kit that allows students to quickly assemble autonomous robotic systems by programming with their hands [12–14]. For example, students can construct a line following robot by combining cubelets with wheels, cubelets that can sense light, a power cubelet, and a structural blocking cubelet. Cubelets do not require an external computer to work and allow creative discovery with little to no instruction.

This paper describes our experiences with two half-hour sessions with 8th graders at a middle school. Students were surveyed both on their perceptions on the role of robotics in their daily lives, computer science as a field of study, and on their prior interest to engage in computer science activities in their professional or personal lives. Data shows that an hour engagement with Cubelets spread over two days has significantly altered their perceptions and could therefore be delivered in a single, 1-hour session as an outreach activity, or form the basis for a multi-session teacher-led curriculum, much like [6], but without requiring a computer lab or setup of a programming environment.

This paper is organized as follows. Section 2 describes the curriculum as well as the information given to the students before and during the activity. Section 3 describes the background of the student population and the survey questions. Section 4 provides results from pre- and post tests. Section 5 discusses these results and Section 6 concludes the paper.

(a) Alarm consisting of sound (transparent), distance sensor (black) and battery block (gray).

(b) Runner consisting of distance sensor, battery, and two wheel blocks (transparent).

(c) Chaser consisting of "Runner" with inverted (red block) distance sensor.

(d) Light-tower consisting of a flash-light block, rotator, knob (to adjust the speed), and power block.

(e) Max-speed Runner made of a runner with a knob and "min" block (pink).

(f) Line Follower made of two light sensors (black blocks), two wheels (transparent blocks) and a passive connector block (green).

Fig. 2 Robots built by the students during the first session

2 Curriculum

The course curriculum was divided into two half-hour activities that were presented to the students during two 40-minute class periods with 10 minutes reserved for assessment, and was designed to teach students about several concepts at the core of computer science: decomposing problems into sub-problems and composing solutions to partial problems to solve a larger problem. The Cubelets robotics kit facilitated this lesson by allowing the students to break down the logical problems (what are the robot's inputs and how does it respond to them) the same way that they decomposed the task of building a complex physical robot.

2.1 Cubelets

Cublets are a set of cubic building blocks that connect via magnets and a proprietary unisex-connector. They are about an inch wide. Connectors not only withstand torque and orthogonal forces (magnets are strong enough to withstand the gravitational pull of up to three other Cubelets), but also allow information and energy to flow between Cubelets. Cubelets each implement a specific sensing, actuation, or logic function. Sensing Cubelets emit sensory information to their neighbors, actuator Cubelets consume information, and logic Cubelets manipulate information flow. There exist also passive Cubelets, which exclusively block or forward information, as well as a Power Cubelet, which includes rechargable batteries. These batteries can power the Cubelets for multiple hours of activity, depending on the actual use. Actuation Cubelets include Cubelets with a single wheel (Drive), a rotating face (Rotate), a lamp (Flashlight), and a bargraph (Bar). Sense Cubelets include brightness, infrared distance sensor, a potentiometer (Knob), and temperature sensors. Think Cubelets include Inverter, which performs a mathematical operation equivalent to 1-value, a Maximum Cubelet, which forwards only the maximum value that it receives on any of its faces, as well as the Blocker, which only forwards energy, and the passive Cubelet, which forwards both energy and power it receives. Cubelets used in this curriculum are shown in Figures 1 and 2.

2.2 First Part

The first part focusses on using the Cubelets to accomplish simple tasks and begins with an explanation of the basic functions of the individual cubes. This explanation includes how they are powered, how the cubes fit together, their categorization as Think, Sense, and Actuation blocks, what values could be produced by sensor blocks, how these values propagate across the blocks, and how Actuator blocks are controlled by the values they receive. The students are then shown how the blocks could be assembled to produce simple behavior using the Power, Knob, Bar Graph, and Inverse cubes.

Next, the class is divided into groups of 4-7 students and each group is given a set of Cubelets. In the study presented in this paper, the class with 17 students was divided into three groups and each group was given the following cubes: 2 Power, 3 Passive, 2 Brightness, 3 Distance, 2 Drive, 2 Flashlight, 1 Bar, 1 Knob, 1 Temperature, 1 Blocker, 1 Rotate, 1 Inverse, and 1 Maximum. The class with 26 students was divided into four groups and each group was given the following cubes: 1 Power, 2 Passive, 1 Brightness, 2 Distance, 2 Drive, 1 Flashlight, 1 Blocker, and 1 Inverse. In each case the remaining cubes were available for the students to use, and each group was encouraged to lend cubes to other groups when they were not using them.

The students were then given *only* the following descriptions of several robots and asked to build them (see Figure 2 for corresponding pictures):

1. Alarm - Makes noise when something approaches it.
2. Runner - Drives away from objects that try to approach it.
3. Chaser - Moves forward at maximum speed until it gets close to another object.
4. Lighttower - Rotating flashlight whose speed can be adjust with a dial.
5. Max-speed Runner - Same as the Runner with the added ability to set a maximum speed.
6. Line Follower - Follows a black line on a white surface.

During the class period all groups were able to complete the first five tasks, most were able to build a runner robot with the ability to set a maximum speed, and at least one group in each class was able to build the line following robot. While all groups had sufficient blocks to complete almost all tasks (in some cases they had to use some blocks from those available to the entire class) additional blocks would facilitate group participation. With the limited supply of blocks most groups would have one or two students who did most of the hands on interaction while the rest of the students commented or observed.

2.3 Second Part

The second part includes both a mini-lecture and discussion. While the first 30 minutes are spent using the blocks to build the simple reactive robots described above, the main focus of the second part is on composing groups of blocks to build more complicated behavior. Specifically, the instructor motivates a scenario such as "tracking down a source of nuclear radiation using a Geiger counter" and discusses with the students what the basic functions are that such a robot must have. These include a pivoting sensor to track down the source of radiation and logic that controls the robot in this direction.

During the lecture several collections of blocks that are relevant to this task were passed around the classroom (Figure 3) and students were asked to postulate as to the functions the partial robots. These included a simple robotic arm, a logical AND gate (here the students were introduced to the concept of binary input), a distance-, and direction-sensing 'head', and a large driving mechanism. While there were not enough blocks to assemble all of these components at the same time, they could be combined to form a robot capable of looking around, finding the nearest object, and moving to it.

The subjects of problem decomposition and solution composition were discussed in the context of the partial robots the students were studying. Once the students were aware of the function of these robotic fragments, they were able to identify how they might be able to re-use the same design when building robots in the future. The idea that each functional unit's behavior could be reasoned about independently, based on their inputs, was then presented.

Up to this point, the physical structures that were being built in class or presented to them determined the behavior of the system. This is not the case for most computing systems, including robots, and so the idea of separating the physical and logical behaviors of a system was introduced. Students learned about *for* loops, *if-then-else* statements, and *functions* and how they were the logical 'blocks' that computer programs were constructed from. A concrete example of a chat application that would allow students to communicate with each other using their phones was then used to show how logical problems could be sub-divided in the same way that a robot could be divided into its functional units. The different components identified for this application were authentication (logging in), communication (sending data over the network), input handling, and displaying output. Lastly, the students were asked to identify several sub-problems in building robots and how they might be solved and re-used. These sub-problems included movement and balance, vision and sensing, and interacting with the physical world. Here, the robots were used to introduce the students to the idea of functional units and *functions* (in the programming language sense), and how the behavior of components of larger systems could be reasoned about based on their inputs and outputs.

(a) (b) (c)

Fig. 3 Robot parts examined during the second part of the activity. *a)* Simple arm mechanism capable of approaching a block and stopping when it connects. *b)* An AND gate that receives binary input from the knob blocks and produces a binary output from the *min* block at the lower-right. *c)* A robotic 'head' capable of detecting the distance and direction of objects around it.

3 Student Background and Survey Content

3.1 Student Background

Angevine Middle School (AMS) in Lafayette, CO, serves grades 6-8. AMS has a high population of students of Latin origin (42.3%), and 72% of the 8th graders score "unsatisfactorily" or "partially" in math, 48% in reading, 59% in writing, and 61% in science in the Colorado Student Assessment Program (CSAP) [1]

The curriculum described here was taught to four cohorts of 8th graders within their social studies course. This setting was chosen to prevent any bias from a STEM-based setting such as science and math classes, both in terms of preconceptions to the field and to the activity itself. All four classes met consecutively on Tuesday and Friday. A pre- and post assessment was administered to two of the cohorts ("Group 1" and "Group 2") before their first exposure to the Cubelets and after the second session on Friday, respectively.

Group 1 consisted of 17 students, 8 male, 9 female. Group 2 consisted of 26 students, 6 male, 20 female. Each group included one special needs student who was accompanied by a paraprofessional. The paraprofessionals were present in the class room, but did not interfere with their student's engagement in the curriculum.

3.2 Questionnaire

The questionnaire consisted of a pre-test and a post-test administered before the first (Tuesday) and after the last (Friday) activity, respectively. Both pre- and post-test consist of identical questions, with the post-test including one additional quantitative question.

The questions are geared to explore conceptions on robotics, the role of computer science in robotics, and the level of interest to engage in computer science as a field of study and/or as a hobby. The questions were designed by the teacher to be age and student appropriate. Questions were projected to a white-board one-by-one and narrated by the teacher and responses collected immediately. This approach required the students to write down answers spontaneously without time for discussion or deliberation. Questions asked were:

1. What is the purpose of a robot?
2. How can Computer Science "build" robots?
3. How can a robot help humans solve problems?
4. How interested are you to pursue Computer Science in High School and beyond? (1=not at all interested, 5=very interested)
5. How interested are you to purse Computer Science in your spare time, for example reading magazines? (1=not at all interested, 5=very interested)
6. *Have you done any research on robotics since last time?*

[1] 2010 data, available from `http://www.schoolview.org`, last retrieved April 25, 2012

Questions 1-5 were administered for both pre- and post-test. Question 6 was administered only at the post-test.

4 Results

4.1 Pre-test

Response rate in group 1 were 17/18 and 26/26 in group 2. Two ESL (English as a second language) female students required additional help by their teacher to answer the questionnaire (group 1). All results are aggregated over both groups, whereas students that did not participate in the post test (3) were removed, leading to 40 samples.

The dominating pattern (25/40) in answering *Question 1* was that robots make the life of humans easier by being assistive companions in everyday live (“robots ease human life”, “robots help humans with things”,“to help or entertain”), followed (5/40) by robots being tools for performing dull or dangerous jobs (“robots do things that humans don’t want to do”), and (4/40) robots are more efficient than humans (“robots are more efficient”,“robots are meant to eliminate human error”). Other comments (6/40) were tongue-in-cheek (indicated by drawn emoticons) of the kind “robots creep people out” and “to take over the world”. *Question 3* lead to similar patterns and outcomes, with students mentioning specific applications such as surgery, manufacturing, repairing, or outperforming the human mind such as in Chess or Jeopardy.

Answers to *Question 2* fell into three groups: students (11/40) able to identify a computer as significant component of a robotic system, which allows the robots to execute plans and reason (“Computer Science helps by telling the robot what to do”, “It helps to make the robot ‘think’”, “To make the robot smart”, “Because a robot is basically a computer that can move and talk”); students (13/40) that understand Computer Science as a discipline involving programming, but do not clearly articulate the role of programming in robotics (“because they program the robot”, "It helps with programming”); and students (16/40) who seem to have little to no understanding of Computer Science (“It uses a robot to make a robot”, “because they understand everything”, “By showing and telling us how to build one”, “because it takes a lot of science to build one”, “to visualize how a robot will function”, “to design a robot”).

Quantitative results to Questions 4 and 5 are provided in Section 4.3. Answers provided by male and female students are significantly different with a p-value of 0.019 for Question 4, and p-value of 0.06 for Question 6.

4.2 Post-test

Of the 17 students in the first group, only 14 were present during the post-test. These 14 students consisted of 7 male and 7 female students. All students in the second group were present for both the pre- and post-test.

Answers to *Question 1* on the post-test followed many of the same patterns as during the pre-test. The most common answers (24/40) stated that robots make life on humans easier by assisting in everyday tasks. The two next most common answers were that robots do tasks that are too difficult, dull, or dangerous for humans (6/40), and that robots do whatever they are programmed to (6/40). Several students (3/40) stated that robots are machines that receive input and act upon it. One student (1/40) answered that robots do tasks that they are more efficient at than humans. No students gave tongue-in-cheek answers to this question on the post-test.

The same three groups of answers to *Question 2* identified in the pre-test are also present in answers to the post-test. However, the majority of students (24/40) were now able to identify the role of a computer in a robot, and the role of computer scientists in programming it. The remaining students either correctly identified programming as something that computer scientists do (9/40) or demonstrated a continued lack of understanding of how Computer Science relates to robotics (7/40).

Quantitative results on Questions 4 and 5 for pre- and post test are compared in Section 4.3. Answers provided by males and females were significantly different with a p-value of 0.014 for Question 4, and a p-value of 0.11 for Question 5.

When asked if the students had researched robotics between the first and second session, six students answered in the affirmative.

4.3 Qualitative and Quantitative Improvements

In answering *Question 2* approximately half of the students (19/40) showed an increased understanding of the role of computers and computer scientists in the creation and operating of robots. The other half of the students (21/40) showed no improvement in their understanding, although 9 of these students had previously identified the role of computers in robotics.

Of those students who showed an increased level of understanding 6 students who had not previously been able to identify the role of programming in robotics were able to indicate in their answers during the post-test that it was involved in their operation. An additional 2 students who answered similarly, and 8 students who identified Computer Science and programming as involved in the creation of robots, were able to state that a program controlled the actions of a robot. None of the students showed a decreased understanding of the role of Computer Science and programming in the field of robotics.

In answering *Question 4* ("How interested are you to pursue Computer Science in High School and beyond?"), all means show increasing trends, from 2.67 ± 1.22 to 2.91 ± 1.3 overall, from 2.36 ± 1.05 to 2.58 ± 1.24 for females, and from 3.25 ± 1.34 to 3.53 ± 1.18 for males. Although the increases are consistent among all groups, the means are only weakly significant (p value ≈ 0.16 for all three paired distributions, all, male and females). This analysis, however, forgoes the polarizing effect the exercise had and which is reflected in the increasing variance: 16 students ranked Question 4 higher than at the pre-test, whereas 12 ranked it lower and 12 remain unchanged. (10 females ranked higher, 8 worse, 6 males ranked better, 4 worse.)

For *Question 5* ("How interested are you to pursue Computer Science in your spare time, for example reading magazines?"), all means show a slight decrease, from 2.36 ± 1.38 to 2.3 ± 1.23 overall, from 2.10 ± 1.31 to 2.10 ± 1.11 for females, and from 2.86 ± 1.40 to 2.68 ± 1.38 for males. These distributions are not significantly different, paired t-test return p-values of 0.74 overall, 0.30 for males and 0.48 for females. Also here, the activity has been polarizing: 11 students increase their ranking, while 11 decrease it, and 22 remain unchanged. (7 females ranked higher and 7 worse, 4 males ranked higher and 4 worse.)

5 Discussion

The response to the questions "What is the purpose of a robot?" and "How can robots help humans to solve problems?" show that robotics is overwhelmingly received as a positive by the 8th graders and is perceived to improve the life of humans one way or the other. This is important as it shows that robotics can indeed serve as a strong motivator to engage in STEM. Whereas the observation that tongue-in-cheek answers to this question vanished in the post-test can be attributed little statistical significance due to the low number of samples, it could be attributed to students being more confident on the subject than before, which is supported by data.

In response to "How can Computer Science help to build robots", a little more than half show a more or less solid understanding how Computer Science can contribute to robotics and therefore to the societal benefits identified by most. Around 40% of the students show little to no understanding of Computer Science as a field. The fact that around half of the students do (subjectively) improve their understanding of the role of computation in robotics is encouraging, however.

As expected given the low enrollment by females in Computer Science, interest of girls in studying computer science (both in school and at home) is significantly lower than that of boys. Although an increase in chosing Computer Science as a field of study due to the activity is only marginal (from 2.36 to 2.58 on a scale from 1 to 5) for this group, the data shows a polarizing effect (also for boys) of the activity. By actively engaging into a Computer Science-related activity, students can make up their mind: 10 girls that are more interested than before (more than a third), are 10 potential future CS students, whereas 8 that are even less interested, are 8 more students that might have avoided a potential mistake and engage in a disappointing activity.

Interest in "Computer Science" or "Robotics" expressed by the students, as well as an increase of their interest level after a playful activity such as the one described here, should not be overestimated, however. Computers and robotics are clearly exciting, especially when offered as extra-curricular activity in a class-room setting, but science and learning how to make them might not. As anecdotal evidence we note the response of one student who expressed that he wants to pursue Computer Science "very much" (Question 6), but cannot imagine studying Computer Science ("not at all on Question 5). He annotated "Because I don't like working with Computers about Science". Conversely, drawing students into a field of study without presenting them with the technical rigor it requires might even be detrimental for their success.

As much as interest in Computer Science as a field of study increases, results on interest in Computer Science as an afterschool activity show little change. A possible explanation might be that the proposed activity is not very well suited on enabling this transitions. Cubelets are relatively new and little available (the 3 kits used in this activity were part of the first 3000 Cubelets shipped to Beta-customers) and are expensive at $299 per kit. In order to serve as a vehicle to bring the classroom activity home, both its availability and affordability need to increase, e.g., as demonstrated in [6] who allow participants to take the robot home for a nominal price of around $60 (59 CHF). Alternatively, lessons could be extended with "action items" that students could pursue at home and/or in follow-up after-school activities.

Although the test group have been middle schoolers, the proposed activity might equally well extend to 4th or 10th graders. While the complexity of possible construction is open-ended (the platform is Turing universal), younger children can potentially be reached by increased scaffolding of the activities. Exploring these opportunities is subject to future work.

6 Conclusion

This paper presents a 1-hour curriculum that allows outreach activities that increase awareness and understanding for computer science and robotics in a middle school setting. The activity is centered around Cubelets, a modular robotics construction kit that allows students to program with their hands, and which can be deployed with little to no preparation. The impact of the proposed activity has been examined by testing it on two classes of 8th graders within two 30 minute sessions plus a pre-test and post-test. Interest and understanding of computer science increases both qualitatively and quantitatively. Students are able to more precisely articulate the role of computing in robots after building a series of primitive, reactive vehicles, a mini-lecture and examining a set of more complex robotic building blocks including an AND gate. Students also show a significant increase in interest to pursue computer science as a field of study, whereas interest to pursue Computer Science as an after school activity remains unchanged. The proposed curriculum has therefore the potential to allow quick interventions by computer science graduate students or professors in local schools as well as rural areas as little to no setup time is required and the time of engagement is low. Here, it can provide encouragement to pursue STEM-related topics in the future for some, while leading to a more informed choice for others.

Acknowledgements. This work has been supported by NSF CAREER award #1150223. C. Wailes is supported by a NSF GK-12 fellowship. We are grateful for this support.

References

1. A modular and extendable robotics platform for education. In: 39th IEEE Frontiers in Education Conference, FIE 2009, pp. 1–4 (October 2009)
2. Brand, B., Collver, M., Kasarda, M.: Motivating students with robotics. In: The Science Teacher, pp. 44–49 (April/May 2008)
3. Bruder, S., Wedeward, K.: Robotics in the classroom. In: IEEE Robotics and Automation Magazine, pp. 25–29 (2003)
4. Carbonaro, M., Rex, M., Chambers, J.: Using lego robotics in project-based learning environment. Interactive Multimedia Electronic Journal of Computer-Enhanced Learning 6(1) (2004)
5. Goff, M., Vernon, M.: Using LEGO RCX bricks as the platform for interdisciplinary design projects. In: Proceedings of the ASEE Annual Conference and Exposition, number Session 3425 (2001)
6. Magnenat, S., Riedo, F., Bonani, M., Mondada, F.: A programming workshop using the robot "thymio ii": The effect on the understanding by children. In: IEEE Workshop on Advanced Robotics and its Social Impacts (ARSO), pp. 24–29 (2012)
7. Matson, E., DeLoach, S.: Using Robots to Increase Interest of Technical Disciplines in Rural and Underserved Schools, Salt Lake City, UT (June 2004)
8. Mondada, F., Bonani, M., Raemy, X., Pugh, J., Cianci, C., Klaptocz, A., Magnenat, S., Zufferey, J., Floriano, D., Martinoli, A.: The e-puck, a robot designed for education in engineering. In: Proceedings of the 9th Conference on Autonomous Robot Systems and Competitions, vol. 1, pp. 59–65 (2009)
9. Nourbaksh, I.: Robot diaries: Creative technology fluency for middle school girls. IEEE Robotics & Automation Magazine, 16–18 (March 2009)
10. Petre, M., Price, B.: Using robotics to motivate 'back door' learning. Education and Information Technologies 9(2), 147–158 (2004)
11. Pomalaza-Raez, C., Groff, B.: Retention 101: Where robots go ... students follow. Journal of Engineering Education 92(1), 85–90 (2003)
12. Schweikardt, E.: Modular robotics as tools for design. In: Proceedings of the 6th ACM SIGCHI Conference on Creativity & Cognition, Washington, DC, USA (June 2007)
13. Schweikardt, E., Gross, M.: roBlocks: a robotic construction kit for mathematics and science education. In: ICMI 2006 Proceedings of the 8th International Conference on Multimodal Interfaces, pp. 72–75 (2006)
14. Schweikardt, E., Gross, M.: The robot is the program: interacting with roBlocks. In: Proceedings of the 2nd International Conference on Tangible and Embedded Interaction, Bonn, Germany (February 2008)
15. Spencer, D., Jaksic, N.: A multidisciplinary robotics learning environment: what Mindstorms and DARPA urban challenge have in common. ASEE Computers in Education Journal 1(3), 32–39 (2010)
16. Welch, A.: The Effect of the FIRST Robotics Competition on High School Students' Attitudes Toward Science. PhD thesis, University of Kansas. Curriculum and Teaching (2007)

A Fast Coalition Structure Search Algorithm for Modular Robot Reconfiguration Planning under Uncertainty

Ayan Dutta, Prithviraj Dasgupta, Jose Baca, and Carl Nelson

Abstract. We consider the problem of reconfiguration planning in modular robots. Current techniques for reconfiguration planning usually specify the destination configuration for a modular robot explicitly. We posit that in uncertain environments the desirable configuration for a modular robot is not known beforehand and has to be determined dynamically. In this paper, we consider this problem of how to identify a new 'best' configuration when a modular robot is unable to continue operating efficiently in its current configuration. We build on a technique that enumerates all the possible partitions of a set of modules requiring reconfiguring as a coalition structure graph (CSG) and finds the 'best' node in that graph. We propose a new data structure called an uncertain CSG (UCSG) that augments the CSG to handle uncertainty originating from the motion and performance of the robot. We then propose a new search algorithm called searchUCSG that intelligently prunes nodes from the UCSG using a modified branch and bound technique. Experimental results show that our algorithm is able to find a node that is within a worst bound of 80% of the optimal or best node in the UCSG while exploring only half the nodes in the UCSG. The time taken by our algorithm in terms of the number of nodes explored is also consistently lower than existing algorithms (that do not model uncertainty) for searching a CSG.

1 Introduction

Over the past few years, modular robots have been proposed as an attractive paradigm for building highly dexterous robots that are capable of maneuvering in

Ayan Dutta · Jose Baca · Prithviraj Dasgupta
Computer Science Department, University of Nebraska at Omaha, NE 68182, USA
e-mail: {adutta,jbacagarcia,pdasgupta}@unomaha.edu

Carl Nelson
Mechanical Engineering Department, University of Nebraska-Lincoln
e-mail: cnelson5@unl.edu

M.A. Hsieh and G. Chirikjian (eds.), *Distributed Autonomous Robotic Systems*, Springer Tracts in Advanced Robotics 104,
DOI: 10.1007/978-3-642-55146-8_13,

environments that are difficult to move in. The major advantage offered by a modular robot is that it consists of individual modules which can be dynamically configured into a shape or configuration that enables the robot to perform tasks under its current conditions. One of the principal computational challenges in designing modular robots is to solve the problem of *reconfiguration planning* - given a set of modules in a certain configuration how to reconfigure those modules to achieve a desired configuration while reducing the time and cost expended to achieve the new configuration. This problem becomes non-trivial if the target configuration is not known *a priori*, and, consequently, the set of all possible configurations has to be explored on-the-fly to find the best possible configuration. Because the space of possible configurations is exponential in the number of modules, conventional search algorithms are unsuitable to solve the reconfiguration planning problem within a reasonable amount of computation time and space. The problem becomes further complicated if we include uncertainty in the mobility and connections of modules, which are practical considerations for any physical robot. In this paper, we address this reconfiguration planning problem for modular robots under uncertainty, using a representation from coalition game theory called coalition structure graph (CSG). We augment the basic CSG to handle uncertainty in modules' movement and in the environment using parameters derived from physical characteristics of a modular robot called ModRED. We formulate the reconfiguration planning problem as an uninformed search problem on the UCSG and propose a modified branch-and-bound algorithm called *searchUCSG* to solve it. We have simulated our algorithm for reconfiguration planning of ModRED and shown that it explores much fewer nodes (about 50%) and its solution quality is within 80% of the optimal node in UCSG. And also the runtime of our algorithm is relatively less than existing algorithms(that do not include uncertainty) to find the optimal coalition structure.

2 Related Work

Modular self-reconfigurable robots (MSRs) are a type of self-reconfigurable robots that are composed of several modules. These modules can change their connections with each other to manifest different shapes of the MSR and select a shape that enables the MSR to perform its assigned task efficiently [12]. An excellent overview of the state of the art MSRs and related techniques is given in [13]. Out of the three types of MSRs — chain, lattice and hybrid - we have used a chain-type MSR to illustrate the experiments in this paper although our techniques could be used for other types too. The self-reconfiguration problem in MSRs has been solved using search-based [1, 2] and control-based techniques [10]. However, both these techniques require the initial and goal configuration to be determined before the reconfiguration process starts. A third technique called task-based reconfiguration has recently shown considerable success [5]. Here the goal configuration of an MSR doing reconfiguration is not determined *a priori*, but is determined as the configuration that helps the MSR perform its task efficiently. Our work in this paper is targeted towards task-based reconfiguration techniques; we do not explicitly specify a goal configuration but allow the reconfiguration algorithm to select a new configuration that reduces the reconfiguration cost i.e. cost for going from one configuration to another which includes operations like docking, undocking, aligning, crawling, etc. and thus selects the best configuration. Coalition game theory gives a set of techniques that can be used by a group of agents to form teams or coalitions with each other [7]. A coalition can be loosely defined as a set of agents that remain together with the intention of cooperating with each other, possibly to perform a task. In terms of MSRs a coalition represents a set of MSR-modules that are connected together. Within coalition games, the coalition structure generation problem that deals with partitioning the agents has received significant attention. This problem is NP-complete, and Sandholm [11] and Rahwan [8] have proposed anytime algorithms to find near-optimal solutions. In contrast to these works, we incorporate uncertainty into the CSG and propose a new algorithm with branch and bound -based pruning to find the best coalition structure.

3 ModRED MSR

We have used an MSR called ModRED [3] that is currently being developed by us, for implementing and testing the techniques in this paper. Unlike most other MSRs, it has 4 DOF (3 rotational and 1 translational); this allows each module to rotate along its long axis as well as extend along that same axis. Single ModRED module is shown in Figure 1(a). This combination of DOF enables the MSR to achieve a greater variety of gaits to possibly maneuver itself out of tight spaces. For the simulated version of each module, we have used a GPS node that gives global coordinates on each robot[1], an accelerometer to determine the alignment of the robot with the ground, in addition to the Xbee modules in the physical robot. Two ModRED modules performing an inchworm motion and its major components are shown in Figure 1[2]. The movement of the MSR in fixed configuration is enabled through gait tables [12]. Videos showing the movement of the MSR in different configurations (e.g., chain, ring) using gait tables are available at

Parameter	Description
Processing	Arduino Fio (Atmel ATmega328P)
Motor Driver	Easy Driver Stepper Motor Driver (8 step microstepping, 750 mA perphase current rating)
Navigation and sensing	Infrared with a range of 4-30 cm, Bump switches for tactile sensing, 9-DOF Razor Inertial Measurement Unit (triple-axis gyro-ITG-3200, triple-axis accelerometer-ADXL345, and triple-axis magnetometer-HMC5843)
Communication	XBee radio modems (Unobstructed range: 120m, transmission power: 1mW)
Power source	3.7 V Lithium-polymer (Li-Po) battery packs

Fig. 1 (a) Single module of the MSR. (b) Two modules doing inchworm motion (c) Major components of the MSR

[1] In the physical MSR, relative positioning is planned to be calculated by combining IMU and IR sensors.

[2] The order of connectivity does not alter the operation of modules.

http://cmantic.unomaha.edu/projects/modred/. While moving in a fixed configuration, if the MSR's motion gets impeded by an obstacle or an occlusion in its path, it needs to reconfigure into a new configuration so that it can continue its movement efficiently. In the next section, we formalize the MSR self-reconfiguration problem and then provide a coalition structure graph-based approach for finding the best configuration.

4 Dynamic Self-Reconfiguration in MSRs

Let A be the set of modules or agents that have been deployed in the environment. Let $\Pi(A)$ be the set of all partitions of A and let $CS(A) = \{A_1, A_2, ..., A_k\} \in \Pi(A)$ denote a specific partition of A. We call $CS(A)$ a configuration, and A_i as the i-th MSR in that configuration. $A_i = \{a_{i_1}, a_{i_2}, a_{i_3}, ..., a_{i_{|A_i|}}\}$ where a_{i_1} and $a_{i_{|A_i|}}$ are the leading and trailing modules of A_i respectively and $\{a_{i_j}, a_{i_{j+1}}\}, j = 1...|A_i| - 1$ is the set of physically coupled modules in A_i. When $\mid A_i \mid = 1$ the MSR is a single module. We define $Val : \Pi(A) \rightarrow \mathbb{R}$, a value function that assigns each partition $CS(A) \in \Pi(A)$ a real number. $Val(CS(A))$ is a metric that gives a virtual reward or benefit obtained by the robots when they perform their assigned tasks while in the configuration $CS(A)$. Evidently, the most suitable configuration for a set of modules is the one that maximizes $Val(CS(A))$. The problem that we study in this paper is the following:

Definition 1 *MSR Reconfiguration Problem. Given a set of modules A and an arbitrary configuration $CS_{old}(A) = \{A_1^{old}, A_2^{old}, ..., A_k^{old}\}$ in which they are deployed, find a new configuration $CS_{new}(A) = \{A_1^{new}, A_2^{new}, ..., A_{k'}^{new}\}$ such that the following constraint is satisfied:*

$$\max_{CS_{new}(A) \in \Pi(A)} Val(CS_{new}(A))$$

Note that k and k' in the above definition may be different. Such reconfigurations can happen, for example, when a set of modules is deployed into the environment from an aircraft and the modules need to get into a configuration that maximizes their value. Another instance of reconfiguration could happen when an MSR gets stuck at an obstacle while navigating during an exploration task and needs to get into a new configuration so that it can continue performing its navigation. The objective of the MSR reconfiguration problem is to get the MSR into a new configuration that allows it to continue its task while giving the highest value over its current partitions.

In our previous work [9, 4], we have shown that a systematic way to go about analysing the configurations in $\Pi(A)$ is provided by a hierarchical graph structure called a coalition structure graph(CSG) [7]. In a CSG, each partition $CS(A) \in \Pi(A)$ is called a coalition structure and appears as a node in the CSG. The parts or subsets of a partition are called coalitions, denoted by S. A CSG with 4 agents is shown in Figure 2. CSG nodes are organized into levels. *Level l* indicates that every node in level l in CSG has exactly l subsets or coalitions as its members. CSGs offer a structured way of exploring coalition structures because a node at level $l - 1$ can be generated by combining a pair of coalitions from a node at level l. Let $succ$:

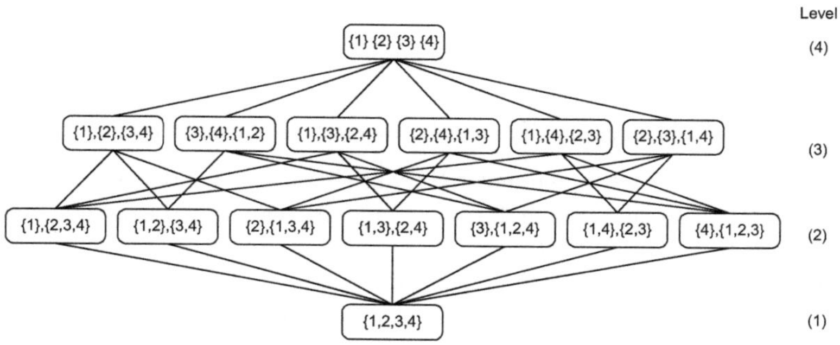

Fig. 2 Coalition structure graph with 4 agents

$\Pi(A) \rightarrow \Pi(A)$ denote a successor function that takes a node at level l and generates a node at level $l-1$. $succ^{(k)}(CS(A))$ denotes the node that is reached from $CS(A)$ by applying the $succ(\cdot)$ function k times. Each node or coalition structure $CS(A)$ is associated with a value $Val(CS(A))$ that corresponds to the value of the partition or coalition structure $CS(A)$. For the context of MSR reconfiguration, an agent a_i corresponds to a single MSR-module, a coalition S corresponds to an MSR A_j, while a coalition structure $CS(A)$ corresponds to a set of MSRs or a configuration of A. The $succ(CS(A))$ operation corresponds to a pair of MSRs which are part of the configuration $CS(A)$ moving close to each other, aligning and docking with each other to give a new configuration $CS'(A)$. To solve the MSR reconfiguration problem given in Definition 1, we have to find the coalition structure in the CSG that corresponds to the maximum value, i.e., find $CS^*(A) = arg \max_{CS(A) \in \Pi(a)} Val(CS(A))$. In the rest of the paper, for the sake of legibility, we have dropped the argument A from CS^* and CS, assuming that it is appropriately defined based on the context. For convenience with the CSG traversal algorithm in Section 4.2, we define the depth of a node at level l as $d = \mid A \mid - l$. The depth of the root node of the CSG is 1 and that of the bottommost node is $\mid A \mid$.

4.1 Uncertainty in Reconfiguration of Modular Robots

Uncertainty in the operations required to reconfigure modules in ModRED is an important consideration to extract the desired behavior of the robot. Uncertainty is caused by inexact or unexpected operation by the robot, which cannot be calculated accurately *a priori*. We have considered two sources of uncertainty in ModRED, *viz.*, (a) motion uncertainty from robot physics and environment, and, (b) performance uncertainty from robot operation, which are discussed below.

4.1.1 Motion Uncertainty

Unexpected motion and alignment of the robot modules can cause ModRED's behavior to deviate from ideal operation. We have considered three major sources of uncertainty under this category that could affect the reconfiguration process:

(i) Distance uncertainty is the uncertainty arising out of the distance required to be traversed by a pair of MSRs before docking with each other. It is modeled as a half-Gaussian distribution $\mathcal{N}(\mu_{du}, \sigma_{du})$.

(ii) Alignment uncertainty is the uncertainty arising out of the angle each MSR in a pair of MSRs needs to rotate before they can align with each other prior to docking. It is modeled as a Gaussian distribution, $\mathcal{N}(\mu_{au}, \sigma_{au})$.

(iii) Environment uncertainty is the uncertainty arising from the operational conditions in the environment due to factors such as obstacles, terrain conditions, surface friction, etc. that affect the movement of a pair of MSRs while moving towards and docking with each other. We consider three discrete values for environment uncertainty, $e_u = \{hi, med, lo\}$. The uncertainty is modeled as a multi-variate half-Gaussian distribution $\mathcal{N}(\mu_{eu}, \sigma_{eu})$. Modeling it as a half-Gaussian distribution allows us to represent it as a folded standard normal distribution where the fold occurs at a cumulative probability of 0.5 i.e. at the mean μ_{eu}.

To combine the Gaussian representing the motion uncertainty, we consider their weighted mean with variance [6]. We associate with each Gaussian a weight that denotes the Gaussian effect on the total motion uncertainty of the robot. These weights are denoted by w_{du}, w_{au}, and w_{eu} respectively, and each weight is given by the inverse of the corresponding Gaussian variance. The weighted mean of the three Gaussian then gives the total motion uncertainty, expressed as a probability, when two MSRs A_i and A_j attempt to connect with each other, as given below:

$$prob(A_i, A_j) = \frac{1}{w_{du} + w_{au} + w_{eu}} (w_{du} \cdot p_{du} + w_{au} \cdot p_{au} + w_{eu} \cdot p_{eu}), \tag{1}$$

where $p_{du} \in \mathcal{N}(\mu_{du}, \sigma_{du})$, $p_{au} \in \mathcal{N}(\mu_{au}, \sigma_{au})$ and $p_{eu} \in \mathcal{N}(\mu_{eu}, \sigma_{eu})$, and, $w_{du} = \frac{1}{\sigma_{du}^2}$, $w_{au} = \frac{1}{\sigma_{au}^2}$, $w_{eu} = \frac{1}{\sigma_{eu}^2}$.

4.1.2 Performance Uncertainty

When a robot moves in a certain configuration, its performance might vary depending on several factors such as how well the modules are physically connected, how well the modules can lift each others weight, how much battery the modules have, how well multiple MSRs coordinate with each other while operating, etc. To model these operational uncertainties, we assume that the utility received by the set of MSRs has a certain variance around the ideal value given by the value $Val(CS)$. We denote this variance as a lower and upper bound on $Val(CS)$ denoted by $EV_{lb}(CS)$ and $EV_{ub}(CS)$, where $EV_{lb}(CS) = (1 - p_l) \cdot Val(CS)$, $EV_{ub}(CS) = (1 + p_u) \cdot Val(CS)$ and $p_l, p_u \in [0, 1]$. For legibility, from now on we use the term value to refer to EV, without loss of generality.

Uncertain CSG. To integrate the reconfiguration uncertainties into the CSG, we propose an uncertainty augmented structure called the *Uncertain Coalition Structure Graph*(UCSG). Formally, $UCSG = (V, E, w)$, where $V = \Pi \ni CS$ is the set of vertices of the UCSG corresponding to the set of all partitions of the agents or modules, $e_{i,j} \in E : e_{ij} = (v_i, v_j), v_i, v_j \in V, v_j = succ(v_i)$ is the set of edges in the UCSG, and, $w : E \to [0,1]$ is a weight associated with edge e_{ij}. We have taken $w_{ij} = w(e_{ij}) = prob(A_i, A_j)$ so that it represents the motion uncertainty involved in forming the coalition structure corresponding to v_j starting from the coalition structure corresponding to v_i. Within the context of the UCSG, we formulate the MSR reconfiguration problem as finding the configuration or coalition structure CS^* that has the maximum expected worst-case (lower-bound) value, that is, $CS^* = \arg\max_{CS \in V} EV_{lb}(CS)$.

Because the number of nodes in the UCSG is unchanged from that in the CSG, and is still exponential, exhaustive search techniques are computationally expensive to implement. To address this problem, we describe an intelligent pruning technique to find the node CS^* in the UCSG, as described in the next section.

4.2 Generating and Pruning the UCSG

Our pruning strategy for the UCSG is based on the insight that a pair of MSRs that are likely to incur a high amount of motion uncertainty to get connected with each other will lead to a low-value configuration when included as part of any other configuration. Our objective then is to identify such inefficient pairings as soon as their expected value is calculated and prevent further exploration of the nodes of the UCSG that include those pairings. Starting from singleton modules at the root of the UCSG, the earliest such inefficient pairings can be determined is at depth 2, when two singletons come together to form a two-module MSR. We mark such inefficient 2-module configurations as a *bad coalition (BC)*, as defined below:

Definition 2 Bad Coalition. *Let v_1 denote the root of the UCSG and $\mathbf{v_2}$ denote the children of the root node. Then, any $v \in \mathbf{v_2}$ is marked as a bad coalition iff $prob(a_i, a_j) \leq bc_{thr} : (a_i, a_j) \subset v$, where $bc_{thr} \in [0,1]$.*

Nodes generated while exploring the UCSG which include any bad coalitions are pruned immediately. We assume that all modules are within communication range of each other, and, at the beginning of the reconfiguration process, modules communicate their position and angle to each other within a global reference frame.

The basic algorithm for searching the UCSG for CS^* is a uniform cost search as shown in Algorithm 1. The search starts at the root node, where every module is a singleton, and checks its children for bad coalitions. Bad coalitions, if any, are stored in the set BC. The nodes that do not contain any bad coalitions are placed in the set called $OPEN$. The nodes in $OPEN$ are partitioned into two sets, called *unpromising* and *promising* nodes ($\bar{v}^{unprom}$ and $\bar{v}^{prom}$ respectively) based on whether the nodes upper bound lies below or above the highest value of the lower bound, EV^*_{lb} seen thus far. Nodes in each of these partitions are sorted according to their upper bound value EV_{ub} to ensure that within each partition, less promising nodes are inspected

Algorithm 1. Algorithm to search for best coalition structure in UCSG

searchUCSG(root)
Input: *root* // set of singleton agents represented as root of UCSG
Output: CS^*: coalition structure in UCSG with max. expected value
Calculate $EV_{lb}(root)$ and $EV_{ub}(root)$;
$EV_{lb}^* \leftarrow EV_{lb}(root)$;
$\bar{v}_{root}^{children} \leftarrow$ Generate children of root;
$BC \leftarrow IdentifyBadCoalition(v_{root}^{children})$;
$OPEN \leftarrow \bar{v}_{root}^{children} \setminus BC$;
Sort nodes in $OPEN$ based on EV_{ub};
Partition $OPEN$ into $\bar{v}^{uprom}$ and $\bar{v}^{prom}$ given by:
    $\bar{v}^{unprom} = \{v \in OPEN : EV_{ub}(v) \leq EV_{lb}^*\}$
    $\bar{v}^{prom} = \{v \in OPEN : EV_{ub}(v) > EV_{lb}^*\}$
Prune($\bar{v}^{uprom} \cup \arg\min_{\bar{v}^{prom}}(EV_{lb}(\bar{v}^{prom}))$);
$d = 2$;
while $d < targetdepth$ **do**
    $EV_{lb}^* \leftarrow \max(EV_{lb}^*, \max(EV_{lb}(OPEN))$;
    for *every node* $v \in OPEN$ **do**
        $\bar{v}^{children} \leftarrow$ Generate children of v
            after removing any children nodes with coalitions in BC
        $OPEN \leftarrow OPEN \setminus \{v\}$;
        $OPEN \leftarrow OPEN \cup \bar{v}^{children}$;
    Sort nodes in $OPEN$ based on EV_{ub};
    Partition $OPEN$ into $\bar{v}^{uprom}$ and $\bar{v}^{prom}$ given by:
        $\bar{v}^{unprom} = \{v \in OPEN : EV_{ub}(v) \leq EV_{lb}^*\}$
        $\bar{v}^{prom} = \{v \in OPEN : EV_{ub}(v) > EV_{lb}^*\}$
    Prune($\bar{v}^{uprom} \cup \arg\min_{\bar{v}^{prom}}(EV_{lb}(\bar{v}^{prom}))$);
    $d \leftarrow d + 1$;
$EV_{lb}^* \leftarrow \max(EV_{lb}^*, \max(EV_{lb}(OPEN))$;
$CS^* \leftarrow$ node with EV_{lb}^*;
return CS^*;

IdentifyBadCoalition($\bar{v}$)
Input: $\bar{v}$: set of nodes belonging to UCSG
Output: BC: set of bad coalitions
//v only contains nodes at depth 2 of the UCSG which consist of
//exactly one 2-agent coalition and remaining singleton coalitions.
$BC \leftarrow \{\emptyset\}$;
for *every* $v \in \bar{v}$ **do**
    if $\exists (i,j) \in v : u(i,j) << u_2$ **then**
        $BC \leftarrow BC \cup (i,j)$;
return BC;

Algorithm 2. Algorithm to prune a set of nodes from the OPEN list in a depth first manner.

Prune$(\bar{v})$
Input: $\bar{v}$: set of nodes belonging to *OPEN*
Output: *winner*: node from $\bar{v}$ with highest expected future utility
for $l = 1$ *to* k **do**
  //for k levels expand branch starting from node v_j
  for *every* $v_j \in \bar{v}$ **do**
    $\bar{v}^{children}_{j,l} \leftarrow succ^{(l)}(v_j)$ (after removing any bad coalitions);
    $v^{child}_{j,l} \leftarrow \arg\max_{v \in \bar{v}^{children}_{j,l}} EV_{ub}(v)$;
    if $\exists j' : v^{child}_{j',l} = v^{child}_{j,l}$ **then**
      $\bar{v} \leftarrow \bar{v} \setminus \{v_j\}$
      $pruned \leftarrow pruned \cup \{v_j\}$;
    else
      $grad_j \leftarrow \frac{(EV_{lb}(v^{child}_{j,l}) - EV_{lb}(v_j)) + grad_j \cdot (l-1)}{l}$;

//Project each line up to *targetdepth* and select winner
$v^{max}_k \leftarrow \max_{v \in v^{child}_{j,k}} EV_{lb}(v)$;
$grad^{max}_k \leftarrow$ gradient corr. to v^{max}_k;
$winner \leftarrow v^{max}_k$;
//find intersection depth of v^{max}_k with every unpruned child of v_j at depth k
for *every* $v^{child}_{j,k}$ **do**
  $inter_j \leftarrow \frac{EV_{lb}(v^{max}_k) - EV_{lb}(v^{child}_{j,k})}{grad_{j,k} - grad^{max}_k}$;
if $\min_j |(targetdepth - inter_j)| \leq thresh$ **then**
  $winner \leftarrow \arg\min_j | targetdepth - inter_j |$;
  //break ties in *winner* by selecting *winner* with $(targetdepth - inter_j)$
$pruned \leftarrow pruned \cup (\bar{v} \setminus \{winner\})$;
$OPEN \leftarrow OPEN \setminus pruned$;
return *winner*;

and possibly pruned earlier. During each iteration, the algorithm partially prunes the nodes in $\bar{v}^{unprom}$ and then recursively expands the un-pruned nodes in *OPEN*. The best expected lower and upper bound values in the entire UCSG, EV^*_{lb} and EV^*_{ub}, are updated at the end of each iteration. The algorithm explores UCSG nodes up to a certain depth called *targetdepth* and returns the node with the highest value of lower bound EV^*_{lb} that has been encountered so far.

The pruning mechanism used for nodes in the unpromising partition, $\bar{v}^{unprom}$, requires some insight. Consider a situation where there is only one node in $\bar{v}^{unprom}$ and $\bar{v}^{prom}$ respectively. Denote these nodes by $v_k \in \bar{v}^{unprom}$ and $v_l \in \bar{v}^{prom}$ and their lower bound values by $EV_{lb}(v_k)$ and $EV_{lb}(v_l)$ respectively. By definition of the promising and unpromising partition, $EV_{lb}(v_k) < EV_{lb}(v_l)$. In the conventional branch and bound algorithm, v_k can be pruned right away as the successor nodes in its subtree cannot improve on the values in v_l's subtree. However this might not be the case in the UCSG because of the operational uncertainty between modules. Let $p_{k1}, p_{k2}, p_{k3}...p_{kd'}$ denote the operational uncertainty denoted by the weights (w_{ij}) encountered starting from v_k up to a node at depth d' away from v_k's depth. Similarly, let $p_{l1}, p_{l2}, p_{l3}...p_{ld'}$ denote the corresponding operational uncertainty starting from node v_l up to a node at depth d' below v_l. The values p_{km} and $p_{ln}(m,n = 1...d')$ are probabilities determined by the operational conditions. Consider a case where for every m,n $p_{km} >> p_{ln}$. In such a scenario, we can have $EV_{lb}(succ^{(d')}(v_k)) > EV_{lb}(succ^{(d')}(v_l))$ implying that d' levels below the current depth, v_k's lower bound might be higher than v_l's lower bound. Therefore, it might be incorrect to make a decision about pruning node v_k as soon as it is placed in $\bar{v}^{unprom}$.

To address this problem, we propose a lookahead-based technique for making a decision to prune a node from the unpromising partition. The main idea of the lookahead technique is to check whether the lower bound of any node in the unpromising partition might exceed the lower bound of the worst node (lowest EV_{ub}) in the promising partition at a lower depth in the UCSG called *targetdepth*. However, expanding all the successor nodes of every node in $\bar{v}$ up to *targetdepth* is a computationally expensive operation as the number of successor nodes grows exponentially. Therefore, we divide the lookahead process into two steps - (i) a *gradient calculation phase* that expands a node for k successive levels from the current level, retains only the best successor node at each level, and, calculates the change in the value or gradient of EV_{lb} from the current level upto k levels below; (ii) a *projection phase*, that calculates the expected value of EV_{lb} at *targetdepth* by projecting the last calculated value of EV_{lb} using the gradient. As shown in Algorithm 2, in the gradient calculation phase, each node $v_j \in \bar{v}$, is expanded for k successive levels and only, $v_{j,l}^{child}$ the successor node at each level l with highest value of the upper bound EV_{ub} is retained. After removing duplicate nodes, the change in the value of the lower bound EV_{lb} from v_j up to $v_{j,l}^{child}$ is calculated as $grad_j$. In the projection phase, we first identify the best node (with highest value of EV_{lb}) at the last calculated level $(current\,level + k)$. We call this node *winner* and its associated gradient for EV_{lb} as $grad_k^{max}$. We then inspect each of the remaining nodes $v_j \neq winner$ at $(current\,level + k)$ and calculate the level, $inter_j$, at which v_j's EV_{lb} value exceeds *winner*'s EV_{lb} value. If $inter_j$ is at *targetdepth* or near (*i.e.* within *thresh* levels of *targetdepth*), the node $v_j \in OPEN$ having the highest positive difference in EV_{lb} at or near *targetdepth* with *winner* is marked as the new *winner*. All nodes except the final *winner* are then pruned, while *winner* is added to *OPEN* for expansion in the next iteration step.

5 Experimental Results

To verify the performance of our proposed reconfiguration planning technique, we implemented the searchUCSG algorithm in C++ on a desktop PC (Intel Core i7 - 960 3.20GHz, 12GB DDR3 SDRAM).

Experimental Setting. We consider a setting where a set of $n = 4...25$ ModRED modules are placed randomly within a 4×4 m^2 environment. Initially none of the modules are connected with each other. The objective of the modules is to find the configuration that gives the highest value. To do this, each module runs the *searchUCSG* algorithm given in Algorithm 1. The value of an MSR $A_i \in CS$ is defined as:

$$Val(A_i) = \begin{cases} Val(1), \text{ if } |A_i| = 1; \\ Val(1) \times (|A_i| + \frac{|A_i|^2}{10}), \text{ if } |A_i| \leq A_{max}; \\ Val(1) \times e^{-(|A_i| - A_{max}) \times 10}, \text{ if } |A_i| > A_{max}. \end{cases} \quad (2)$$

This value function[3] causes value of an MSR to monotonically increases from a singleton up to a certain size A_{max}, beyond which it decreases exponentially. In other words, it gives preference to forming larger coalitions or MSRs up to certain maximum size A_{max}, which is given by the physical limitations of MSR modules to connect with each other and maneuver while remaining connected. The value of a configuration or coalition structure is given by: $Val(CS) = \sum_{A_i \in CS} Val(A_i)$. For our experiments, we have used $Val(1) = 5$. The values of different parameters used by our algorithm are shown in Table 1. The environment uncertainty is set to *medium* for most experiments, unless otherwise stated.

Table 1 Different parameters used for our simulations

Parameter	Symbol	Values
Number of agents	n	$\{4....25\}$
Maximum desired coalition size	A_{max}	$2(n=4), 4(n=6), 6(n>6)$
Max. depth explored in UCSG	$targetdepth$	$\{\frac{n}{2}, \frac{2n}{3}, \frac{3n}{4}, n-1\}$
Look-ahead depth	k	$\frac{n-currentdepth}{2}$
Mean and std. dev. for distance uncertainty	p_{du}, σ_{du}	0, $\frac{17}{3}$(comm. range of modules = 17cm)
Mean and std. dev. for angle uncertainty	p_{au}, σ_{au}	$\frac{\pi}{2}, \frac{\pi}{6}$
Mean and std. dev. for env. uncertainty	$p_{eu=hi}, \sigma_{eu=hi}$ $p_{eu=med}, \sigma_{eu=med}$ $p_{eu=lo}, \sigma_{eu=lo}$	$0.66, 0.1$ $0.33, 0.1$ $0, 0.1$
Prob. for estimating EV_{lb} from V_{CS}	p_l	0.5 for $n \leq 12$ 0.2 for $n > 12$
Prob. for estimating EV_{ub} from $V(CS)$	p_u	0.2
Bad coalition threshold probability	bc_{thr}	0.1

5.1 *Simulation Results*

In the first set of experiments, we analyzed the effect of the main concept of our algorithm i.e. finding the best coalition structure possible from UCSG with intelligent pruning. For $4 \leq n \leq 12$, we were able to do an exhaustive search in the space of

[3] Value or *reward* can be determined from the history of past performances of a coalition.

all coalition structures to find the optimal coalition structure and see that our algorithm is able to find the optimal coalition structure for all values of n. The time taken to find the optimal value with our intelligent pruning technique is given in Figure 3(a). For $n > 12$, exhaustive search becomes prohibitive as its complexity is $O(n^n)$. So for more than 12 agents, the highest coalition structure value that our algorithm finds with $targetdepth = n - 1$ (exploring all depths, but with pruning) is used as the *best* value. The ratio between the optimal ($n \leq 12$) or best ($n > 12$) value and the value found by our algorithm, for different values of $targetdepth$, are shown in Figures 3 (b) and (c). Figure 3(b) shows that for $n \leq 12$ if we vary the exploration depth, $targetdepth$ in the UCSG, then for $targetdepth = \frac{n}{2}$, on an average we can get 80% of the optimal value[4]. If we increase the limit to $targetdepth = \frac{2n}{3}$, then the achieved value is almost 90% or above of the optimal value and if we further increase $targetdepth$ to $\frac{3n}{4}$, then it is 95% of the optimal value. Empirically, we can say that our algorithm provides the *worst bound* of 80% if we explore till depth $\frac{n}{2}$ and this *bound* increases as we go explore deeper. For $n = 15...25$ agents, in Figure 3(c), we can see that if we explore upto $targetdepth = \frac{n}{2}$, then the value obtained by our algorithm is 90% of the best value obtained. And for $targetdepth = \frac{2n}{3}$, this ratio increases to almost 95%, whereas for $targetdepth = \frac{3n}{4}$ it is more than 95%. From this set of results, we can say that our algorithm provides the *worst bound* of 90% if we go till depth $\frac{n}{2}$ (for more than 12 agents) and this *bound* increases as we further explore lower depths.

No. of agents	Ratio to opt. value	Runtime (mins)
4	1	0.0006
6	1	0.005
7	1	0.04
8	1	0.2
9	1	1.08
10	1	4.22
12	1	6.35

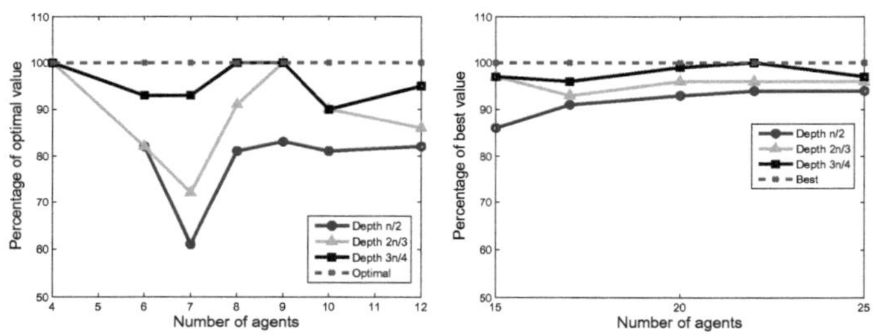

Fig. 3 a) Ratio of values of coalition structure found using searchUCSG algorithm with $targetdepth = n - 1$ to the optimal value and corresponding running times taken by searchUCSG algorithm, (b) Ratio of value found by searchUCSG to the optimal found (expressed as percentage) for 4 to 12 agents, (c) Ratio of value found by searchUCSG to the optimal found (expressed as percentage) for 15 to 25 agents

The quality of the solution found by our algorithm is also dependent on the ratio between exploration depth $targetdepth$ and A_{max}. Recall that the value function we have defined, gives the highest value for an MSR (coalition) that is of size A_{max}. The first time a coalition of size A_{max} occurs in the UCSG is at depth A_{max}. This implies

[4] As we have fixed A_{max} at 6, for 7 agents we can find coalitions of size 6 only at depth 6 (i.e. d= n-1). That is why, if we go for $targetdepth = \frac{n}{2}$, then obtained value is only 60% of the optimal value.

that $targetdepth \geq A_{max}$, for our algorithm to be able to find the coalitions with good values, and the higher the value of $\frac{targetdepth}{A_{max}}$ is, the closer the solution is to the optimal or best value. For example, in Figure 3 (a), with $n = 12$ and $targetdepth = \frac{n}{2} = 6$ which is equal to A_{max}, the value found by our algorithm is on an average within 80% of the optimal. But when $n = 25$, and $targetdepth = \frac{n}{2} = 12$ which is greater than A_{max}, the algorithm has already seen several nodes with coalitions of size $A_{max} = 6$ and consequently finds a better solution which is within 94% of the best value. We have performed 10 simulation runs with different values for medium noise settings and found that variance in value found at different *targetdepths* is within $1-3\%$.

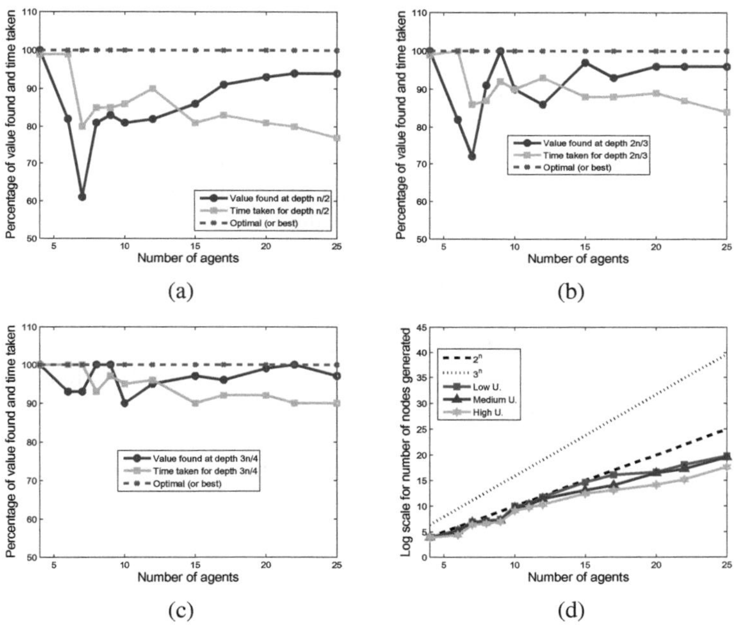

Fig. 4 Comparison of value found and time taken for different *targetdepth* values. (a) $targetdepth = \frac{n}{2}$, (b) $targetdepth = \frac{2n}{3}$, (c) $targetdepth = \frac{3n}{4}$; (d) Comparison of number of nodes explored, in log scale, between our algorithm (using low, medium and high environment uncertainty) and 3^n and 2^n times obtained in previous works.

We have also compared the percentage of optimal(or *best*) value we are getting till a certain *targetdepth* using our algorithm and percentage of the time taken to find the optimal value. The results are shown in Figures 4 (a)-(c) for up to 25 agents. We see that as as the number of agents increases, we get values that are closer to the optimal or best value, while taking less time. Also as we explore deeper into the UCSG, this relative difference between the best found value and the time taken increases. As can be seen from the graphs, as *targetdepth* increases from $\frac{n}{2}$ to $\frac{3n}{4}$, the relative difference between the percentage value found and percentage time taken also increased - the time got successively lower and the best value found got successively higher. This implies that as our algorithm proceeds, the improvement in value of the node found is more than the cost (time) incurred to find the node, that is, our algorithm takes less time to find a better node, as it proceeds further.

We have also compared our result with previously established bounds for the CSG search problem [11, 8], where the complexity of the algorithms were $O(3^n)$ and $O(2^n)$ respectively. As can be seen from Figure 4(d) (shown with log scale on y-axis), using our algorithm, for all three of the environment uncertainty types ($eu = \{lo, med, hi\}$), the number of nodes generated is lower than the other algorithms. For $n < 12$ agents, although the curves for our algorithm's time appear very close to the 2^n line on the log scale, on a linear scale they take an average time of 75% ($eu = lo$), 62.5% ($eu = med$) and 40% ($eu = hi$) of the worst case time of 2^n.

6 Conclusion and Future Work

In this paper we have proposed a MSR reconfiguration planning technique that models the problem as a new data structure called a UCSG and then described an algorithm to intelligently prunes the search within the UCSG to find the best configuration of a set of MSR modules. Currently, our technique starts with all singletons or individual modules and finds the best configuration or partition among them. We are currently extending our algorithm to enable it to start from any configuration of modules and change to the 'best' possible configuration. We are also looking at more structured ways to incorporate the performance uncertainty of modules using agent types within a Bayesian game framework. Yet another direction we are investigating is to automatically determine the optimal $targetdepth$ based on the values of n and A_{max}. Finally, we are working on implementing this algorithm on physical ModRED modules.

References

1. Butler, Z., Brynes, S., Rus, D.: Distributed motion planning for modular robots with unit decompressable modules. In: IEEE/RSJ Intl. Conf. Intell. Rob. and Sys., pp. 790–796 (2001)
2. Chirikjian, G., Pamecha, A., Ebert-Uphoff, I.: Evaluating efficiency of self reconfiguration in a class of modular robots. Robotics Systems 13, 317–338 (1996)
3. Chu, K., Hossain, S.G.M., Nelson, C.: Design of a four-dof modular self-reconfigurable robot with novel gaits. ASME Intl. Design Engg. Tech. Conf, DETC2011-47746 (2011)
4. Dasgupta, P., Ufimtsev, V., Nelson, C., Mamur, S.M.G.: Dynamic reconfiguration in modular robots using graph partitioning-based coalitions. In: Intl. Conf. on Auton. Agents and Multi-Agent Systems (AAMAS), Valencia, Spain (2012)
5. Kamimura, A., Yoshida, E., Murata, S., Hurokawa, H., Tomita, K., Kokaji, S.: Distributed Self-Reconfiguration of M-TRAN III Modular Robotic System. Intl. J. of Robotics 27(1), 373–386 (2008)
6. Meier, P.: Variance of a weighted mean. Biometrics 9(1), 59–73 (1953)

7. Myerson, R.: Game Theory: Analysis of Conflict. Harvard University Press, Cambridge (1997)
8. Rahwan, T., Ramchurn, S., Jennings, N., Giovannucci, A.: An anytime algorithm for optimal coalition structure generation. J. Artif. Intell. Res (JAIR) 34, 521–567 (2009)
9. Ramaekers, Z., Dasgupta, R., Ufimtsev, V., Hossain, S.G.M., Nelson, C.: Self-reconfiguration in modular robots using coalition games with uncertainty. In: Automated Action Planning for Autonomous Mobile Robots (2011)
10. Rosa, M., Goldstein, S., Lee, P., Campbell, J., Pillai, P.: Scalable shape sculpturing via hole motions. In: IEEE Intl. Conf. Rob. and Auton., Orlando, FL, pp. 1462–1468 (2006)
11. Sandholm, T., Larson, K., Andersson, M., Shehory, O., Tohme, F.: Coalition structure generation with worst case guarantees. Artificial Intelligence 111(2), 209–238 (1999)
12. Stoy, K., Brandt, D., Christensen, D.: Self-Reconfigurable Robots: An Introduction. The MIT Press, Cambridge (2010)
13. Yim, M., et al.: Modular self-reconfigurable robot systems: Challenges and opportunities for the future. IEEE Robotics and Automation Magazine 14(1), 43–53 (2007)

Flexible Self-reconfigurable Robots Based on Thermoplastic Adhesives

Fumiya Iida, Liyu Wang, and Luzius Brodbeck

Abstract. The paper introduces a concept of flexible self-reconfiguration that makes use of thermoplastic adhesives (TPAs) in robotic systems. TPAs are polymer-based materials that exhibit several interesting mechanical properties beneficial for self-reconfiguration. For example, thermoplasticity enables robots to flexibly fabricate a number of different mechanical structures, while temperature-dependent adhesion allows systems to make robust connection and disconnection. This paper introduces robotic self-reconfiguration by using three TPA handling processes, i.e. structure formation, connection and disconnection. These processes are then examined in a few practical application scenarios, i.e. pick and place operations of a variety of objects, autonomous body extension of robotic manipulators, and robots climbing on uneven surfaces. And finally we discusses challenges and perspectives of this approach.

1 Introduction

Capabilities to significantly vary body structures play an essential role in the adaptability of biological systems. For example, animals start their lives smaller and simpler and then gradually develop more complex systems, replace old body parts such as hairs and nails, and heal unanticipated mechanical fracture of bones and muscles to survive in uncertain complex environments.

Previously a number of robotics engineers have been fascinated by such functions in nature, and attempted to implement them through self-reconfiguration and self-assembly in modular robots [4, 17]. Typically self-reconfigurable robots consist of a number of predefined mechatronic modules that can form lattice-type structures, self-disassemble the structures, and transform the structures into another shape

Fumiya Iida · Liyu Wang · Luzius Brodbeck
Bio-Inspired Robotics Laboratory, Institute of Robotics and Intelligent Systems,
ETH Zurich, Leonhardstrasse 27, CH-8092, Zurich Switzerland
e-mail: {fumiya.iida,liyu.wang,luzius.brodbeck}@mavt.ethz.ch

M.A. Hsieh and G. Chirikjian (eds.), *Distributed Autonomous Robotic Systems*,
Springer Tracts in Advanced Robotics 104,
DOI: 10.1007/978-3-642-55146-8_14, © Springer-Verlag Berlin Heidelberg 2014

by using actuators in the modules [12, 17]. Based on this basic concept, various mechatronic modules were designed to demonstrate the use of reconfigurable body structures for a wheel-like structure transforming into a legged robot to perform locomotion on different kinds of terrain [8], for example. In order to form larger structures, the self-assembling approach has been intensively studied, in which stochastic physical interactions are exploited as the main drive. This approach usually employs mechatronic modules that take advantage of the stochastic physical processes such as water flows, and the structures can be developed by controlling the connection and disconnection between the modules [5, 18, 19]. The variations of this approach can also be found in the 'passive self-assembly' in chemical engineering [1], or robotic assembly and disassembly of passive micro components [6]. There has been also an increasing interest in the use of unconventional materials to overcome some of the underlying problems of conventional approaches, in which, among others, the variations of freeform fabrication techniques were usually employed [11, 3]. While the significant progress has been made in this field recently, there are still a number of challenges in order to substantially increase the adaptability of our robotic systems.

The goal of this paper is to introduce recent development in our self-reconfigurable robot project which makes use of thermoplastic adhesive material. By explaining the basic technologies and case studies, we discuss the benefits and challenges in this approach. Note that this paper focuses only on the conceptual design principles of this approach, with a rather restricted description of technical details. For those interested, technical details can be found in publications [2, 3, 10, 13, 14, 15].

2 Automated Handling of TPAs

Flexible self-reconfiguration exploits physical characteristics of TPAs. TPAs are solvent-free polymer-based materials that exhibit a phase transition within a relatively small range of temperature [9]: From room temperature up to around 60-80 $^{\circ}C$, they are viscoelastic solid, while they become liquid at higher temperatures over 150 $^{\circ}C$. The transition is bi-directional and repeatable such that a solid TPA cube at room temperature, for example, can be heated up and deformed into arbitrary shapes, and the deformed shape can be maintained by being cooled down. Another important mechanical characteristic of TPAs is their adhesion property that can be controlled through the bi-directional phase transition. In the liquid phase, TPAs are highly adhesive to almost any solid materials and bonds can be formed with a high strength when being cooled down. Quantitatively, the bonding strength using TPAs is in the range of 0.01-5 MPa (depending on the type of TPAs and bonded solids) at room temperature, and it can be dramatically reduced by increasing temperature of the bonds (Fig. 1A).

To make use of the unique material properties of TPAs for robotic self-reconfiguration, automated handling technologies for TPAs are necessary onboard a robotic system. Below we explain three important TPA handling processes, i.e. structure formation, connection and disconnection processes.

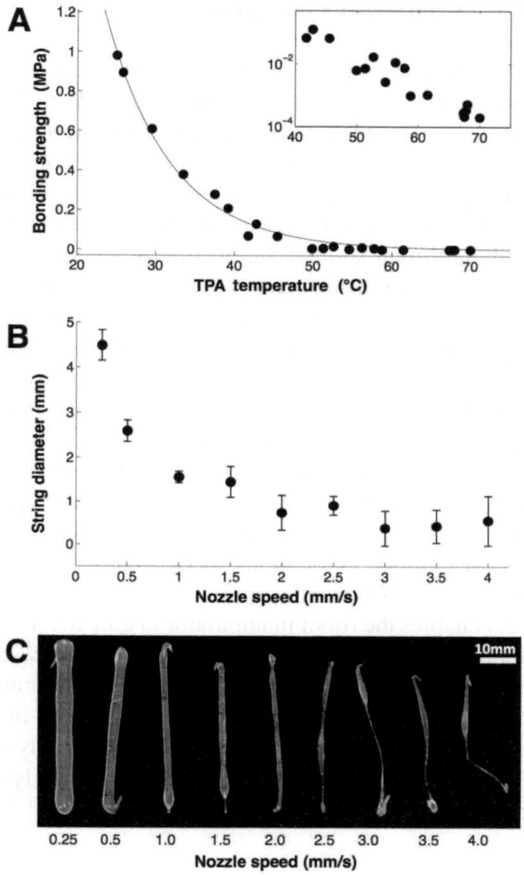

Fig. 1 Exploitation of material properties of TPAs. (A) Bonding strength changes over temperature (reproduced from data in [13]). (B) Diameter of TPA threads formed by FFF with a robotic arm moving at horizontally at different speeds. (C) Examples of TPA threads formed by FFF corresponding to data in (B).

2.1 Active Connection and Disconnection

In order to make use of the unique material properties of TPA in a robotic system, we developed a manipulator that is equipped with a specifically tailored end-effector. The robot manipulator can exploit these mechanical characteristics of TPA for active connection and disconnection processes [13, 14]: By installing a "heating plate" at an end-effector of robotic manipulator (Fig. 2), it is able to liquefy a solid-phase TPA to induce adhesion and bonding to the heating plate when it is cooled down. Similarly, bonding strength can be reduced dramatically by increasing the surface temperature of the heating plate, and by utilizing this characteristic, an attached TPA structure can be easily separated. The cooling processes can be usually achieved passively because the TPA material is solid in room temperature, although the speed can be significantly improved when a robot employs active heat pumps such as Peltier elements.

2.2 Structure Formation

The adhesive property of TPA can also be exploited in another process, i.e. additive fabrication [3]. Here we consider using TPA in the so-called Fused Filament Fabrication [7], in which variations of mechanical structures can be fabricated by placing a liquid-phase TPA flow and solidifying it. For example, when a thread is placed to form a spiral trajectory on a flat table, it cools down into a solid flat disk at room temperature, and an upright wall can be formed by accumulating a thread vertically. Although such additive fabrication processes usually employ other thermoplastics such as acrylonitrile butadiene styrene, we found it possible to use TPA for the same purpose by specifically developing a TPA handling device (TPA Supplier shown in Fig. 2) that regulates the flow of liquid-phase TPA under the kinematic control of a robot manipulator. So far our experimental setup is capable of continuously and smoothly extruding TPA thread with the minimum diameter of 1mm, and the process is destabilized when fabricating with a smaller diameter (Fig. 1B and C). It is also important to mention that both the storage modulus and tensile strength of TPA are as high as 10MPa at room temperature, which allows to flexibly develop mechanical components with practical structural strength.

The additive fabrication process can be further enhanced by using both the TPA Supplier and Connector, which we call a "bonding assembly processes". This process enables the robot manipulator to glue together TPA structures that are independently fabricated, thus more complex structures can be produced. Bonding assembly is also used to connect the fabricated TPA structures to the end-effector of the robot manipulator for task execution. As exemplified in these processes, owing to the mechanical characteristics of TPA, the robotic body extension enriches the variation of mechanical structures that can be autonomously developed on the fly, when compared to the other conventional approaches.

3 Flexible Self-reconfiguration

The proposed TPA handling processes can be used in a variety of self-reconfigurable robot applications, owing to the simplicity of mechanisms. In this section, we introduce three case studies in which robotic systems make use of TPAs for self-fabrication of body parts, connection and disconnection with mechatronic modules, and self-reconfiguration with random objects.

3.1 Self-fabrication of Body Parts

The TPA-based additive fabrication and active connection-disconnection processes can provide a significant flexibility in self-reconfiguration of robotic systems. Here we consider the so-called "robotic body extension" method, which was physically realized based on a position-controlled five-axis robot manipulator that is equipped with TPA Connector and Supplier (Fig. 2A). An external host computer sends operation commands of joint trajectories of the robot manipulator as well as of the TPA

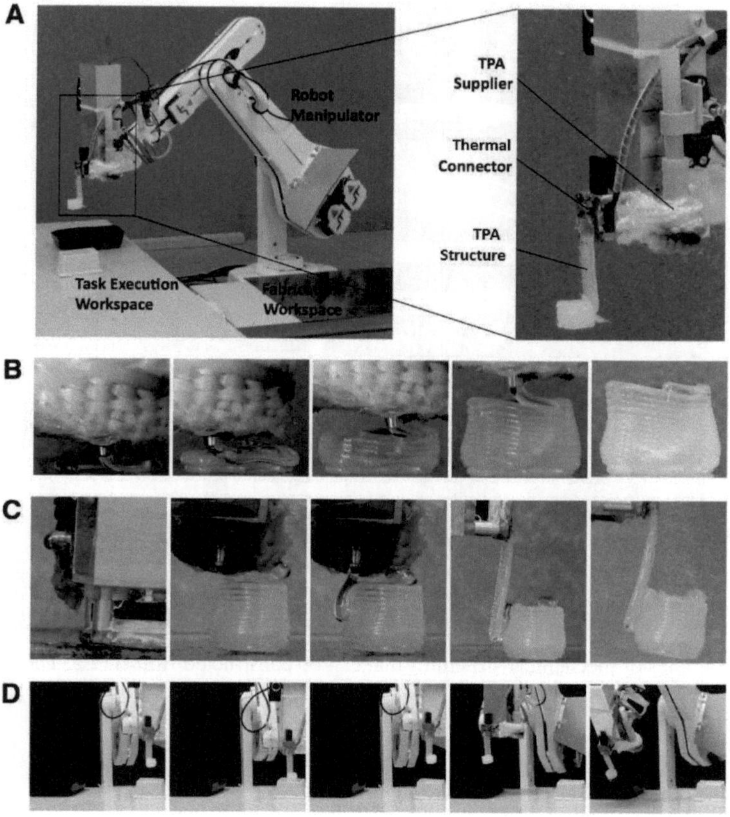

Fig. 2 A robotic manipulator equipped with TPA Supplier and Thermal Connector. (A) A photograph of the robot manipulator and a magnified look of its end effector. (B) The fabrication process. (C) The active connection process. (D) Snapshots of water transportation after body extension with a self-made TPA scoop.

handling processes in TPA Connector and Supplier. For continuous operations over an extended period of time, the robot manipulator is installed in a large experimental arena consisting of fabrication and task-execution workspaces, and all devices are externally powered.

Figure 2 shows one of the representative body extension experiments, in which the robot manipulator fabricated a "spoon" through additive fabrication and bonding assembly processes. The manipulator was initially controlled to fabricate a cup-like structure by using TPA Supplier (Fig. 2B), and then it was bonded to another fabricated stick-like structure to form a spoon (Fig. 2C). When the spoon is completed and connected to the end-effector, the same robot manipulator can start another operation to transport water from a cup to another (Fig. 2D). These processes of robotic body extension were fully automated through a control sequence, and the

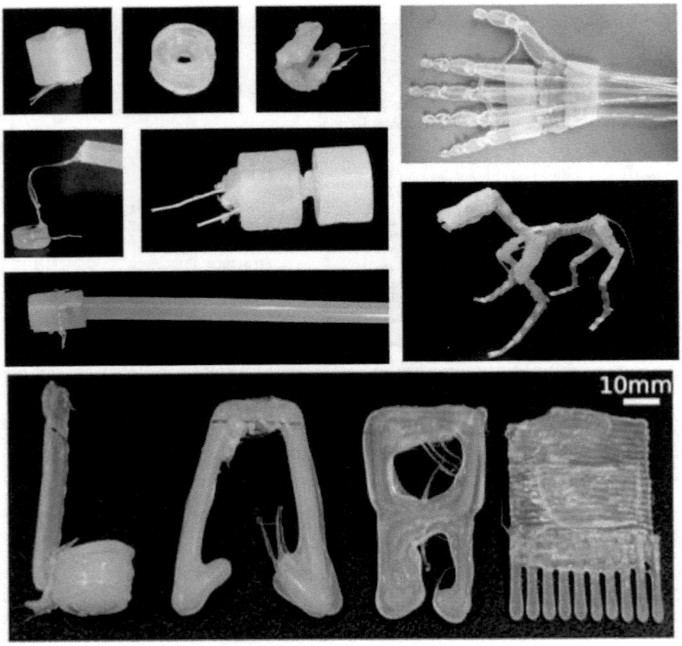

Fig. 3 Examples of mechanical structures that can be constructed from TPAs. The top structures were manually constructed. The bottom four structures were fabricated by the robot manipulator in Fig. 2.

robotic manipulator can autonomously switch between processes such as fabrication of body components, assembly of mechanical structures, or execution of practical tasks.

Compared to the conventional self-reconfigurable robots developed in the past (e.g. [8, 17]), the proposed robotic body extension has a few distinctive characteristics in its self-reconfiguration operations. First, because of the additive fabrication processes based on the TPA material, this approach has a significantly more variety of reconfiguration possible. As shown in Fig. 3, we have conducted a number of different structures that can be useful for the robotic applications by using the proposed approach. Second, owing to the bi-directional and repeatable transitions between adhesive and solid phases, the fabricated structures can be re-structured or re-assembled by controlling the temperature, which is usually not the case with the conventional rapid prototyping methods. This characteristics also enables the robot to connect and disconnect the self-fabricated mechanical structures to its own body. And third, the TPA adhesive property can be applied to many different materials including metal, wood, and other plastics, thus the approach can be used in many variations of reconfiguration scenarios as shown in the next subsections.

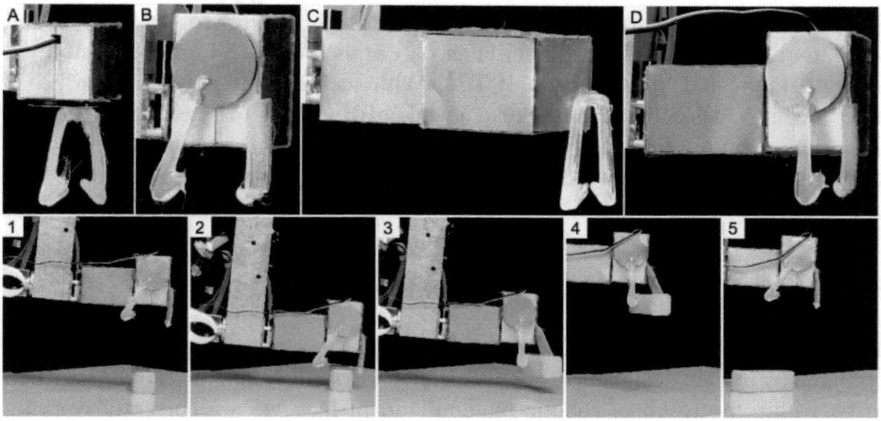

Fig. 4 Demonstration of body extension through TPAs with simple building blocks. (A-D) Four different combinations of passive and active building blocks, as well as self-made TPA parts. Snapshots 1-5 show execution of a pick-and-drop task after body extension with the biggest combination out of the four.

3.2 Self-reconfiguration with Mechatronic Modules

One of the main challenges in the proposed self-reconfigurable robots is that self-fabrication of body parts requires significant amount of time, energy and TPA material in order to fabricate large mechanical structures. For example, the fabrication of a 60mm-long spoon in our robotic setup generally requires approximately 30-40 minutes. Moreover, an additional challenge of the proposed approach is the physical limitations that are originated in the mechanical properties of TPA material. Because the structural strength of TPA material is limited around 10MPa and it cannot be easily increased, the robotic systems are not always able to fabricate the shapes and sizes of structures as they need.

In order to account for this challenge, we have been exploring a solution in which the robotic manipulator could take advantage of existing mechanical structures (Fig. 4; [2]). This case study assumes that, within the reach of robotic manipulator, there exist mechanical structures such as box-like blocks that can be actively connected and disconnected. This method fist employs the TPA Supplier which provide liquified TPA on the target object, and second, uses Active Connector to bond to the object. This method can be also applied to bond an object to another object (instead of the Active Connector), with which the robotic manipulator is able to construct arbitrary beam structures. The limitation of such beam structures depends on the bonding strength of TPA material, while we have so far demonstrated a significant extension of body structures in this approach as shown in Fig. 4.

The benefit of such self-reconfiguration through modules lies also in the fact that the blocks attached to the robotic manipulator can be 'robotized' (e.g. a block containing remotely controlled sensors and motors), and other types of random objects as long as they are made of materials being able to attached by TPA. For example,

Fig. 4 shows a simple block that contains a remotely controlled servomotor, and the output shaft of the motor (the circular part in the object in Fig. 4) is connected to a self-fabricated 'finger' made of TPA. With such a configuration, the robot is able to achieve tasks that cannot be achieved only by using the passive mechanical structures. Figure 4 for example demonstrates an active pick-and-drop operation in which the servomotor is move the TPA fingers to pick a plastic box and drop it in the air.

3.3 *Self-reconfiguration with Random Objects*

The adhesive property of TPA can not only be used for self-reconfiguration of blocks prepared by human designers in advance, but also for random objects that happened to be within the reach of robotic manipulator. In fact, we are currently working on the visually guided robotic manipulator that is capable of recognize stones and wooden pieces on the floor to autonomously construct mechanical structures. This capability is mainly due to the TPA's adhesive property that can be varied over a large range with many different materials. Because its relatively large bonding strength against variations of materials, the proposed self-reconfiguration process can be applied to attachment/detachment of very large modules, and ultimately it becomes climbing locomotion of self-reconfigurable robots.

Figure 5 and 6 show two prototypes of climbing robot based on the TPA handling devices that we developed in our laboratory [10, 15, 16]. The first prototype (Fig. 5) is equipped with an onboard TPA Supplier that is capable of providing liquid TPA between the robot's feet and climbing surface for attaching the body on a

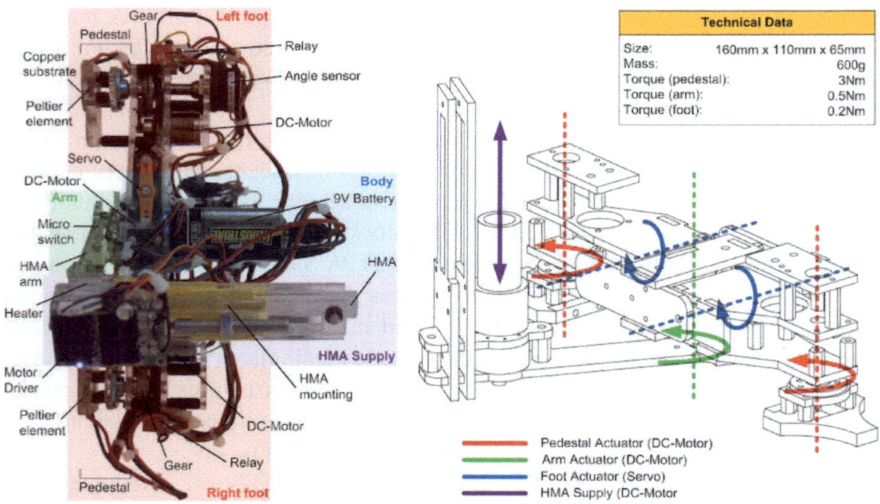

Fig. 5 The first prototype of a mechatronic climbing module that is equipped with Thermal Connectors and TPA Supplier

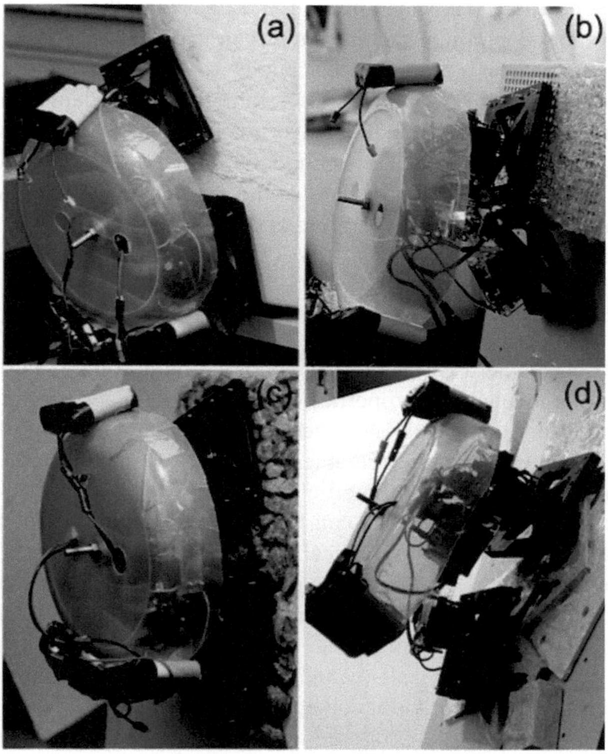

Fig. 6 The third prototype of a mechatronic climbing module that is self-contained and optimized for a large payload and vertical locomotion on complex surfaces

vertical wall. The feet contain a pair of Peltier elements that increase and decrease the connecting surface for attachment and detachment. The second and third prototypes [15, 16] have no TPA Suppliers onboard, thus they rely on preloaded TPAs on the feet or on the surfaces of other modules. Fig. 6 shows the latest prototype that was developed for testing climbing performance in various vertical surfaces. The robot has a dimension of $30 \times 30 \times 15$ cm^3, and it is capable of moving vertically with its own mass of 1.25 kg including the batteries and control electronics onboard.

When employing TPA's adhesive property, there are two important characteristics that need to be considered. First, bonding strength of TPAs usually varies depending on material properties of bonding surfaces: bonding strength to aluminum surface is much lower than wooden one, for example. This characteristics can be exploited when we carefully select materials in the robot. One of our climbing robots, for example, is capable of climbing on aluminum surface without leaving TPA residuals behind, because the connecting surfaces of robot feet are made of copper which exhibits larger bonding strength than that of aluminum. Therefore, when the robot peels off a bonded foot, TPA stays on the foot rather than the climbing aluminum wall [15]. And the second important design principle of TPA-based connection and disconnection mechanism is related to the characteristics of heat conductivity on the connecting surfaces. Because TPA's bonding strength is dependent on material temperature, connection and disconnection performances (such as speed and remaining residuals) largely depend on the heat conductivity of the materials. Usually it is more difficult to disconnect from materials with larger conductivity because it requires more energy to increase temperature at the connecting surface.

4 Conclusion and Perspectives

This paper introduces an approach to self-reconfigurable robots that takes advantage of the unique mechanical properties of thermoplastic adhesive material. The material has both thermoadhesive and thermoplastic characteristics, thus, by controlling material temperature, we are able to regulate connectivity of the material to bonding surfaces as well as geometric structure of the material itself. Furthermore, because these TPA properties can be bi-directional and repeatable, we are able to cyclically exploit them in practical self-reconfiguration tasks. Owing to the unique mechanical characteristics of this material, we were able to develop three TPA handling processes (i.e. active connection, disconnection, and structure forming), which enabled three distinctive case studies of self-reconfigurable robots, i.e. self-fabrication of body parts, self-reconfiguration through mechatronic modules, and self-reconfiguration with random objects.

The main contribution of the proposed approach is largely originated in the simplicity of mechanisms for freeform fabrication and connectivity control. Because the TPA material allows us to control both structural plasticity and connectivity by using the simple processes, we were able to develop small and portable devices that can be implemented into different robotic platforms. The processes of shape and connectivity control are not entirely new as there are a number of rapid prototyping devices and well-established mechanisms for connectivity control available nowadays. The innovations of this particular approach, however, lies in the fact that, because of the simplicity of mechanisms due to the TPA material properties, we are able to implement both plasticity and connectivity control processes in a compact robotic system, which are essential functions in self-reconfigurable robotic systems.

Although our exploration is still in a nascent stage, we are able to foresee a few important challenges based on the case studies so far. First, one the most important challenges is the improvement of TPA material properties because most of the capabilities of our robots rely on the mechanical characteristics of the material. The behavioral performances of our robots such as the sizes, structural strength, and precision of fabricated objects are mostly determined by the tensile modulus of TPA itself, and there are clear limitations as long as we use the material in the same processes. In this sense, it is very challenging for the proposed TPA-based mechanisms to generate connection strength comparable to the conventional mechanical fixation approaches. Second, along with the material-related research, it is also necessary to optimize further the handling technologies of TPA in order to improve the scalability of the proposed approach. In particular, our robotic platforms have relatively limited capabilities in thermo and pressure control of TPA material, which restrict the speed, precision, and size of self-reconfiguration processes. In general, it is a significant challenge to improve speed and precision in the temperature-dependent control approaches. Another important research direction is to consider the technology for 're-morphing' of TPA-based objects: in our current setup, the TPA-based objects fabricated by the robot cannot be modified later, although, with an additional TPA handling device, the robot should be able to partially liquify the previously fabricated structures to shape into another. And third, another exciting challenge will be the design and implementation of more enhanced computational processes. All of the demonstrations shown in our project so far were fully automated although they had significant biases of human designers such as the pre-determined trajectories in fabrication processes and highly structured task-environments. These detailed constraints should be relaxed through the investigations of computer optimization, signal processing, and system identification.

Acknowledgements. This research was supported by the Swiss National Science Foundation Professorship Grant No. PP00P2123387/1, and the ETH Zurich Research Grant ETH-23-10-3.

References

1. Boncheva, M., Bruzewicz, D.A., Whitesides, G.M.: Millimeter-scale self-assembly and its applications. Pure Appl. Chem. 75(5), 621–630 (2003)
2. Brodbeck, L., Iida, F.: Enhanced robotic body extension with modular units. In: Proc. 2012 IEEE/RSJ IROS (2012) (in press)
3. Brodbeck, L., Wang, L., Iida, F.: Robotic body extension based on hot melt adhesives. In: Proc. 2012 IEEE ICRA, pp. 4322–4327 (2012)
4. Chirikjian, G.S.: Reviewing the issues of robotic self-X. IEEE Robot Autom. Mag. 14(4), 6–7 (2007)
5. Christensen, A.L., O'Grady, R., Dorigo, M.: Morphology control in a multirobot system. IEEE Robot Automat. Mag. 14(4), 18–25 (2007)
6. Diller, E., Pawashe, C., Floyd, S., Sitti, M.: Assembly and disassembly of magnetic mobile micro-robots towards deterministic 2-D reconfigurable micro-systems. Int. J. Robot Res. 30, 1667–1680 (2011)
7. Jones, R., Haufe, P., Sells, E., et al.: RepRap: The replicating rapid prototyper. Robotica 29, 177–191 (2011)
8. Kurokawa, H., Tomita, K., Kamimura, A., et al.: Distributed self-reconfiguration of M-TRAN III modular. Int. J. Robot Res. 27(3-4), 373–386 (2008)
9. Li, W., Bouzidi, L., Narine, S.S.: Current research and development status and prospect of Hot-Melt-Adhesives: A review. Ind. Eng. Chem. Res. 47, 7524–7532 (2008)
10. Osswald, M., Iida, F.: A climbing robot based on hot melt adhesion. In: Amato, N. (ed.) Proc. 2011 IEEE/RSJ IROS, pp. 5107–5112 (2011)
11. Revzen, S., Bhoite, M., Macasieb, A., et al.: Structure synthesis on-the-fly in a modular robot. In: Amato, N. (ed.) Proc. 2011 IEEE/RSJ IROS, pp. 4797–4802 (2011)

12. Stoy, K., Brandt, D., Christensen, D.J.: Self-reconfigurable robots: An introduction. MIT Press, Cambridge (2010)
13. Wang, L., Iida, F.: Physical connection and disconnection control based on hot melt adhesives. IEEE-ASME Trans Mechatron (2012a), doi:10.1109/TMECH.2012.2202558
14. Wang, L., Iida, F.: Towards "soft" self-reconfigurable robots. In: Proc. 4th IEEE RAS/EMBS BioRob., pp. 593–598 (2012b)
15. Wang, L., Neuschaefer, F., Bernet, R., et al.: Design considerations for attachment and detachment in robot climbing with hot melt adhesives. In: Proc. 2012 IEEE ICRA, pp. 1181–1186 (2012)
16. Wang, L., Graber, L., Iida, F.: Climbing vertical terrains with a self-contained robot. In: Proc. 2011 IEEE/RSJ IROS 2011 (2012) (in press)
17. Yim, M., Shen, W.M., Salemi, B., et al.: Modular self-reconfigurable robot systems: Challenges and opportunities for the future. IEEE Robot Autom. Mag. 14(1), 43–52 (2007)
18. Yun, S., Rus, D.: Optimal self-assembly of modular manipulators with active and passive modules. Auton. Robot 31(2-3), 183–207 (2011)
19. Zykov, V., Mytilinaios, E., Desnoyer, M., et al.: Evolved and designed self-reproducing modular robotics. IEEE Trans. Robot 23(2), 308–319 (2007)

Self-assembly and Self-reproduction by an M-TRAN Modular Robotic System

Haruhisa Kurokawa, Akiya Kamimura, and Kohji Tomita

Abstract. A feasibility study was made for self-assembly and self-reproduction using an M-TRAN modular robotic system. Presuming that many stand-alone modules are scattered on a flat plane, various models of self-organizing systems have been examined; (1) randomly driven self-assembly, (2) self-assembly accelerated by worker, seed, and mold robots, and (3) self-reproduction based on a universal constructor. Basic functions for the universal constructor were tested by experimentation.

1 Introduction

Self-assembly and self-reproduction are fundamentally important phenomena in biology for maintaining an individual's life (growth and health) and for continuing the evolution of species. Although such phenomena and their underlying principles do not work in general on macro-scale kinetic systems, studies of various kinds have been conducted to simulate or make use of them [1, 4].

One such approach is to simulate self-assembly of mechanical components agitated by a randomly vibrating container [2, 5, 6, 8, 16]. With such a setting, it is not easy to realize natural force among scattered components, comparable to electromagnetic force, to induce accumulation and cohesion.

Swarms of mobile robots cooperating and forming spatial and logical conformations can be viewed as another approach. In this case, each robot has sufficient sensing devices and there can be environmental measuring systems to find the relative location and orientation of robots. With locomotion capability of robots, mutual interaction can be made to simulate either particle dynamics or self-assembly. A few of them, however, have dealt with self-assembly of robots, such that separated robots physically connect to others [3].

Haruhisa Kurokawa · Akiya Kamimura · Kohji Tomita
National Institute of Advanced Industrial Science and Technology (AIST), Tsukuba, Ibaraki, 305-8568 Japan
e-mail: {kurokawa-h,kamimura.a,k.tomita}@aist.go.jp

M.A. Hsieh and G. Chirikjian (eds.), *Distributed Autonomous Robotic Systems*, Springer Tracts in Advanced Robotics 104,
DOI: 10.1007/978-3-642-55146-8_15,

Studies of self-reconfigurable modular robots constitute another approach. There have been, however, few examples of actual hardware realization compared to swarms of mobile robots [12, 19]. Such modules generally work only when they are connected, and an individual module in most systems is not locomotive by itself. In [13, 20], however, separated modular robots reconnected after detecting their relative location using image sensing. Those trials can be regarded as simple modular robotic self-assembly.

Compared to self-assembly as above, self-replication and self-reproduction are much less studied on a macro-scale [1]. In some studies such as [11, 10], the environment in which a parent robot works seems so structured that components to be assembled are supplied in a predefined manner. The environment, not the robot, possesses all necessary information, hence the term 'self' is vague. Moreover, the von Neumann's well-known principle of a self-reproducing automaton in [18] is little considered. In the case of modular robotic approaches such as in [21], modules cooperate as components with a constructor robot, performing comprehensive self-reconfiguration. Therefore, the distinction between a constructor, components, and an instruction is not clear.

We intend to design several types of self-assembly and self-reproduction using M-TRAN modules, to incorporate all processes together in the hope of drawing a feasible evolutional history from self-assembly to self-reproduction. Compared to each study described above where components and the environment are designed independently for each process, we assume an environment common to all processes and finally propose self-reproduction using a universal constructor. Though actual controllers and algorithms were not developed for most of the following systems, basic experiments were made to verify kinetic functions for self-reproduction.

2 M-TRAN Modular Robot

The M-TRAN system is based on the cubic lattice with each module composed of a pair of semi-cubic blocks (half-cylinder and half-cube), called passive and active blocks, connected by a link block via two rotary actuators (Fig. 1(a)) [9, 14, 15].

Modules connect via connection interfaces (flat surfaces), where only active (dark color) and passive (white) blocks are mutually attached (connection polarity). The modules can be assembled into robots of various configurations. Examples of robotic forms that are capable of locomotion are shown in Fig. 1(b) and Figs. 2(d), (f), (g), and (i) [7]. Robots in various configurations can self-reconfigure into others as shown in Fig. 1 and Fig. 2. All of those were verified through experimentation [9].

Generally a modular robot does not separate out, which has advantages: 1) relative location of a module is in principle obtainable, and is easily controlled fundamentally in the case of a lattice-based system; 2) communication between any two modules is possible; and 3) synchronization, cooperation, and centralized control is easily attainable as long as not too many modules are used, etc.

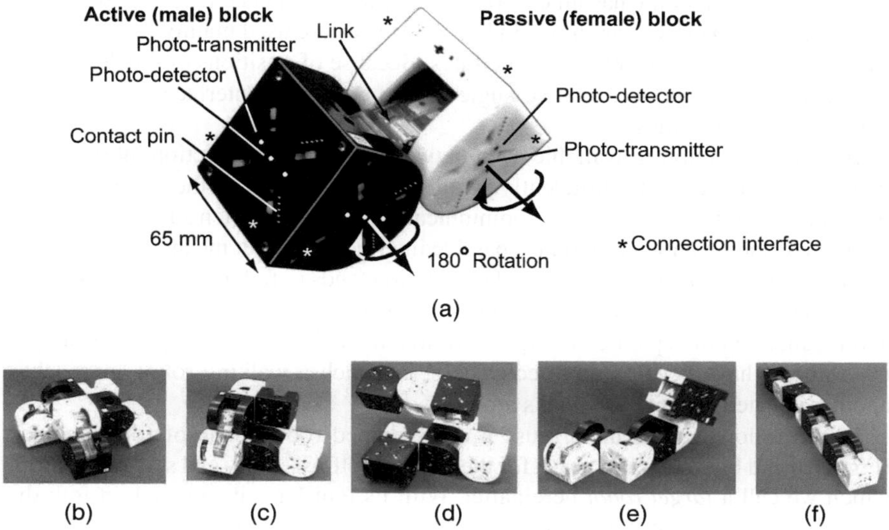

Fig. 1 M-TRAN module and self-reconfiguration. (a) Single hardware module. (b)–(f) Experiment of self-reconfiguration from a four legged robot to a linear robot. A reversal process of this, from (f) to (b), was also verified by experiments.

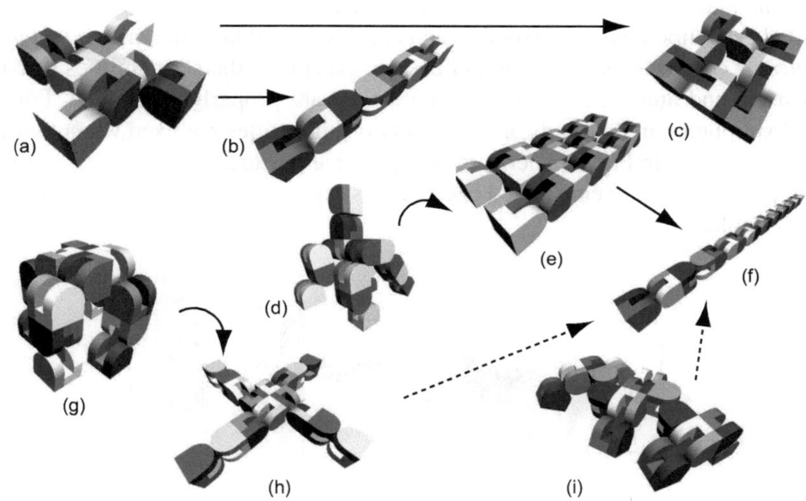

Fig. 2 M-TRAN configuration and self-reconfiguration: (a), (d), (g), and (i) show robotic forms capable of locomotion. Curved arrow: transformation simply by joint motion. Straight solid arrow: self-reconfiguration. Broken arrow: reconfiguration to serial forms similar to (f).

An M-TRAN module has nine infrared (IR) transmitters and detectors placed on six connection interfaces as shown in Fig. 1(a). They are used mainly for communication between directly connected modules. Because of positions of the devices on each connection interface, relative angle of two connected interfaces is detectable among four possibilities.

IR communication is available even between detached connection interfaces, if they are proximate to and sufficiently parallel to each other. With the current built-in program, separate modules try to communicate and connect when a module's active interface is facing another module's passive interface. Without means of sensing the position and alignment, such a trial of connection tends to fail, but failure can be detected and retries can be made automatically. The IR signal is also used for detection and localization of separated modules. Although localization is not precise, simple experiments have verified its effectiveness, in which a walking robot approaches another isolated module and walks around it.

In the following, we mainly use a cross-shaped robot made of four modules (Fig. 1(b) and Fig. 2(a)) as a target for the tasks of self-assembly and self-replication, which we call a *target robot* hereinafter. With its four legs, it can walk in four directions and make a turn using simple gate patterns. Each flat surface of the legs is equipped with IR devices for detection of and communication with another module. Additionally, it is presumed that its joint controller can detect an object that obstructs joint motion.

We presume in the following that there is an ideal flat ground on which target robots and many stand-alone modules are scattered. Those stand-alone modules are regarded as components for assembly and are designated hereinafter as *component modules* or *components*. They remain passive except for their capabilities of communication and automatic connection until they are properly assembled. For the sake of simplicity, in this study, all of their two-joint angles are fixed when they are inactive as shown in Fig. 3. There are four possible postures, but only two of them, p1 and p2 in Fig. 3, are considered because of simplicity and because the other two are less stable.

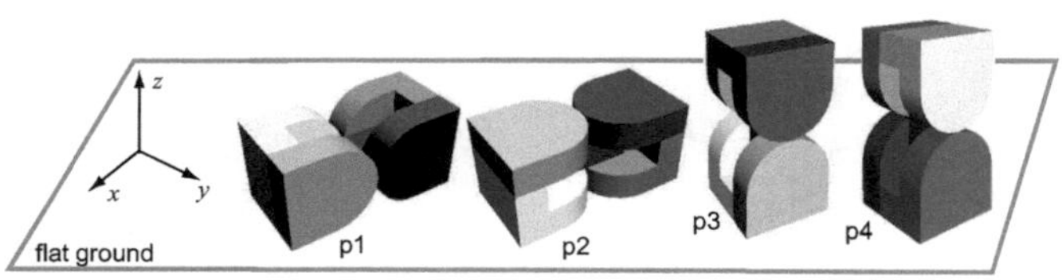

Fig. 3 Components on flat ground. There are four possible postures on the ground.

3 Self-assembly

3.1 Stochastic Assembly

As a preliminary study, consider self-assembly similar to [2, 5, 6, 8], i.e., component modules are placed on a flat bed, which is shaken randomly to give components perturbation and agitation. Two components might accidentally collide, and though less likely, connect to form several configurations (Fig. 4). When the formed configuration does not match a part of the target, it should be dissolved. Consequently, the yield of this stochastic assembly will be very small.

Instead of using a vibrating bed, a target robot is useful to assist and to accelerate assembly as a worker. It is sufficiently dexterous to walk, to approach, to push, and to transfer a component module (Fig. 5(a) and 5(b)). Of course a component module must be detected and the relative position and orientation are properly measured. Such measurement can be done, in principle, by walking around and sensing an IR signal from a component by many detectors equipped with the walking robot.

Because the components are assumed to be scattered randomly, the relative position and orientation of components must be adjusted. Pushing by a robot can change the position and orientation of a component on the flat surface. Changing the posture of a component between p1 and p2 in Fig. 3 is not straightforward, but with sequences of self-reconfiguration, the worker (target) robot can make such a change (Figs. 5(c), (d)).

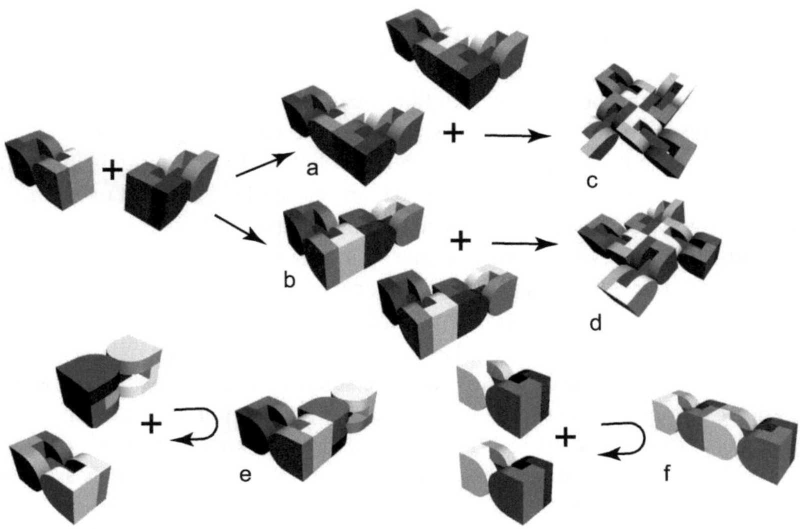

Fig. 4 Examples of random assembly. Configurations e and f are not included in the target and to be dissolved. Note that the robots d is a mirror image of c, and at the same time equivalent to c turned upside down. In this paper, they both are regarded as targets.

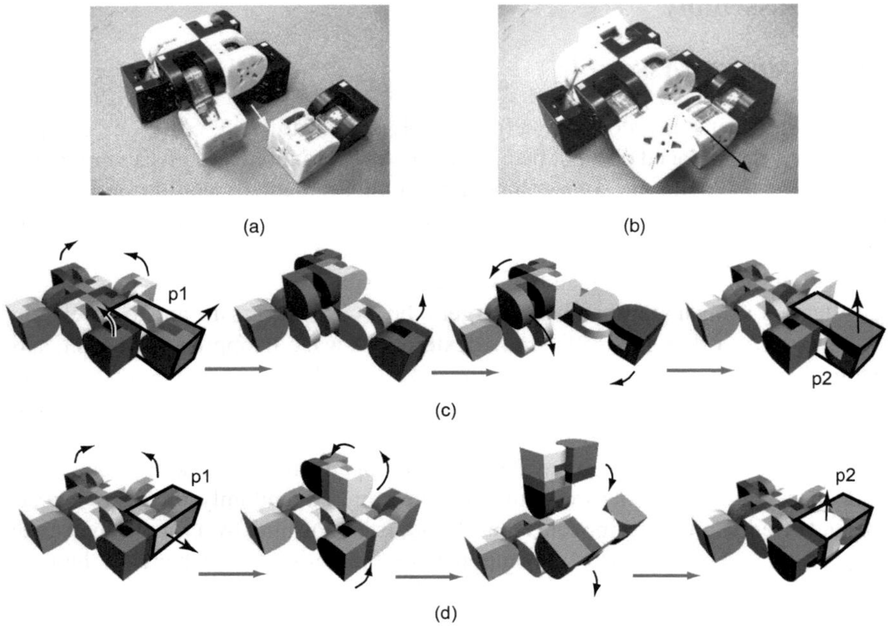

Fig. 5 Worker robot: (a) approach, (b) transportation, (c), (d) posture change

3.2 Seed and Mold

In the above process, the robot works like a catalyst or an enzyme in a chemical or biological reaction [17]. The success rate, i.e., the yield of reaction process, will be still low, because the robot's sensing ability is low and positioning is inaccurate. The following two ideas, a seed and a mold, are more like a catalyst or an enzyme utilizing their geometric affinities in accelerating the assembly process.

Presuming that there are two robots, one acting as a worker and another as a seed for assembly. A worker carries and connects components to a seed one by one, then the seed grows as shown in Fig. 6. The assembly process can be controlled such that the seed sends a message to the worker, which includes the stage of assembly and which is distinct in each direction to guide the subsequent assembly by the worker.

Geometry of the seed is crucial for a successful connection. In the assembly path from (a) to (c) or (d) via (b) in Fig. 6, for example, connection of A or B in (b) might mostly fail. Rather more successful will be a new module, such as C or D in (f), placed at and pushed to a concave corner of (a). Therefore, guiding along the latter path from (a) to (g) via (f) will be more successful than the former from (a) to (c), though the products (e), (h), and (j) need to reconfigure to the target configuration.

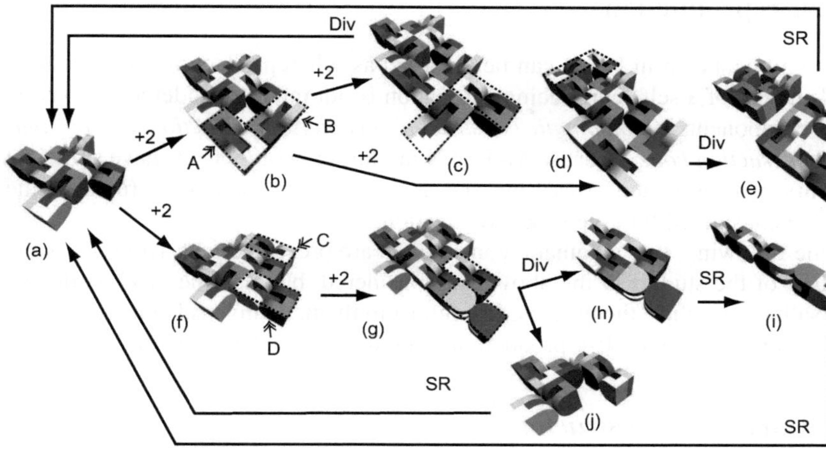

Fig. 6 Assembly pathes using a seed robot. At each arrow, '+2' implies that two modules are added, and 'Div' and 'SR' respectively imply division and self-reconfiguration.

Such concavity, as used in [13, 16], can be utilized by another structure shown in Figs. 7(a)–(e), which works as a universal, variable shape mold for assembly. This chain can also work as a guide along which a worker robot travels, maintaining physical contact. If a part of modules in a chain is twisted as in Fig. 7(f), then more effective guidance by IR communication is possible, although molding becomes less universal.

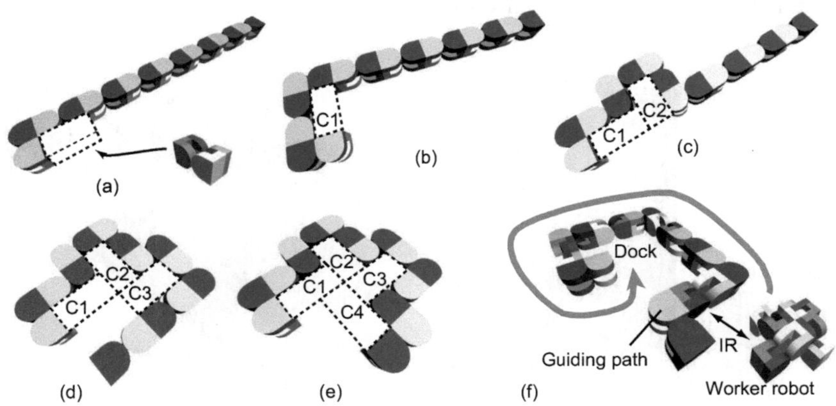

Fig. 7 Two-dimensional universal mold. (a)–(e) The chain needs to change its shape as components are properly assembled. (f) Similar chain capable of emitting IR signal to guide a worker robot.

4 Self-reproduction

The assembly shown in Fig. 6 can be regarded as self-reproduction, but von Neumann's model of a self-reproducing automaton is not much considered. It consists of four components: an *automatic constructor*, a *copier of instruction*, a *controller*, and an *instruction code*. Because M-TRAN modules as components have sufficient capability of information processing and control, an automatic constructor is the first to be considered for kinetic self-replication.

In the following, two automatic constructors are presented first. Not only is the condition of the studies in the above not considered, but also they lack sufficient universality to produce themselves. In contrast to them, a universal constructor in a reduced dimension and self-reproduction using it will then be proposed.

4.1 Automatic Constructor

Making an automatic constructor is straightforward using a lattice-based self-reconfigurable modular robotic system, if components are the same modules as the constructor and if they are supplied on the lattice grids. One such example is a three-dimensional manipulator as shown in Fig. 8(a).

A three-dimensional manipulator, however, requires hardware capability in powerful actuation and precision control, which is beyond most modular robots currently available. More realistic is two-dimensional construction, which suffices for universal construction, because many three-dimensional robots can be transformed from two-dimensional ones as in Fig. 2.

Fig. 8(b) is an example, which can construct most two-dimensional structures within its size. The process of construction is the following: 1) Components to be assembled are supplied on the base structure. 2) Each component (module) moves

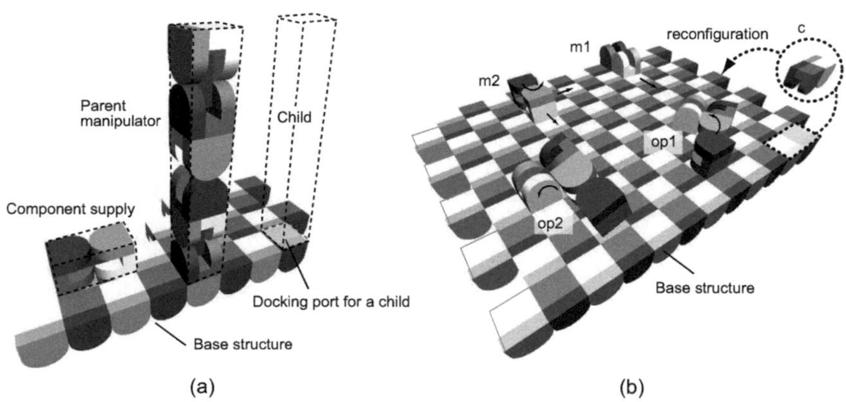

Fig. 8 Automatic constructor: (a) three-dimensional, (b) two-dimensional. In (b), op1 changes the line of m1's motion and op2 converts between m1 and m2.

by itself (basic motions m1 or m2 in Fig. 8(b)) or a pair of modules work in cooperation (basic motions op1, and op2 in Fig. 8(b)) to reach the desired position and in the desired orientation [14].

For these two cases of 3D and 2D constructors, the environmental conditions in the previous sections are not considered, and components may be either floating in the air, and approach and connect to the base structure, or they might be supplied from the base structure made of modules by reconfiguration (Fig. 8(b)). In addition, in the construction process, components do not remain passive but instead actively participate in construction. Also, they are not useful for self-reproduction because they only assemble structures that are smaller than themselves.

4.2 Universal Constructor

An example of a universal constructor is portrayed in Fig. 9(a), which can, in principle, make any serial structure in any length. It has two manipulators M1 and M2, and a port where an assembled serial chain of components are supported. This constructor works on a flat ground, and components remain passive during the assembly process.

Assume that there is a robot, i.e., the worker robot in the previous section, which carries a component to the universal constructor one by one. Once a component is in its reach, the universal constructor tries blind actions of dragging and pushing to set the component to either position A or B in Fig. 10. There are four possible orientations of the component at each position as in Fig. 10(c). By other trial motions of the manipulators M1 and M2 such as those shown in Fig. 11, current orientation and posture of the component are determined. Then the component is manipulated and aligned to be at the position A in Fig. 10(a) and in either the p1 or p2 posture in a proper polarity order. Finally, the component is connected to a semi-assembled product supported at the port in Fig. 9(a), and pushed at the place of the former semi-assembled product.

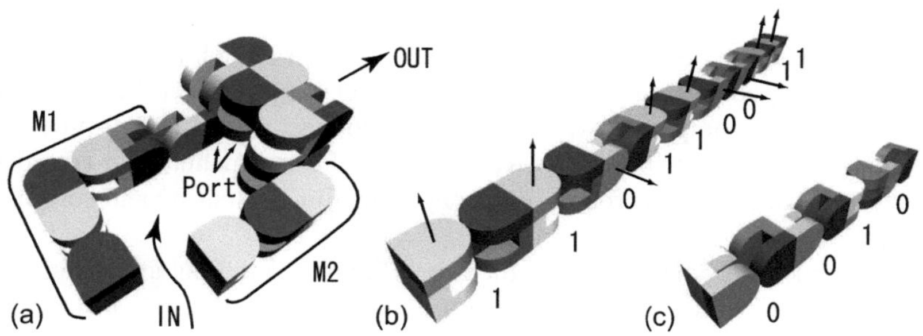

Fig. 9 1-D universal constructor: (a) Manipulators M1 and M2, and a port used to hold a product. (b) Instruction code for construction, with 0 and 1 respectively for the posture p1 and p2 in Fig. 3. (c) Code for the robot in Fig. 1(f).

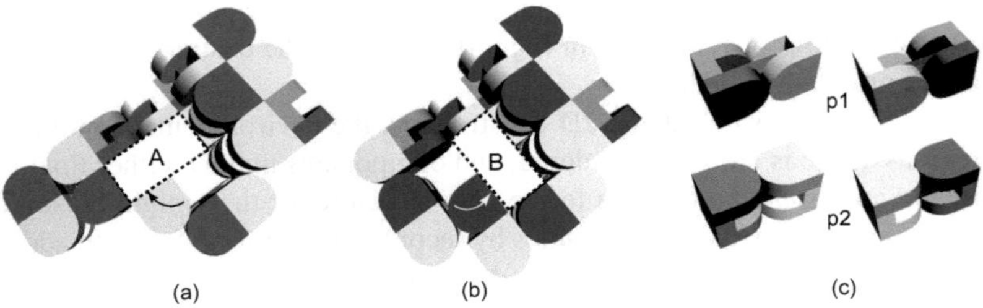

Fig. 10 Sensing and aligning a component. (a)(b) By pushing a component using two manipulators (M1 and M2 in Fig. 9), a component, if any, is detected and aligned. (c) Four possible orientations of a component for each position A in (a) or B in (b).

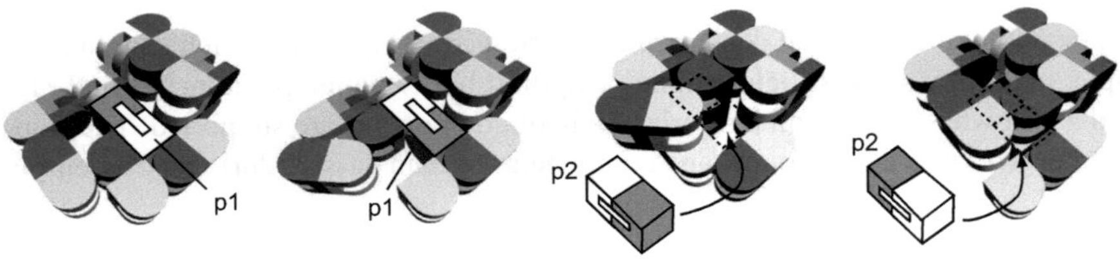

Fig. 11 Orientation sensing and manipulation. When a module is at B in Fig. 10(b), four configurations of the manipulators are tried, each of which can connect to a module in one of four orientations in Fig. 10(c). After connection, the component can be manipulated to the desired position and orientation.

A product of the construction is a serial chain of modules, each in the desired posture, either p1 or p2, which can be represented by a sequence of two numerals, 0 and 1, as presented in Figs. 9(b) and 9(c). For example, the code 0010 represents a serial robot in Fig. 9(c), which is transformable to a cross shaped, target robot as in Fig. 1.

Two experiments were conducted successfully, which verify the feasibility of the construction process. Fig. 12(a) shows an experiment in which a component was dragged, aligned, connected, and pushed. In another experiment in Fig. 12(b), the posture of the new component was changed and connected to the semi-product of two components. These experiments were conducted using manual, not automatic, control. For an autonomous process, elaborate programs for sensing and control are necessary.

Consequently, this constructor has the capability of assembling any serial structure as long as the code, which is a binary number, is provided. Considering the self-reconfiguration capability described in Sect. 2, it can assemble not only a target robot but also any robot in Fig. 2. Moreover, it can reproduce itself because this constructor is indeed a serial chain made of nine modules as shown in Fig. 9 (b). This serial robot can also work as a worker robot with its snake-like form or with another

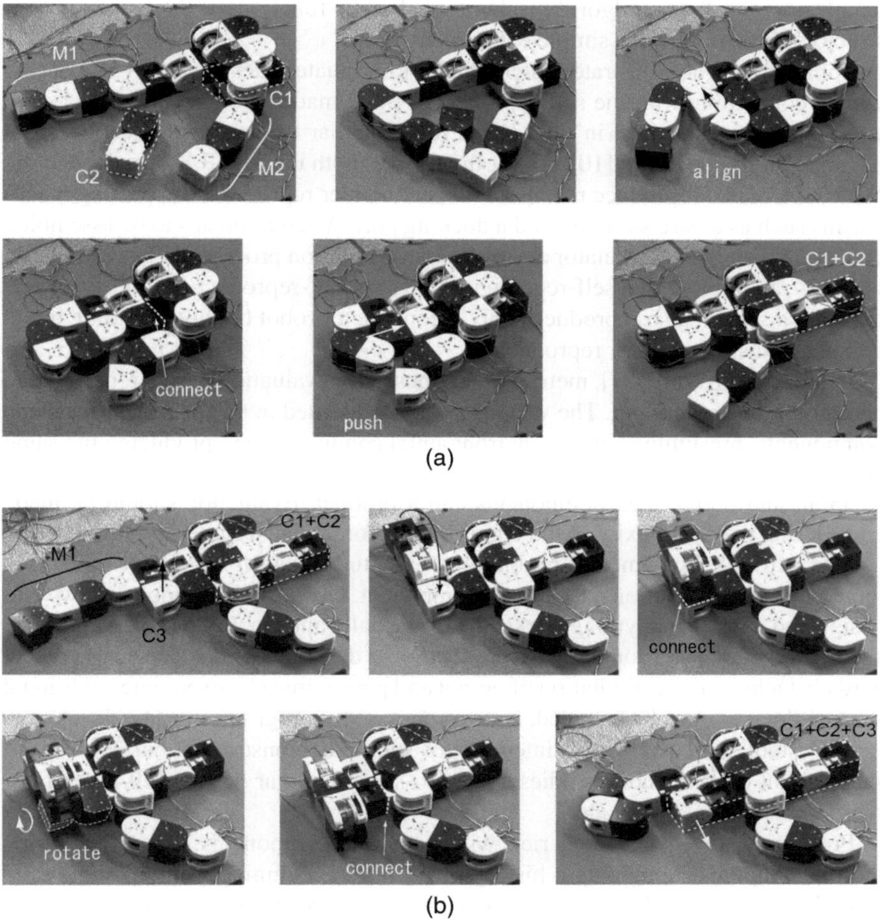

Fig. 12 Experiments of serial construction: (a) the component C2 is dragged, aligned, connected to C1, and pushed to a new position; (b) Posture of C3 is changed by M1. (Video available at http://staff.aist.go.jp/kurokawa-h/self-repMTRAN.mp4)

dexterous form after self-reconfiguration. Considering two such robots, a worker robot and a constructor, as a group, this group, as an individual, self-reproduces.

5 Discussion and Conclusion

Several models and scenarios of self-assembly and self-reproduction using an M-TRAN modular robotic system are presented. Most of the models are originated from other studies, but the final self-reproducing modular system based on a universal constructor is original and unique among others. All of them are outlined without details of controllers or algorithms, but basic kinetic functions of the universal constructor are verified using simple experiments.

Some models are illustrated intentionally to evaluate other works. Figure 8(a) is almost compatible with the self-reproducing robot made by Molecube in [11, 21], and the system presented in Fig. 7(f) is made similar to the self-replicating robot having a guiding track in [10]. As for the former, both the M-TRAN model and the Molecube model reproduce manipulators, but neither reproduces supporting mechanisms such as a base structure and a docking port. As components to be assembled cooperate with the manipulator during their construction process, the operation as a whole is rather more of a self-reconfiguration than self-reproduction. As for the latter, such a system also reproduces only a constructor robot but not a guiding track, which is indispensable for reproduction.

In the studies as [10, 21], metrics are proposed for evaluation of self-reproduction based on input and output. The various models presented in this paper use the same components, presuming similar environmental conditions, and producing the same target robot. Evaluation of these different models, therefore, cannot be made merely by input and output, i.e., components and a product. Evaluation should be made considering the complexity of the controller and/or the environment.

It is surely a fundamental problem of this study that an M-TRAN module is too complex and functional as a mere component. Given such components, we can design and construct anything physically or virtually and we will be able to produce a hardware demonstration of self-reproduction as described in this paper. Of course, hard obstacles, such as limitation of output and power, must be overcome, additional sensing devices must be installed, and various control programs must be developed.

This study, however, is not intended for such a demonstration, but is meant to shed a light on how future studies of macro-scale modular robotic self-assembly / reproduction should be made.

Presuming an environment rich with primitive components, similar to a primordial soup, and imagining a history of evolution beginning from a simple self-assembling compound of a few components to a complex self-reproducing entity. The question is how such evolution is feasible. If there is a universal constructor in the last section, evolution of serial robots, as its product, is easily understood because the code directly represents the robot. As described in Sect. 3.2, however, a serial chain seems less feasible to appear in the early stage, because of its simple geometry. Therefore, all the models in this paper might not be sufficient to make up a likely evolution history.

A more difficult problem is evolution of controllers, which need to take place simultaneously with the evolution of morphology and function. Some controllers, such as a reproduction process as a lattice-type self-reconfiguration as in [21] and a locomotion controller based on Central Pattern Generator as in [7], are by themselves suitable for evolutional computation such as Genetic Algorithm. For simple controllers based on 'if-then' rules, other methods such as Genetic Programming also can be useful tools. Simply combining all such tools, however, is not appropriate, because controllers for positioning, manipulation, and locomotion, and algorithms including decision making are all to be included, and each tool requires its own objective function. We need an intrinsic, unified system model with a unique objective function, which is as simple as possible to reduce calculation time, but is general enough to cover necessary functions.

Even with the above difficulties, not only M-TRAN but also any other self-reconfigurable modular robot, either homogeneous or heterogeneous, will be a useful platform for constructive studies seeking a feasible scenario, partially at least, of artificial evolution in morphology, function, and control.

References

1. Freitas Jr., R., Merkle, R.: Kinematic Self-Replicating Machines. Landes Bioscience (2004)
2. Griffice, S., Goldwater, D., Jacobson, J.M.: Self-replication from random parts. Nature 636, 437–7059 (2005)
3. Gros, R., Bonani, M., Mondada, F., Dorigo, M.: Autonomous Self-Assembly in Swarm-Bots. IEEE Trans. Robotics 22(6), 1115–1130 (2006)
4. Gross, R., Dorigo, M.: Self-assembly at the macroscopic scale. Proc. IEEE 96(9), 1490–1508 (2008)
5. Hosokawa, K., Shimoyama, I., Miura, H.: Two-dimensional micro-self-assembly using the surface tension of water. Sensors and Actuators 57, 117–125 (1996)
6. Kaloutsakis, G., Chirikjian, G.S.: A stochastic self-replicating robot capable of hierarchical assembly. Robotica 29, 137–152 (2011)
7. Kamimura, A., Kurokawa, H., Yoshida, E., Murata, S., Tomita, K., Kokaji, S.: Automatic locomotion design and experiments for a modular robotic system. IEEE/ASME Trans. Mechatron 10(3), 314–325 (2005)
8. Klavins, E., Ghrist, R., Lipsky, D.: Graph Grammars for Self Assembling Robotic Systems. In: ICRA 2004, pp. 5293–5300 (2004)
9. Kurokawa, H., Tomita, K., Kamimura, A., Kokaji, S., Hasuo, T., Murata, S.: Distributed Self-reconfiguration of M-TRAN III Modular Robotic System. Intl J. Robotics Res. 27(3-4), 373–386 (2008)
10. Lee, K., Chirikjian, G.S.: Robotic Self-Replication. IEEE Robotics Automat. Mag., 34–43 (December 2007)
11. Lipson, H., Pollack, J.B.: Self-reproducing Machines. Nature 435, 163–164 (2005)
12. Murata, S., Kakomura, K.: Self-Reconfigurable Robot: Shape-Changing Cellular Robots Can Exceed Conventional Robot Flexibility. IEEE Robotics Automat. Mag., 71–78 (March 2007)
13. Murata, S., Kakomura, K., Kurokawa, H.: Toward a Scalable Modular Robotic System. IEEE Robotics Automat. Mag., 56–63 (December 2007)
14. Murata, S., Yoshida, E., Kamimura, A., Kurokawa, H., Tomita, K., Kokaji, S.: M-TRAN: Self-Reconfigurable Modular Robotic System. IEEE/ASME Trans. Mechatron 7(4), 431–441 (2002)
15. M-TRAN (Modular Transformer), `http://unit.aist.go.jp/is/frrg/dsysd/mtran3/`
16. Penrose, L.S.: Self-reproducing Machines. Sci. Amer. 200, 105–114 (1959)

17. Rubenstein, M., Krivokon, M., Shen, W.M.: Robotic Enzyme-Based Autonomous Self-Replication. In: Proc. IROS, pp. 2661–2666 (2004)
18. von Neumann, J.: Theory of Self-Reproducing Automata. Univ. Illinois Press (1966)
19. Yim, M., et al.: Modular Self-Reconfigurable Robot Systems: Challenges and Opportunities for the Future. IEEE Robotics Automat. Mag., 43–52 (March 2007)
20. Yim, M., Shirmohammadi, B., Sastra, J., Park, M., Dugan, M., Taylor, C.J.: Towards Robotic Self-reassembly After Explosion. In: IROS 2007, pp. 2767–2772 (2007)
21. Zykov, V., Desnoyer, M., Lipson, H.: Evolved and Designed Self-Reproducing Modular Robotics. IEEE Trans. Robotics 23(2), 308–319 (2007)

Self-assembly in Heterogeneous Modular Robots

Wenguo Liu and Alan F.T. Winfield

Abstract. This paper describes the distributed self-assembly of a multi-robot 'organism' from a swarm of autonomous heterogeneous modular mobile robots. A distributed self-assembly strategy based on a symbol sequence representation is proposed. Constructed from a tree representation of an organism, the symbol sequence is presented as a well organised nested structure in a compact format. It includes not only information on the topology of the organism but also how the organism will be self-assembled. The proposed approach has been tested with real robot prototypes. Results show that robots can successfully self-assemble to required target body plans within certain time frames.

1 Introduction

Self-reconfigurable modular robots has became a popular research topic since late 1980's. From the CEBOT [1] to the SYMBRION project [2], more than 30 systems have been developed over 2 decades. Although earlier studies focused mainly on solving the mechanical engineering challenges for designing the hardware, recent research has paid more attention to high level algorithms toward greater controllability of self-assembly, self-reconfigurable and self-repair of the system. The complexity of the modules and the architecture of the system vary in different studies. Some use a lattice architecture where docking/undocking can only occur at points within some virtual cells. The lattice architecture requires a simpler mechanical design and simplifies the computational representation, thus the reconfiguration planning is more achievable and scalable. Some other systems do not use the virtual cell as the docking point for their units. Instead, a number of modules can form a chain to reach any point in the operating space; the so called chain architecture. To

Wenguo Liu · Alan F.T. Winfield
Bristol Robotics Laboratory, University of the West of England, Bristol, UK
e-mail: wenguo.liu@brl.ac.uk, alan.winfield@uwe.ac.uk

M.A. Hsieh and G. Chirikjian (eds.), *Distributed Autonomous Robotic Systems*,
Springer Tracts in Advanced Robotics 104,
DOI: 10.1007/978-3-642-55146-8_16, © Springer-Verlag Berlin Heidelberg 2014

get to a specific point and carry out reconfiguration, the chain architecture requires a more complicated design with additional sensing. To overcome the limitation of both approaches, a hybrid architecture is favoured by many recent designs. Among these, a singular design of individual module is normally used, *i.e.*, a homogeneous design. Apart from the essential functionality requirement of the modules, *e.g.*, mechanical docking, some degree of freedom of rotation, robots have been provided with a very limited range of sensors as appropriate to the design challenge and research interests. Also, few designs have considered providing individual robots with autonomous motion capabilities. This typically requires that modules have to be manually attached to each other prior to any reconfiguration process. The lack of sensing and mobility limits the possibility of re-assembling after pre-assembled structures have fallen apart either purposely or accidentally.

To close the gap between modular robotic systems and multiple/swarm robotic systems, researchers are seeking new designs that are capable of autonomous self-assembly and self-reconfiguration. The SamBot [3] project consists of a group of identical robots. Each robot is equipped with two differential driven wheels and several IR sensors for autonomous self-assembly. As with many other modular robotic systems, a rotation arm is used in SamBot to provide one extra degree of rotation freedom thus enabling self-reconfiguration capability. In contrast, the SYMBRION project (www.symbrion.eu) takes a heterogeneous approach, and robots with more sensing, computation and actuation. Various modules with complementary motion capabilities have been developed for this project. The main focus of SYMBRION is to investigate the controllability and evolvability of artificial multi-cellular organisms from both engineering and scientific perspectives. The SYMBRION robots can either work fully autonomously with their own sensing and locomotion capabilities to explore the environment, or physically dock with each other to form different organism structures to accomplish more complex tasks that a single robot is not capable of, for example, climbing a wall or moving over a gap.

One of the fundamental requirements of the SYMBRION project is that robots must be able to self-assemble into a given structure (body plan) starting from a single module. To achieve this, apart from the autonomous docking/undocking capabilities, we need the right morphology control mechanism. A bio-inspired gradient based process has been widely used to study the pattern growth problem in agent-based cell systems [5, 4, 6], and later adapted to the modular robotic systems [7]. This approach has however been tested only in simulation and for homogeneous systems. In addition, the uncertainty of the final generated body shapes imposes another challenge for controlling the macro-locomotion of the whole structure at a later stage. An alternative is to build the target shapes, which are known to be controllable, from some stored pre-defined body shapes. A SWARMMORPH-script language has been proposed by Christensen for arbitrary morphology generation for a group of *s-bot* robots in a 2D environment [8]. The morphologies are pre-specified as sets of rules stored in scripts which can be communicated and subsequently executed on the newly connected robot. Note that unlike the *s-bot* robots, the SYMBRION robots will initially form a 2D planar structure and then lift itself from a 2D planar configuration to a 3D configuration and, with respect to locomotion, will function as a macroscopic whole. The aggregated organism will also be able to disassemble and reassemble into different morphologies to fit the requirements of the task.

In our previous work, an ID-based strategy was proposed to self-assemble the SYMBRION robots into certain predefined 2D structures [9]. The ID-based approach assumed that all robots in the swarm are identical and each robot stores the same set of predefined structures information. In addition, this strategy allows only one robot to join the organism at a time, using a fixed docking face, more specifically, the Front side, which constrains the topology of the shapes the swarm can self-assemble into. This paper will extend the previous work by removing all these assumptions and limitations. A new strategy will be developed to enable parallel recruiting and any-side-docking in order to improve the efficiency and robustness of the self-assembly strategy. In particular, the new propsed strategy will take the heterogeneity of the system into account.

The rest of the paper is organised as follows: Section 2 introduces the robots of the SYMBRION project and the hardware configuration for autonomous docking. Section 3 proposes the internal representation of organism body plan for the self-assembly process. Section 4 outlines the controller framework and discusses the morphology control strategy. The proposed approach is validated in Section 5 using simulation and real robot experiments. The paper ends by drawing some conclusions and further work in Section 6.

2 The Robots

Figure 1 shows three types of robots developed in the project. These robots have different shapes and are equipped with different locomotion actuators. The Backbone robot has two specially designed wheels which allow the robot to move forwards, backwards and sideways; this robot requires a flat surface. With its tracked locomotion the Scout robot can move across uneven surfaces and is suitable for exploration tasks. Like many other modular robots, both Backbone and Scout robots have cubical shapes with four docking faces and one degree of freedom of bending. The ActiveWheel robot is designed to carry and transport an organism consisting of several Scout or Backbone robots in the most energy-efficient way. It consists of two symmetrically arranged arms, connected via a 180° turning hinge, and 4 omniwheels. Two docking elements are placed on the same axis of the hinge for docking with other robots.

Some unified docking elements are placed on all three types of robots to allow stable physical connections between robots. In addition, electrical contacts next to the docking units can be coupled automatically to provide inter-robot communication and power sharing busses between two connected robots. As shown in Figure 2, two versions of docking elements are installed on robots: an active docking unit and a passive docking unit. By removing the locking mechanism, passive docking units can reduce the size of the assembly while still providing all other mechanical and

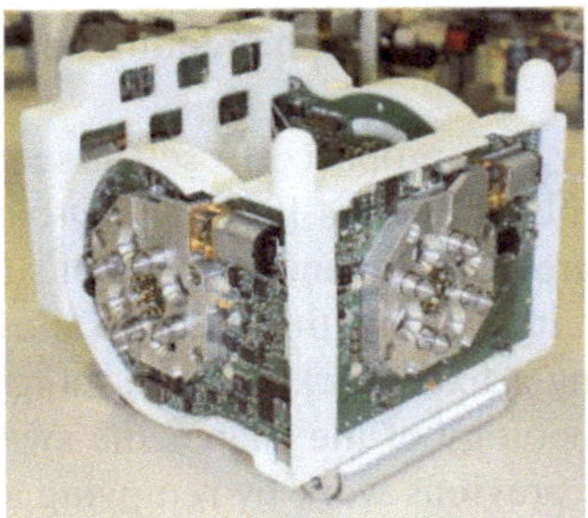

(a) BackBone robot

(b) Scout Robot

(c) ActiveWheel Robot

Fig. 1. Robot prototypes. The BackBone robot has 4 active docking faces. The Scout robot has 2 active docking faces and 2 passive docking faces, while the ActiveWheel has only 2 passive docking faces.

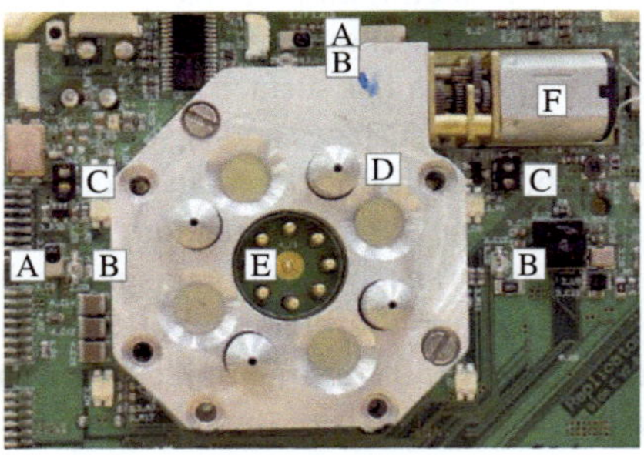

(a) Active docking unit

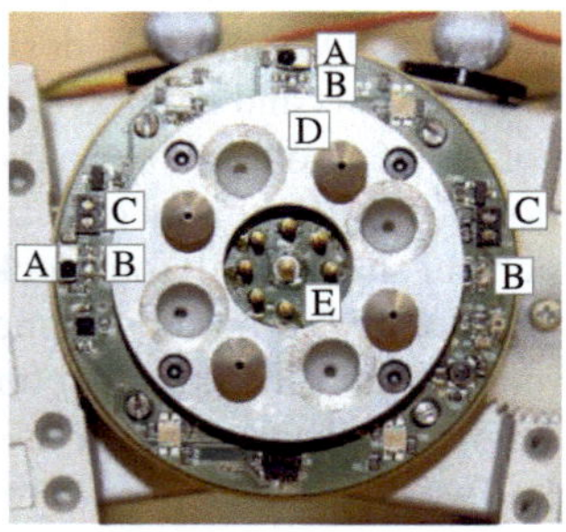

(b) Passive docking unit

Fig. 2. Unified docking mechanisms: A – IR receivers, B – IR LEDs, C – IR sensors, D – docking mechanics, E – electrical docking contacts, F – locking motor electrical features. Note that for any valid robot-robot connection, at least one active docking unit needs to be present so that two docking units can be locked securely.

Infrared (IR)-based sensing - including proximity detection, docking alignment detection and local communications circuits - has been developed for the SYMBRION robot to achieve autonomous docking in a 2D planar environment [10]. These sensors have similar placement on each docking face of the robot. More specifically, two IR sensors have been placed symmetrically above and on either side of the docking unit; one IR LED is placed directly above the docking unit, while the other two LEDS are located on either side of the docking unit. These LEDs are used to emit different frequency signals for obstacle detection, docking alignment and communication. The IR sensors work for both obstacle detection and docking alignment detection. IR remote control receivers are placed next to the IR LED on each docking face for communications.

3 Internal Representation of Organism Body Plan

To investigate how specified organism shapes can be self-assembled from a swarm of freely mobile robots, a common representation of the target shape must first be defined. The representation is also essential for the topology exploration of the organism in the macro-locomotion and the self-reconfiguration tasks. This study assumes that the self-assembly process always occurs in a 2D planar environment between a partially formed organism and a group of freely moving robots where individual robots join the growing organism using their own locomotion. The hardware constraints of the robot platform indicate that at any time one robot can attach to the partially formed organism using only one docking port. Thus no circular paths can appear among several connected modules in the target organism shapes. The body plan of an organism in a 2D planar environment can therefore be represented as a tree structure T. If we let $T = (V,C)$, then V is a set of nodes and C represents a set of connections. Each node $v \in \{r_1, r_2, ..., r_m\}$ corresponds to a robot module in the organism, where r_i is the type of robots and m is the total number of types in the system. Each connection $c = ((v_1, o_1), (v_2, o_2))$ includes a pair of nodes (v_1, v_2) and the orientation of corresponding connection ports, o_1 and o_2. Note that each type of robot r_i can have k_i connection ports. Clearly, unlike the common representation of a tree, there is no root node in the organism body plan and each node may have a different type.

There are many different ways to represent trees in a computer system. The design of data structures depend on the algorithm used for the tree. As there is no central control module in the complete organism, information on the body shape will have to be transferred and stored across all modules. Besides a smaller memory requirement, a data structure which can be easily manipulated and transferred via common robot-robot communication means is preferred. To satisfy these requirements, this paper proposes a well formatted parenthesis symbol sequence to represent the body plan of the organism. Let T be a tree representation of an organism with n robots, a symbol sequence S of tree T is defined as a sequence of $2(n-1)$

symbols s generated from a depth first traversal of T. There are two kinds of symbols in a sequence, s_c and s_p, where $s_c = f(c)$ is a "connection" symbol, which corresponds to a connection $c = ((v_1, o_1), (v_2, o_2))$ in T using a mapping function $f()$, while s_p is a special "pairing" symbol. The size of "connection" symbol $|s_c|$ equals $(\sum k_i)^2$ and "pairing" symbol $|s_p|$ is 1.

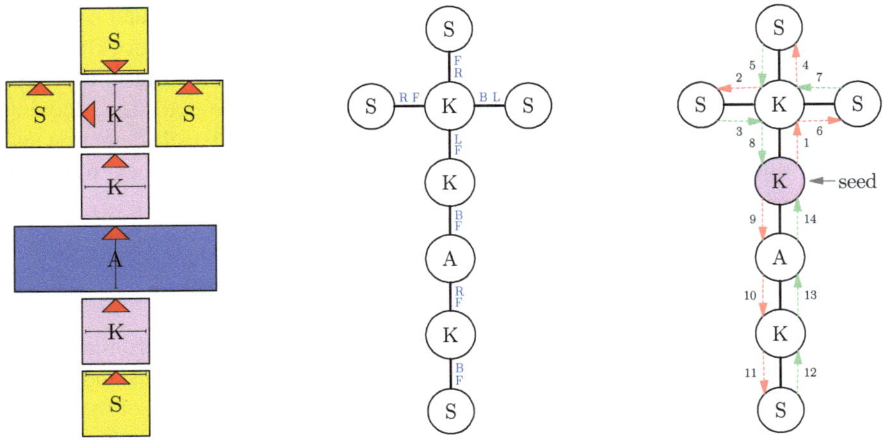

(a) Organism body plan (b) Tree represention (c) Order of traversal

Fig. 3. An example of organism body plan and its tree representation. 'K' represents BackBone robot, 'A' for ActiveWheel robot and 'S' for Scout robot. 'F', 'R', 'B', 'L' denotes the Front, Right, Back and Left docking ports of a robot respectively.

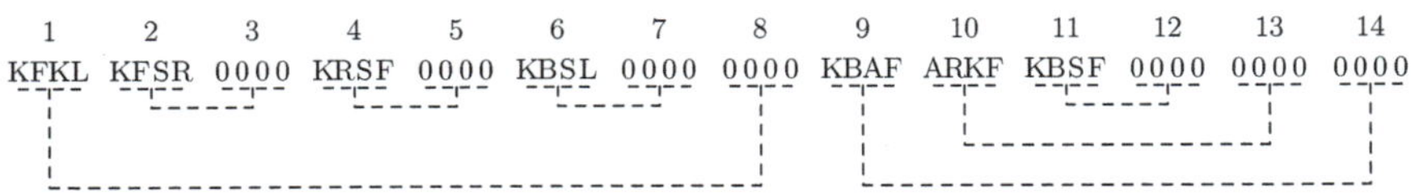

Fig. 4. An example symbol sequence of the organism. Each pair of s_c and s_p symbols, annotated and connected with dashed lines, represent one edge in its tree representation.

To obtain a symbol sequence from a tree representation, a node, denoted as v_o, must first be chosen as the starting point of the traversal. Depending on the starting node, different symbol sequences can be generated for the same tree. Compared to the general tree representation, a symbol sequence representation is encoded with the information where the organism starts to grow, which is indeed desirable for the morphology control mechanism introduced in the next section. Take the organism body plan shown in Fig.3(a) as an example. There are 3 Backbone, 4 Scout and 1 ActiveWheel robots in the organism where the headings are indicated with triangles and the rotation axes are marked with a bar. Its corresponding tree representation is shown in Fig.3(b) and all connected docking ports are labelled along the edges. Starting from a node marked with the dark colour, Fig.3(c) depicts the order of each edge being visited in a depth first traversal, which follows the travel order of "Front

→ Right → Back → Left" for each node. Through the traversal, a s_c symbol is added to a symbol sequence if an edge is first traversed and a s_p symbol is added each time an edge is traversed in the opposite direction. For simplification, let s_c to be a string in a format of "parent node type | parent connection side | child node type | child connection side", where each node type or connection side is denoted with one character, and s_p to be a string of "0000", then the generated symbol sequence can be presented as shown in Fig. 4. Here the node types are abbreviated as 'K', 'S' and 'A' for Backbone, Scout and ActiveWheel robot respectively, while connection sides 'F', 'R', 'B' and 'L' stand for Front, Right, Back and Left respectively. For instance, a symbol of "KFKL" can be read as a traversal along the edge from a Backbone robot's front side to another Backbone robot's left side. Also shown in Fig. 4, the symbol sequence can be annotated as a nested structure. For each s_c symbol, there is a corresponding pairing symbol s_p in the sequence. Each pair of s_c and s_p symbols corresponds to one edge in the tree representation. Their relative positions then depict the topology of the tree structure.

A symbol sequence can be obtained recursively from a depth first traversal of the tree. By carefully choosing the mapping function $f(c)$ and pairing symbol s_p, the generated symbol sequence requires only a small amount of memory to be stored and transferred. For example, in this study each s_c symbol is constructed using only 1 byte, with every two bits storing the robot type or the orientation of the connected docking port as shown below. If let 1 - 3 denote the robot type 'K', 'S' and 'A'

7	6	5	4	3	2	1	0
parent node type		parent node docking port		child node type		child node docking port	

respectively and 0 - 3 correspond to the connection ports 'F', 'R', 'B' and 'L' respectively, then a symbol of "KFKL" can be depicted as 0b01000111 = 0x47, while symbol s_p can be identified using 0.

Because of the well organised format and nested structure of a symbol sequence, its tree representation can be reconstructed using Algorithm 1.. This is important

Algorithm 1. symbol sequence to graph representation

Precedure: OgSequenceTraversal (symbol_sequence S, tree T, node v)

```
for each branch sequence S_b in S do
    create new node w from S_b.symbols[0];
    insert w to T
    if v is not empty node then
        connect w with v;
    end if
    create a new symbol sequence S_1 by removing the first and last symbol of S_b;
    OgSequenceTravelsal( S_1, T, w);
end for
```

as the topology information can be more easily accessed from a tree representation for macro-locomotion control of a complete organism, while the symbol sequence representation is more convenient to be shared among the robots. Meanwhile, it provides a trival way to check any geometrical conflict of the body shape that a symbol sequence stands for. For example, if the geometrical distance between any two non-neighbouring robots in the reconstructed tree is less than a certain value, the body plan can not be self-assembled because of the physical interference between robots.

4 Self-assembly Strategy

In SYMBRION scenarios, depending on whether physically connected to other robots, a robot works in one of two modes: swarm mode and organism mode, and switches between them accordingly. The transition from swarm mode to organism mode is determined by the morphology control strategy. In general, successful docking requires the collaboration between two robots: one remains stationary, called the recruiter, which emits some guiding signals; the other moves and aligns along these signals and then docks to the recruiter. To build an organism shape, the process needs to be initialised by a robot, called the seed. During the self-assembly process, one freely moving robot can dock and join to the partially formed organism only when it is signalled. To grow the correct body shape, it is critical that the right robots in the organism mode become recruiters at the right time, and the right type of freely moving robots respond to the recruiters.

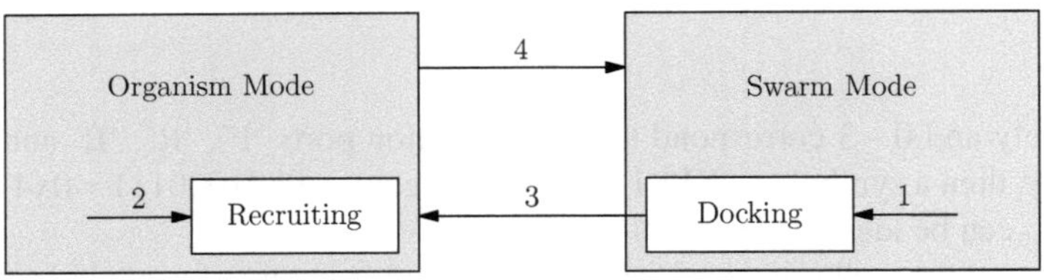

Fig. 5. Controller framework for the SYMBRION robots. Transition conditions: 1 – recruiting signals are received and request type matches; 2 – recruiting is required as per morphology control strategy; 3 – docking is accomplished; 4 – disassembly is required.

This paper adopts a finite state machine as the controller framework for all robots. Fig. 5 shows part of the controller which is the focus of this paper, a complete version can be found in [9]. Once the self-assembly process starts, the seed robot becomes the first recruiter in the organism and thus moves to the *Recruiting* state. Assuming that the target organism shape has already been stored as the symbol sequence in its memory, the seed robot extracts the branch symbol sequences from the complete symbol sequence. A branch symbol sequence S_b of S is defined as a sub symbol sequences between one of the starting node v_o associated s_c symbol and corresponding paired s_p symbol. The number of branch symbol sequences equals the number of associated s_c symbols of S. For example, the symbol sequence shown in Fig.4 has two branch symbol sequences: "KFKL KFSR 0000 KRSF 0000 KBSL 0000 0000" and "KBAF ARKF KBSF 0000 0000 0000". Each branch sequence is in fact endowed with connection information to its descendants in the target organism body shape. The seed robot can therefore use this information to decide which types of robot need to be recruited from which docking port. The behaviours in the *Recruiting* state can be grouped into different stages:

stage 1 broadcast recruiting message from corresponding docking ports, indicating the types and connection side of the robot d recruited;
stage 2 emit guiding signals to help docking robots to align to them;
stage 3 synchronise the locking process if required, as both active docking unit and passive docking unit may be involved;
stage 4 send the branch symbol sequences to newly docked robots.

Any robot in swarm mode may become docking robots when recruiting messages are received and their type matches with that requested. The docking robot then moves close to the recruiter guided by the signals and docks to it using the correct docking port. Upon receiving the branch symbol sequence S_b from the recruiter, it extracts a new symbol sequence S_1 by removing the first s_c symbol and the last s_p symbol in S_b. This new symbol sequence represents the sub tree which is rooted from itself. If there are branch sequences in the symbol sequence S_1, the newly joined robot becomes a recruiter and moves into the *Recruiting* state to recruit more robots into the organism. Otherwise it changes state to organism mode and hence stops any further growth of the organism from itself. A recruiter exits the *Recruiting* state once all its required docking ports are joined by other robots. Algorithm 2. outlines the behaviours for the recruiter and docking robots. Note that the exact behaviours in each state can have different implementations because of the heterogeneity of robots.

The morphogenesis process completes when all recruiters exit the *Recruiting* state. As the self-assembly process is fully distributed and no single module acts as the coordinator for the growth of the organism, once the process has been initialised by a seed robot, it is important that all robots are aware of completion of the morphogenesis process and then behave accordingly to transform the organism from 2D to 3D to initialise the macro-locomotion controller. One solution to check the self-assembly process is to pass a message token across the developing organism periodically to request all robots to register if they are recruiters or not. This message token travels across the organism following some rules, for example in the same way a symbol sequence is generated, and back to the original sender. Once it shows no recruiters are present in the organism, the self-assembly process is assumed finished and the sender will issue another message to notify the success of morphogenesis to all robots. The selection of such a sender remains open. It can be any robot in the organism. A good candidate is however the most recently joined robot that has no other robots to be recruited, *i.e.*, a leaf node robot in the tree representation. In other words, whenever a leaf node robot joins the organism, a progress check of self-assembly will be performed.

Algorithm 2. Self-assembly strategy for recruiter and docking robots

```
Behaviour: in state Docking
if docking is NOT accomplished then
    locate and align to the recruiter;
else if new symbol sequence information received then
    extract the branches from received symbol sequence;
    if branches exisit then
        enable the corresponding docking ports;
        start to emit beacon signals;
        move to state Recruitment;
    else
        issue a message token to check progress of self-assembly;
        if message token is returned then
            if No recruiters are presented then
                notify the completion of self-assembly;
            end if
            move to organism mode;
        end if
    end if
end if
Behaviour: in state Recruitment
if All required docking port is docked then
    move to organism mode;
else
    for Each recruiting docking port do
        if new robot is docked then
            send branch symbol sequence;
            stop emitting guiding signals;
        else if currenttime % RECRUITMENT_SIGNAL_INTERVAL then
            broadcast recruiting message;
        end if
    end for
end if
```

Communications, both wireless (IR) and wired (Ethernet), are very important in synchronising the high level self-assembly strategy and also the low level autonomous docking process between two robots. Due to the interference of IR signals and the low bandwidth, messages are more likely to get lost when transferred wirelessly. To address this issue, IR messages should normally be kept as simple (short) as possible. Meanwhile, each IR message may need to be repeated several times until it has been acknowledged. A full analysis of effect of communication failures on the algorithm is presented in [11].

5 Experiments and Discussion

The proposed controller and self-assembly strategy has been implemented and tested with a small number of real prototype SYMBRION robots. Each robot runs the same controller framework as described in the paper and utilises the IR signals for recruiting and docking alignment. Autonomous docking between the recruiter and a docking robot is synchronised via local sensing and wireless communication. The locomotion control for each type of robots has been carefully designed according to their unique hardware specification. As the initialisation of the self-assembly process is out the scope of this study, each time a seed robot is chosen manually with a predefined symbol sequence to start the self-assembly process. Fig. 6 shows the progress of autonomous self-assembly of an organism with 2 ActiveWheel and 2 BackBone robots. The approach was successfully and repeatably demonstrated for a number of configurations and seed robot positions. In all cases we see that the target organism shape is formed within a certain time frame and all robots in the organism are notified of the success of the process using the progress checking mechanism. Videos of different experiments are available online at http://goo.gl/VPGGm.

(a) 24s (b) 32s (c) 36s

(d) 1m32s (e) 1m44s (f) 1m56s

Fig. 6. Self-assembly experiment with 2 Backbone and 2 ActiveWheel robots. The experiment is initialised by the ActiveWheel in the middle with a symbol sequence of "AFKF0000ABKFKBAF00000000".

Since a symbol sequence representation includes also the location information of the seed robot in the organism, a selection of different locations for the seed, thus different symbol sequences, do indeed result in building the same organism. Although robots have to join the organism one after another, the potential for multiple recruiters in the self-assembly process shows that the growth of the organism is in fact a parallel process. It follows that the position of the seed robot (in the organism) can affect the completion time of the self-assembly process. To evaluate this, experiments are performed in Robot3D [12] simulator, instead of real robots as only 4 SYMBRION robots were availabe for testing at the time of writing, for growing the same body plan as shown in Fig.7a. In all cases, 25 robots are initially deployed uniformly in an arena sized 4 m $\times$ 4 m, and the pre-selected seed is always located in the centre of the arena with varying headings in 10 repeated runs.

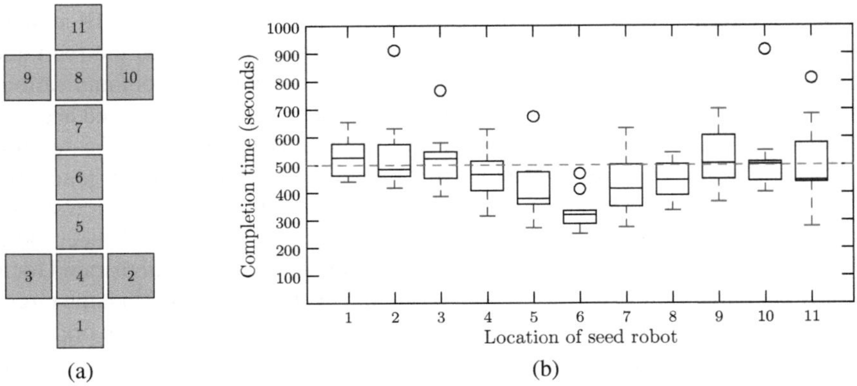

Fig. 7. Comparison of completion time for self-assembly initialised by seed robots in different positions in the organism. Each box in the plot represents the first to the third quartile of the data from 10 experimental runs.

Fig. 7b compares the completion time for self-assembly with seed robots in different positions. We see clearly that the completion time varies with different seed robot positions. It is obvious that the more parallel recruitment in the developing organism, the quicker the growth process. As expected, a seed located in the centre of the organism, *i.e.*, position 6, leads to the fastest growth with an average completion time of 330 seconds. The further the seed is from the centre of the organism, the longer the completion time; for example, a seed from postion 1 has an average completion time of 539 seconds. Since the tested target organism shape has a symmetrical topology, the completion time is very close when the position of the seed robots are similar. For example, the seeds next to the centre of the organism, position 5 and 7, give average completion times of 415 and 425 seconds respectively. In order to optimise the efficiency of the self-assembly process, we therefore require that the process needs to be initialised by a seed robot located as close as possible to the centre of the target body shape. To check whether a seed robot with a particular symbol sequence could result in the most efficient self-assembly process, we can count the number of edges in each branch sequence for this symbol sequence, if all of them are less than half of total number of edges of the parent symbol sequence, then the associated seed robot is located in the center of the body shape that this symbol sequence represents.

6 Conclusion and Future Work

This paper presents a fully distributed algorithm for morphology control in a group of heterogeneous modular self-assembling robots. Each robot in the system is fully autonomous with its own sensing and locomotion, and able to physically join with others using a unified docking mechanism in a 2D planar environment. This study focuses on how specific organism body shapes can be built, starting with a seed robot. A well organised symbol sequence structure has been proposed to represent the target organism body plan and also to form the basis of the self-assembly strategy. The symbol sequence representation encodes not only the topology of the target body shape but also information on how the shape should grow from one robot. Its compact nested structure allows the target body shape to be stored and transferred among robots with minimum memory and low bandwidth communication, which is a requirement of robots with limited resources. Depending on the topology of the body plan, multiple robots may join the partially formed organism from different positions simultaneously. Each time a robot joins the partially formed 2D organism, it receives a branch symbol sequence from its recruiter. Thus, during self-assembly, only the seed robot stores the entire body plan of the target organism; other robots store only parts of the body shape, including itself and its branch. There is a big advantage compared to the ID-based single entry self-assembly strategy proposed in [9]. In case there are failures in the self-assembly process, for example, a newly joined robot malfunctions, only that branch of the body shape will be affected, and organism growth from other points will still go on. If proper fault detection and recovery mechanisms are introduced to detect the faults and remove the malfuction robot, the self-assembly from that point can continue as the shape information can be retrieved directly from the parent recruiter robot.

Since no complex communication protocols and conflict resolution mechanisms are present in the recruitment-docking process, more than one robot can be attracted to the same docking port of a recruiter at the same time. Competition among these robots can inevitably increase self-assembly completion time. We have tested a simple "expelling" message, broadcast by the docking robot as a way of reducing the possible competition; however this has only been tested in simulation with idealised local communication. How efficient this conflict resolution mechanism is and how robustly it works with a large swarm of real robots need to be further investigated. Note that there are many ways in which faults on hardware might disrupt the self-assembly process including, for instance, mechanical failure of the docking mechanism or failure of the power or communications buses across the docking mechanism. Extending and adapting the algorithm to compensate for such faults during self-assembly is ongoing work presented in [11].

Acknowledgements. The SYMBRION project is funded by the European Commission within the work programme Future Emerging Technologies Proactive under grant agreement No. 216342.

References

1. Fukuda, T., Nakagawa, S.: Dynamicaly reconfigurable robotic system. In: Proceeding IEEE International Conference on Robotics and Automation, vol. 3, pp. 1581–1586 (1988)
2. Levi, P., Kernbach, S. (eds.): Symbiotic Multi-Robot Organisms: Reliability, Adaptability, Evolution. Springer (2010)

3. Wei, H., Chen, Y., Tan, J., Wang, T.: Sambot: A self-assembly modular robot systems. IEEE/ASME Transactions on Mechatronics 16(4), 745–757 (2011)
4. Doursat, R.: Organically Grown Architectures: Creating Decentralized, Autonomous Systems by Embryomorphic Engineering. Understanding Complex Systems, vol. 21, pp. 167–199. Springer (2008)
5. Nagpal, R.: Programmable self-assembly: Constructing global shape using biologically-inspired local interactions and origami mathematics. Ph.D. thesis, Massachusetts Institute of Technology (2001)
6. Werfel, J.: Biologically realistic primitives for engineered morphogenesis. In: Dorigo, M., Birattari, M., Di Caro, G.A., Doursat, R., Engelbrecht, A.P., Floreano, D., Gambardella, L.M., Groß, R., Şahin, E., Sayama, H., Stützle, T. (eds.) ANTS 2010. LNCS, vol. 6234, pp. 131–142. Springer, Heidelberg (2010)
7. Støy, K.: Using cellular automata and gradients to control self-reconfiguration. Robotics and Autonomous Systems 54(2), 135–141 (2006)
8. Christensen, A., O'Grady, R., Dorigo, M.: SWARMORPH-script: a language for arbitrary morphology generation in self-assembling robots. Swarm Intelligence 2(2), 143–165 (2008)
9. Liu, W., Winfield, A.F.T.: Autonomous morphogenesis in self-assembling robots using IR-based sensing and local communications. In: Dorigo, M., Birattari, M., Di Caro, G.A., Doursat, R., Engelbrecht, A.P., Floreano, D., Gambardella, L.M., Groß, R., Şahin, E., Sayama, H., Stützle, T. (eds.) ANTS 2010. LNCS, vol. 6234, pp. 107–118. Springer, Heidelberg (2010)
10. Liu, W., Winfield, A.: Implementation of an IR approach for autonomous docking in a self-configurable robotics system. In: Proceedings of Towards Autonomous Robotic Systems, pp. 251–258 (2009)
11. Murray, L., Liu, W., Winfield, A., Timmis, J., Tyrrell, A.: Analysing the Reliability of a Self-reconfigurable Modular Robotic System. In: Hart, E., Timmis, J., Mitchell, P., Nakamo, T., Dabiri, F. (eds.) BIONETICS 2011. LNICST, vol. 103, pp. 44–58. Springer, Heidelberg (2012)
12. Winkler, L., Wörn, H.: Symbricator3D – a distributed simulation environment for modular robots. In: Xie, M., Xiong, Y., Xiong, C., Liu, H., Hu, Z. (eds.) ICIRA 2009. LNCS (LNAI), vol. 5928, pp. 1266–1277. Springer, Heidelberg (2009)

Error-Tolerant Cyclic Sequences for Vision-Based Absolute Encoders

Kevin C. Wolfe and Gregory S. Chirikjian

Abstract. A method for obtaining error-tolerant cyclic sequences similar to de Bruijn sequences is presented. These sequences have a number of potential applications, including use as absolute rotary encoders. This investigation is motivated by the desire to use a vision-based system to obtain the angular position of the wheels of mobile robots as they rotate about their axes. One benefit of this approach is that the actual wheel orientation is observed (as opposed to non-collocated measurements of wheel angles via encoders on the motor shaft). As a result, ambiguities from backlash are eliminated. Another benefit of this system is the ability to apply it quickly to existing systems. Several methods are developed for increasing the robustness of these encoders. An imaging simulator is used to compare the accuracy of a variety of encoding schemes subjected to several levels of image noise.

1 Introduction

The process of transmitting and obtaining state information has been studied from a variety of perspectives. A practical example of this process is determining the orientation of a wheel relative to a fixed frame. A number of rotary encoding strategies have been developed to detect both relative motion and absolute position. For absolute encoding, most rotary encoders fall into one of two categories: single-track and multitrack. Most multitrack systems rely on bits that change in parallel as the wheel turns. Single-track strategies utilize a single code, and segments of this code are often read in a serial fashion similar to a shift register. Fig. 1 provides two examples of common absolute encoders: a Gray code and a de Bruijn sequence. A general overview of rotary encoders can be found in [10], [2], and [11].

Kevin C. Wolfe · Gregory S. Chirikjian
Johns Hopkins University, Baltimore, MD, USA
e-mail: {kevin.wolfe,gregc}@jhu.edu

M.A. Hsieh and G. Chirikjian (eds.), *Distributed Autonomous Robotic Systems*,
Springer Tracts in Advanced Robotics 104,
DOI: 10.1007/978-3-642-55146-8_17, © Springer-Verlag Berlin Heidelberg 2014

(a) 4-bit binary Gray code (b) 4-bit binary de Bruijn sequence

Fig. 1 Above are two examples of codes that can are commonly used for rotary encoders. Small circles indicate sensor locations. The multitrack Gray code (a) is read in a parallel fashion, while the single track de Bruijn sequence (b) is read serially.

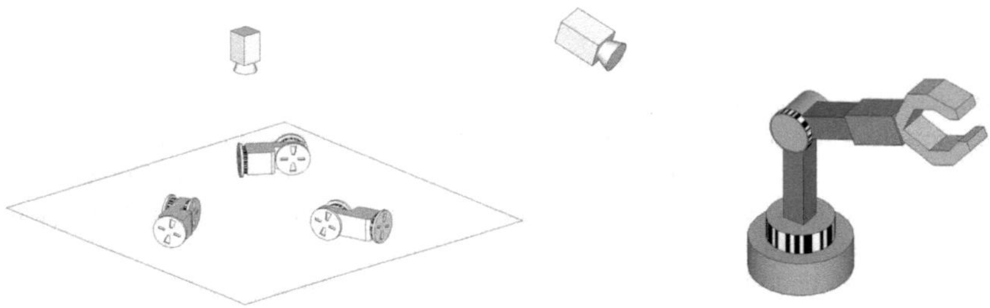

(a) (b)

Fig. 2 Examples of potential applications for vision-based encoders

In this paper, a new method for absolute rotary encoding that can be used in combination with, or in place of, shaft encoders is presented based on single-track encoding. It was developed for use with a new kind of modular reconfigurable robotic platform, the M^3 (Modular-Mobile-Multirobot) system, that is further described in [15], [8], and [14]. In this system, the orientations of the wheels are important for module-to-module docking that occurs on the wheel surfaces. Fig. 2(a) illustrates an initial deployment scenario in which an overhead camera is used to measure the wheel angles, position, and orientation of multiple robotic modules. This ability to easily capture additional state information can be important for nonholonomic motion planning and characterization of backlash. Some issues related to backlash and other uncertainty in nonholonomic robotic systems are discussed in [3], [17], and [1].

The basic approach is to print a cyclic code around the circumference of the wheel. A camera captures a visible portion of this pattern, and a decoding scheme assesses the orientation of the wheel. There are numerous advantages of this approach over traditional shaft encoding: (1) because the measurement is direct, measurement errors that accumulate between the motor shaft and the wheel due to backlash in the drivetrain are eliminated; (2) sensor/payload requirements on the robot are reduced since sensing is external; (3) a single sensor (an overhead camera) can be used to simultaneously assess both pose and wheel angles rather than using multiple sensors; (4) this method can be used to supplement shaft encoders to quantify backlash at any given time; (5) if off-board computing is being used, data bandwidth is reduced as wheel angle information does not need to be sent from each robot to the controlling computer. Additionally, one of the most beneficial advantages of this encoding scheme is its ease of application to existing systems without the need to carefully align and mount collocated encoders and detectors.

Fig. 2(b) illustrates another potential application for the encoders presented. In this type of application, appropriate methods must be developed for dealing with occlusions. The robust nature of the encoding scheme allows for accurate readings despite minor partial occlusion. Occlusions may also be handled using multiple cameras.

The primary focus of this work is on error-tolerant cyclic sequences for use as single track rotary encoders sampled through a noisy channel, computer vision in our case. Error-tolerance here refers to error-detection and/or error-correction. Several groups have investigated error-tolerant single-track sequences that address transition error, errors that arise on the boundary between characters due to slight misalignment of detectors [12, 13, 16]. However, we are concerned with error-tolerance from a more information theoretic standpoint; we would like to improve the certainty with which we make measurements by increasing the Hamming distance between distinct positions along the code. In [6], Heiss developed error-detecting codes for multitrack encoders where the Hamming distance between adjacent locations was one and it was greater than one for all other pairs of distinct locations.

While this work was primarily developed for rotary encoding, portions of it are applicable in a variety of settings. If we consider the problem of phase synchronization of two or more systems through a noisy channel, error tolerant codes such as those described here could be utilized. Hagita et al. further describe this in [5] while also exploring and presenting methods for obtaining error-tolerant sequences similar to the ones developed here. However, they only provide a method for generating error-correcting binary sequences of period $2^{2^m-m-2}-1$ with 2^m-2-length subsequences for integer values of m [5]; such sequences may not be suitable for rotary encoders given specific design considerations.

2 Error-Tolerant Cyclic Sequences

Let $S = (a_1, a_2, a_3, \ldots a_{K-1}, a_K, a_1, a_2, \ldots)$ be a cyclic sequence with a period of K whose elements are taken from an alphabet, $\mathbb{A}$, with q distinct elements. Now let $s_n(i)$ be a subsequence of S of length n starting at the ith element of S. For example, $s_5(3) = (a_3, a_4, a_5, a_6, a_7)$.

S is a *de Bruijn sequence* for subsequences of length n if it has a period of q^n and all possible n-length combinations of the elements of $\mathbb{A}$ appear. De Bruijn sequences have been studied and used as absolute rotary encoders. This is done by "writing" q^n consecutive elements of the sequence around a circular object, such as a wheel.

Taking n-sequential readings then allows the angular position to be determined to within $\pm\frac{\pi}{q^n}$ radians. This is illustrated in Fig. 1(b) for $q = 2$ and $n = 4$.

Any cyclic sequence, S, can be used as an absolute rotary encoder in a similar fashion provided that all n-length subsequences are unique. This can be formalized further if we let $d_H(s_n(i), s_n(j))$ be the Hamming distance[1] between subsequences $s_n(i)$ and $s_n(j)$. Now consider using the sequence S as an absolute encoder; $d_H(s_n(i), s_n(j))$ represents the distance between the message sent for position i and position j. For convenience we can further define the minimum Hamming distance between two distinct positions as

$$\delta(S,n) = \min\{d_H(s_n(i), s_n(j)) \,|\, i, j \le K, i \ne j\}.$$

Thus, for S to be a suitable cyclic sequence for use as an absolute encoder in the manner described above, $\delta(S,n)$ must be at least 1 (i.e., no n-length subsequence appears more than once).

If this process is viewed as transmitting an actual angular position through a noisy channel, it may be desirable to have this minimum distance be greater than 1, providing some degree of error-tolerance. If we assume that we only transmit and receive elements from $\mathbb{A}$, we are able to detect up to $\delta(S,n) - 1$ errors. For error-correction, if $\delta(S,n) \ge 2e - 1$, we are able to correct up to e errors. Using vision as a transmission channel, the data received may differ from $\mathbb{A}$ and thus it is less obvious the "number" of errors we are able to detect or correct; we are more concerned with the maximizing the mutual information. In either case, it should be clear that our goal is to increase $\delta(S,n)$ while decreasing n for a fixed or bounded K.

For the remainder of the paper we will refer to a cyclic sequence S as an (n,d)-sequence if $\delta(S,n) \ge d$. Using this notation it is clear that all de Bruijn sequences are $(n,1)$-sequences; as such, they are not well suited for use as encoders in the presence of noise. In Sections 3 and 4 we will explore solutions to this problem.

3 Obtaining (n,d)-Sequences and the De Bruijn Graph

A *de Bruijn graph*, $G = (V,E)$, for n-length subsequences on an alphabet, $\mathbb{A}$, of q characters is a directed graph where each vertex, $v_i \in V$, represents one of the q^n possible n-length code words. The edges connect vertices that represent possible sequential subsequences. Thus, a directed edge $(v_i, v_j) \in E$ exists if v_j can be attained by appending an element of $\mathbb{A}$ to the right of a left shifted version of v_i. Inasmuch, if

$$v_i = (a_{v_i 1}, a_{v_i 2}, \ldots, a_{v_i n}),$$

then

$$(v_i, v_j) \in E \Leftrightarrow v_j = (a_{v_i 2}, \ldots, a_{v_i n-1}, b) \quad \text{for some} \quad b \in \mathbb{A}.$$

For further information regarding de Bruijn graphs see [4].

[1] The Hamming distance for two sequences of equal length is taken to be the number of corresponding positions at which the sequences differ.

Table 1 Maximum length of binary (n,d)-sequences $(q=2)$

n	$d=1$	$d=2$	$d=3$	$d=4$
3	8	3		
4	16	4		
5	32	10		
6	64	15	7	
7	128	31	14	7
8	256	63	16	8
9	512	≥100	31	11
10	1024	≥188	62	22

Cycles[2] of length K such that $K \geq n$ within a de Bruijn graph are analogous to $(n,1)$-sequences. Given a cycle, C, of length $K \geq n$ from a de Bruijn graph we can construct a cyclic sequence, S_C, by sequentially taking the last element of the code words represented by its first K vertices. Therefore, let

$$C = (v_i, v_{i+1}, \ldots, v_{i+m}, v_i)$$

then

$$S_C = (a_{v_i n}, a_{v_{i+1} n}, \ldots, a_{v_{i+m} n})$$

and $C \approx S_C$. Thus, a Hamiltonian cycle[3] in G is analogous to a de Bruijn sequence.

While any cycle in G that is at least n in length represents an $(n,1)$-sequence, some cycles in G may correspond to sequences whose minimum distance is greater than 1. Given some n and $d > 1$, we could obtain (n,d)-sequences by enumerating all cycles in G using methods given in [7] and [9], and removing those that do not satisfy the constraint on d. However, for $d > 1$ the number of cycles with minimum distance d is a relatively small subset of all of the cycles in G. Consequently, enumerating all cycles may be inefficient; for example, is known that on q characters there exist $(q!)^{q^{n-1}} q^{-n}$ unique de Bruijn sequences each of which does not need to be considered when looking for such (n,d)-sequences. Therefore, a modified cycle finding algorithm was developed. This is given in Algorithm 1 which takes in a de Bruijn graph and a minimum distance and, through a breadth-first search, finds all cycles such that all vertices have a pairwise Hamming distance of at least d. Enforcement of this constraint during each search step eliminates the need to enumerate unsuitable paths or cycles and to remove unsuitable cycles at termination. Table 1 provides a summary of the maximum length (n,d)-sequences that have been found for a variety of n and d combinations using a binary alphabet.

[2] Throughout the paper, the term *cycle* will refer only to *simple cycles*, those with no repeated vertices other than the first and last.

[3] A Hamiltonian cycle is a cycle in which every vertex in a graph is visited exactly once.

Algorithm 1 Find all (n,d)-sequences

```
Require: De Bruijn digraph G = (V,E) with vertices for subsequences of length n
procedure AllCyclesWithMinDistance(G,d)
    k ← 0
    U ← {}
    for all v ∈ V do
        k ← k+1
        P_k ← (v)
        C_k ← false
        T_k ← true
    Note: P_k represents a path. C_k = true if P_k is a cycle. T_k = true if P_k should be further
    explored.
        while T(j) = true for some j do
            i ← arg min_{1≤j≤k}(T_j = true)
            y ← last element in P_i
            T_i ← false
            for all x ∈ Children(y,G) do
                if HammingDist(x,P_i) ≥ d then
                    k ← k+1
                    P_k ← Append(P_i,x)
                    C_k ← false
                    T_k ← true
                    if x ∈ P_i ⋃ U then
                        T_k ← false
                        if x = v then
                            C_k ← true
                        end if
                    end if
                end if
            end for
        end while
        U ← U ⋃ v
    end for
    Return all P_j such that C_j = true
end procedure

function Children(v,G)
    Let G = (V,E)
    Return all w ∈ V such that (v,w) ∈ E
end function

function HammingDist(v,P)
    Return minimum Hamming distance between v and all vertices in P
end function

function Append(P,v)
    Let P = (p_1, p_2, ... p_m)
    P' ← (p_1, p_2, ... p_m, v)
    Return P'
end function
```

4 Multiple Parallel Sequences

Given a set of constraints on the length of the sequence, K, and the length of code words, n, using a single sequence may not provide enough uniqueness between all distinct positions. We can provide increased certainty by introducing additional sequences that are written in parallel to the first.

Let us assume that we want to use m sequences. For pairs of positions in one sequence whose Hamming distances are close, we can attempt to ensure that those pairwise distances are greater in the remaining sequences. Thus for two sequences $S^{(1)}$ and $S^{(2)}$, for all i and j such that $d\left(s_n^{(1)}(i), s_n^{(1)}(j)\right) = \delta\left(S^{(1)}, n\right)$ we desire $d\left(s_i^{(2)}, s_j^{(2)}\right) > \delta(S^{(2)}, n)$. To this end, we can define a $K \times K$ distance matrix $\mathbf{D}(S^{(l)}, n)$ as

$$\left[\mathbf{D}\left(S^{(l)}, n\right)\right]_{ij} = d\left(s_n^{(l)}(i), s_n^{(l)}(j)\right).$$

Using this notation we can easily sum distance matrices to obtain a distance matrix whose entries represent the total Hamming distance between two positions across all sequences.

To that end, we would like to determine the cyclic shifting or phasing that maximizes the following:

$$\max_{\mathbf{P}_1,\ldots\mathbf{P}_k \in \mathscr{P}} \left(\min_{i,j|i \neq j} \left[\sum_{l=1}^{k} \mathbf{P}_l \left(\mathbf{D}\left(S^{(l)}, n\right)\right) \mathbf{P}_l^{-1} \right]_{ij} \right) \tag{1}$$

Here the cyclic shifts are represented by shift permuation matrices $\mathbf{P}_1, \ldots \mathbf{P}_k \in \mathscr{P}$. $\mathscr{P}$ can be defined as

$$\mathscr{P} = \{\mathbf{P} | \mathbf{P} = (\mathbf{e}_{n-j}, \mathbf{e}_{n-j+1}, \ldots \mathbf{e}_n, \mathbf{e}_1, \ldots \mathbf{e}_{n-j-1}), \\ 0 \leq j \leq n-1\}$$

where $\mathbf{e}_i$ is the ith standard basis vector. It should be noted that the maximization in (1) can be simplified by letting P_1 equal the $K \times K$ identity matrix.

It is likely that a number of shifts will result in the same value of (1). To choose from among these we consider the standard deviation between all pairs of positions; by minimizing this standard deviation, we seek to provide a uniformity to the Hamming distance between positions.

5 Reading Wheel Position

For our application, a single overhead camera and binary sequences (i.e. $q = 2$ and $\mathbb{A} = \{0(\text{black}), 1(\text{white})\}$) are utilized. Using this camera and fiducials on the robots, we are able to locate the position and orientation of each robot. From this pose information and the known geometry of a robotic module, we determine where to read the sequences. The intensity images of the sequences are sampled along line

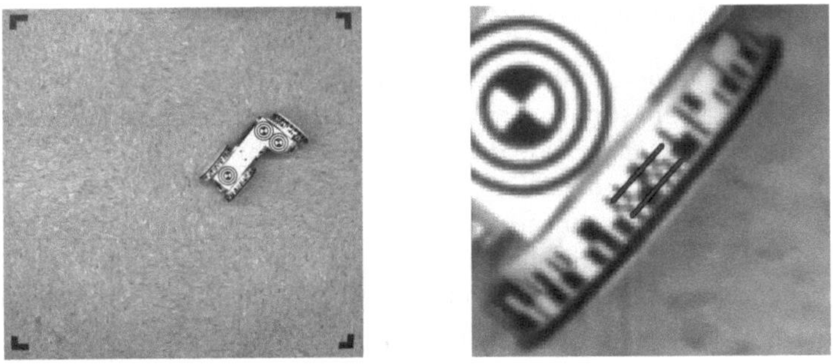

(a) View from overhead camera (b) Close-up of reading locations

Fig. 3 Knowing the relative position of the robot with respect to camera allows the position to be read. In (a) the full camera image is shown. The two short line segments in (b) indicate where sampling of the sequences is performed in the image.

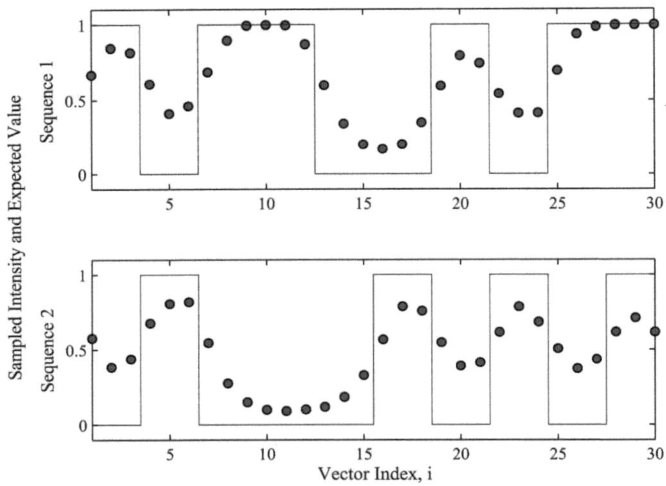

Fig. 4 Intensity readings and expected values sampled from Fig. 3.

segments; an example of this sampling is shown in Fig. 3 and Fig. 4. We take pn interpolated samples where p is the number of samples per bit. This sampling strategy enables us to increase accuracy beyond $\pm\frac{\pi}{K}$ radians to approximately $\pm\frac{\pi}{pK}$ for K-length sequences. Let $\mathbf{x}(S,t) \in [0,1]^{pn}$ be a reading taken from the image of a sequence S at time t. Let $\mathbf{y}(s_n(i)) \in \{0,1\}^{pn}$ such that

$$\mathbf{y}(s^n(i)) = (y_1, y_2, \dots y_p, y_{p+1}, \dots y_{pn})^T = (a_i, a_i, \dots a_i, a_{i+1}, \dots a_n)^T .$$

Let $\mathbf{h} = \{-0.5\}^{pn}$ be a constant vector used to shift $\mathbf{x}(S^{(l)},t)$ and $\mathbf{y}(s_n^{(l)}(i))$. This shift is needed to ensure that all vectors have the same 2-norm.

We can find the position using the summed cross-correlation across all m sequences. This can be done by letting b equal

$$b = \arg\max_{i} \left(\sum_{l=1}^{k} \left(\mathbf{x}(S^{(l)}, t) + \mathbf{h} \right)^{\mathsf{T}} \left(\mathbf{y}\left(s_n^{(l)}(i) \right) + \mathbf{h} \right) \right).$$

Then the angular reading can be taken as

$$\theta = \frac{2b\pi}{pK}.$$

6 Design Considerations and Imaging Simulator

For a given application, there are a number of design considerations that need to be taken into account. First, the radius and width of the cylindrical surface on which the code is to be written on will impose a number of design constraints including the number of sequences to write in parallel and the length of the sequences. These geometric factors must be coupled with the imaging resolution, position of the camera, and image quality to determine a suitable set of coding sequences. For a fixed cylinder geometry, increasing the number of sequences written in parallel will increase the Hamming distance between distinct angular positions, but will also increase the noise and blurring experienced in reading the position. Similarly, increasing sequence length reduces the size of an individual bit leading to less certainty when attempting to read the bit.

In an attempt to quantify the relative quality of various coding strategies, an imaging simulator was developed to easily test a variety of encoders. The simulator creates a virtual cylinder with binary sequences written on the outer surface. It then sets the cylinder's pose and angular position. A virtual image is captured of the cylinder at a specified resolution and common image noise is added at specified levels. The forms of image noise considered are Gaussian blurring, additive white Gaussian noise, and a reduction in the dynamic range. In addition to image noise, minor noise was added to the position of the cylinder. This has the effect of measurement uncertainty of the position of a robotic module. Once we have an image, we use the same decoding technique described in Section 5.

For a given coding scheme, the simulator was used to test reading a wheel with a given geometry in a variety of poses and at numerous noise levels. The amount of noise considered for each test was determined by a noise coefficient used to scale all of the noise sources simultaneously (see Fig. 6). The percentage of correct readings were then used to determine the effectiveness of a particular encoding. A reading was considered correct if it was within a set tolerance limit; for the test results presented in Fig. 7, the tolerance limit was set at $\pm\frac{2\pi}{K}$ for K-length sequences. Parameters were chosen that closely match those found in the M^3*Express* system [15]. The cylinder had a radius of 55mm and a width of 12mm. The image resolution was chosen so that the each pixel represents approximately one square millimeter.

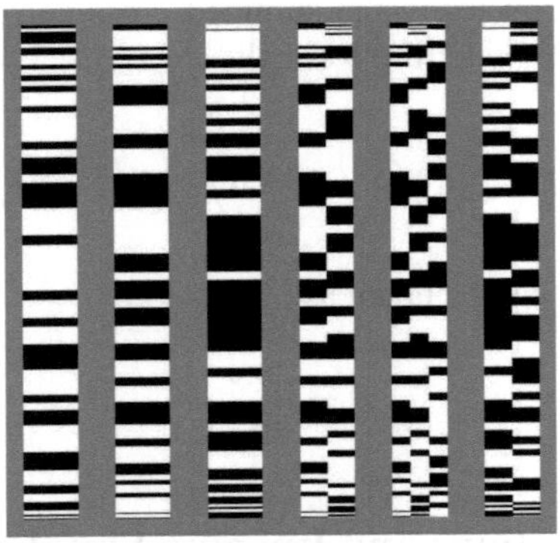

Fig. 5 Six example encoders that were tested using the image simulator. Each encoded cylinder is shown from above at high resolution. From left to right: a single $(10,1)$-sequence, a single $(8,1)$-sequence, a single $(10,2)$-sequence, two parallel $(8,1)$-sequences, three parallel $(8,1)$-sequences, and two parallel $(10,2)$-sequences.

Several examples of encoding scenarios that were tested are provided in Fig. 5. These include a single $(8,1)$-sequence, a single $(10,1)$-sequence, a single $(10,2)$-sequence, two parallel $(8,1)$-sequences, three parallel $(8,1)$-sequences, and two parallel $(10,2)$-sequences. All sequences tested have a length of 170. The $(n,1)$-sequences were generated randomly and then checked to ensure a minimum Hamming distance of 1. These $(n,1)$-sequences are similar to a standard de Bruijn sequence encoding scheme. The $(10,2)$-sequences were generated using Algorithm 1. The phasing of the multiple sequence encodings was performed as described in Section 4.

7 Discussion and Future Work

We have presented a new method for encoding and reading the angle of a wheel using vision based detection. The encoding strategy was inspired by the use of de Bruijn sequences as single track rotary encoders. In Sections 3 and 4, we describe methods for developing encoders that are more tolerant to uncertainty and noise. These methods include increasing the Hamming distance between distinct positions for single cyclic sequences and writing multiple sequences in parallel.

In addition to developing encoders, six example encoding strategies were tested using an imaging simulator described in Section 6. The results shown in Fig. 7 reveal several things. It is clear that the multi-sequence encoders perform better than single sequence encodings. This can be attributed to the redundancy provided by adding additional sequences. However, the performance difference between using one and two $(8,1)$-sequences is much more dramatic than the difference between using two and three. This may be due to blurring of the borders of bits or positioning error. As additional sequences are added, the width of each strip is reduced. For a fixed resolution image, this reduced size can lead to additional ambiguity for each bit.

(a) Noise Coefficient = 0.0 (b) Noise Coefficient = 0.5 (c) Noise Coefficient = 0.75

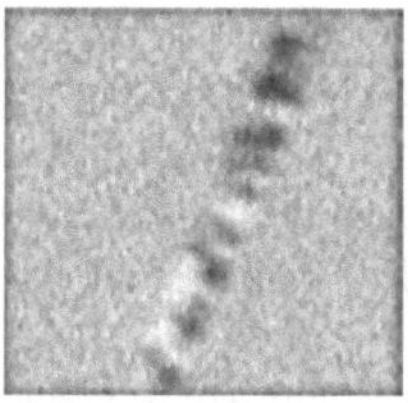

(d) Noise Coefficient = 1.0

Fig. 6 Four sample images demonstrating the amount of image noise introduced in the image simulator.

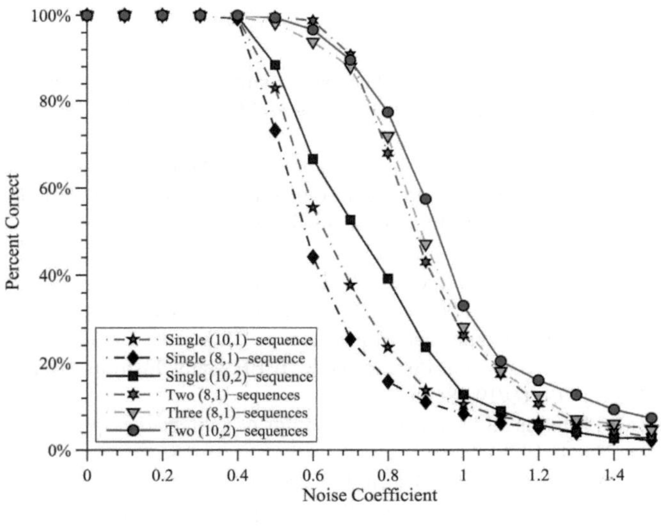

Fig. 7 The percentage of correctly read encoder positions at various noise levels for the encoder examples shown in Fig. 5.

Thus, there is a limit to the added benefit of adding additional sequences. Also evident is that for the single sequence encoders, the $(10,2)$-sequence outperforms both the $(10,1)$ and $(8,1)$-sequences. We believe that this is due to the increased Hamming distance between any two positions. Finally, it should be noted that the encoder which employed both strategies had the highest success rate at nearly all noise levels.

While the methods described here may provide more robust encoding strategies, there is still additional cataloging of sequences for $n \geq 10$ and $d > 1$ that can be performed. Moreover, future investigation can be performed to attempt to enumerate sequences for larger alphabets, $q > 2$. These could be useful as a replacement for writing two or more codes in parallel. For example, if $q = 4$, one could imagine representing the four symbols using two bits side-by-side (i.e. the alphabet, $\mathbb{A}$, could be $\{(0,0),(0,1),(1,0),(1,1)\}$). Writing such a sequence on a wheel would then look very similar to writing two sequences in parallel; however, a greater minimum Hamming distance may be achievable for a smaller number of total readings. Finally, in addition to investigating additional encodings, future work may include testing and experimental validation of the accuracy of the encoders on the $M^3Express$ and development of strategies for handling partial occlusions.

Acknowledgements. The authors would like to graciously thank Dr. Matthew Moses for his helpful insights and discussions.

This work was suported in part by NSF grant IIS-0915542 RI: Small: Robotic Inspection, Diagnosis, and Repair.

References

1. De Luca, A., Oriolo, G., Samson, C.: Feedback control of a nonholonomic car-like robot. In: Laumond, J. (ed.) Robot Motion Planning and Control. LNCIS, vol. 229, pp. 171–253. Springer, Heidelberg (1998)
2. Everett, H.R.: Sensors for Mobile Robots: Theory and Application. A K Peters (1995)
3. Fierro, R., Lewis, F.L.: Robust practical point stabilization of a nonholonomic mobile robot using neural networks. Journal of Intelligent and Robotic Systems 20, 295–317 (1997)
4. Golomb, S.W.: Shift Register Sequences. Holden-Day (1967)
5. Hagita, M., Matsumoto, M., Natsu, F., Ohtsuka, Y.: Error correcting sequence and projective de bruijn graph. Graphs and Combinatroics 24(3), 185–194 (2008)
6. Heiss, M.: Error-detecting unit-distance code. IEEE Trans. Instrum. Meas. 39(5), 730–734 (1990), doi:10.1109/19.58616
7. Johnson, D.B.: Finding all the elementary circuits of a directed graph. SIAM J. on Computing 4(1), 77–84 (1975)
8. Kutzer, M.D.M., Moses, M.S., Brown, C.Y., Scheidt, D.H., Chirikjian, G.S., Armand, M.: Design of a new independently-mobile reconfigurable modular robot. In: Proc. IEEE International Conf. on Robotics and Automation (2010)
9. Mateti, P., Deo, N.: On algorithms for enumerating all circuits of a graph. SIAM J. on Computing 5(1), 90–99 (1976)
10. Mayer, J.R.R.: The Measurement, Instrumentation, and Sensors Handbook, ch. 6.8. CRC Press (1999)
11. Stein, D., Scheinerman, E.R., Chirikjian, G.S.: Mathematical models of binary spherical-motion encoders. IEEE/ASME Trans. Mechatronics 8(2), 234–244 (2003)
12. Tomilnson, G.H.: Absolute type shaft encoder using shift register sequences. Electronics Letters 23(8), 398–400 (1987)

13. Tomlinson, G.H., Ball, E.: Elimination of errors in absolute position encoders using m-sequences. Electronics Letters 23(25), 1372–1374 (1987)
14. Wolfe, K.C., Kutzer, M.D., Armand, M., Chirikjian, G.S.: Trajectory generation and steering optimization for self-assembly of a modular robotic system. In: Proc. IEEE International Conf. on Robotics and Automation (2010)
15. Wolfe, K.C., Moses, M.S., Kutzer, M.D., Chirikjian, G.S.: M3express: A low-cost independently-mobile reconfigurable modular robot. In: Proc. IEEE International Conf. on Robotics and Automation (2012)
16. Yamaguchi, K., Imai, H.: Periodic sequences for absolute type shaft encoders. In: Sakata, S. (ed.) AAECC 1990. LNCS, vol. 508, pp. 36–45. Springer, Heidelberg (1991)
17. Zhang, Q., Shippen, J., Jones, B.: Robust backstepping and neural network control of a low-quality nonholonomic mobile robot. International Journal of Machine Tools and Manufacture 39, 1117–1134 (1999)

Arbitrary Lattice Formation with Flocking Algorithm Applied to Camera Tracking System

Sho Yamauchi, Hidenori Kawamura, and Keiji Suzuki

Abstract. Flocking algorithms for a multi-agent system are distributed algorithms that only have simple rules for each agent but generate complex formational movement. These algorithms are known as swarm intelligence and are robust and disaster tolerant for most cases. We consider that flocking algorithms that have these characteristics are the way to generate homeostasis in a system. We expect that by making use of this algorithm the system can tune its self parameters and thus maintain a high performance. First, to apply a flocking algorithm to a system, we extended the flocking algorithm to form an arbitrary lattice for further flexibility. We then applied the extended flocking algorithm to a position tracking camera system as an example.

1 Introduction

In a multi-agent system, several distributed algorithms such as the work of Reynolds [12] are known to have agents flock[14]. Systems using these techniques are robust and has resistance to mechanical breakdown even when some of them are destroyed. Such systems are known to possess swarm intelligence[1][2] and there are multi-agent robotic systems that make use of these techniques[10].

Other examples that use swarm intelligence include the hexapod robot OSCAR[5] which is designed to be fault tolerant by applying the boids algorithm. Each of OSCAR's legs is assumed to be a boids agent, and although any of OSCAR's legs may be destroyed, OSCAR can keep walking by balancing its body using a boids formation control algorithm.

These types of multi-agent systems can be viewed in terms of the consensus problem and such an approach have been applied to the study of flocking dynamics for birds and fishes to engineering systems[8]. There are several types of

Sho Yamauchi · Hidenori Kawamura · Keiji Suzuki
Graduate School of Information Science and Technology, Hokkaido University,
Kita 14, Nishi 9, Kita-ku Sapporo, Hokkaido 060-0814, Japan
e-mail: sho-yamauchi@complex.ist.hokudai.ac.jp

M.A. Hsieh and G. Chirikjian (eds.), *Distributed Autonomous Robotic Systems*,
Springer Tracts in Advanced Robotics 104,
DOI: 10.1007/978-3-642-55146-8_18,

application such as, under the situation with restricted communication data rate[15], event-based communication type[13], gossip algorithm type[4][3] and agents with complex dynamics[6][7][9].

However, there is no formal method to apply these techniques to generate homeostasis in an autonomous system. We introduce a way to apply a flocking algorithm to a system that tracks the position of a robot with its camera to make the system adapt to changes in brightness in a room. We assume that each control parameter of the camera is an agent of in a flocking algorithm. We expect that these techniques to be capable of enabling autonomous systems to achieve homeostasis.

2 Flocking Algorithm

2.1 *Analyzable Flocking Algorithm*

Without a general theoretical framework, it is hard to extend existing algorithms so we employ the flocking algorithm presented by Olfati-Saber[11]. This work presents a theoretical framework for the design and analysis of distributed flocking algorithms and our proposed extension is as follows.

Each agent tries to keep a fixed distance with other agents and moves keeping a lattice alignment (Fig. 1a). We assume the presence of a virtual agent to indicate the ideal movement and each agent follows this leader's movement (Fig. 1b). We consider a graph G that consists of a set of vertices $V = \{1,2,...,n\}$ and a set of edges $E \subseteq V \times V$. Also, the adjacency matrix is denoted as $A = [a_{ij}]$. We assume that each node is an agent and V is an agent set.

Let $q_i, p_i, u_i \in \mathbb{R}^m$ denote the position, velocity, input of agent i for all $i \in V$ respectively. Each agent has dynamics the following dynamics

$$\begin{cases} \dot{q}_i = p_i, \\ \dot{p}_i = u_i. \end{cases} \tag{1}$$

The set of neighbors of agent i is denoted by

$$N_i = \{j \in V : \|q_j - q_i\| < r\} \tag{2}$$

where $r > 0$ is the interaction range and $\|\cdot\|$ is a Euclidean norm in $\mathbb{R}^m$.

Agents are designed to maintain the same distance to the other agents. In other words, agents must be aligned such that their positions achieves a uniformed lattice that satisfies equation (3).

$$\|q_j - q_i\| = d, \forall j \in N_i. \tag{3}$$

We then have to determine the input u_i to enable the agent to follow the virtual agent and while satisfying the constraints given by equation (3).

The input u_i for agent i is determined by three terms as follows.

$$u_i = f_i^g + f_i^d + f_i^\gamma \tag{4}$$

where the term f_i^g is a gradient-based term of potential among agents, the term f_i^d is a velocity consensus/alignment term that acts as a damping force, and the term f_i^γ is a navigational feedback term. We now determine each term of the right side of an equation (4). Here, σ-norm is defined as

$$\|z\|_\sigma = \frac{1}{\varepsilon}[\sqrt{1+\varepsilon\|z\|^2} - 1] \tag{5}$$

where $\varepsilon > 0$ and this norm is differentiable everywhere. Also, gradient $\sigma_\varepsilon(z) = \nabla\|z\|_\sigma$ is defined as follows

$$\sigma_\varepsilon(z) = \frac{z}{\sqrt{1+\varepsilon\|z\|^2}} = \frac{z}{1+\varepsilon\|z\|_\sigma}. \tag{6}$$

The bump function $\rho_h(z)$ is continuous at $[0,1]$ and is defined as

$$\rho_h(z) = \begin{cases} 1 & z \in [0,h) \\ \frac{1}{2}[1+\cos\pi\frac{(z-h)}{(1-h)}] & z \in [h,1] \\ 0 & otherwise \end{cases} \tag{7}$$

where $h \in (0,1)$. With this bump function, each element of the adjacency matrix is given by

$$a_{ij} = \rho_h(\|q_j - q_i\|_\sigma / \|r\|_\sigma) \tag{8}$$

where $a_{ii}(q) = 0$ for all i and q. Let $V(q)$ be defined as the the collective potential function and is given by

$$V(q) = \frac{1}{2}\sum_i \sum_{j \neq i} \psi_\alpha(\|q_j - q_i\|_\sigma) \tag{9}$$

where ψ_α is the action function $\phi_\alpha(z)$ defined as follows

$$\phi(z) = \frac{1}{2}[(a+b)\sigma_1(z+c) + (a-b)] \tag{10}$$

$$\phi_\alpha(z) = \rho_h(z/\|r\|_\sigma)\phi(z - d_\alpha) \tag{11}$$

$$\psi_\alpha(z) = \int_{\|d\|_\sigma}^{z} \phi_\alpha(s)ds \tag{12}$$

where $0 < a \leq b, c = |a-b|/\sqrt{4ab}, \sigma_1(z) = z/\sqrt{1+z^2}$. Then,

$$f_i^g = -\nabla_{q_i} V(q) = \sum_{j \in N_i} \phi_\alpha(\|q_j - q_i\|_\sigma)\sigma_\varepsilon(q_j - q_i) \tag{13}$$

$$f_i^d = \sum_{j \in N_i} a_{ij}(q)(p_j - p_i) \tag{14}$$

$$f_i^{\gamma} = -c_1(q_i - q_r) - c_2(p_i - p_r) \tag{15}$$

where q_r,p_r are the position and velocity of the virtual agent respectively and c_1,c_2 are positive constant. Here, f_i^g is an attractive/repulsive force as a function of distance between agents, f_i^d is a force that even out the speed of agent and f_i^{γ} is a force that have agent follow the virtual agent.

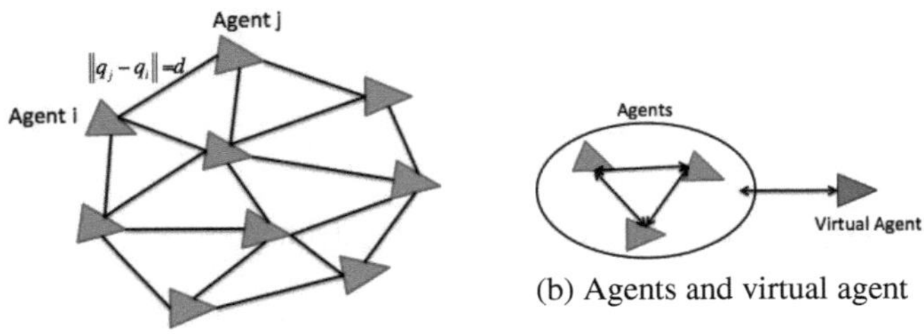

(a) Agent lattice

(b) Agents and virtual agent

Fig. 1 Flock of agents

2.2 Application of Flocking Algorithm for Autonomy in a System

To apply this flocking algorithm to autonomous system control, we assign system parameters to agent positions of a flocking algorithm. However, the value of the parameter that indicates the state of the system is different for each state and the distance among agents, *i.e.*, the difference between the parameter values are not uniform as given by equation (3). Furthermore, ideal values of parameter are all different. For this reason, we cannot directly apply the flocking algorithm that enables multiple agents to form a uniformed lattice while following a single virtual agent. Hence, we have to extend the flocking algorithm to be able to form any lattice and have the ability to follow several virtual agents.

2.3 Extended Flocking Algorithm

The original flocking algorithm is to have agents configure along a uniform lattice that satisfies equation (3). However, this algorithm cannot form any lattice that has an arbitrary distance among agents, so it is hard to apply this algorithm to general systems. Therefore, we extended the flocking algorithm to be able to form arbitrary lattice and follow several virtual agents.

We consider a matrix D. This matrix D has elements given by the distance d in equation (3) between each agent. The distance between agent i and j is denoted as d_{ij}. Now we assume that there are n agents.

$$D = \begin{bmatrix} d_{11} & d_{12} & \cdots & d_{1n} \\ d_{21} & \ddots & & \vdots \\ \vdots & & \ddots & \vdots \\ d_{n1} & \cdots & \cdots & d_{nn} \end{bmatrix} \tag{16}$$

where $d_{ii} = 0, (i = 1, 2, ..., n)$, $d_{ij} = d_{ji}, (i, j = 1, 2, ..., n)$.

Also, we assume that each agent has its own virtual agent. Some virtual agents can be shared among a number of agents and we consider a total of s virtual agents. The distance among virtual agents, or the lattice of virtual agents, is denoted by matrix R as

$$R = \begin{bmatrix} r_{11} & r_{12} & \cdots & r_{1s} \\ r_{21} & \ddots & & \vdots \\ \vdots & & \ddots & \vdots \\ r_{s1} & \cdots & \cdots & r_{s} \end{bmatrix} \tag{17}$$

where $r_{ii} = 0, r_{ij} = r_{ji}, (i, j = 0, 1, ..., s)$.

We define r in equation (2) for each agent as follows

$$J = \{1, 2, ..., s\} \tag{18}$$

$$J_{(-k)} = J - \{k\} \tag{19}$$

$$r_i^{min} = \min_{j \in J_{(-k)}} r_{kj} \tag{20}$$

with r_i the value of r for agent i with destination k. Therefore, (2),(8),(11) are updated as follows

$$N_i = \{j \in V : \|q_j - q_i\| < r_i^{min}\}, \tag{21}$$

$$a_{ij} = \rho_h(\|q_j - q_i\|_\sigma / \|r_i\|_\sigma) \in [0, 1], j \neq i, \tag{22}$$

$$\phi_\alpha(z) = \rho_h(z / \|r_i\|_\sigma)\phi(z - \|d_{ij}\|_\sigma). \tag{23}$$

Also, if agent i has a virtual agent k, then (15) is given by

$$f_i^\gamma = -c_1(q_i - q_{rk}) - c_2(p_i - p_{rk}) \tag{24}$$

where q_{rk} and p_{rk} are the position and velocity of virtual agent k respectively.

Fig.3 shows the example of the flocking configuration achieved by this extended algorithm. In Fig.3, three circles indicate the virtual agent. Agents are represented as a small triangle. The virtual agent for each agent is the circle which is the same color. In this experiment, the distance d_{ij} is determined as

$$d_{ij} = \begin{cases} const & s_i = s_j \\ r_{ij} & s_i \neq s_j \end{cases} \tag{25}$$

where s_i is the virtual agent of agent i and s_j is that of agent j.

Fig.3 shows the sequential results of the blue circle's movement.

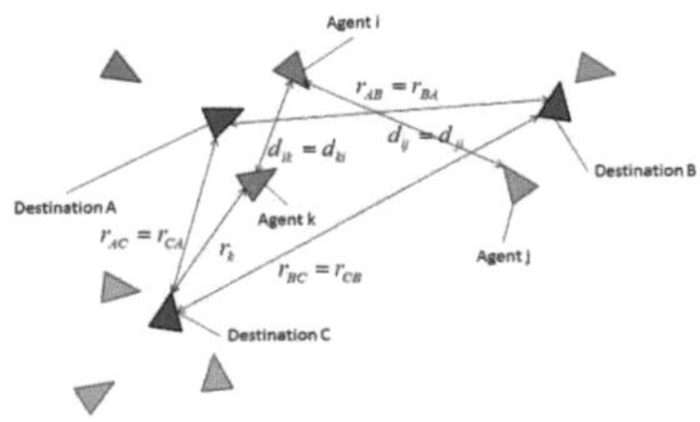

Fig. 2 Relationship among agents

(a) initial position (t=0) (b) formation at t = 1000

Fig. 3 2 dimensional flocking experiment

2.4 1-dimensional Flocking Algorithm and Probabilistic Fluctuation

By systematically varying the various parameters and understanding the outcomes, it might be possible to improve robustness and adaptivity of whole system. We assume that a system is a flock of parameters and the system can change its parameters in accordance with sensor inputs. To apply proposed algorithm to the control of the system, we consider the state of agent as a scalar. We assign the system parameter and specific sensor to agents, so each value of system parameters and specific sensors such as an angle of a servo and input of a pressure sensor is the state of the agent. We thus assume the whole system as a flock of agents.

However, the state value of the agent that is the assigned sensor is determined only by sensor input, and thus cannot be changed by the agent. This agent is not able to change its own state in accordance with the equation (4) in the same way as other agents. Therefore, the objective of the agents is to reduce the error between the state value of an agent that is the assigned sensor and that of its virtual agent by the movement of the others. We refer to the virtual agent as an ideal value if the state of the virtual agent is scalar.

On the other hand, each agent of a flocking algorithm is designed to determine its position only from the relationship with neighborhood agents, but is not designed to adjust the whole flock by moving itself. For this reason, if the state value of the agent is close to its ideal value, whole flock stagnate even though the state value of the other agent is far away from its ideal value.

To avoid such stagnation, we also apply probabilistic fluctuation to agents. We added a probabilistic fluctuation term to input u_i. We changed equation (4) as follows

$$u_i = f_i^g + f_i^d + f_i^\gamma + f_i^{prob} \tag{26}$$

$$f_i^{prob} = c_{prob} \varepsilon v_{random} \tag{27}$$

where c_{prob} is a positive constant and $v_{random} \in (-0.5, 0.5)$.

2.5 Motion of the Flock

When applying proposed flocking algorithm to the problem such as optimization of parameter. If the position of a virtual agent is fixed based on his local situation, it is not able to search wide a area because agents tend to aggregate to the virtual agent's neighborhood. These characteristics cause the stagnation in the neighborhood of local minimum point. To avoid such situation, we have to search better answer by moving virtual agents. We thus apply hill climbing like method to this flocking algorithm. We introduce a mechanism that has a whole flock move in accordance with the error between agents that represents the sensor and their virtual agents.

Let $q_G(t)$, $q_G(t+1)$ be the center of gravity of flock at a given time $t, t+1$ respectively. Also, let $\varepsilon(t)$, $\varepsilon(t+1)$ be the error between agents that represent sensor and their virtual agent at given time $t, t+1$ respectively. Now, we define the flock movement vector Δq as

$$\Delta q = \left(1 - \frac{\varepsilon(t+1)}{\varepsilon(t)}\right) \{q_G(t+1) - q_G(t)\}. \tag{28}$$

We determine the position of virtual agents at given time $t+2$ by

$$q_r(t+2) = q_r(t+1) + \Delta q. \tag{29}$$

This mechanism make a flock move in adirection that reduces the error value between agents that represent the sensor and their virtual agents.

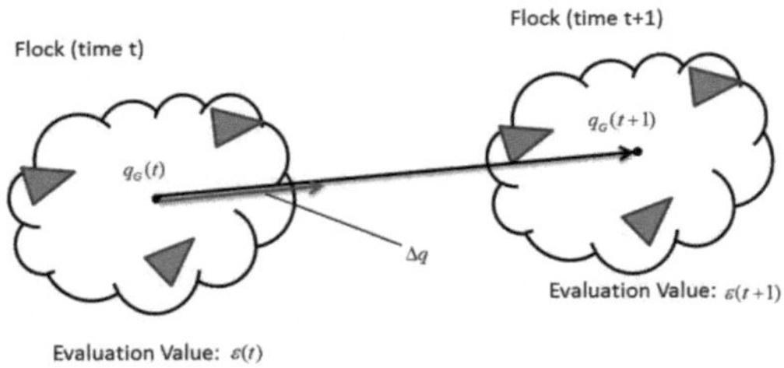

Fig. 4 Flock shift mechanism

3 Application of Flocking Algorithm to Camera Control

By applying the proposed flocking algorithm to a system, we hypothesize that the system will be able to adapt to changes in environment. To show this, we applied the proposed flocking algorithm to an actual system and evaluated its performance. As an example, we apply the flocking algorithm to a position tracking camera system. A tracking camera often has a difficult time tracking objects without parameter adjustments due to changes of in lighting conditions in the environment, *e.g.*, changes in brightness, direction of light, and other factors. Such a system has many control parameters and the system can adapt changing the various parameters.

3.1 Experiment Environment

In this experiment, we used a position tracking camera system that tracks a line follower with 3 types of marker, red, blue, and green. The tracking system employs both color extraction and outline extraction to the images obtained from the camera. For each color, it assumes each line follower's position as the centroid of a region enclosed by an outline that has maximum area. Also, we measured the error between tracked and actual positions for evaluation.

The line followers are Lego Mindstorms NXT robots (Fig. 5). The camera used for the position tracking system is UCAM-DLY300TAWH (Fig. 5) and the system can control brightness, contrast, saturation, sharpness, as well as gamma and white balance. The range of values are as follows: Brightness ∈ [-10, 10], Contrast ∈ [0, 20], Saturation ∈ [0, 10], Sharpness ∈ [0, 10], Gamma ∈ [100, 200], White balance ∈ [2800, 6500].

We assume that each agent has one of these parameters as the state value. Also, the maximum areas of outlines for each colors are the sensor value. Therefore, we

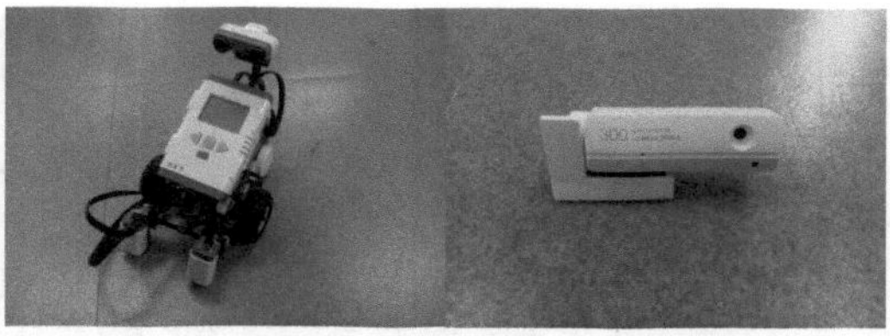

Fig. 5 Lego Mindstorms NXT as line tracing car and camera UCAM-DLY300TAWH

prepared a flock made of 9 agents and assigned each value to each agent position q_i as follows.

1. Brightness q_1
2. Contrast q_2
3. Saturation q_3
4. Sharpness q_4
5. Gamma q_5
6. White balance q_6
7. Detected area of red marker q_7
8. Detected area of blue marker q_8
9. Detected area of green marker q_9

When the actual value was x, we determined the position of agent q_i as

$$q_i = (x - minimum\ value)/(maximum\ value - minimum\ value) \tag{30}$$

to standardize it to $[0,1]$.

We defined the ideal value of each parameter and sensor value is the value at the situation that the system tracks all the positions properly in a bright condition (situation 1) and measured them first. After that, we turned off the lights (situation 2) and experimented using the same parameters.

The coefficients of system are determined as follows : $h = 0.5$, $a = 0.5$, $b = 0.5$, $c_1 = 0.5$, $c_2 = 0.5$, $c_{prob} = 0.01$. Each default camera parameter is given by: 1. Brightness = 0, 2. Contrast = 3, 3. Saturation = 5, 4. Sharpness = 5, 5. Gamma = 150, 6. White balance = 4588.

We compared the result of three types of settings, the case with no adjustment mechanism (Default), adaptation using the flocking algorithm without a flock movement mechanism given by equation (29) (Algorithm 1) and adaptation using the flocking algorithm with a flock movement mechanism given by equation (29) (Algorithm 2).

3.2 Experimental Result

The experimental result in situation 1 and situation 2 is shown in Fig. 8. Also, the average error value in each trial is shown in Table. 1. Error value e for each step is determined as follows.

[1] Default [2] Algorithm 1 [3] Algorithm 2

Fig. 6 Camera tracking experiment in situation 1

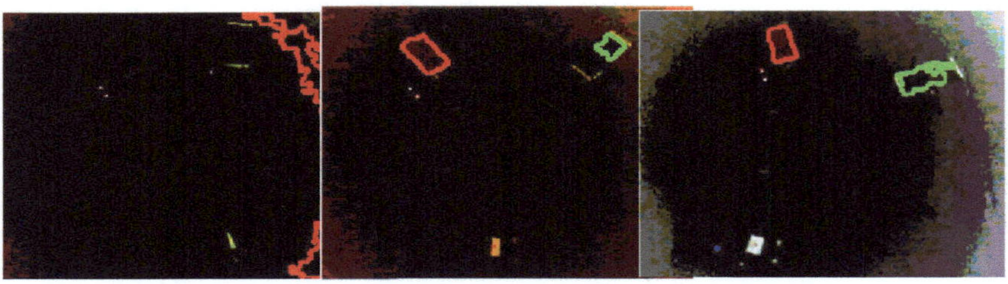

[1] Default [2] Algorithm 1 [3] Algorithm 2

Fig. 7 Camera tracking experiment in situation 2

$$e_1 = (actual\ position\ of\ red\ marker) - (tracked\ position\ of\ red\ marker) \quad (31)$$

$$e_2 = (actual\ position\ of\ green\ marker) - (tracked\ position\ of\ green\ marker) \quad (32)$$

$$e_3 = (actual\ position\ of\ blue\ marker) - (tracked\ position\ of\ blue\ marker) \quad (33)$$

$$e = \sqrt{e_1^2 + e_2^2 + e_3^2} \quad (34)$$

In situation 1, Default, Algorithm 1 and Algorithm 2 indicate almost the same result.

On the other hand, in situation 2, Default is not able to track the line follower's position properly and the error become greater. However, Algorithm 1 and Algorithm 2 could track the position by changing the camera parameter correctly. Also, from the aspect of average error value, the results of Algorithm 2 provided the best results for all the situations. From this result, by applying a flocking algorithm, the system can adjust itself to follow the change of environment that the camera picks up even though the ideal values at the changed environment are unknown.

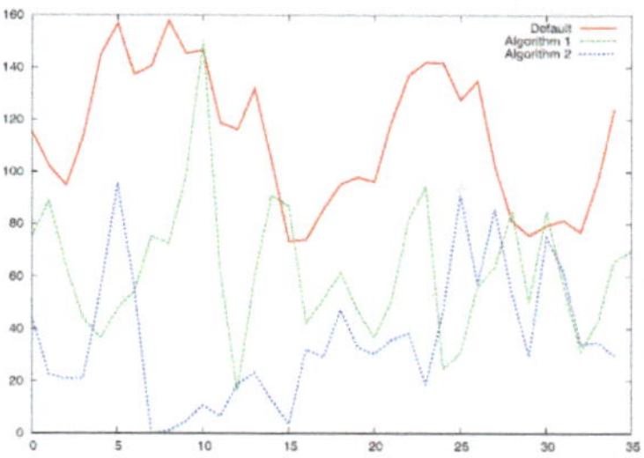

Fig. 8 Error value of tracking in bright room

Table 1 Experimental results of average error values

	situation 1	situation 2
Default	52.315	167.374
Algorithm 1	53.218	114.901
Algorithm 2	52.468	88.114

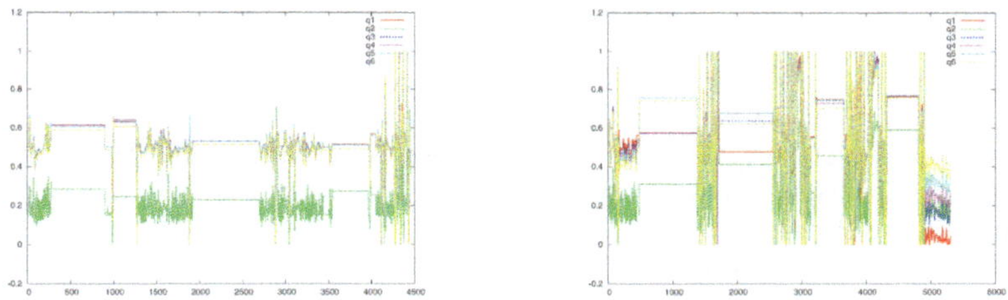

[1] Algorithm 1 [2] Algorithm 2

Fig. 9 State of each agent in situation 2

Also, without a flock movement mechanism, there were cases when the accumulated errors increased and the result were unstable. Movement of each agent is shown in Fig. 9.

4 Conclusion

We extended the flocking algorithm to construct an arbitrary lattice. Also, we introduced probabilistic fluctuation term and adjusted the sensor values by controlling the parameters autonomously. In addition, we introduced flock movement mechanism by applying evaluation value using sensor values. We showed how a flocking algorithm can be applied to a position tracking camera system in a fluctuating environment and showed how this application worked smoothly.

References

1. Bonabeau, E., Dorigo, M., Théraulaz, G.: From natural to artificial swarm intelligence (1999)
2. Bonabeau, E., Dorigo, M., Theraulaz, G.: Inspiration for optimization from social insect behaviour. Nature 406(6791), 39–42 (2000)
3. Boyd, S., Ghosh, A., Prabhakar, B., Shah, D.: Randomized gossip algorithms. IEEE Transactions on Information Theory 52(6), 2508–2530 (2006)
4. Cai, K., Ishii, H.: Average consensus on general digraphs. In: 2011 50th IEEE Conference on Decision and Control and European Control Conference (CDC-ECC), pp. 1956–1961. IEEE (2011)
5. Jakimovski, B., Meyer, B., Maehle, E.: Self-reconfiguring hexapod robot oscar using organically inspired approaches and innovative robot leg amputation mechanism. In: International Conference on Automation, Robotics and Control Systems, ARCS 2009, Orlando, USA (2009)
6. Kaizuka, Y., Tsumura, K.: Consensus via distributed adaptive control. In: World Congress, vol. 18, pp. 1213–1218 (2011)
7. Liu, L., Tsumura, K., Hara, S.: Adaptive consensus for a class of uncertain nonlinear multi-agent dynamical systems? In: Proceedings of the 18th IFAC World Congress, pp. 1219–1224 (2011)
8. Mesbahi, M., Egerstedt, M.: Graph theoretic methods in multiagent networks. Princeton University Press (2010)
9. Miyasato, Y.: A design method of model reference adaptive formation control. In: Proceedings of the 18th IFAC World Congress, pp. 1198–1203 (2011)
10. Mondada, F., Pettinaro, G.C., Guignard, A., Kwee, I.W., Floreano, D., Deneubourg, J.L., Nolfi, S., Gambardella, L.M., Dorigo, M.: Swarm-bot: A new distributed robotic concept. Autonomous Robots 17(2), 193–221 (2004)
11. Olfati-Saber, R.: Flocking for multi-agent dynamic systems: Algorithms and theory. IEEE Transactions on Automatic Control 51(3), 401–420 (2006)
12. Reynolds, C.W.: Flocks, herds and schools: A distributed behavioral model. In: ACM SIGGRAPH Computer Graphics, vol. 21, pp. 25–34. ACM (1987)
13. Seyboth, G.S., Dimarogonas, D.V., Johansson, K.H.: Control of multi-agent systems via event-based communication. In: World Congress, vol. 18, pp. 10086–10091 (2011)
14. Williams, A., Glavaski, S., Samad, T.: Formations of formations: Hierarchy and stability. In: Proceedings of the 2004 American Control Conference, vol. 4, pp. 2992–2997. IEEE (2004)
15. Zhang, Q.Z.J.F., Zhang, J.F.: Distributed quantized averaging under directed time-varying topologies. In: World Congress, vol. 18, pp. 2356–2361 (2011)

Part IV

Formation Control and Motion Planning for Robot Teams

A Stochastic Optimal Enhancement of Feedback Control for Unicycle Formations

Ross P. Anderson and Dejan Milutinović

Abstract. We consider an optimal feedback control approach for multiple nonholonomic vehicles to achieve a distance-based formation with their neighbors using only local observations. Beginning with a non-optimal feedback formation control, each agent determines an additive correction term to its non-optimal control based on an elliptic Hamilton-Jacobi-Bellman equation so that its actions are optimal and robust to uncertainty. In order to avoid offline spatial discretization of the stationary, high-dimensional cost-to-go function, we exploit the stochasticity of the distributed nature of the problem to develop an equivalent estimation problem in a continuous state space using a path integral representation. Consequently, each agent independently computes its optimal feedback control using a discrete-time Unscented Kalman smoother. Our approach is illustrated by a numerical example in which five agents achieve a pentagon with aligned and equal velocities.

1 Introduction

Nonholonomic vehicle formations, in which each agent is tasked with attaining and maintaining pre-specified distances from its neighbors, are beginning to demonstrate its significance and potential impact in a variety of applications in both the public and private sector [21]. The control approaches in relation to this problem have typically relied on stability analyses, or artificial potential functions [1, 4, 5, 6, 22, 23, 25].

Although previously-considered control approaches lead to satisfying results, they are non-optimal, and stability is usually only proven in the deterministic case. Because of this, we think it fruitful to examine the additional control input necessary to drive the non-optimal system into a formation *optimally* and in a manner that is

Ross P. Anderson · Dejan Milutinović
University of California, Santa Cruz, 1156 High Street, Santa Cruz, CA 95064
e-mail: {anderson,dejan}@soe.ucsc.edu

M.A. Hsieh and G. Chirikjian (eds.), *Distributed Autonomous Robotic Systems*, Springer Tracts in Advanced Robotics 104,
DOI: 10.1007/978-3-642-55146-8_19,

robust to uncertainty. In this work, we begin with a non-optimal feedback control policy [25], which provides an artificial potential function for distributed formation control of deterministic nonholonomic vehicles with feedback-controlled turning rate and acceleration. Then we introduce a stochastic optimal feedback control for each agent defining an additive control input in order to reach the formation in an optimal way. Consequently, our control approach is optimal, and, due to the adopted artificial potential function [25], it provides collision-free agent trajectories.

To compute an optimal feedback control, one must solve the Hamilton-Jacobi-Bellman (HJB) equation, which is a nonlinear partial differential equation (PDE). However, the state space dimension of multi-agent systems makes this conventional approach to stochastic optimal control impossible. In this work, we exploit the distributed nature of the problem at hand in order to make its solution tractable. The distributed formation control problem is inherently stochastic – from the perspective of one agent, neighbors' control inputs are unknown, as well as the consequences of these inputs to agent trajectories due to agent model uncertainties. Along these lines, this work considers the problem of controlling one agent based on observations of its neighbors *and the probability of their future motion*. This probability distribution arises from an assumption that a prior for the unknown control input of an agent can be robustly described as Brownian motion [12], which, for the agent model considered in this paper, results in a so-called "banana distribution" prior [17]. Based on our prior and the system kinematics, we can induce a probability distribution of the relative state $\mathbf{x}$ to all neighbors in an interval $(\mathbf{x}, \mathbf{x}+d\mathbf{x})$ at a particular future time [28]. More importantly, this probability distribution over future system trajectories can be used to statistically infer the probability distribution of the *control*, and, hence, the optimal control. In particular, the relations between the solutions to optimal control PDEs and the probability distribution of stochastic differential equations [8, 19, 31], allows certain stochastic optimal control problems to be written as an estimation problem on the distribution of optimal trajectories in continuous state space, in a manner known as the path integral (PI) approach (see [13, 14, 26, 27] as well as [15] and references therein for a more recent review of results, and see [18, 20] for an analogous approach in the open-loop control case).

Multiagent systems have previously been studied using the PI approach [2, 3, 30, 29], but in these works, the agents cooperatively compute their control from a marginalization of the joint probability distribution of the group's trajectory. In this paper, we develop a method by which agents independently compute their controls without explicit communication. Moreover, previous works using the PI approach have formulated a receding horizon optimal control problem for which stability is difficult to guarantee [11]. We therefore consider an elliptic control problem over a planning horizon that ends only when the formation is reached. In this sense, each agent is estimating both their optimal control and the time that the formation will be achieved. Finally, our work differs from previous PI implementations in that the optimal feedback control is computed independently by each agent from a nonlinear Kalman smoothing algorithm.

This paper is organized as follows. Section 2 introduces the formation control problem as viewed by a single agent in the group, followed by a derivation of a path integral representation in Section 3. Section 4 presents a Kalman smoother method for computing individual agent control. Section 5 illustrates our method with a simulated five-agent formation, and we conclude with Section 6.

2 Control Problem Formulation

We consider a team of agents, each described by a Cartesian position (x_m, y_m), a heading angle θ, a speed v_m, and the kinematic model:

$$\begin{aligned} \mathrm{d}x_m(t) &= v_m \cos\theta_m \mathrm{d}t \\ \mathrm{d}y_m(t) &= v_m \sin\theta_m \mathrm{d}t \\ \mathrm{d}\theta_m(t) &= \omega_m \mathrm{d}t + \sigma_{\theta,m}\mathrm{d}w_{\theta,m} \\ \mathrm{d}v_m(t) &= u_m \mathrm{d}t + \sigma_{v,m}\mathrm{d}w_{v,m}, \end{aligned} \tag{1}$$

where ω_m and u_m are the feedback-controlled turning rate and acceleration, respectively, and where $\mathrm{d}w_{\theta,m}$ and $\mathrm{d}w_{v,m}$ are mutually independent Wiener process increments with corresponding intensities $\sigma_{\theta,m}$ and $\sigma_{v,m}$, respectively.

To achieve a distributed control, the problem is formulated from the perspective of just one agent, which we call the agent-in-focus, or AiF for short. We denote the state (x, y, θ, v) of the AiF *sans* subscript. The AiF observes M neighbors with subscripts $m = 1, \ldots, M$, irrespective of the total number of agents in the population. The AiF computes an optimal feedback control to reach a formation with respect to observed neighbors, and the formation is achieved when the inter-agent distances $r_m = \sqrt{(x-x_m)^2 + (y-y_m)^2}$, $m = 1, \ldots, M$ reach a set of predefined nominal distances δ_m within a tolerance ϵ_r, the agents' heading angles are aligned within a tolerance ϵ_θ, and the speeds are equal within a tolerance ϵ_v. For the AiF, this occurs when its perspective of the system state $\mathbf{x}$ belongs to a set $\mathscr{F}$:

$$\begin{aligned} \mathscr{F} = \{\mathbf{x} : &|r_m - \delta_m| \leq \epsilon_r, \\ &|\theta - \theta_m| \leq \epsilon_\theta, \\ &|v - v_m| \leq \epsilon_v, \ \ m = 1, \ldots, M\}. \end{aligned} \tag{2}$$

The AiF further assumes no information about the observations made by its neighbors. In other words, the AiF assumes that its observed neighbors are only observing the AiF. This type of scenario is illustrated in Fig. 1.

We introduce collision avoidance by adding to the original kinematic model the artificial potential function [25] for collision-free velocity vector alignment of groups of vehicles described by a noiseless version of (1). The evolution equations for $\theta(t)$ and $v(t)$ become

$$\mathrm{d}\theta(t) = \omega_{\mathrm{D}}\mathrm{d}t + \omega\mathrm{d}t + \sigma_\theta \mathrm{d}w_\theta, \qquad \mathrm{d}v(t) = u_{\mathrm{D}}\mathrm{d}t + u\mathrm{d}t + \sigma_v \mathrm{d}w_v,$$

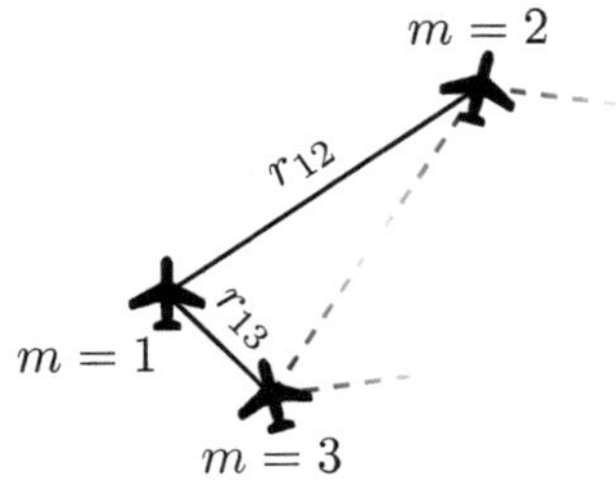

Fig. 1 In this scenario, agent 1 (the AiF) observes neighbors 2 and 3 (solid lines) and attempts to achieve the inter-agent spacings $r_{12} = \delta_{12}$ and $r_{13} = \delta_{13}$, as well as alignment of heading angles and speeds. Agent 1 is unaware of any other observation connectivity (dashed lines).

so that the controls ω and u are interpreted as the optimal *correction* terms to the deterministic turning rate feedback control ω_{D} and acceleration feedback control u_{D}, respectively, in the presence of uncertainty. For brevity, the reader is directed to [25] for the relevant equations for ω_{D} and u_{D}. Suffice it to say that in the deterministic case ($\sigma_{\theta,m} = \sigma_{v,m} = 0$), the non-optimal feedback control ω_{D} and u_{D} ensures collision avoidance, that is, the inter-agent distances remain strictly positive[1], and it helps to align the vehicle velocity vectors. Moreover, it guarantees that the group will tend toward a minimum of the artificially-constructed potential energy $V(r_m)$. Our specific choice of $V(r_m) = \delta_m^2 ||r_m||^{-2} + 2\log||r_m||$ (see [25]) causes the potential to reach minimum value when all inter-agent distances $r_m \to \delta_m$. Note that from the perspective of the AiF, the non-optimal controls executed by a neighboring agent m are based on that neighbor's sole observation, i.e., the observation of the AiF, and independent of others in the population, as before.

Finally, we look to the evolution equations for the heading angle $\theta_m(t)$ and speed $v_m(t)$ of a neighbor m to the AiF. In our formulation, the optimal controls $\theta_m(t)$ and $v_m(t)$ will be computed by the AiF, while assuming that agent m is only observing the AiF. The distributed nature of this problem is preserved, since during simulation, the control executed by agent m will differ from what is expected from it. Therefore, the control the AiF computes for agent m is modeled as the mean value of a Gaussian random variable:

$$\mathrm{d}\theta_m(t) = \omega_{\text{D},m}\mathrm{d}t + N\left(\omega_m \mathrm{d}t, \sigma_{\theta,m}^2 \mathrm{d}t\right), \qquad \mathrm{d}v_m(t) = u_{\text{D},m}\mathrm{d}t + N\left(u_m \mathrm{d}t, \sigma_{v,m}^2 \mathrm{d}t\right),$$

where $\sigma_{\theta,m}$ and $\sigma_{v,m}$ take into account both kinematic uncertainty *and control uncertainty*, so that $\sigma_{\theta,m} > \sigma_\theta$ and $\sigma_{v,m} > \sigma_v$. In summary, we have a kinematic model for the AiF of the form

[1] Ensuring collision avoidance among agents with non-zero collision radii would require a different deterministic control or artificial potential function than is considered in this work.

$$dx(t) = v\cos\theta dt \tag{3}$$
$$dy(t) = v\sin\theta dt \tag{4}$$
$$d\theta(t) = \omega_{\mathrm{D}} dt + \omega dt + \sigma_\theta dw_\theta \tag{5}$$
$$dv(t) = u_{\mathrm{D}} dt + u dt + \sigma_v dw_v \tag{6}$$
$$dx_m(t) = v_m \cos\theta_m dt \tag{7}$$
$$dy_m(t) = v_m \sin\theta_m dt \tag{8}$$
$$d\theta_m(t) = \omega_{\mathrm{D},m} dt + \omega_m dt + \sigma_{\theta,m} dw_{\theta,m} \tag{9}$$
$$dv_m(t) = u_{\mathrm{D},m} dt + u_m dt + \sigma_{v,m} dw_{v,m}, \qquad m = 1,\ldots,M. \tag{10}$$

This model can be written in a general form:

$$d\mathbf{x}(t) = f(\mathbf{x})dt + B\mathbf{u}dt + \Gamma d\mathbf{w}, \tag{11}$$

where the state vector $\mathbf{x}$ includes the system state from the perspective of the AiF, $f(\mathbf{x})$ captures the kinematics including the deterministic, collision-avoiding controls, and $\mathbf{u} = [\omega, u, \omega_1, u_1, \ldots, \omega_M, u_M]^T$ is a vector of optimal feedback controls to be computed by the AiF. The Wiener process $d\mathbf{w}$ captures the uncertainty due to model kinematics for each agent, as well as the uncertainty due to the control executed by neighboring agents $m = 1,\ldots,M$.

Our goal is to compute the feedback controls $\mathbf{u}(\mathbf{x})$ that minimize the total accumulated cost until the formation is reached (a problem sometimes called control until a target set is reached). We define the following cost functional for the AiF:

$$J(\mathbf{x}) = \min_{\mathbf{u}} \mathbb{E}\left\{\int_0^\tau \frac{1}{2}\Big(k(\mathbf{x}) + \mathbf{u}^T R\mathbf{u}\Big) ds\right\}, \tag{12}$$

where $\tau = \inf\{t > 0 : \mathbf{x}(t) \in \mathscr{F}\}$ is a (finite) first exit time, i.e., the first time that the state reaches a formation in $\mathscr{F}$ defined in (2). We note that, unlike previous works that either use a receding horizon approach or fix a final time, the final time τ is not known in advance. The positive semi-definite matrix R provides a quadratic control penalty, and the instantaneous state cost $k(\mathbf{x})$,

$$k(\mathbf{x}) = (h(\mathbf{x}) - \boldsymbol{\mu})^T Q(h(\mathbf{x}) - \boldsymbol{\mu}), \tag{13}$$

is a quadratic that reaches minimum value when the distances to each of the AiF's M neighbors equal the nominal distances δ_m, the heading angles are equal, and the speeds are equal with:

$$h(\mathbf{x}) = [r_1,\ldots,r_M, \theta-\theta_1,\ldots,\theta-\theta_M, v-v_1,\ldots,v-v_M]^T, \tag{14}$$
$$\boldsymbol{\mu} = [\delta_1,\ldots,\delta_M, 0,\ldots,0]^T, \tag{15}$$

and Q is a diagonal positive definite matrix. Note that this instantaneous state cost reaches minimum value when the potential energy associated with noiseless control also reaches minimum value. However, the noiseless controls also prevent collisions using an infinite potential energy when inter-agent distances approach $r_m = 0$. Since the correction control $\mathbf{u}$ is penalized, it will not overcome this barrier, and collision avoidance is still ensured.

3 Path Integral Representation

In this section we show how the optimal control problem can be represented as a path integral over possible system trajectories. The derivation is similar to that used in previous works, but the new type of cost functional used in this paper warrants a new derivation. The (stochastic) Hamilton-Jacobi-Bellman equation for model (11) and cost functional (12) is

$$0 = \min_{\mathbf{u}} \left\{ (f + B\mathbf{u})^T \partial_{\mathbf{x}} J + \frac{1}{2} \mathrm{Tr}\left(\Sigma \partial_{\mathbf{x}}^2 J\right) + \frac{1}{2} k(\mathbf{x}) + \frac{1}{2} \mathbf{u}^T R \mathbf{u} \right\}, \tag{16}$$

where $\Sigma = \Gamma\Gamma^T$. We have chosen boundary conditions for this PDE as $J(\mathbf{x}(\tau)) = 0$ for $\mathbf{x} \in \mathscr{F}$. The HJB equation must typically be solved numerically in a discretized state space until a steady state is reached (see [16], for example). However, the structure of the problem at hand allows us to avoid this through a suitable transformation.

The optimal control $\mathbf{u}(\mathbf{x})$ that minimizes (16) is

$$\mathbf{u}(\mathbf{x}) = -R^{-1} B^T \partial_{\mathbf{x}} J, \tag{17}$$

which, when substituted back into the HJB equation, yields:

$$0 = f^T \partial_{\mathbf{x}} J - \frac{1}{2} (\partial_{\mathbf{x}} J)^T B R^{-1} B^T \partial_{\mathbf{x}} J + \frac{1}{2} \mathrm{Tr}\left(\Sigma \partial_{\mathbf{x}}^2 J\right) + \frac{1}{2} k(\mathbf{x}(t)). \tag{18}$$

Next, we apply a logarithmic transformation [7] $J(\mathbf{x}) = -\lambda \log \Psi(\mathbf{x})$ for constant $\lambda > 0$ to obtain a new PDE

$$\begin{aligned} 0 = \frac{k(\mathbf{x})}{2\lambda} - \frac{f^T}{\Psi} \partial_{\mathbf{x}} \Psi - \frac{1}{2} \frac{1}{\Psi} \mathrm{Tr}\left(\Sigma \partial_{\mathbf{x}}^2 \Psi\right) \\ - \frac{1}{2} \frac{\lambda}{\Psi^2} (\partial_{\mathbf{x}} \Psi)^T B R^{-1} B^T \partial_{\mathbf{x}} \Psi + \frac{1}{2} \frac{1}{\Psi^2} (\partial_{\mathbf{x}} \Psi)^T \Sigma \partial_{\mathbf{x}} \Psi. \end{aligned} \tag{19}$$

In the model (3)-(10), it can be seen that the optimal controls $\mathbf{u}(\mathbf{x})$ act as a correction term to the deterministic controls and the stochastic noise. Penalizing this control (12) suggests that the optimal control is that which is, in some sense, "close" to the passive, deterministic process (see [27] for a more precise definition in terms of Kullback-Leibler divergence). Moreover, this implies that the possibility of a large stochastic disturbance (either due to neighbors' unknown controls or model kinematics) requires the possibility of a greater control input. Because of this, we assume that the noise in the controlled components is inversely proportional to the control penalty, or $\Sigma = \Gamma\Gamma^T = \lambda B R^{-1} B^T$. This selects the value of the control penalty that we shall use in the following and is given by $R = \lambda\ diag\left(\sigma_\theta^{-2}, \sigma_v^{-2}, \sigma_{\theta,1}^{-2}, \sigma_{v,1}^{-2}, \dots, \sigma_{\theta,M}^{-2}, \sigma_{v,M}^{-2}\right)$ and also causes the quadratic terms on the second line of (19) to cancel, so that the remaining PDE for Ψ is linear:

$$0 = f^T \partial_{\mathbf{x}} \Psi(\mathbf{x}) + \frac{1}{2}\mathrm{Tr}\left(\Sigma \partial_{\mathbf{x}}^2 \Psi\right) - \frac{k(\mathbf{x})}{2\lambda}\Psi(\mathbf{x}), \tag{20}$$

$$\Psi(\mathbf{x}) = 1, \qquad \mathbf{x} \in \mathscr{F} \tag{21}$$

As before, this could be solved numerically until a steady state is reached. However, the Feynman-Kac equations [19, 31] connect certain linear differential operators to adjoint operators that describe the evolution of a *forward* diffusion process beginning from the current state $\widetilde{\mathbf{x}}(0) = \widetilde{\mathbf{x}}_0 = \mathbf{x}$. From the Feynman-Kac equations, the solution to (20) is [8]:

$$\Psi(\mathbf{x}) = \mathbb{E}_{\widetilde{\mathbf{x}},\tau|\widetilde{\mathbf{x}}_0}\left\{\Psi(\widetilde{\mathbf{x}}(\tau))\exp\left(-\frac{1}{2\lambda}\int_0^\tau k(\widetilde{\mathbf{x}}(s))ds\right)\right\}. \tag{22}$$

where $\widetilde{\mathbf{x}}(t)$ satisfies the path integral-associated, uncontrolled dynamics (cf. (11)),

$$\mathrm{d}\widetilde{\mathbf{x}}(t) = f(\widetilde{\mathbf{x}}(t))\mathrm{d}t + \Gamma \mathrm{d}\mathbf{w}, \tag{23}$$

with initial condition $\widetilde{\mathbf{x}}(0) = \mathbf{x}$. The expectation in (22) is taken with respect to the joint distribution of $(\widetilde{\mathbf{x}}, \tau)$ of sample paths $\widetilde{\mathbf{x}} = \widetilde{\mathbf{x}}(t)$ that begin at $\widetilde{\mathbf{x}}_0 = \mathbf{x}$ and evolve as (23) until hitting the formation $\widetilde{\mathbf{x}}(\tau) \in \mathscr{F}$ at time τ. Unlike previous works, where the terminal time is fixed and known in advance, this stopping time is a property of the set of stochastic trajectories $\widetilde{\mathbf{x}}(t)$.

The distribution $(\widetilde{\mathbf{x}}, \tau)$ is difficult to obtain. Monte Carlo techniques may be used to sample trajectories $\widetilde{\mathbf{x}}$, but hitting the formation is a rare event unless there is an artificial mechanism to "guide" the trajectory into the formation. In this work, we determine the trajectory $\widetilde{\mathbf{x}}|\tau, \mathbf{x}_0$ conditioned on its hitting time. From the law of total expectation,

$$\Psi(\mathbf{x}) = \mathbb{E}_{\tau|\mathbf{x}_0}\left\{\mathbb{E}_{\widetilde{\mathbf{x}}|\tau,\widetilde{\mathbf{x}}_0}\left[\Psi(\widetilde{\mathbf{x}}(\tau))\exp\left(-\frac{1}{2\lambda}\int_0^\tau k(\widetilde{\mathbf{x}}(s))ds\right)\right]\right\} \tag{24}$$

$$= \mathbb{E}_{\tau|\mathbf{x}_0}\left\{\Psi(\mathbf{x}_0|\tau)]\right\}. \tag{25}$$

In practice, we find that the inner distribution $\Psi(\mathbf{x}_0|\tau)$ exhibits small tails for most τ and has high probability for just a small range of τ. Moreover, the range of τ with higher likelihood $\Psi(\mathbf{x}_0|\tau)$ is that which appears to equally balance state and control costs. Therefore, we consider a discrete set $(\tau_1, \dots, \tau_{N_\tau})$ of N_τ possible values for τ with non-informative, uniform prior probabilities. This implies that the distribution of the hitting times is implicitly encoded in the length (and the ensuing cost) of the path $\widetilde{\mathbf{x}}|\tau_i, \mathbf{x}_0$. Since $\Psi(\mathbf{x}(\tau_i)) = 1$ from (21), the solution (24) can be expanded as:

$$\Psi(\widetilde{\mathbf{x}}) = \frac{1}{N_\tau} \sum_{i=1}^{N_\tau} \mathbb{E}_{\widetilde{\mathbf{x}}|\widetilde{\mathbf{x}}_0,\tau_i} \left\{ \Psi(\widetilde{\mathbf{x}}(\tau_i)) \exp\left(-\frac{1}{2\lambda} \int_0^{\tau_i} k(\widetilde{\mathbf{x}}(s)) ds \right) \right\} \tag{26}$$

$$= \frac{1}{N_\tau} \sum_{i=1}^{N_\tau} \mathbb{E}_{\widetilde{\mathbf{x}}|\widetilde{\mathbf{x}}_0,\tau_i} \left\{ \exp\left(-\frac{1}{2\lambda} \int_0^{\tau_i} k(\widetilde{\mathbf{x}}(s)) ds \right) \right\}. \tag{27}$$

By discretizing the interval $[0, \tau_i]$ into N_i intervals of equal length Δt, $t_0 < t_1 < \ldots < t_{N_i} = \tau_i$, we can consider a sample of the discretized trajectory $\widetilde{\mathbf{x}}^N|\mathbf{x}_0, \tau_i = (\widetilde{\mathbf{x}}_1, \ldots, \widetilde{\mathbf{x}}_{N_i})$. Under this discretization in time, the solution (24) can be written as

$$\Psi(\widetilde{\mathbf{x}}) = \frac{1}{N_\tau} \lim_{\Delta t \to 0} \sum_{i=1}^{N_\tau} \int d\widetilde{\mathbf{x}}^N P(\widetilde{\mathbf{x}}^N|\widetilde{\mathbf{x}}_0, \tau_i) \exp\left[-\frac{\Delta t}{2\lambda} \sum_{k=1}^{N_i} k(\widetilde{\mathbf{x}}_k) \right], \tag{28}$$

where $d\widetilde{\mathbf{x}}^N = \prod_{k=1}^{N_i} d\widetilde{\mathbf{x}}_k$ and where $P(\widetilde{\mathbf{x}}^N|\widetilde{\mathbf{x}}_0, \tau_i)$ is the probability of a discretized sample path, conditioned on the starting state $\widetilde{\mathbf{x}}_0$ and hitting time τ_i, given by

$$P(\widetilde{\mathbf{x}}^N|\widetilde{\mathbf{x}}_0, \tau_i) = \prod_{k=0}^{N_i-1} p(\widetilde{\mathbf{x}}_{k+1}|\widetilde{\mathbf{x}}_k, \tau_i). \tag{29}$$

Since the uncontrolled process (23) is driven by Gaussian noise with zero mean and covariance $\Sigma = \Gamma\Gamma^T$, the transition probabilities may be written as

$$p(\widetilde{\mathbf{x}}_{k+1}|\widetilde{\mathbf{x}}_k, \tau_i) \propto \exp\Big(-\frac{1}{2} (\widetilde{\mathbf{x}}_{k+1} - \widetilde{\mathbf{x}}_k - f(\widetilde{\mathbf{x}}_k)\Delta t)^T \times (\Delta t \lambda B R^{-1} B^T)^{-1} (\widetilde{\mathbf{x}}_{k+1} - \widetilde{\mathbf{x}}_k - f(\widetilde{\mathbf{x}}_k)\Delta t) \Big) \tag{30}$$

for $k < N_i - 1$, and $p(\widetilde{\mathbf{x}}_{k+1}|\widetilde{\mathbf{x}}_k, \tau_i) = \mathbb{1}_{h(\widetilde{\mathbf{x}}_{k+1})=\boldsymbol{\mu}}(\widetilde{\mathbf{x}}_{k+1})$ for $k = N_i - 1$.

The path integral representation of $\Psi(\widetilde{\mathbf{x}})$ is obtained from equations (28)-(30), and can be written as an exponential of an "action" $S(\widetilde{\mathbf{x}}^N|\widetilde{\mathbf{x}}_0, \tau_i)$ [10] along the time-discretized sample trajectories $(\widetilde{\mathbf{x}}_1, \ldots, \widetilde{\mathbf{x}}_N)$:

$$\Psi(\widetilde{\mathbf{x}}) \propto \lim_{\Delta t \to 0} \sum_{i=1}^{N\tau} \int d\widetilde{\mathbf{x}}^N \exp\left(-S(\widetilde{\mathbf{x}}^N|\widetilde{\mathbf{x}}_0, \tau_i) \right) \tag{31}$$

$$S(\widetilde{\mathbf{x}}_1, \ldots, \widetilde{\mathbf{x}}_N|\widetilde{\mathbf{x}}_0, \tau_i) = \sum_{k=1}^{N_i} \frac{\Delta t}{2\lambda} k(\widetilde{\mathbf{x}}_k) + \sum_{k=0}^{N_i-1} \frac{1}{2} (\widetilde{\mathbf{x}}_{k+1} - \widetilde{\mathbf{x}}_k - \Delta t f(\widetilde{\mathbf{x}}_k))^T \times \left(\lambda \Delta t B R^{-1} B^T\right)^{-1} (\widetilde{\mathbf{x}}_{k+1} - \widetilde{\mathbf{x}}_k - \Delta t f(\widetilde{\mathbf{x}}_k)). \tag{32}$$

Differentiating (31) with respect to $\widetilde{\mathbf{x}}_0$, we can obtain the optimal control (17) as [2]

$$\begin{aligned}\mathbf{u}(\widetilde{\mathbf{x}}) &= \lim_{\Delta t \to 0} \lambda R^{-1} B^T \partial_{\widetilde{\mathbf{x}}} \log \Psi \\ &= \lim_{\Delta t \to 0} \sum_{i=1}^{N_\tau} \int d\widetilde{\mathbf{x}}^N P(\widetilde{\mathbf{x}}^N | \widetilde{\mathbf{x}}_0, \tau_i) \mathbf{u}_L(\widetilde{\mathbf{x}}^N | \widetilde{\mathbf{x}}_0, \tau_i) \qquad (33) \\ &= \lim_{\Delta t \to 0} \sum_{i=1}^{N_\tau} \mathbb{E}_{P(\widetilde{\mathbf{x}}^N | \widetilde{\mathbf{x}}_0, \tau_i)} \left\{ \mathbf{u}_L(\widetilde{\mathbf{x}}^N | \widetilde{\mathbf{x}}_0, \tau_i) \right\} \qquad (34)\end{aligned}$$

where $\lim_{\Delta t \to 0} P(\widetilde{\mathbf{x}}^N | \widetilde{\mathbf{x}}_0, \tau_i) = P(\widetilde{\mathbf{x}} | \widetilde{\mathbf{x}}_0, \tau_i)$ is the probability of an *optimal* trajectory conditioned to hit the formation at time τ_i,

$$P(\widetilde{\mathbf{x}}^N | \widetilde{\mathbf{x}}_0, \tau_i) \propto e^{-S(\widetilde{\mathbf{x}}^N | \widetilde{\mathbf{x}}_0, \tau_i)}, \qquad (35)$$

which weights the local controls $\mathbf{u}_L(\widetilde{\mathbf{x}}^N | \widetilde{\mathbf{x}}_0, \tau_i)$ in (33), defined by

$$\mathbf{u}_L(\widetilde{\mathbf{x}}^N | \widetilde{\mathbf{x}}_0, \tau_i) = \frac{\widetilde{\mathbf{x}}_1 - \widetilde{\mathbf{x}}_0}{\Delta t} - f(\widetilde{\mathbf{x}}_0). \qquad (36)$$

Although the resulting control law is stationary, the state space is too large for it to be computed offline. Because of this, after computing $\mathbf{u}(\mathbf{x}) = \mathbf{u}(\widetilde{\mathbf{x}}_0)$, each agent executes only the first increment of that control, at which point the optimal control is recomputed. Then (34) is

$$\begin{aligned}\mathbf{u}(\mathbf{x}) &= \frac{\mathbb{E}_{P(\widetilde{\mathbf{x}}^N | \widetilde{\mathbf{x}}_0)} \{\widetilde{\mathbf{x}}_1\} - \mathbf{x}}{\Delta t} - f(\mathbf{x}) \\ &= \frac{\mathbb{E}_\tau \mathbb{E}_{P(\widetilde{\mathbf{x}}^N | \widetilde{\mathbf{x}}_0, \tau)} \{\widetilde{\mathbf{x}}_1\} - \mathbf{x}}{\Delta t} - f(\mathbf{x}). \qquad (37)\end{aligned}$$

In other words, the control (37) applied by an agent in state $\mathbf{x}$ is constructed from a realization of the unknown or random dynamics of the system that maximizes the probability of the trajectory that starts from $\mathbf{x}$ and evolves until hitting the formation. This probability is weighted by the cost accumulated along the path. One may compute the optimal control (37) once the path probability $P(\widetilde{\mathbf{x}}^N | \widetilde{\mathbf{x}}_0, \tau_i)$ has been computed, a nontrivial task to be discussed in the following section.

4 Computing the Control with Kalman Smoothers

In this section we present our approach to compute the control in (37). Although Monte Carlo techniques can be used to generate samples of the maximally-likely trajectory $P(\widetilde{\mathbf{x}}_1, \ldots, \widetilde{\mathbf{x}}_N | \widetilde{\mathbf{x}}_0)$, we find them to be slow in practice due to the high dimension of this problem ($\widetilde{\mathbf{x}}^N \in \mathbb{R}^{4NM}$). Moreover, when sampling a trajectory $\widetilde{\mathbf{x}}^N | \widetilde{\mathbf{x}}0, \tau_i$, the trajectory must be conditioned to hit the formation at τ_i. Finally, it is not necessary to sample the entire distribution $P(\widetilde{\mathbf{x}}^N | \widetilde{\mathbf{x}}_0)$ since only the estimate $\widehat{\mathbf{x}}_1 \equiv \mathbb{E}_{P(\widetilde{\mathbf{x}}^N | \widetilde{\mathbf{x}}_0)} \{\widetilde{\mathbf{x}}_1\}$ is needed.

Therefore, in this work, we treat the temporal discretization of the optimal trajectory $\widetilde{\mathbf{x}}^N$ as the hidden state of a stochastic process, where appropriately-chosen measurements of this hidden state are related to the system goal $\boldsymbol{\mu}$ (15). The optimal control can then be computed from the optimal estimate $\widehat{\mathbf{x}}_1$ given the process and measurements over a fixed interval $t_1, \ldots, \tau_i$. We define the following nonlinear smoothing problem.

Nonlinear Smoothing Problem:
Given measurements $\mathbf{y}_k = \mathbf{y}(t_k)$ for $t_k = t_1, \ldots, t_N = \tau_i$, where $t_{k+1} - t_k = \Delta t$, compute the estimate $\widehat{\mathbf{x}}_{1:N}$ of the hidden state $\widetilde{\mathbf{x}}_{1:N}$ from the nonlinear hidden state-space model:

$$\widetilde{\mathbf{x}}_{k+1} = \widetilde{\mathbf{x}}_k + \Delta t f(\widetilde{\mathbf{x}}_k) + \varepsilon_k \tag{38}$$

$$\mathbf{y}_k = h(\widetilde{\mathbf{x}}_k) + \eta_k, \tag{39}$$

where $f(\cdot)$ and $h(\cdot)$ are as in Section 2, and ε_k and η_k are independent multivariate Gaussian random variables with zero mean and covariances:

$$\mathbb{E}\left(\varepsilon_k \varepsilon_k^T\right) = \lambda \Delta t B R^{-1} B^T \tag{40}$$

$$\mathbb{E}\left(\eta_k \eta_k^T\right) = \begin{cases} \frac{\lambda}{\Delta t} Q^{-1} & k = 1, \ldots, N-1 \\ 0 & k = N \end{cases}. \tag{41}$$

The smoothing is initialized from $\widetilde{\mathbf{x}}_0 = \mathbf{x}$, the current state of the system as viewed by the AiF. Measurements $\mathbf{y}_k$ are always exactly $\mathbf{y}_k = \boldsymbol{\mu}$. □

To show the relation between the nonlinear smoothing problem and the stochastic optimal control problem, we write the probability of a hidden state sequence $\widetilde{\mathbf{x}}^N$ given measurements $\mathbf{y}_k$ and the initial state $\widetilde{\mathbf{x}}_0$, which is [9] $P\left(\widetilde{\mathbf{x}}^N | \widetilde{\mathbf{x}}_0, \mathbf{y}_1, \ldots, \mathbf{y}_N\right) \propto \prod_{k=1}^N p(\mathbf{y}_k | \widetilde{\mathbf{x}}_k) p(\widetilde{\mathbf{x}}_k | \widetilde{\mathbf{x}}_{k-1})$, where

$$\begin{aligned} p(\mathbf{y}_k | \widetilde{\mathbf{x}}_k) &\equiv p(\boldsymbol{\mu}_k | \widetilde{\mathbf{x}}_k) = N\left(h(\widetilde{\mathbf{x}}_k), \eta_k \eta_k^T\right) \\ &\propto \exp\left\{ -\frac{\Delta t}{2\lambda} (h(\widetilde{\mathbf{x}}_k) - \boldsymbol{\mu})^T Q (h(\widetilde{\mathbf{x}}_k) - \boldsymbol{\mu}) \right\} \end{aligned} \tag{42}$$

$$\begin{aligned} p(\widetilde{\mathbf{x}}_k | \widetilde{\mathbf{x}}_{k-1}) &= N\left(\widetilde{\mathbf{x}}_{k-1} + \Delta t f(\widetilde{\mathbf{x}}_{k-1}), \Delta t \Sigma\right) \\ &\propto \exp\left\{ -\frac{1}{2} \left(\widetilde{\mathbf{x}}_k - \widetilde{\mathbf{x}}_{k-1} - \Delta t f(\widetilde{\mathbf{x}}_{k-1})\right)^T \right. \\ &\quad \left. \times \left(\lambda \Delta t B R^{-1} B^T\right)^{-1} \left(\widetilde{\mathbf{x}}_k - \widetilde{\mathbf{x}}_{k-1} - \Delta t f(\widetilde{\mathbf{x}}_{k-1})\right) \right\}. \end{aligned} \tag{43}$$

Comparing the right hand sides of (42-43) with (31)-(32), it can be seen that they are identical to those in the stochastic optimal control problem.

Since the optimal control (37) is based on the probability of a full trajectory of fixed length and values $\boldsymbol{\mu}$ are available in advance, the expected value of the trajectory originating from state $\widehat{\mathbf{x}}_0$ conditioned to hit the formation at time τ_i, that is, the hidden states $\widehat{\mathbf{x}}_k$, $k = 1,\ldots,N$, can be found by filtering and then smoothing the process given the values $\boldsymbol{\mu}_k$ using a nonlinear fixed-interval Kalman smoother. Such an algorithm assumes that the increments given by (42) and (43) are Gaussian to some extent, but the algorithm is sufficiently fast to be applied in *real-time* by each unicycle in a potentially large group with an even larger state space, motivating its use in this work.

Once this estimated trajectory has been computed for each τ_i, the expectation over τ_i may be computed using (32) and (35). This would result in an average of the controls $u_L(\widetilde{\mathbf{x}}|\widetilde{\mathbf{x}}_0, \tau_i)$ to be applied, weighted by the probability of the optimal trajectory for each τ_i. In other words, each agent would estimate both the optimal system trajectory (from its perspective) given the time the formation will hit *and* the hitting time of the formation. Hitting the target sooner would save on state costs, but may cause an increase in control costs, and vice versa.

When the smoothing is complete and agents have applied their computed control, each agent must then observe the actual states of its neighbors so that the next iteration begins with the correct initial condition. We provide a pseudo code for our computations in Algorithm 1. In practice, the controller/smoother must be capable of efficiently smoothing over the horizon $[t_0, \tau_i]$. The computational complexity of the smoother used in this work is analyzed in [24], where it is seen that the number of operations required by the smoother roughly scales with the number of neighbors as M^3.

Algorithm 1. Formation control algorithm applied by each agent

```
x(t) ← measured state of system from AiF viewpoint
μ ← nominal distances
while x(t) ∉ 𝓕 do                                                    ▷ defined by (2)
    for i = 1,...,N_τ do
        E{x̃_1},...,E{x̃_N} ← KalmanSmoother (initial state = x_0, horizon = [0, τ_i],
                                              measurements = μ)
        E{u_L(x̃^N | x_0, τ_i)} ← (E(x̃_1) − x)/Δt − f(x)                 ▷ from (36)
        S^(i) ← action (E(x̃_1),...,E(x̃_N))                              ▷ using (32)
    end for
    u(x) = (1/N_τ) Σ_{i=1}^{N_τ} E{u_L(x̃^N | x_0, τ_i)}
    Apply computed corrective control u(x)                              ▷ using (11)
    x(t) ← measured state of system from AiF viewpoint
end while
```

5 Results

In this section, the Kalman smoothing method is employed so that five agents achieve the formation of a regular pentagon, where each agent is individually estimating the hidden optimal trajectory based on the relative kinematics of all of its neighbors. The agents observe all others, but, as described in Section 2, only the inter-agent connections *known* by an agent are used when computing the control. The instantaneous state cost (13) penalizes the mean squared distance from the unicycle to all of its $M = 4$ neighbors in excess of the side length of the pentagon (5

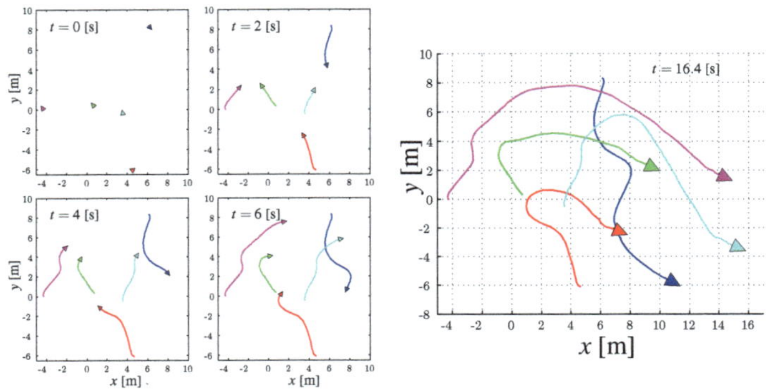

Fig. 2 Five agents, starting from random initial positions and a common speed $v = 2.5$ [m/s], achieve a regular pentagon formation by an individually-optimal choice of acceleration and turning rate, without any active communication. The frames at 2 [s] and 4 [s] show an example of collision avoidance between the two upper-left agents.

[m]) or the diagonal of the pentagon, depending on the relative configuration of the pentagon encoded in δ_m, $m = 1, \dots, 4$.

The system and algorithm parameters were chosen as $\lambda = 100$, $\sigma_\theta = 0.1, \sigma_v = 0.05, \sigma_{\theta,m} = 1, \sigma_{v,m} = 1, N_\tau = 10$, $\tau = 1, \dots, 10$, $Q = 100I$, $\epsilon_r = \epsilon_v = 0.1$, $\epsilon_\theta = 10°$, and $\Delta t = 0.1$, with all units relative to meters and seconds. The control was computed using a Discrete-time Unscented Kalman Rauch-Tung-Striebel Smoother [24]. Fig. 2 shows the trajectories of all agents, while the inter-agent distances and agents' angles and speeds can be seen in Fig. 3. The actual stopping time was $\tau = 16.1$ [s], and the agents did not collide. In the last second of simulation, a minor deviation in the agents' heading angles and speeds was seen. This was due to a final correction in agents' relative distances that was needed after having converged in heading angle and speed. Without the addition of the optimal controls, the agents formed a loose pentagon, but the collision-avoiding controls acting alone led to oscillatory trajectories, and the formation tolerances (2) were not reached in the first 100 [s] of simulation (Fig. 3).

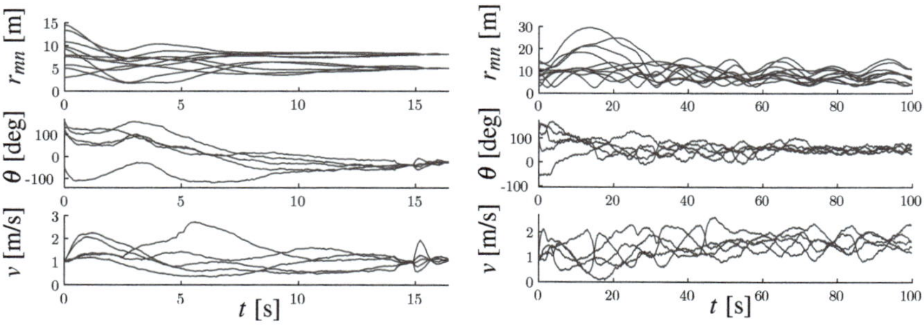

Fig. 3 Inter-agent distances r_{mn}, agent heading angles θ, and agent speeds v as a function of time using the stochastic optimal control (left) and the deterministic feedback control (right).

6 Discussion

This work considers the problem of unicycle formation control in a distributed optimal feedback control setting. Since this gives rise to a system with huge state space, we exploit the stochasticity inherent in distributed multi-agent control problems in order to apply a path integral method.

Each agent computes its optimal control using a nonlinear Kalman smoothing algorithm. The measurement and process noise of the smoothing problem are created using the structure of the cost function and stochastic kinematics. Aside from instantaneous observations of neighbors, the formation is created and maintained without any communication among agents. The possibility to extend the model to three spatial dimensions will be pursued.

In order to prevent collisions among agents, the optimal turning rate and acceleration controls affect the system alongside a non-optimal feedback control law based on an artificial potential function. This suggests that the type of stochastic optimal control problem considered in this work may provide a way to improve other deterministic feedback control laws used for multi-robot systems in the presence of uncertainties.

Acknowledgements. This work was supported by NSF GRFP 0809125. We thank the reviewers for their helpful comments and the Symposium attendees for their valuable feedback.

References

1. Anderson, B., Fidan, B., Yu, C., Walle, D.: UAV formation control: theory and application. In: Blondel, V., Boyd, S., Kimura, H. (eds.) Recent Advances in Learning and Control. LNCIS, vol. 371, pp. 15–34. Springer, Heidelberg (2008)
2. van den Broek, B., Wiegerinck, W., Kappen, B.: Graphical model inference in optimal control of stochastic multi-agent systems. Journal of Artificial Intelligence Research 32(1), 95–122 (2008)
3. van den Broek, B., Wiegerinck, W., Kappen, B.: Optimal control in large stochastic multi-agent systems. In: Adaptive Agents and Multi-Agent Systems III. Adaptation and Multi-Agent Learning, pp. 15–26 (2008)
4. Bullo, F., Cortes, J., Martinez, S.: Distributed control of robotic networks: A mathematical approach to motion coordination algorithms. Princeton University Press, Princeton (2009)
5. Dimarogonas, D.: On the rendezvous problem for multiple nonholonomic agents. IEEE Transactions on Automatic Control 52(5), 916–922 (2007)
6. Elkaim, G., Kelbley, R.: A Lightweight Formation Control Methodology for a Swarm of Non-Holonomic Vehicles. In: IEEE Aerospace Conference. IEEE, Big Sky (2006)
7. Fleming, W., Soner, H.: Logarithmic Transformations and Risk Sensitivity. In: Controlled Markov Processes and Viscosity Solutions, ch. 6. Springer, Berlin (1993)
8. Freidlin, M.: Functional Integration and Partial Differential Equations. Princeton University Press, Princeton (1985)
9. Gelb, A.: Applied Optimal Estimation. The MIT Press, Cambridge (1974)
10. Goldstein, H.: Classical Mechanics, 2nd edn. Addison-Wesley (1980)

11. Jadbabaie, A., Hauser, J.: On the stability of unconstrained receding horizon control with a general terminal cost. In: Proceedings of the 40th IEEE Conference on Decision and Control, vol. 5, pp. 4826–4831. IEEE, Orlando (2001)
12. van Kampen, N.G.: Stochastic Processes in Physics and Chemistry, 3rd edn. North Holland (2007)
13. Kappen, H.: Linear Theory for Control of Nonlinear Stochastic Systems. Physical Review Letters 95(20), 1–4 (2005)
14. Kappen, H.J.: Path integrals and symmetry breaking for optimal control theory. Journal of Statistical Mechanics, Theory and Experiment 2005, 21 (2005)
15. Kappen, H.J., Gómez, V., Opper, M.: Optimal control as a graphical model inference problem. Machine Learning 87(2), 159–182 (2012)
16. Kushner, H.J., Dupuis, P.: Numerical Methods for Stochastic Control Problems in Continuous Time, 2nd edn. Springer (2001)
17. Long, A.W., Wolfe, K.C., Mashner, M.J., Chirikjian, G.S.: The Banana Distribution is Gaussian: A Localization Study with Exponential Coordinates. In: Proceedings of Robotics: Science and Systems, Syndey (2012)
18. Milutinović, D.: Utilizing Stochastic Processes for Computing Distributions of Large-Size Robot Population Optimal Centralized Control. In: Proceedings of the 10th International Symposium on Distributed Autonomous Robotic Systems, Lausanne, Switzerland (2010)
19. Oksendal, B.: Stochastic Differential Equations: An Introduction with Applications, 6th edn. Springer, Berlin (2003)
20. Palmer, A., Milutinović, D.: A Hamiltonian Approach Using Partial Differential Equations for Open-Loop Stochastic Optimal Control. In: Proceedings of the 2011 American Control Conference, San Francisco, CA (2011)
21. Parker, L.E.: Multiple Mobile Robot Systems. In: Sciliano, B., Khatib, O. (eds.) Springer Handbook of Robotics, ch. 40, pp. 921–941. Springer (2008)
22. Ren, W., Beard, R.: Distributed consensus in multi-vehicle cooperative control: Theory and applications. Springer, New York (2007)
23. Ryan, A., Zennaro, M., Howell, A., Sengupta, R., Hedrick, J.: An overview of emerging results in cooperative UAV control. In: 2004 43rd IEEE Conference on Decision and Control, vol. 1, pp. 602–607 (2004)
24. Särkkä, S.: Continuous-time and continuous-discrete-time unscented Rauch-Tung-Striebel smoothers. Signal Processing 90(1), 225–235 (2010)
25. Tanner, H., Jadbabaie, A., Pappas, G.: Coordination of multiple autonomous vehicles. In: IEEE Mediterranean Conference on Control and Automation. IEEE, Rhodes (2003)
26. Todorov, E.: General duality between optimal control and estimation. In: 47th IEEE Conference on Decision and Control, vol. 5, pp. 4286–4292. IEEE, Cancun (2008)
27. Todorov, E.: Efficient computation of optimal actions. Proceedings of the National Academy of Sciences of the United States of America 106(28), 11,478–11,483 (2009)
28. Wang, M.C., Uhlenbeck, G.: On the theory of Brownian Motion II. Reviews of Modern Physics 17(2-3), 323–342 (1945)
29. Wiegerinck, W., van den Broek, B., Kappen, B.: Optimal on-line scheduling in stochastic multiagent systems in continuous space-time. In: Proceedings of the 6th International Joint Conference on Autonomous Agents and Multiagent Systems, p. 1 (2007)
30. Wiegerinck, W., Broek, B., Kappen, H.: Stochastic optimal control in continuous space-time multi-agent systems. In: 22nd Conference on Uncertainty in Artificial Intelligence, Cambridge, MA (2006)
31. Yong, J.: Relations among ODEs, PDEs, FSDEs, BDSEs, and FBSDEs. In: Proceedings of the 36th IEEE Conference on Decision and Control, pp. 2779–2784. IEEE, San Diego (1997)

Maximum-Leaf Spanning Trees for Efficient Multi-Robot Recovery with Connectivity Guarantees

Golnaz Habibi and James McLurkin

Abstract. This paper presents a self-stabilizing distributed algorithm for the recovery of a large population of robots in a complex environment—that is, to gather them all in a goal location. We assume the robots do not have a map of the environment, but instead use short-range network communications and local sensing to physically route themselves towards the goal location. Since the robot's motion can disrupt robots network communication and localization in complex environments, we desire an algorithm that can maintain connectivity while preserving efficient operation. Our approach constructs a spanning tree for physical routing, but only allows the leaves of this tree to navigate the goal location. This distributed maximum-leaf spanning tree (DMLST) ensures connectivity, while providing an efficient recovery by allowing the maximum number of robots to be mobile. We present empirical results on the competitive ratio of the DMLST and it is very good, approaching the optimal solution for our communication network. DMLST recovery has been tested in simulation, and implemented on a system of thirteen robots. While a basic recovery fails in all experiments, the DMLST recovery succeeds efficiently in most trials.

1 Introduction

Many practical applications of multi-robot systems, such as search-and-rescue, exploration, mapping and surveillance require robots to disperse across a large geographic area. Often overlooked is the need to *recover* the robots to a goal location after the application is complete. Recovery can be challenging in complicated environments. We focus on large populations of simple robots operating in large environments. We assume that constructing and sharing a global map of the en-

Golnaz Habibi · James McLurkin
Rice University, 6100, Main St., Houston, TX
e-mail: {gh4,jmclurkin}@rice.edu

M.A. Hsieh and G. Chirikjian (eds.), *Distributed Autonomous Robotic Systems*,
Springer Tracts in Advanced Robotics 104,
DOI: 10.1007/978-3-642-55146-8_20,

vironment is beyond the limits of their processing and communications. We also assume that GPS navigation and infrastructure communication are unavailable in indoor environments. Additionally, central communication is not practical for large populations, because of non scalability. *i.e.* the required bandwidth does not scale with the population size. We also assume that the communication range is much smaller than the size of environment. Therefore, a multi-hop network is required for communication. We need to measure the neighboring positions of a robot to be able to navigate. Because there is no map or GPS, we assume that each robot has access to the *local network geometry*, *i.e.* communications with nearby robots and position information of neighbors. This local network geometry supports multi-hop communications and distributed algorithms for a configuration control.

While sonar and radio positioning perform poorly in cluttered environments, line-of-sight communication such as IR communication and camera vision can provide us appropriate local geometry information. However, visual obstructions can block views of neighboring robots. These limited sensing abilities make the recovery more challenging in complicated environments. We desire a distributed algorithm that can function with the knowledge of only *local network geometry* and overcome recovery challenges.

This paper presents a distributed algorithm to *safely* and *efficiently* recover large populations of robots with limited sensing from complex and unstructured environments. Informally, we define safety as the ability to recover the robots without leaving any robot behind. We define efficiency as the comparison between actual and optimal execution time. We define the execution time as the time that is required for all the robots to go to the goal location. In addition, we desire a *self-stabilizing* distributed algorithm; an algorithm that produces a desired configuration from any connected configuration in a bounded time.

The primary contribution of our work is a distributed algorithm to recover a large population of robots efficiently while maintaining network connectivity. We present a distributed algorithm for efficiently computing an approximation of Maximum Leaf Spanning Tree(MLST). We combine this tree with mid-angle navigation to recover a large number of robots in a variety of environments.

The paper is organized as follows: The motivation of the paper and related work are presented in Section 2. Section 3 presents our model, assumptions and some preliminary concepts which include network communications assumptions and a dynamic model of the robot. Section 4 describes our distributed MLST algorithm (DMLST). The recovery of robots based on DMLST is explained in Section 5. Section 5 also analyzes the time efficiency and correctness of our recovery algorithm. We present simulation and experimental results in Section 6 and Section 7 respectively. We conclude the paper with a brief discussion of the work in Section 8.

2 Motivation and Related Work

Existing approaches do not solve the recovery problem for multi-robot systems—that is, these algorithms do not recover all of the robots in a goal location. For example, Batalin [1] used stationary vertices as navigation guides, but these stationary nodes remain stationary and are not recovered. By using a simple quantized control law, a group of agents with limited sensing achieve rendezvous and gather in a location [5]. However robots gather anywhere in the space and there is no specific location as a goal location for the recovery. Moreover, the sensor model is not realistic in [5]. Li and Rus [6] develop a distribute algorithm for sensor networks to guide robots to a target, but they do not guarantee inter-robot connectivity during the recovery.

We are looking for an algorithm to recover all the robots in a goal location while maintaining connectivity. For this purpose, we use the subset of robots as guides *Guide* for physical navigation. In order for the robots to travel along an efficient path to the goal location, we require that each robot u has at least one neighbor $N(u) \in Guide$ such that $N(u)$ is closer to the goal location. An spanning tree with a robot in the goal location as the root provides exactly this desired structure for navigation. The spanning tree provides a connected network in which each robot has a path to the root. We assume only the leaves of the tree can move during the recovery and internal nodes become stationary. We require an algorithm that converges quickly and provides an spanning tree in which each robot has a stationary parent to act as a navigation guide to lead the robot back to the goal location. we are looking for a spanning tree with maximum number of leaves to increase the number of moving robots during a recovery.

A Maximum Leaf Spanning Tree is a tree that spans a graph with the smallest possible number of internal vertices, producing a maximum number of leaf vertices [16]. Having a larger number of leaves allows that a larger number of robots are moving at any given time, decreasing the average physical time for all robot to get the goal location. Finding MLST is well-studied and is a NP-hard problem. There are different versions of MLST that are centralized [11], distributed [4] and self-stabilizing [4, 9]. However, all of these approaches constructs MLST in statics networks, but we seek an algorithm to generate an MLST in a dynamic networks.

We design a Distributed MLST(DMLST), that is an approximation of the Maximum-leaf Spanning Tree, to improve the efficiency of the recovery such that the largest number of robots move toward the goal location. This spanning tree is built based on the information of the depth of the graph of robot network which is provided by the broadcast tree and ad-hoc multi-hop communication. In the second part of the paper, we introduce a mid-angle navigation algorithm to navigate robots to the goal location, while reducing physical interference from inter-robot collisions.

3 Model and Assumptions

We assume that the network starts from a dispersed but connected state. First, we assume that there is a distinguished goal robot R_G located at the goal location. We use R_G as the root of a tree. There are no assumptions on the size or shape of the network or of the environment. The communication network is an undirected graph $G = (V,E)$. Each robot is modeled as a vertex, $u \in V$, where V is the set of all robots and E is the set of all robot-to-robot communication links. The neighbors of each vertex $u \in V$ are the set of robots within communication range r of robot u with a line-of-sight connection, denoted $N(u) = \{v \mid \{u,v\} \in E\}$. Each robot sits at the origin of its local coordinate system, with the $\hat{x}$-axis aligned with its current heading. Each robot can measure the angle with respect to its own $\hat{x}$-axis to each of its neighbors, but with a limited resolution of $\frac{\pi}{8}$.

Each robot is modeled as a small disk with position of $u.pos = (x, y, \theta)$. The robot has a differential drive, and can rotate in place and translate up to some maximum speed v_{max}. Obstacle sensors detect any collision with the environment and other robots. There is an *obstacle avoidance behavior* that can effectively maneuver the robot away from any collision. We do not model the *physical interference* [12] caused by these collisions with robots in the simulations, hence the dimension of the robot is ignored. However, this interference affects the experimental results in Section 7.

Algorithm execution occurs in a series of synchronous *rounds*. A Synchronizer models an asynchronous distributed system to a synchronous distributed system [13] . This syncronizer simplifies analysis and is straightforward to implement in a physical system. At each round, every robot $u \in V$ broadcasts the unique message $u.message$ to all of its neighbors. Therefore, robot u receives a $v\ .message$ from each neighbor in $v \in N(u)$. Each message contains a tuple of integers of the form: $u.message = (u.id, hops, u.ChildrenCount, u.SelectedParent, u.Stationary)$. The meaning of these fields will be described in Section 4. Here we simply note that the magnitude of each field for a fixed maximum number of robots n, is smaller than $log_2 n$, , i.e. the number of bits required to identify each robot. This produces a total message of constant size.

We assume robot platforms with limited sensors and capabilities. In particular, we assume that robots do not have a map of the environment, nor the ability to localize itself relative to the environment geometry. The problem of localization and mapping is typically solved with SLAM [18, 3] or Monte Carlo Localization (MCL) [8, 17, 7]. However, both of these techniques are beyond the sensing and processing capabilities of our platform. Because, the sensor model of our experimental platform is limited to measuring the angles to neighboring robots *i.e.* bearing-only sensing.

4 Distributed Maximum Leaf Spanning Tree Algorithm

Our distributed algorithm for finding the maximum leaf spanning tree has three stages: 1) the robots construct a spanning tree rooted in the goal robot; 2) Each robot counts its children and shares this information with its neighbors; 3) Each robot selects a parent in a fashion intended to maximize the number of children per parent. We label these *SelectedParent* robots as *internal* nodes, and the rest as *leaves*. This section describes this process in detail. The results of our algorithm is compared with the solution of brute force algorithm as an optimal solution. We use this comparison to compute the competitive ratio.

Algorithm 1. Algorithm for counting children of robot *u*

1: **Do forever**
2: *ChildrenCount* $\leftarrow 0$
3: **for** $\forall v \in N(u) \ni v.hop > u.hop$ **do**
4:     **if** $(v.SelectedParent = u.id) \vee (v.SelectedParent = undecided)$ **then**
5:         $u.ChildrenCount \leftarrow u.ChildrenCount + 1$
6:     **end if**
7: **end for**
8: BroadCast(*ChildrenCount*)

After broadcast tree is constructed by using the *goal robot* as a source [10]. The source become the root of the tree and all robots calculate their *hop* value. The hop value implies the shortest path network route to the source.

Algorithm 1 shows the procedure for counting the children of robot *u*. Robot *u* reads the messages *v.message* from its neighbors $v \in N(u)$ and calculates *u.ChildrenCount*. *u.ChildrenCount* contains the number of children that have actually selected robot *u* as *SelectedParent*, plus UNDECIDED children (line 4 - 6). *ChildrenCount* of robot *u* is broadcasted through the network. Algorithm 2 shows how a robot *u* selects its parent. Lines 6 to 14 search for the parent with *MaxChildrenCount*. All the parents with *MaxChildrenCount* are saved in the *PotentialParent*

Algorithm 2. Algorithm for selecting parent of robot *u*

1: **Do forever**
2: *PotentialParent* $\leftarrow \varnothing$
3: *MaxChildrenCount* $\leftarrow 0$
4: *SelectedParent* $\leftarrow$ *undecided*
5: **if** $N(u)$ has not been changed since the last round **then**
6:     **for all** $v \in N(u)$ **do**
7:         **if** $ChildrenCount > MaxChildrenCount$ **then**
8:             *PotentialParent* $\leftarrow \varnothing$
9:             Add *v* to *PotentialParent*
10:             $MaxChildrenCount \leftarrow ChildrenCount$
11:         **else if** $ChildrenCount = MaxChildrenCount$ **then**
12:             Add *v* to *PotentialParent*
13:         **end if**
14:     **end for**
15:     **if** (size(*PotentialParent*) > 1) **then**
16:         $Prob \leftarrow rand(0,1)$
17:         **if** $Prob \leq K$ **then**
18:             *SelectedParent* $\leftarrow$robot with minimum *id* is selected from *PotentialParent*
19:         **end if**
20:     **else**
21:         *SelectedParent* $\leftarrow$ *PotentialParent*
22:     **end if**
23: **end if**

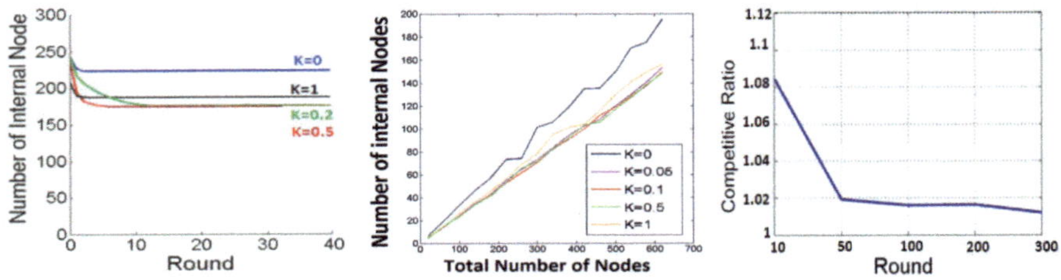

(a) (b) (c)

Fig. 1. Comparison of DMLST algorithm for different values of K . (a): The average number of internal nodes vs. rounds for 50 trials and the population of 500 robots is compared for different values of probability K. This plot shows that $K = 0.5$ gives the best result. (b): The plot of the average number of internal nodes vs. the total population of nodes for the same varying K in (a). (c): The average of competitive ratio of DMLST against the number of iterations.

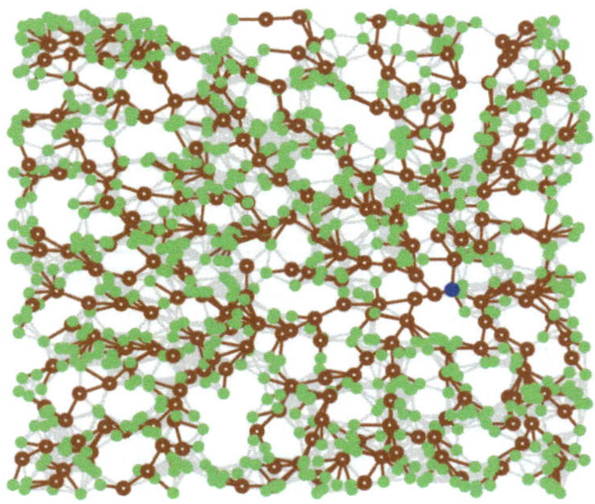

Fig. 2. Distributed MLST for a graph with 1000 nodes. Blue circle is the root of the tree. Leaves and internal nodes in the tree are in green and brown colors respectively. Edges of the original graph which are selected in the tree are colored in brown. The edges of the graph that are not selected in the tree are in gray. Note: nodes are generated randomly and sometimes they overlap. Overlapping nodes, especially leaves, may look like cycles, but there is no loop in the generated tree.

set. If there are two or more parents with *MaxChildrenCount*, or if the parents of the robot changes, the robot defers its decision to the next round, and sends UNDECIDED as *SelectedParent*. Without this constraint, a newly added potential parent is never selected, because robot's *ChildrenCount* is initially set to zero; therefore, the newly added robot always loses in the competition with other potential parents, even if it has the most potential children. The condition in line 5 in Algorithm 2 guarantees that robots do not ignore new potential parents. The situation when a robot has more than one parent with *MaxChildrenCount* is called *symmetrical* state. In *symmetrical* state, robot does not have enough information to make a decision. In this case, a robot chooses parent with minimum *id* with probability K. Otherwise, the robot remains UNDECIDED and waits for the next round to get information from its neighbors decision and make a choice of parent (line 15-22 of algorithm 2). Using the probability K eliminates the possibility of the symmetry.

Figure 1 shows the convergence of DMLST algorithm over rounds. For $K = 1$, the algorithm always chooses the parent with minimum *id*; when $K = 0$, the robot always waits for the next round to decide and a selection is never made. In other words, the symmetry is never broken, and robots select all the potential parents with *MaxChildrenCount*. Since the result may not be a tree, we use $K = 0$ only for the analysis here and we choose $K > 0$ for our algorithm. Symmetry also happens in the very beginning of the algorithm. Initially robots do not have any information about their neighbors and send UNDECIDED, *i.e.* robots select all their potential parents as parent. In other words, all robots are internal nodes initially. Some nodes become leaves as the algorithm is converging and a tree is gradually formed. Figure 1(b) shows the average number of internal nodes versus the robot network size in 12 trials for different values of K. These results show that the best choice for K is 0.5, because the algorithm converges the fastest with the minimum number of internal nodes compared to other choices. We define *Competitive Ratio* as the ratio of number of internal nodes in optimal solution to number of internal nodes in DMLST algorithm. We use brute force algorithm as an optimal solution to compute the *Competitive Ratio*. Figure1 (c) illustrates how the competitive ratio decreases as the number of iterations of DMLST increases.

5 DMLST Recovery Algorithm

DMLST recovery has three main parts that run concurrently. 1)A BFS tree is built from the communication broadcast, rooted from the *goal robot* as a source. 2)The approximation of MLST is generated based on BFS tree. 3) Internal robots become stationary and leaf robots navigate towards the source robot by selecting *NavigationGuides* from the set of neighbors which are closer to the source robot or in the same distance.

As robots move towards the source and the tree updates, internal robots become leaves and move towards the goal location. In this way, the number of moving robots increases so that all the robots move toward the source and ultimately get the goal location. The proposed recovery algorithm depends on frequent updates of the broadcast communication tree. We assume that the tree updates faster than the robot's motions [14]. This assumption will also be enforced our hardware experiments to prevent dis-connectivity. Algorithm 3 shows how robots become stationary or start moving during recovery algorithm.

5.1 MidAngle Navigation Algorithm

There are many existing approaches for navigation in multi-robot systems [6, 1]. We use the basic formulation of Li [6] and Batalin [1] . However, we modify these methods for the bearing-only sensors on our robot platform [15]. In bearing-only systems, there is no information about the distance between two robots. A robot only knows the orientation to its neighbors. Algorithm 4 describes the function of Mid-angle navigation algorithm. This algorithm is a slight improvement over

Algorithm 3. DMLST Recovery algorithm

Do forever
Construct BFS tree by IRCommBroadCast()
Construct a tree by DMLST algorithm
for each Robot u **do**
    if $u.ChildrenCount = 0$ **then**
        $u.Stationary \leftarrow 0$
        MidAngleNavigation(u)
    else
        $u.Stationary \leftarrow 1$
        Robot u stops moving
    end if
end for

Algorithm 4. Mid-angle Navigation for robot u

Do forever
$NavigationGuide \leftarrow$ all neighbors $v \in N(u) \ni v.hop < u.hop$
if $NavigationGuide.size = 1$ **then**
    $S \leftarrow v \in N(u) \ni v.hop = u.hop$
    if $S \neq \varnothing$ **then**
        $NewGuide \leftarrow S$ that has minimum bearing with the robot in $NavigationGuide$
        Add $NewGuide$ to $NavigationGuide$
    end if
end if
$MidAngle \leftarrow$ Average of bearing $(u,v), v \in NavigationGuide$
Turn by $theta = MidAngle$

simply navigating directly to a parent in the communications tree. In Mid-angle navigation, a robot moves on the bisector of the angles that robot builds with its *NavigationGuide*. Therefore, the robot navigates between its *NavigationGuides* in order to prevent collisions. This fact can be seen, since the angle bisector of a triangle always intersects the opposite side with the intersection point between two ends of that side. In the case that robot has only one *NavigationGuide*, robot moves directly to its *NavigationGuide*.

5.2 Correctness of DMLST Recovery Algorithm

We show the correctness of recovery algorithm in three terms, safety, progress and self-stabilization. **Safety:** The correct operation of recovery algorithm requires that the network does not disconnect. We can show this by induction as follows: in DMLST-based recovery, each robot has at least one parent that is stationary. The edges in the tree is traversable, because of line-of-sight communication. Every edge between two nodes in the tree implies the physical path between two robots . Therefore, if we assume that communication between stationary robots is reliable, each robot has a reliable path through the stationary robots in the network to the source. and the network is always connected. **Progress:** In order to guarantee forward progress of our algorithm, we need to show two things: first, our tree always has at least one moving robot; second, moving robots are moving towards the source. Leaves are moving robots, and it follows from the definition of a tree that there will always be at least one leaf. Demonstrating motion in the correct direction requires us to show that robots with fewer hops in the tree are geometrically closer to the source. The proof in Li [6] shows that this type of geometric, distributed BFS algorithm cannot have a local minimum in the number of hops, or basins of attraction for navigating robots. Therefore, robots in such approaches with fewer hops are always closer to the source, and always move toward the source. Our Mid-angle navigation algorithm is slightly different in that robots can move towards a single parent or a point between two parents. However, robots always move towards guides with the same or fewer hops. Therefore, they never move away from the goal location.

Self-Stabilization: The DMLST algorithm is self-stabilizing, meaning that you can start the robots from any arbitrary configuration, or provide a disturbance in position, state or population, but algorithm will eliminate the effects of this perturbation after a predictable number of communication rounds [2]. Changes in initial configuration do not affect in the algorithm performance, because the tree is built outward from the source, and all robots ultimately recover the source by using the constructed maximum leaf spanning tree as a line path and moving towards the goal location. hops counters and redirect their paths to the new location of the source. In addition, if a robot is removed, different outcomes happen: 1) If the removed robot is a leaf robot, other network maintain connectivity, because the leaf node is located at the end of the tree and does not affect on the connectivity of the network. Therefore, the recovery process continues without any disconnections. 2) If the stationary robot is removed, as long as the network connectivity remains, the child of the removed robot updates its guides by selecting a new parent or become stationary to combat the change. Updating the guides or switching to stationary takes only one round. Our algorithm is self-stabilizing as long as the network remains connected after changing configuration. However, removal of a internal robot can disconnect the network. This connection failure happens when the removed robot's child does not have any other neighbor with fewer hop to select as its parent.

5.3 *Time Complexity and Path Efficiency*

Time complexity of recovery algorithm depends on two separate components: complexity of the DMLST algorithm and complexity of navigation. The time complexity of DMLST depends on the topology of the robot network, especially the configuration of symmetries in the graph of a robot.We remind that a symmetry happens when a robot has multiple parents with the maximum *ChildrenCount*. Therefore, robot should wait for the next round to pick its parent. In the worst case, network symmetry extends horizontally to at most the diameter of the graph at each depth from the source. Time complexity is bounded by $O(d^2)$, where d is the depth of the graph. Another complexity component arises from the motion of the robots as they navigate to the goal location. Obviously, the lower bound for the recovery is the depth of the graph, because the furthest robot determines the time for recovering all robots at the goal location. In practice, the arrival time is always more than this lower bound, because of interference of robots. Moreover, mid-angle navigation usually gives a longer path than the shortest path to the goal location. To measure the efficiency of our algorithm, we use *path efficiency*. Path efficiency (PE) is a number, $0 \leq PE \leq 1$, and is defined as the ratio of the shortest path to the length of the path which is created by the algorithm. We have computed the PE of DMLST recovery for 50 trials, when 100 robots recover in a quarter-circle environment (Figure 3). The calculated PE has a mean of 0.78 and the standard deviation of 0.08.

6 Simulated Experiments

We assume robots are uniformly distributed over a quarter-circle of radius R, and the source robot is located at the corner of the quarter-circle. The robots are modeled as points and the effects of physical interference from collisions are ignored. In this ideal case robots are uniformly distributed in the workspace and no limitation exists in sensing and movements, the number of robots that are at the goal location can be computed by equation (1). In other words, t is defined as the time it takes for robots located initially at distance d to get the goal location. If the velocity of all robots is a constant v, the distance to the goal location(d) can be expressed as $d = vt$. In equation (1), n is the total number of robots, and $m(t)$ is the average number of robots that arrive the goal location in time t. The furthest robot's distance to the source is R. Furthermore, we assume a holonomic chassis, so that each robot can move in a straight line towards the source robot. Roughly speaking, this is expected to be $m(t) = O(t^2)$, as the number of robots at any given radius is $O(d^2)$.

$$m(t) = \frac{n\pi d^2}{\pi R^2} = \frac{n(vt)^2}{R^2} = \frac{nv^2t^2}{R^2} \tag{1}$$

Local Coordination: In local coordination, we use an inter-robot communications radius that is much smaller than the environment size, $r << R$. In this model, each robot can only communicate with and measure the pose of its neighbors. In this case, each robot has a local coordination whose origin is itself. This coordination reduces the available information for each robot to *local network geometry*.

Each robot is equipped with infrared (IR) sensors that measure the bearing angle to its neighbors.

Basic Recovery: This type of recovery allows all the robots to move to the goal location simultaneously. In a basic recovery each robot calculates the average bearing of its neighbors and moves in that direction. Instead, DMLST recovery chooses stationary parents and uses mid-angle navigation. Figure 3(a) illustrates the flow of

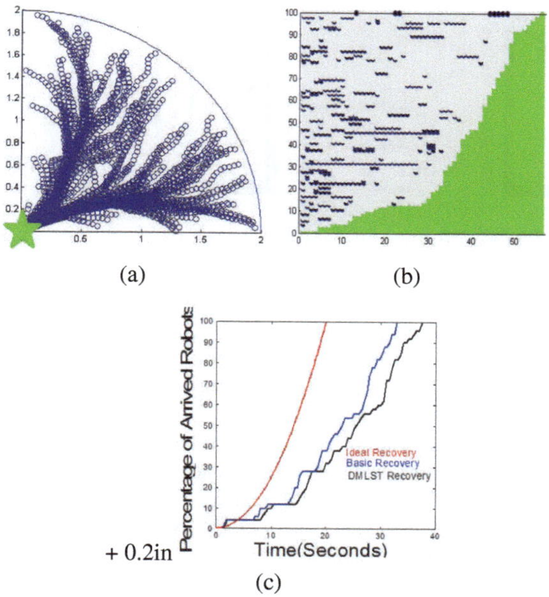

Fig. 3. (a): The trajectory of 100 robots during DMLST recovery algorithm in a quarter-circle environment. Robots are in blue circles and red star shows the goal location. (b): The status of each robot during the recovery in figure (a). Gray, blue and green show moving, stationary and arrived states respectively. The horizontal axis shows the time in seconds and the vertical axis shows the robots sorted by arrival time. (c): The comparison of recovery algorithms performance in a quarter-circle environment which is illustrated in figure (a). Ideal recovery, DMLST Recovery and basic recovery are shown in red, black and blue color respectively.

robot during DMLST recovery. Robots are recovered to the corner of the quarter-circle which has been defined as the goal location. Figure 3(b) shows three statuses of each robot during the recovery that are stationary, moving and arrived at the goal location. Figure 3(c) compares different recoveries for 100 robots in a quarter-circle environment. In this simulation, we assigned $v = 0.1m/s$ and sampling time $t = 0.05s$. As we expect, recovery from global coordination is very similar to the curve that arises from equation (1).

While basic recovery is efficient when recovering in a open environment, connectivity can fail in a complex environment as robots drive around corners. Figure 4(a) and Figure 4(c) show the simulation results of DMLST recovery in U-shape and serpentine environments respectively. Figure 4(b) and Figure 4(d) compare basic recovery and DMLST recovery against GPS (ideal) recovery in a U-shape and serpentine environment respectively. ideal recovery uses global coordinates to navigate robots along the shortest path to the goal location. These results illustrate the success of DMLST recovery in maintaining connectivity, even with sharp turns in the workspace. In DMLST recovery, there always exists at least one robot in the corners of the environment which is stationary and maintains connectivity. While,

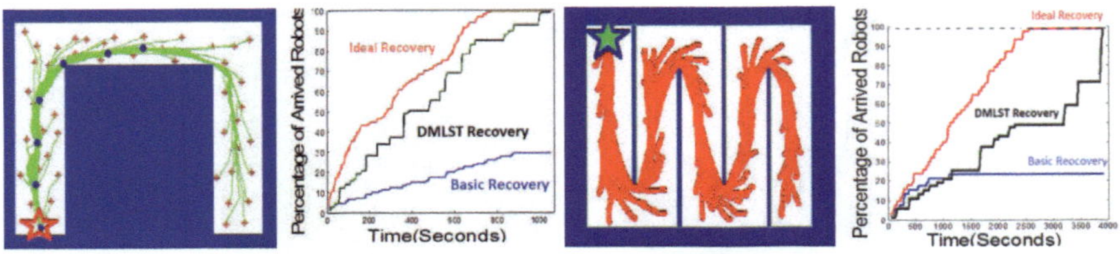

(a) (b) (c) (d)

Fig. 4. (a): DMLST recovery of 50 robots in an U-shape environment with two 90 degrees turns. Red star specifies the goal location. Green curves show the path of the DMLST recovery. Initial positions of the robots are shown in red stars. Additionally, blue points show stationary robots at the half way through the simulation. (b): The comparison of time efficiency of different recovery algorithms in U-shape corridor which is illustrated in (a). Ideal recovery, DMLST Recovery and basic recovery are shown in red, black and blue color respectively.(c): DMLST recovery path of 90 robots in a serpentine environment with four 180 degrees turns. Green star specifies the goal location The path of the recovery is shown in red. (d): The comparison of time efficiency of different recovery algorithms in serpentine-shape corridor which is illustrated in (c). Ideal recovery, DMLST Recovery and basic recovery are shown in red, black and blue color respectively.

basic recovery fails to finish and some robots never reach the goal location, because of the the loss of connectivity. Basic recovery may fail, because all robots always move and there is no guarantee of connectivity and some robot may be lost and unable to reach the goal location.

7 Experiments on Robot Hardware

We have tested DMLST and basic recovery algorithms on 13 real robots for 6 trials in a serpentine corridor(Figure 5). As the goal robot is selected as a source, the tree is generated and robots start moving until they get to the goal location. As illustrated in Figure 5, the basic recovery algorithm is not able to cope with the corners in this environment, so some robots are disconnected during recovery trials. Instead, DMLST recovery is usually able to move all robots to the goal location. The DMLST is updated during the recovery. Some robots become stationary while others move. Stationary parents wait for children, especially in the corners. For better performance, each robot's bump sensors help it move away from the environmental walls and other robots. Figure 6 compares the performance of basic and DMLST recovery for 6 trials. The basic recovery failed in all trials because of loss of network connectivity during the execution. This is not a problem during the straight sections, but was a problem while getting around the corners. In 4 out of 6 trials of DMLST recovery, all of the robots succeeded in rounding corners by maintaining connectivity with a stationary parent(s). The experimental result is very consistent with simulation outputs. However, in other 2 trials, 7 of 13 robots and 6 of 13 robots were disconnected.

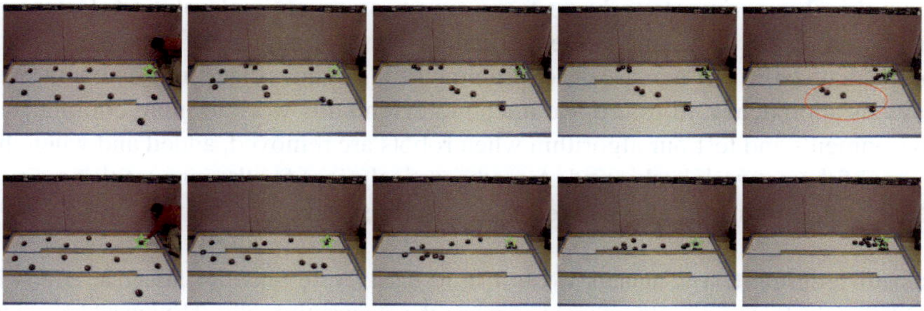

Fig. 5. Real experiment for robots recovery. The environment has two corners with 180 degrees. The dimension of each corridor is 50 cm(width) $\times$ 300 cm(length). Robot diameter: 11 cm, robot communication range:100 cm. First row(left to right):screen-shots of basic recovery; the goal location is a green star. The network becomes disconnected at the second corner, and four robots do not return the goal location.These lost robots are shown in a red elipse. This happens in 100% of 6 trials. Second row(left to right): screen-shots of DMLST Recovery; the goal location is a green star. The network remains connected around the corners, and all of the robots are able to recover to the goal location.

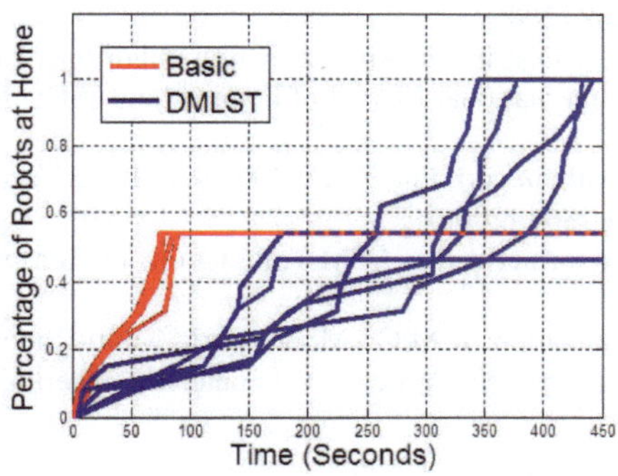

Fig. 6. Real experimental result: the comparison of basic(red) and DMLST(blue) recovery performance in a serpentine environment for 6 trials. The percentage of robots reach the goal location is illustrated against the time. In 4 out of 6 trials, all the robots are able to get the goal location, while in the rest, the DMLST recovery fails and only half of the robots are able to get to the goal location. However, basic recovery fails in all trials.

The errors encountered were largely caused by physical interference between robots, especially around corners. The 180-degree corners in this environment are the worst-case scenario for this kind of recovery; future work might consider a custom navigation behavior tailored to deal with these kinds of corners.

8 Conclusion

Tree algorithm (DMLST) for the robot recovery problem. Our algorithm removes a number of simplifying assumptions made in previous works on robot recovery, most notably providing better guarantees of connectivity and performance in complex environments. We present a simple approximation algorithm to compute the maximum leaf spanning tree in a distributed fashion, which has a very good competitive ratio in empirical experiments to brute force search as an optimal solution. Extensive simulation results and hardware experiments demonstrate the effectiveness of our approach, even in the worst-case environments, and show clear improvement over previous approaches. In future, we intend to extend this work to more complicated environments and test our algorithm when robots are removed, added and when the source of the network is changed. A mathematical proof for the competitive ratio of DMLST algorithm will also be presented in future work.

Acknowledgment. The authors would like to thank Artie Shen for his great efforts in simulating the brute force algorithm to compute the competitive ratio. They would also like to thank Dr. Aaron Becker at Rice University for proof-reading the manuscript.

References

1. Batalin, M., Sukhatme, G., Hattig, M.: Mobile robot navigation using a sensor network. In: Proceedings of the IEEE International Conference on Robotics and Automation, ICRA 2004, vol. 1, pp. 636–641 (2004)
2. Dolev, S.: Self-Stabilization. The MIT Press (2000)
3. Howard, A.: Multi-robot simultaneous localization and mapping using particle filters. In: Robotics: Science and Systems, pp. 201–208 (2005)
4. Kamei, S., Kakugawa, H.: A Self-Stabilizing Distributed Approximation Algorithm for the Minimum Connected Dominating Set. International Journal of Foundations of Computer Science 21(03), 459 (2010)
5. LaValle, S.M., Liberzon, D.: Rendezvous without coordinates. In: 2008 47th IEEE Conference on Decision and Control, pp. 1803–1808 (2008)
6. Li, Q., Rus, D.: Navigation protocols in sensor networks. ACM Transactions on Sensor Networks 1(1), 3–35 (2005)
7. Liu, J., Yuan, K., Zou, W., Yang, Q., Localization, K.M.C.: Monte Carlo Multi-Robot Localization Based on Grid Cells and Characteristic Particles. Computer Engineering, 24–28 (2005)
8. Localization, M.: r., Fox, D., Burgard, W., Kruppa, H., Thrun, S.: A Monte Carlo Algorithm for (March 1999)
9. i Lu, H., Ravi, R., Ravi, R.: The power of local optimization: Approximation algorithms for maximum-leaf spanning tree. In: Proceedings of the Thirtieth Annual Allerton Conference on Communication, Control and Computing, pp. 533–542 (1996)
10. Lynch, N.A.: Distributed Algorithms. Morgan Kaufmann Publishers Inc., San Francisco (1996)
11. Marathe, M.V., Breu, H., Hunt, H.B., Ravi, S.S., Rosenkrantz, D.J.: Simple heuristics for unit disk graphs. Networks 25(2), 59–68 (1995)
12. Maratic: Interference as a Tool for Designing and Evaluating Multi-Robot Controllers. In: Proceedings, AAAI 1997, pp. 637–642. AAAI Press (1997)
13. McLurkin, J.: Analysis and implementation of distributed algorithms for Multi-Robot systems. Ph.D. thesis, Massachusetts Institute of Technology (2008)
14. McLurkin, J.: Measuring the accuracy of distributed algorithms on Multi-Robot systems with dynamic network topologies. In: Proceedings of the International Symposium on Distributed Autonomous Robotic Systems, DARS (2008)

15. McLurkin, J., Lynch, A.J., Rixner, S., Barr, T.W., Chou, A., Foster, K., Bilstein, S.: A low-cost multi-robot system for research, teaching, and outreach. In: Proc. of the Tenth Int. Symp. on Distributed Autonomous Robotic Systems, DARS 2010 (October 2010)
16. Ozeki, K., Yamashita, T.: Spanning Trees: A Survey. Graphs and Combinatorics 27(1), 1–26 (2010)
17. Thrun, S., Fox, D., Burgard, W., Dellaert, F.: Robust Monte Carlo Localization for Mobile Robots (February 2001)
18. Thrun, S., Liu, Y., Koller, D., Ng, A.Y., Ghahramani, Z., Durrant-Whyte, H.F.: Simultaneous localization and mapping with sparse extended information filters. I. J. Robotic Res. 23(7-8), 693–716 (2004)

Multi-Robot Formation Morphing through a Graph Matching Problem

Lantao Liu and Dylan A. Shell

Abstract. We consider the problem of changing smoothly between formations of spatially deployed multi-robot systems. The algorithm presented in this paper addresses scenarios in which gradual and seamless formation transitions are needed, a problem which we term *formation morphing*. We show that this can be achieved by routing agents on a Euclidean graph that corresponds to paths computed on — and projected from— an underlying three-dimensional matching graph. The three-dimensional matching graph is advantageous in that it simultaneously represents a logical assignment problem (for which an optimal solution must be sought) and metric information that comprises the spatial aspects of the Euclidean graph. Together, these features allow one to find concurrent disjoint routing paths for multiple source multiple goal (MSMG) routing problems, for which we prove one may find routing solutions to optimize different criteria. These disjoint MSMG paths efficiently steer the agents from the source positions to the goal positions, the process of which enables the seamless transition from an old formation to a new one.

1 Introduction

Part of multi-robot formation control involves manoeuvring a spatially dispersed system from one formation to another. Formation control has received a great deal of attention and extensive investigation in the past decades (see reviews by Murray [17], Chen and Wang [4]). Most previous studies consider formation control of the whole system, but, in many situations, only parts of the system need to be changed to reach a new formation. For example, sometimes only patches of agents in certain corners need move to other locations, or boundary agents need to fill inner holes. If the majority the system keeps its structure unchanged, while a minority migrate to

Lantao Liu · Dylan A. Shell
Dept. of Computer Science and Engineering, Texas A&M University,
College Station, TX, USA
e-mail: {lantao,dshell}@cse.tamu.edu

M.A. Hsieh and G. Chirikjian (eds.), *Distributed Autonomous Robotic Systems*,
Springer Tracts in Advanced Robotics 104,
DOI: 10.1007/978-3-642-55146-8_21,

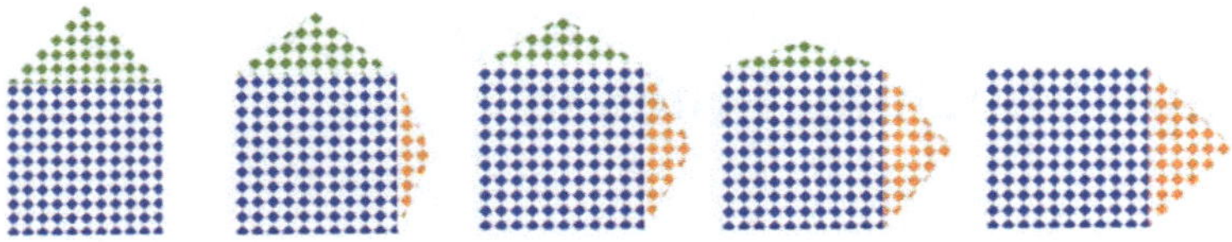

(a) (b) (c) (d) (e)

Fig. 1 Evolution of formation morphing. Source nodes are colored in green (top triangles) and gradually disappear. Goal nodes are colored in orange (right triangles) and gradually grow.

other places, then an incremental variation of the problem is worth addressing. We call formation control in such scenarios *formation morphing* since the formation is changed as if it is gradually "deformed" in places, while the major pattern is unaltered. Fig. 1 shows an example of seamless formation morphing.

Recently we have shown that by exploring the matching graph version of the Hungarian method [10] and interpreting it in three dimensional space, the assignment problem can also be used to deal with the standard routing problems [12]. In this paper we focus on more complicated conditions involving multiple paths generated from the matching graph, and develop an assignment-based formation control method for multi-robot systems for these cases. Specifically, formation morphing is achieved by routing certain agents from their source/initial positions to the defined goal positions along some trajectories, such that all agents involved in the trajectories simultaneously shift to and thus replace their successive neighbors. This is a process which gradually morphs the formation into a new shape. The routing trajectories are a set of interference-free paths on the Euclidean graph in which the nodes are the agents and edges are the traversal links; the paths are projected from the Hungarian augmenting paths in the 3D bipartite graph which we construct. One important contribution of this work is that the formation morphing is carried out in ways which optimize useful criteria, *e.g.*, the overall travel cost is minimized, the total interruptions are minimized (the number of the robots re-deployed is fewest), and the number of disjoint paths that are allowed is maximized. This is because the routing is incorporated in, and projected from, the matching graph from which globally optimal solutions of assignment problems are sought and found.

More specifically, the contributions of this work include:

- The design of a formation control strategy through routing paths projected from a 3D matching graph, which combines the logical description of the matching graph and the spatial embedding of the Euclidean graph.
- An optimal means for producing routing solutions of interest in applications; for example, the global minimal travel distance and shortest hopping distance are analyzed.
- Simultaneous generation of disjoint and conflict-free MSMG paths. Conditions for producing disjoint paths, and the maximal number of such paths are analyzed.

2 Related Work

Formation control is an active topic in the multi-robot research area and many approaches for controlling the formations of various types have emerged during the past decades. To sample from as many distinct taxonomical branches as possible, we acknowledge control theoretic schemes [7, 8], strategies with combinatorial optimizations [16], approaches dealing with geometry and potential fields [5, 22], as well as behavior-based methods [3] and methods employing biological inspired mechanisms [20, 23]. We are particularly interested in the formations of structured and well-aligned patterns, in which agents keep similar distances from each other (this somewhat mimics certain animal behaviors, *e.g.*, schools of fish, cattle herds, inserts swarms, *etc.*). Works dealing with the formations (patterns) that are similar to this work include [2, 11, 18], although their methods differ significantly.

One may also regard formation control as the assignment of robots to goal positions that define the final pattern. Smooth shifting of formation shapes addressed by this paper relates directly to the reassignment of robots to tasks (work specifically addressing reallocation includes that of Karmani *et al.* [9] and Shen and Salemi [21]), and to controllers which enforce some metric (or shape) constraints (notable recent examples of formation control include that of Michael *et al.* [16], Liu *et al.* [13], and Ren and Sorensen [19]). A clear, recent example of reallocation and formation work together is that of Agmon *et al.* [1]. The authors designed a polynomial time graph-based method to extract a subset of the robots from a coordinated group so that this subset can perform a new task while minimizing the cost of interacting with the remaining group.

This paper offers a different perspective: spatial formation transitions of a multi-robot team are achieved by routing agents on the 2D Euclidean graph but doing this by regarding it as a projection from the 3D representation of a corresponding matching graph. The approach has been described in detailed in our recent work [12] where we focused on the analysis of single path properties. In this work we emphasize multi-path conditions which are more complicated and reveal the merits of thinking about formation transitions in this way. The underlying method is the same incremental matching approach, *viz.* execution of stages of the Hungarian Method which produce paths with desired global optimization properties by incorporating both (metric) traversal information and reallocation (logical) costs, simultaneously.

3 Synthesized Matching Graph and Assignment Problem

This work is based on two forms of graphs: the Euclidean graph and the bipartite graph (or *bigraph* for short).

The Euclidean graph is a standard graph $G = (V, E)$ with a metric embedding so that the vertices in V describe locations and edges in E express distances between the vertex pairs. We let each vertex of G denote an agent, and let $w(i, j) = -d(i, j)$ represent the weight of edge $e(i, j) \in E$, where $d(i, j)$ is the travel distance between agent pair (i, j). The negated travel costs transform the problem from minimization to maximization. This transformation does not change the optimization objective but makes the problem consistent with the assignment utility maximization described below. In addition, traversability constraints, limited sensing/communication ranges, and so one, imply that the graph G is likely to be sparse.

Algorithm 1. The Hungarian Algorithm

Require:

An $n \times n$ assignment matrix represented as the complete weighted bigraph $\tilde{G} = (X,Y,\tilde{E})$, where $|X| = |Y| = n$.

Ensure:

A perfect matching M.

1: Generate initial labellings $l(\cdot)$, and an initial matching M in G_e.
2: If M perfect, terminate algorithm. Otherwise, randomly pick an exposed vertex $u \in X$. Set $S = \{u\}, T = \varnothing$.
3: If $N(S) = T$, update labels:
$\delta = min_{\forall x \in S, y \in Y \setminus T} \{l(x) + l(y) - \tilde{w}(x,y)\}$

$$l'(v) = \begin{cases} l(v) - \delta & \text{if } v \in S, \\ l(v) + \delta & \text{if } v \in T, \\ l(v) & \text{otherwise.} \end{cases}$$

4: If $N(S) \neq T$, pick $y \in N(S) \setminus T$.
(a) If y exposed, then $u \rightsquigarrow y$ is an augmenting path, then augment matching M and go to step 2.
(b) If y matched, say to z, extend the tree: $S = S \bigcup \{z\}$, $T = T \bigcup \{y\}$, and go to step 3.

Notes & Definitions:

- $\tilde{w}(x,y)$ is the weight of edge $\tilde{e}(x,y)$.
- Equality graph $G_e = \{\tilde{e}(x,y) : l(x) + l(y) = \tilde{w}(x,y)\}$.
- Neighbor of vertex $u \in X$: $N(u) = \{v : \tilde{e}(u,v) \in G_e\}$.

The Bigraph is the main data structure used in the Hungarian algorithm [10]. In the Hungarian algorithm (see Algorithm 1), is one of the most well-known optimal assignment algorithms and can efficiently solve an $n \times n$ assignment problem in $O(n^3)$ time. In the algorithm, the bigraph $\tilde{G} = (X,Y,\tilde{E})$ is another representation of utility matrix $U = (u_{ij})_{n \times n}$, where X and Y respectively denote the set of agents and tasks, and the set $\tilde{E} = \{\tilde{e}(i,j)\}$ are edges weighted by the utilities $(\tilde{w}(i,j) = u_{ij} = -d(i,j))$ between associated agent-task pairs (i,j), $i,j = 1,\cdots,n$. Since the bigraph represents matching relationships , sometimes it is also called the *matching graph*. The assignment problem is a matching problem where the goal is to find a set of *maximally weighted* and *mutually excluded* edges that constitute a perfect matching M such that each agent in X is uniquely assigned to a task in Y.

The Hungarian algorithm grows a matching by searching for a path, called an *augmenting path*, which consists of an alternating sequence of *matched* and *unmatched* edges but with free end nodes. This means the quantity of unmatched edges

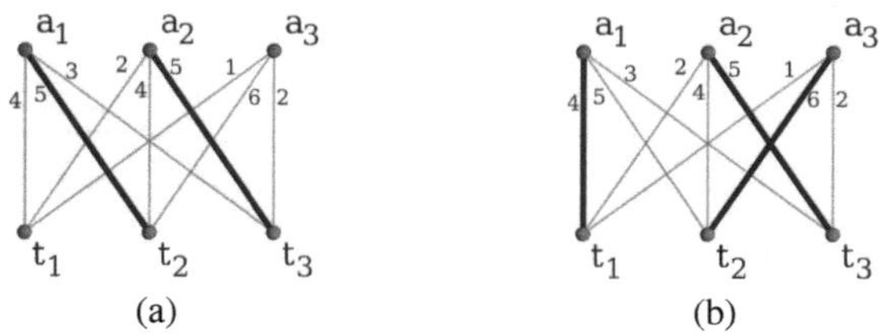

Fig. 2 (a) Two matched edges found after running two stages of the algorithm; (b) A perfect matching consisting of three matched edges is found after one additional stage (by augmenting path $a_3 \to t_2 \to a_1 \to t_1$).

is an odd number and is exactly one more than number of the matched edges. The algorithm augments the set of matched edges by simultaneously flipping the matched and unmatched edges in the augmenting path. (Formal definitions of these operations on matchings are omitted, refer to Lovász and Plummer [15].) In Algorithm 1, steps 2 to 4 describe the procedure of seeking and flipping an augmenting path. We call a single iteration of this procedure a *stage* (see Fig. 2). Note that each stage finds exactly one augmenting path which increases the size of the matching by exactly one. Thus, the algorithm requires at most n stages to obtain all n matched edges with mutually excluded end nodes, thereby forming the optimal assignment solution.

In its conventional use of finding a set of matchings, the bigraph $\tilde{G} = (X, Y, \tilde{E})$ is only interpreted as having meaning in a logical sense; any geometric information used to form the utilities is typically ignored. This differs from the embedded Euclidean graph $G = (V, E)$ which encodes the spatial description directly. But being graphs, both forms have common characteristics, for instance, both the Euclidean graph and the bigraph can be represented with matrices: G can be represented with a symmetric adjacency matrix with $w(i, j)$ as entries, and $\tilde{G}$ can be represented with a non-symmetric utility matrix with $\tilde{w}(i, j)$ as entries. This suggests that perhaps the bigraph may have the potential to express spatial information adequately *if the utility matrix is symmetric*.

All off-diagonal entries of the two matrices have the following relationship:

$$\tilde{w}(i, j) = \tilde{w}(j, i) = w(i, j) = w(j, i), \quad \forall i \neq j, 1 \leq i, j \leq n. \tag{1}$$

This means that if we ignore the diagonal entries, the assignment utility matrix is the form of an adjacency matrix, which is *the basic idea in bringing the two forms of graph together in the construction of a unified one*.

This synthesized graph may be imagined as if an identical copy of the Euclidean graph G had been lifted and placed over G. Via this "extrusion" a three dimensional mesh is formed with two identical layers plus all edges that connect them. Here the two layers correspond to the two partitions of a bigraph, *i.e.*, the bigraph vertex sets satisfy $X = Y = V$. Each edge $e(i, j) \in G$ is replaced with a pair of edges $\tilde{e}(i, j) \in \tilde{G}$

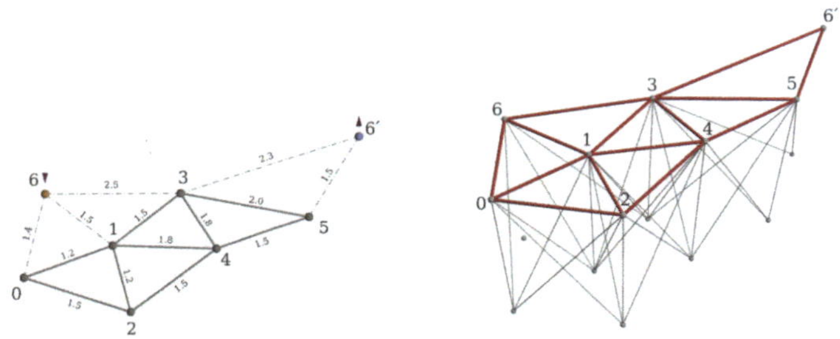

Fig. 3 Mapping from Euclidean graph to bigraph. (a) An Euclidean graph of networked robots with only nearest neighbors connected. This example illustrates the simple case of morphing one agent: vertices 6 and 6′ denote the initial and goal nodes/positions, respectively; (b) The corresponding sparse bigraph visualized in 3D. Bold red edges on top layer do not exist but just show the projection relationship with graph on the left.

and $\tilde{e}(j,i) \in \tilde{G}$. (Note: different from edges in G, in $\tilde{G}$ edges $\tilde{e}(i,j) \neq \tilde{e}(j,i)$ since i,j are nodes from different vertex sets—either X or Y.) An example is illustrated in Fig. 3(b). A more detailed description of this transformation is provided in our preceding work Liu and Shell [12]. In the remainder of the paper we assume that the vertices X in the top layer represent the agent set and vertices Y in the bottom layer denote the task set. Since nodes of either layer are copies from the Euclidean graph, this synthesized graph thus also conveys information about the spatial locations (top nodes describe the agent locations, and bottom nodes describe the task locations). If an agent node is matched to a task node, the agent needs to move from its current location to the newly assigned task location, and when a pair of agent and task nodes are vertically aligned, one can simply imagine that the agent has reached its deployed location and completed the position shift, so need not relocate.

Because this synthesized graph is a matching graph, we call it the *3D bigraph* (*3D matching graph*) and continue to denote it with symbol $\tilde{G}$. Thus, we obtain a mapping (projection) $\Omega : G \rightarrow \tilde{G}$. With known $G = (V,E)$, the projection $\tilde{G} = \Omega(G) = (X,Y,\tilde{E})$. To get $G = \Omega^{-1}(\tilde{G})$, an inverse operation is carried out, analogously.

4 Morphing the Formation

Formation morphing is done by seamlessly transferring agents from the source positions to the predefined goal positions as illustrated in Fig. 1. It can be imagined as cutting a batch of nodes from the source regions and pasting them into the goal regions. Naturally this implies that the two regions must be determined before the morphing operation is begun. Assume that agents in the source positions form a set A, and nodes in the goal positions form set B. Note that a goal location has no agent in it but will be occupied by an agent after the morphing process is finished. For each agent node in A, if it is connected with a unique goal node in B by a routing path, and if we let all agent nodes on this path shift to their successors' locations in a chain, then it looks as if this agent in A is sent to B, and all other nodes in $V \setminus A$ are still occupied by unique agents. This is also the essence of new agent-task insertions to the existing assignment described in our preceding work [12]. Different from that, formation morphing requires multiple interference-free paths to simultaneously steer multiple agents from the source positions to unique goal positions, which is the *multiple sources multiple goals* (MSMG) routing problem.

In this paper, we provide a solid analysis for the generated MSMG paths, as well as the details of applying our method to control formation transitions. We start the analysis by assuming that each node of A and B is directly traversable (with at least one edge connecting) to some nodes in $V \setminus A$. The MSMG routing paths are obtained by projecting the *matched edges* of augmenting paths in 3D bigraph to either planar layer. In order to get these augmenting paths connecting A and B from the Hungarian algorithm, these rules need be followed to construct the 3D bigraph: the top layer is split into two subsets of nodes— A and $X \setminus A$, and the bottom layer is also separated into two subsets— B and $Y \setminus B$. Edges are added between the two layers following the 3D bigraph construction procedure described in Section 3. Nodes of $X \setminus A$ and $Y \setminus B$ are vertically aligned and initialized as matched to represent the stationary intermediate nodes, whereas all other edges are initialized as unmatched. There are only $|A|$ nodes unmatched/un-assigned (assuming $|A| \leq |B|$), and each stage of Hungarian algorithm costs $O(|V|^2)$ time complexity, therefore only $O(|A||V|^2)$ is required to compute all MSMG paths.

4.1 Interference-Free Property of MSMG Paths

Let $P : A \rightsquigarrow B$ denote the set of paths that connect nodes in A and nodes in B. Routing paths P and Q are *disjoint* paths if no node or edge is shared between them. A shared node means that the agent at the intersection of different paths is required to simultaneously replace multiple other agents on corresponding paths, which is impossible.

Theorem 1. *The Hungarian algorithm run on bigraph $\tilde{G} = \Omega(G)$ produces only disjoint paths on graph G.*

Proof. We prove this theorem via contradiction: if the generated paths are not disjoint then there must exist at least two paths with a shared node that crosses between them. An example is illustrated in Fig. 4(a). Since P is also comprised of vertices and edges, it can be denoted by $P = (V^p, E^p)$, alternatively. Assume m paths $P_i = (V_i^p, E_i^p) \subseteq G$ ($m > 1$ and $i = 1, 2, \cdots, m$) are generated from the Hungarian algorithm, and there is shared (crossing) node $v_s \in \bigcup_{\forall i} P_i$ having more than one incoming routing edge and more than one outgoing routing edge. More than two routing nodes must be connected to v_s:

$$|\{u \mid e(v_s, u) \in P_i, \forall i = 1, 2, \cdots, m\}| > 2. \qquad (2)$$

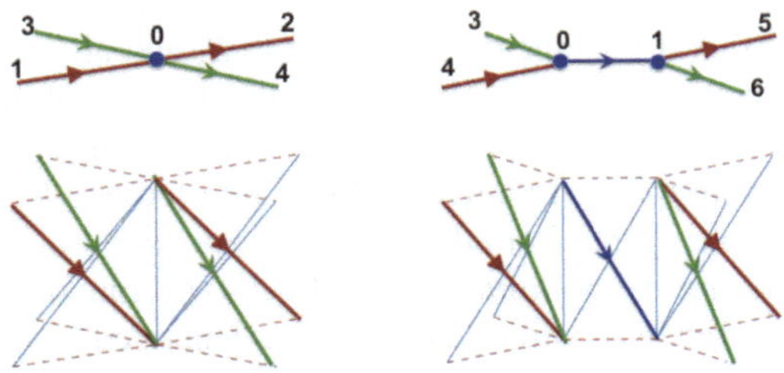

Fig. 4 Neither node nor edge will be shared among multiple paths produced from a matching graph. (a) Assumption of a shared node 0 at the crossing of path $1 \to 0 \to 2$ and path $3 \to 0 \to 4$. The bottom graph is the 3D bigraph showing the violation of mutual exclusion constraint; (b) Assumption of a shared edge $e(0,1)$ belonging to both path $3 \to 0 \to 1 \to 6$ and path $4 \to 0 \to 1 \to 5$.

This means that in bigraph $\tilde{G} = \Omega(G)$ either the corresponding agent node (in top layer of Fig. 4(a)) or the task node (in bottom layer) or both have more than one matched edge, which contradicts the mutual exclusion constraint and violates the feasibility of the assignment solution.

The shared edge case is analogous. □

4.2 Optimality Analysis of MSMG Paths

Thus far we have not described how the diagonal entries in the utility matrix are determined during the construction of a 3D bigraph. The diagonal values are actually the weights of the vertical edges from agents in set $V \setminus A$, see Fig. 3(b) for an example. Producing utility matrices with different diagonal weights, and feeding them to the Hungarian algorithm will produce distinct MSMG paths. Here we show two conditions that yield optimizations of particular interest.

Theorem 2. *The set of MSMG routing paths $P : A \leadsto B$ projected onto the Euclidean graph from the matching computed with the Hungarian algorithm will minimize the global hopping distance*[1] $\mathscr{D}(P)$ *when the weights $\tilde{w}(i,i)$ $(\forall i \in V \setminus A)$ (diagonal utilities) are sufficiently large.*

Proof. Let ζ be the largest absolute value of the utility matrix, *i.e.*,

$$\zeta = \max\{-\min(U), \max(U)\}, \tag{3}$$

and let $\pi = |V|\zeta + \varepsilon$, where ε is a small positive value. Weights of the edges $\tilde{e}(i,i)$ can be made sufficiently large by letting $\tilde{w}(i,i) = \pi$ for all nodes in $V \setminus A$. Now

[1] Hopping distance is also called Geodesic distance, it is the quantity of edges in the path and therefore measures the number of nodes involved and interrupted in the deployment.

assume there exists another path $Q : A \rightsquigarrow B$ with a shorter hopping distance $\mathscr{D}(Q) < \mathscr{D}(P)$. Since all nodes not on the paths themselves maintain their matching, the weight sums $f_s(\cdot)$ for the two matching solutions are

$$f_s(P) = \sum_{\forall_{(i,j)} e(i,j) \in P} \tilde{w}(i,j) + (|V| - \mathscr{D}(P))\pi, \tag{4}$$

and

$$f_s(Q) = \sum_{\forall_{(i,j)} e(i,j) \in Q} \tilde{w}(i,j) + (|V| - \mathscr{D}(Q))\pi, \tag{5}$$

respectively. Since

$$\begin{aligned} f_s(P) - f_s(Q) &= \sum_{\forall_{(i,j)} e(i,j) \in P} \tilde{w}(i,j) - \sum_{\forall_{(i,j)} e(i,j) \in Q} \tilde{w}(i,j) + (\mathscr{D}(Q) - \mathscr{D}(P))\pi \\ &\le \sum_{\forall_{(i,j)} e(i,j) \in P} \tilde{w}(i,j) - \sum_{\forall_{(i,j)} e(i,j) \in Q} \tilde{w}(i,j) - \pi \\ &\le \sum_{\forall_{(i,j)} e(i,j) \in P} \tilde{w}(i,j) - \pi < 0, \end{aligned} \tag{6}$$

contradicting the optimality of the matching from the Hungarian algorithm. □

Theorem 3. *When $w(i,i) = 0$ ($\forall i \in V \setminus A$), the set of MSMG routing paths $P : A \rightsquigarrow B$ computed by projecting the Hungarian algorithm's perfect matching to the Euclidean graph have the globally shortest path length.*

Proof. When $\tilde{w}(i,i) = 0$, $\forall i \in V \setminus A$, the weight sum of the matching solution for the assignment problem is

$$\begin{aligned} f_s(P) &= \sum_{\forall_{(i,j)} e(i,j) \in P} \tilde{w}(i,j) + \sum_{\forall v \notin P} \tilde{w}(v,v) \\ &= \sum_{\forall_{(i,j)} e(i,j) \in P} \tilde{w}(i,j) + 0 = \sum_{\forall_{(i,j)} e(i,j) \in P} w(i,j), \end{aligned} \tag{7}$$

which is essentially the total length of all MSMG routing paths. □

These two path properties are important since the globally shortest paths minimizes the total travel distances for a morphing operation, whereas the globally shortest hopping distance represents the fewest interruptions to the system (an interruption usually bears a cost).

4.3 Concurrent Paths in Narrow Bridges

In investigating paths formed in complex environments, it is important to quantify the maximum number of concurrent paths that can be formed. Narrow spaces may pose a challenge because they can impose a limit on the degree of concurrency that is possible; understanding these limits allows one to decide when sequential treatment (*e.g.*, for subsets of A and B) might be called for.

Definition 1. Let $G = (V,E)$ be a connected graph. A subset $C \subseteq V$ is called a *vertex cut* if $G \setminus C$ (the remaining graph after removing all vertices in C and their incident edges) is disconnected. A *minimal vertex cut* is a vertex cut with the least cardinality.

Definition 2. The *local connectivity* $\kappa(u,v)$ (or $\kappa(A,B)$) is the size of a smallest vertex cut separating non-adjacent vertices u and v (or vertex sets $A \subseteq V$ and $B \subseteq V$).

Theorem 4. (Menger's Theorem) *Let $G = (V,E)$ be a graph and $A,B \subseteq V$, then $\kappa(A,B)$ is equal to the maximum quantity of disjoint A-B-paths* (i.e., *the paths that connect vertices of A and B) in G.*

Proof. Three proofs appear in Diestel [6].

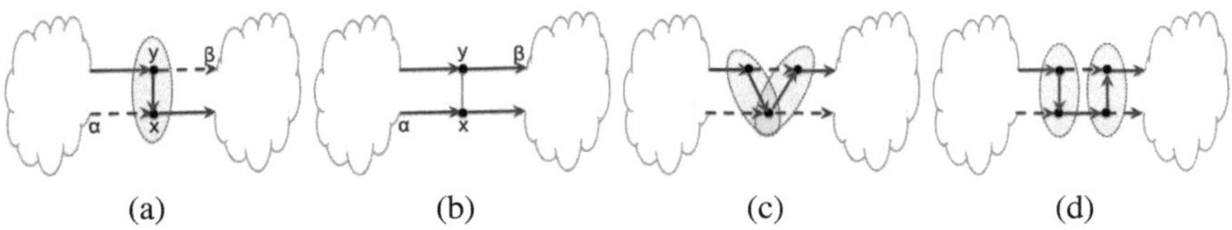

Fig. 5 Conditions on concurrent paths passing through a narrow bridge in the graph. (a) Nodes x and y on a path (solid arrowed edges) are from the same minimal vertex cut circled in an ellipse; (b) Two paths are generated after having found an augmenting path $\alpha \to x \to y \to \beta$; (c)-(d) Vertices of a path are from intersecting and independent minimal vertex cuts, respectively.

Theorem 4 provides the upper bound for the quantity of possible disjoint routing paths. However, we wish to know how close to this bound the disjoint paths produced by the Hungarian algorithm on the corresponding matching graphs are.

Theorem 5. *For connected graph $G = (V,E)$ with $A,B \subseteq V$ and $\min\{|A|,|B|\} \geq \kappa(A,B)$, the number of disjoint paths generated from Hungarian algorithm is equal to $\kappa(A,B)$,* i.e., *Hungarian algorithm running on $\tilde{G} = \Omega(G)$ outputs a set of disjoint A-B-paths, such that each path consists of exactly one cut vertex belonging to a minimal set of cut vertices.*

Proof. Assume a maximal set of disjoint paths $S_p = \{P_i\}, i = 1, \cdots, m$ is output, and assume a minimal vertex cut of G is C. If $|S_p| < |C|$, there must be some path P_l ($l \in [1,m]$) that contains more than one cut vertex from C. Assume these vertices form a set $V_l^{'} \subseteq V$ with $|V_l^{'}| = K$, then there must be $K-1$ edges $E_l^{'} \subseteq E$ (which can also be path segments) connecting these vertices. For an arbitrary edge $e(x,y) \in E_l^{'}$ where $x,y \in V_l^{'}$, x,y must be incident with other edges that are not on routing paths (a property following from the mutual exclusion constraint and the definition of a minimal cut), assume they are $e(\alpha,x)$, $e(y,\beta)$ respectively (illustrated in Fig. 5(a)). Then path $e(\alpha,x)$, $e(x,y)$, $e(y,\beta)$ forms an augmenting path and flipping of matched and unmatched edges cancels edge $e(x,y)$ to effectively bridge two new paths (example shown in Fig. 5(b)). Similarly, other edges in E_l' can also be revised and cancelled, and each such revision will add exactly one new path. Other complex conditions involving multiple intersecting or independent sets of minimal vertex cuts (see. Fig. 5(c) and 5(d)) are treated analogously. There must be $\kappa(A,B)$ disjoint paths generated, each of which consists of exactly one vertex from a minimal vertex cut. □

4.4 *Morphing with Multiple Shifts*

In complex scenarios, we cannot directly generate all MSMG paths in one go. This includes the condition of $\kappa(A,B) < \min\{|A|,|B|\}$, in which at most $\kappa(A,B)$ paths can be generated at one time. Another condition occurs when some nodes in A are not directly traversable to nodes in $V \setminus A$ (no edges directly connect them), these nodes must first morph (move) to some locations near enough to bridge to nodes of $V \setminus A$. In these cases, paths can only be produced in multiple batches and agents may need to carry out several shifts to complete the morphing task.

Generally, those peripheral agents need to be "transferred" first since otherwise inner holes may exist, which also deteriorates the traversability. (Peripheral nodes can be detected in a decentralized way as described in Liu *et al.* [14].) If local connectivity does not allow all peripheral nodes to morph at one time (via conditions given above), then the remaining peripheral nodes are queued and will be processed with high priority in next iteration to first get routing paths. An update of nodes' positions exposes new peripheral agents which can be processed in the following iterations. All future position shifts of an agent are ordered since waypoints require the agent complete movement one at a time. This is exactly the process of formation morphing.

5 Results

We simulated the algorithm with tens to hundreds of robots in order to validate the presented method. The simulator is written in C++ and it runs in a GNU/Linux environment. We assume that all robots are homogeneous and have identical sensing and communication ranges, within which each robot is capable of recognizing its neighbors as well as their distances and bearings. Spatial constraints and sensing/communication ranges may limit each agent's traversablility to only its nearest neighbors. In this work, a traversal link (edge) is added if the distance between a pair of agents is less than a given threshold. As a result, differing thresholds on traversal link length form Euclidean graphs of different sparsity, as illustrated in Fig. 6. To permit easy visualization, robots are simplified as dots, and the formation of the system is aligned in arrays to better reveal the concurrent paths and to show the shifting motions during the morphing. Also, the robots are assumed to autonomously avoid

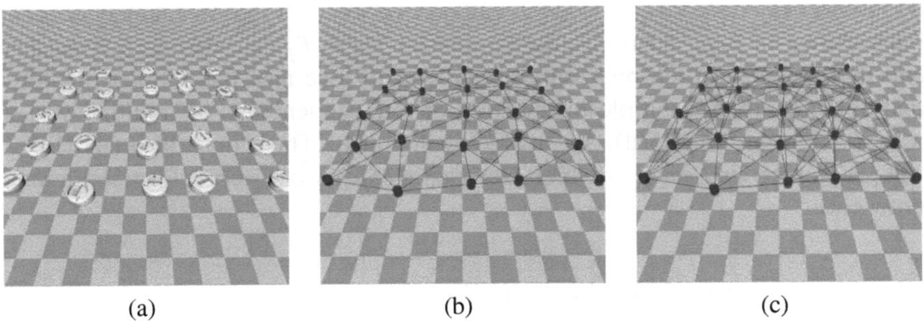

Fig. 6 (a) A formation of robots; (b)–(c) Euclidean graphs with traversal edges of different lengths.

obstacles locally to avoid potential collisions (*e.g.*, making a minimal detour around a obstacle if need be). The algorithm applies to any formation structures so long as the underlying Euclidean graph is connected.

Fig. 7 shows the evolution of an instance of formation morphing. In Fig. 7(b), agents form a triangle pattern, and the goal is to transform into the same size triangle rotated by 180°. Nodes in source positions are identified with green solid dots (forming set A), and the nodes in goal positions are orange circles (forming set B). In this example all nodes in A can traverse directly to nodes in $V \setminus A$.

Fig. 7(b)—7(d) and Fig. 7(e)—7(h) show two morphing processes (the difference lies in different traversal edge lengths, the former have larger thresholds than the latter). Optimization of global hopping distance and travel distances yield the same routing solutions in this example since the formation structure is well-aligned and the traversal link distance threshold is uniform. Since all agents involved carry out the relocation simultaneously, the formation transitions are achieved efficiently.

Fig. 8 illustrates the multiple shifts discussed in Section 4.4. Fig. 8(a) shows the initial and goal formations, and the task is to morph the top and bottom extruded agents to the left and right sides. Unlike the previous example, the traversal edge lengths are short so that the outer layer of green (source) nodes are not directly connected to the inner blue (intermediate) nodes. Two processes with differing thresholds are provided in Fig. 8(a)—8(d) and Fig. 8(e)—8(h), respectively. In particular, Figs. 8(d) and 8(g) show the second round of morphing paths.

Next, we examine the conditions for morphing paths across regions that limit the amount of concurrency. Fig. 9(a) shows a group of agents passing through a small gap. The maximal number of paths that can cross from one side to the other is two; each path passes through exactly one cut vertex in the Euclidean graph. Another example is Fig. 9(b)—9(d) where a square is morphed to a triangle while crossing a waist (one can imagine the waist part is located in between two attenuating walls as in Fig. 9(a), and the agents need to adjust their formations while navigating the confined environment).

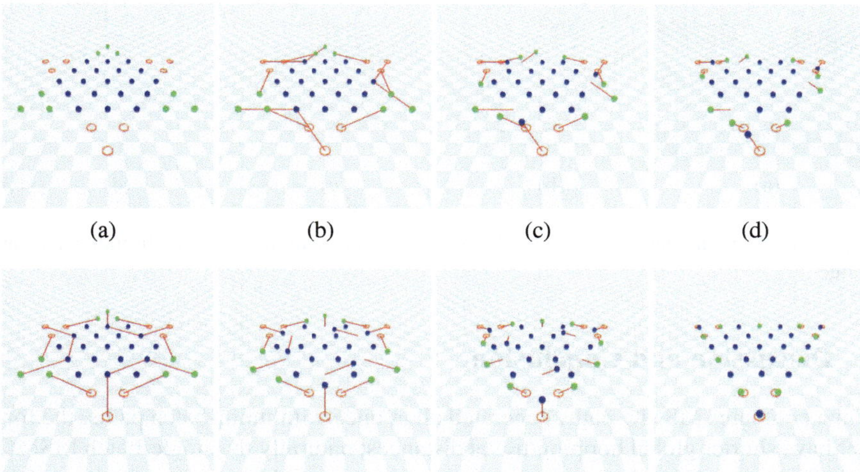
(a) (b) (c) (d)
(e) (f) (g) (h)

Fig. 7 (a) Initial and goal formations; (b)—(h) Processes of formation morphing

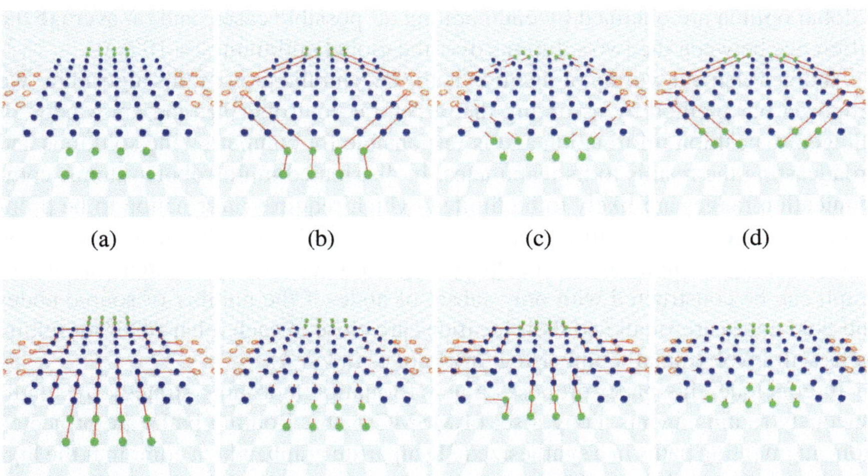
(a) (b) (c) (d)
(e) (f) (g) (h)

Fig. 8 (a) Initial and goal formations; (b)–(h) Processes of formation morphing

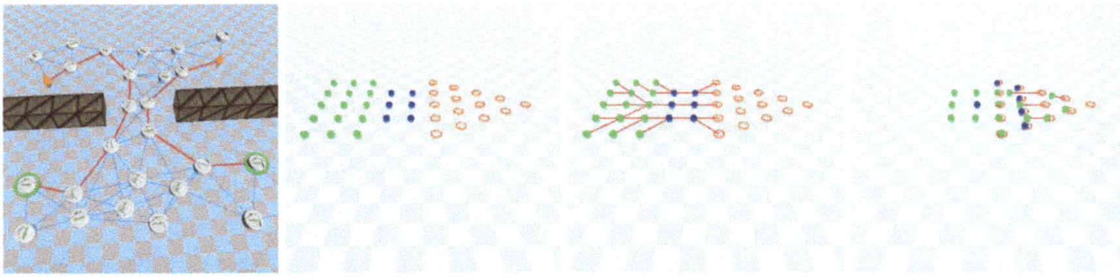

(a) (b) (c) (d)

Fig. 9 (a) Morphing paths across a small gap; (b)–(d) Formation morphing through a narrow bridge

6 Discussion and Conclusion

If one compares MSMG paths generated all at once with multi-batch paths, the former have shorter global hopping/travel distances; naturally the single-batch MSMG paths fail to consider the constraints imposed by limited local connectivity or isolated nodes.In other words, these special conditions either exceed the maximal paths capacity or violate the computation rule of the Hungarian algorithm, as discussed in Section 4.4. Nevertheless, optimality of the multi-batch paths up to the constraints has not been proven. We examined these conditions and ran experiments with $\sim$50 agents, and the results show that our solutions are very close to the global optima. (Global optima are obtained by enumerating all possible cases, and on average the difference between the two solutions over the global optimum $\leq\sim 10\%$).

Another element worthy of mention is the decentralization of this algorithm. We proposed our method by assuming that a central controller has the knowledge of all agents' position information and is responsible for computing the solutions and broadcasting the results to its teammates. In distributed multi-robot systems, communication ranges are likely to be limited, which requires information to be propagated using existing multi-hop methods. Additionally, not all agents need to be involved and communicated with. Morphing can occur in a local area and 3D bigraph can be constructed with only subset of nodes if the number of source nodes and goal nodes are small and their locations are close to each other (their morphing paths will involve small number/region of nodes in $V \setminus A$).

To conclude, this paper proposes a new formation morphing strategy by simultaneously routing agents along a set of MSMG paths. Routing paths are projected from augmenting paths in a synthesized 3D bipartite graph, which combines the logical description of the matching graph and the spatial embedding of the Euclidean graph. Since the paths are computed from the optimal assignment algorithm, useful optimized properties on the paths are revealed, including the characteristic of disjointedness, the global optimality of hopping/travel distances for all same-batch paths, and the maximal number of paths given connectivity constraints. We provided different formation morphing scenarios in simulation to illustrate and validate the various distinct conditions identified in the theoretical analysis.

References

[1] Agmon, N., Kaminka, G.A., Kraus, S., Traub, M.: Task Reallocation in Multi-Robot Formations. J. of Physical Agents 4(2) (2010)

[2] Alonso-Mora, J., Breitenmoser, A., Rufli, M., Siegwart, R., Beardsley, P.: Multi-robot system for artistic pattern formation. In: IEEE International Conference on Robotics and Automation, pp. 4512–4517 (2011)

[3] Balch, T., Arkin, R.C.: Behavior-based formation control for multi-robot teams. Trans. on Robotics and Auto. 14(6), 926–939 (1997)

[4] Chen, Y.Q., Wang, Z.: Formation control: A review and a new consideration. In: IEEE/RSJ Int. Conf. Intellig. Robots and Syst., pp. 3181–3186 (2005)

[5] Consolini, L., Morbidi, F., Prattichizzo, D., Tosques, M.: A geometric characterization of leader-follower formation control. In: 2007 IEEE International Conference on Robotics and Automation, pp. 2397–2402 (2007)

[6] Diestel, R.: Graph Theory, 3rd edn. Graduate Texts in Mathematics, vol. 173. Springer, Heidelberg (2005)

[7] Fax, J.A., Murray, R.M.: Information flow and cooperative control of vehicle formations. IEEE Transactions on Automatic Control 49(9), 1465–1476 (2004)

[8] Hsieh, M., Kumar, V., Chaimowicz, L.: Decentralized controllers for shape generation with robotic swarms. Robotica 26(5), 691–701 (2008)

[9] Karmani, R.K., Latvala, T., Agha, G.: On scaling multi-agent task reallocation using market-based approach. In: Intl Conf. on Self-Adaptive and Self-Organizing Systems (2007)

[10] Kuhn, H.W.: The Hungarian Method for the Assignment Problem. Naval Research Logistics Quarterly 2(1), 83–97 (1955)

[11] Kurabayashi, D., Choh, T., Cheng, J., Funato, T.: Adaptive formation transition among a mobile robot group based on phase gradient. In: IEEE International Conference on Robotics and Biomimetics, pp. 2001–2006 (2009)

[12] Liu, L., Shell, D.A.: Tunable routing solutions for multi-robot navigation via the assignment problem: A 3d representation of the matching graph. In: Proceedings of IEEE International Conference on Robotics and Automation (May 2012)

[13] Liu, L., Wang, Y., Yang, S., Watson, G., Ford, B.: Experimental studies of multi-robot formation and transforming. In: Proc. of UKACC Intl. Conf. on Control (2008)

[14] Liu, L., Fine, B., Shell, D.A., Klappenecker, A.: Approximate characterization of multi-robot swarm shapes in sublinear-time. In: Proceedings of IEEE International Conference on Robotics and Automation, pp. 2886–2891 (2011)

[15] Lovász, L., Plummer, M.D.: Matching Theory. North-Holland (1986)

[16] Michael, N., Zavlanos, M.M., Kumar, V., Pappas, G.J.: Distributed multi-robot task assignment and formation control. In: IEEE Intl. Conf on Robotics and Automation, pp. 128–133 (2008)

[17] Murray, R.M.: Recent Research in Cooperative Control of Multi-Vehicle Systems. In: Intl. Conf. on Advances in Control and Optimization of Dynamical Systems (2007)

[18] Ravichandran, R., Gordon, G., Goldstein, S.C.: A Scalable Distributed Algorithm for Shape Transformation in Multi-Robot Systems. In: Proceedings of the IEEE International Conference on Intelligent Robots and Systems (2007)

[19] Ren, W., Sorensen, N.: Distributed coordination architecture for multi-robot formation control. In: Robotics and Autonomous Systems, pp. 324–333 (2008)

[20] Reynolds, C.W.: Flocks, herds and schools: A distributed behavioral model. SIGGRAPH Comput. Graph. 21(4), 25–34 (1987)
[21] Shen, W.-M., Salemi, B.: Distributed and dynamic task reallocation in robot organizations. In: IEEE Intl. Conf. on Robotics and Automation, pp. 1019–1024 (2002)
[22] Song, P., Kumar, V.: A potential field based approach to multi-robot manipulation. In: IEEE Intl. Conf. on Robotics and Automation, pp. 1217–1222 (2002)
[23] Tanner, H.G., Jadbabaie, A., Pappas, G.J.: Flocking in Fixed and Switching Networks. IEEE Transactions on Automatic Control 52(5), 863–868 (2007)

Hexagonal Lattice Formation in Multi-Robot Systems

Sailesh Prabhu, William Li, and James McLurkin

Abstract. We present an algorithm that arranges a multi-robot system into a regular hexagonal lattice. This configuration provides continuous coverage with the fewest number of robots required. It also has a bounded stretch over a fully-connected graph, producing an efficient multi-hop communications network. Our algorithm uses artificial forces to move each robot to local potential energy wells. A local error correction algorithm detects and corrects most local lattice errors. Both algorithms are fully distributed, requiring *local network geometry* information, but no global coordinates. We present analysis of the potential energy wells that form the lattice, a proof of the upper bound on the spanning ratio of a hexagonal packing, and the error detecting and correcting algorithm. Simulation results demonstrate the effectiveness of the approach for large populations of robots.

1 Introduction

Key applications of multi-robot systems, like mapping, exploration, search-and-rescue, and surveillance, require robots to disperse over large geographical area while maintaining a communications network. Some of these applications require continuous coverage of areas, efficient network connectivity, and the efficient allocation of each robot. This work accomplishes this task with a distributed algorithm that positions the robots at the vertices of a hexagonal lattice. Figure 1 shows the results of our algorithm run in simulation with 600 robots.

The hexagonal configuration requires the fewest robots per unit area of any regular tessellation pattern and produces a connected communication network with a bounded stretch. Figure 2 shows four alternatives for coverage, a line of robots, and the three regular tessellations: hexagonal, square, and triangular lattices. All three

Sailesh Prabhu · William Li · James McLurkin
Rice University, 6400, Main St., Houston, TX
e-mail: {snp3,jmclurkin}@rice.edu, wl11@mit.edu

M.A. Hsieh and G. Chirikjian (eds.), *Distributed Autonomous Robotic Systems*,
Springer Tracts in Advanced Robotics 104,
DOI: 10.1007/978-3-642-55146-8_22,

regular tessellations produce a communication graph with a bounded *stretch*, which is the ratio between the distance a message must travel through the network between two robots and the Euclidean distance between these same two robots. A constant stretch ensures that the distance a message has to travel depends only on the Euclidean distance. Therefore, we can increase the number of robots without increasing the distance the message has to travel. The line of robots provides full coverage with the smallest number of robots, but produces a communications network with a stretch that is $O(n)$, where n is the total number of robots in the network. Of these, the hexagonal lattice configuration provides total coverage of an area, uses the least number of robots, and produces a network with a bounded stretch.

There are four contributions in this paper. (1) We present an algorithm that uses *virtual forces* between two species of robots to produce a hexagonal cell, and (2) we show that these same forces will allow the hexagonal cell to propagate into a hexagonal lattice. (3) We show that the potential energy field around the robots has a local minimum that causes errors in the lattice, but the error can be detected and corrected with a second distributed algorithm. (4) We present a proof sketch of the

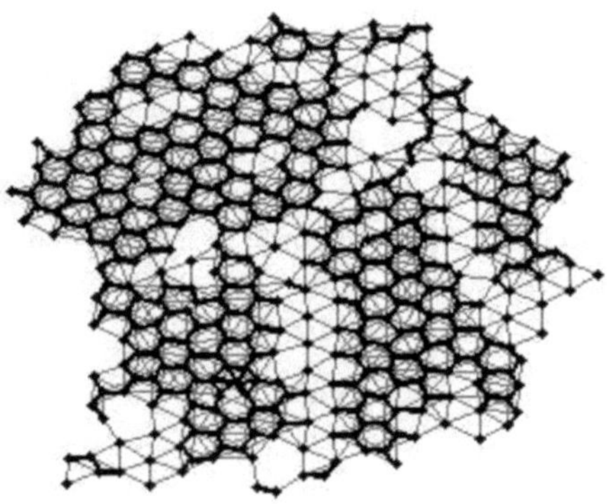

Fig. 1 An example hexagonal lattice from a simulation run of our algorithm.Video

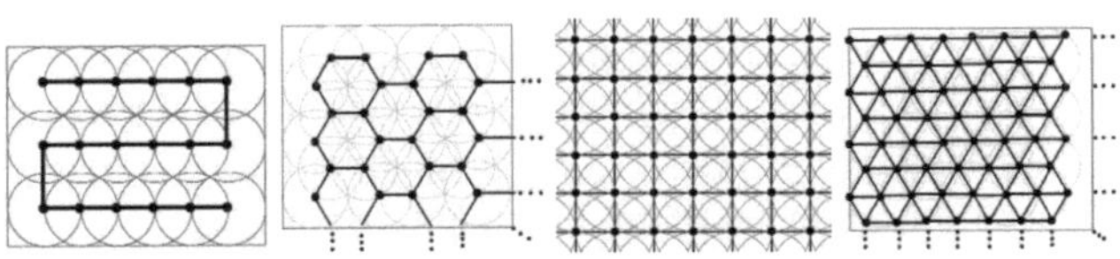

(a) Line Graph (b) Hexagonal Lattice (c) Square Lattice (d) Triangular Lattice

Fig. 2 Four different communication network topologies for full coverage, sorted by the density per unit area. The line graph uses the smallest number of robots, but has graph stretch of $O(n)$. The triangular and square lattice have bounded stretch, but uses more robots than necessary. The hexagonal lattice also has bounded graph stretch, but requires the fewest number of robots for full coverage, only $\frac{2}{3}$ of the robots from a triangular lattice.

maximum stretch of the spanning graph induced by the lattice, and empirical data that the average stretch is actually much lower. Finally, we demonstrate the results of simulations with large populations of 400 robots. All of the algorithms are fully distributed, scale to large numbers of robots, and require only the *local network geometry* - *i.e.* the positions of neighboring robots in the communications network. The algorithms require limited inter-robot communication, and are designed to run on simple, low-cost mobile robot platforms.

1.1 Related Work

Our hexagonal lattice algorithm builds on the work of Spears et al. Using artificial forces, they developed formation control algorithms to construct triangular[1] and square lattices [1, 2, 3, 4]. We modified their framework to make hexagonal cells stable, and we developed an algorithm to remove lattice imperfections.

Triangular lattices provide complete coverage but incur a great deal of overlap in the communication ranges of neighboring robots. As shown in Figure 2, a hexagonal lattice can be formed from a triangular lattice by removing every third robot from each line. A virtual robot could replace a robot in this position as suggested by Mullen et al. [5], but we chose not to use this method because it would require coordination among several robots to determine the center of the hexagonal cells. Our approach uses two species of robots, similar to the "spin up/down" idea of Spears et. al. to produce a potential energy field that has energy wells located at the vertices of a hexagonal cell.

Unfortunately, a system of artificial, which Spears et Al. termed the *physicomimetic* framework, often leads to robots settling down in unfavorable local minima [4]. Martinson and Payton were able to avoid local minima in square lattices by using the robot's on-board compass to draw several parallel lines and making each robot attracted to the parallel lines [6]. Once on the parallel line, they only move on the line. This scheme avoids several local minima that may occur off the lines parallel to the two orthogonal axis in a square lattice. We could not employ this method in avoiding local minima while creating the hexagonal lattice because the hexagonal lattice is not formed by orthogonal lines, and, instead we developed another algorithm to remove a robot from a local minimum.

Previous work by Tucker Balch and Maria Hybinette shows the possibility of using potential fields to maintain a geometric formation while also avoiding obstacles. The robots constructed "social potentials," where each robot influenced the potential energy in its vicinity [7]. Desai et al. created an algorithm to maintain relative positions and construct a formation [8]. Our work differs from that of Balch and Hybinette and Desai et al. because they created a set of geometric formations; however, we create a repeating hexagonal lattice. Balch and Hybinette showed that by placing repulsive forces near obstacles, the robots could successfully avoid them. In future work, we can have the robots in the repeated pattern maneuver around obstacles.

[1] There is an unfortunate habit in the literature of referring to triangular lattices as hexagonal lattices. In this work, we name a lattice by the polygon that forms it.

While we focus on constructing a repeated hexagonal lattice, our algorithm shares some similarities to previous work in flocking. Blake Eikenberry et al. created an algorithm that allows a swarm to construct a formation. The robots detected nearby robots, constructed trajectories, and positioned themselves in the proper relative positions [9]. Our work differs from work done by Eikenberry et al. because they had to prescribe the relative positions for each robot. In our work, the robots determine their relative positions using a distributed algorithm [10]. Hanada et al. constructed algorithms for separating triangular lattices at an obstacle and then reunifying them. However, our work is novel because they did not create a self-organized hexagonal lattice. Turgut et al. created algorithms for self-organizing the heading and proximity of the robots. However, our algorithm not only maintains a certain proximity, but it also constructs a repeated hexagonal lattice formation.

2 Problem Statement

We assume that we have n randomly distributed robots that can measure their *local network geometry*, the positions of their neighbors in each robot's reference frame. A robot is able to move freely in any direction in the plane. We assume that each robot has a sensing radius of R and a local communication radius of $1.5\sqrt{3}R$. Therefore, the robots can sense a radius of R around them with their camera, microphone, etc., but they can communicate via antennae to a radius of $1.5\sqrt{3}R$. Our algorithm will provide full-coverage of the area with the sensing radius. We wish to implement this on low-cost systems with limited communications and processing, so care is taken to implement solutions that are simple and scalable. We seek to position the robots onto the vertices of a hexagonal lattice with hexagons that have a side length of R.

For communication, we assign $\log_2(n)$ bits to provide unique IDs to each robot, 1 bit to identify the type of robot, and 32 bits to communicate the error the robot senses. The total number of bits required is $\log_2(n)+33$.

3 Graph Search in Hex Lattice

Let K be a complete graph where the nodes are arranged as the vertices of a hexagonal lattice. Let H be the edges of the hexagonal lattice, and G be the vertices of the hexagonal lattice. Define $d_h(A,B)$ as the shortest distance A and B on H and $d(A,B)$ as the shortest distance between A and B on K.

Define the stretch as $\frac{d_h(A,B)}{d(A,B)}$. First, we use computational methods to produce a distribution of the stretch. Then, we prove the stretch is no greater than 1.5.

3.1 *The Distribution of Stretch*

To compute the distribution of stretch, we constructed a hexagonal lattice that is ten hexagons by ten hexagons and calculated the distribution of the stretch shown in

Figure 3(c). The maximum ratio is 1.5, and this maximum corresponds to traveling between opposite sides of a hexagonal cell. Furthermore, the histogram indicates that the distribution of stretch resides mostly in the range 1.2-1.4.

3.2 Analytical Solution for Stretch

We wish to show

$$\frac{d_h(A,B)}{d(A,B)} \leq 1.5.$$

Let (a,b) be the index of an arbitrary lattice point. WLOG choose one point to be the origin $(0,0)$, restrict $a,b \geq 0$, and let side lengths equal one. Assign (a,b) as shown in Figure 3(a).

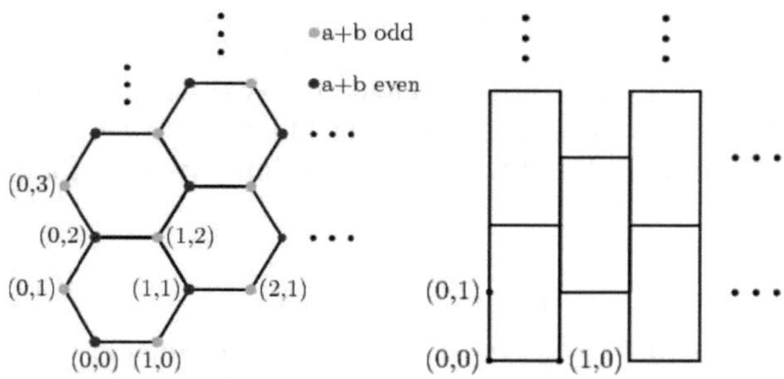

(a) Indexing Hexagonal Lattice Nodes (b) Brick Lattice

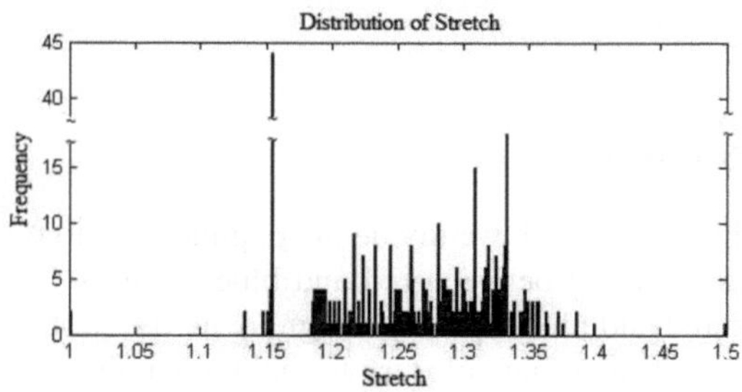

(c) Distribution of Stretch

Fig. 3 **a:** The hexagonal lattice can be compressed into a "brick" lattice without changing walk lengths. The stretch is at most 1.5. **b:** The "brick" lattice conserves the indexes and edges from the hexagonal lattice **c:** We computed the ratio between the shortest path traveled and the Euclidean distance based on a large lattice. This lattice has a size of ten hexagons positioned vertically and horizontally for a total of 420 vertices. The largest ratio of 1.5 between the distance traveled along the lattice and the Euclidean distance occurs in traveling between opposite sides of the same lattice as shown in the. There is a spike in the histogram at a stretch of 1.15, corresponding to the most common stretch between any two nodes.

Let $d(a,b)$ denote the Euclidean norm of (a,b). Then

$$d(a,b) = \sqrt{x(a,b)^2 + y(a,b)^2}.$$

The y-position of point (a,b) is

$$y(a,b) = y(b) = \frac{\sqrt{3}}{2}b.$$

The x-position depends on the parity of $a+b$:

$$x(a,b) = \left\{ \begin{array}{ll} \frac{3}{2}a - \frac{1}{2}, & \text{if } a+b \text{ odd} \\ \frac{3}{2}a, & \text{if } a+b \text{ even} \end{array} \right\}.$$

Let $d_h(a,b)$ denote the shortest walk from $(0,0)$ to (a,b) on the hexagonal lattice. Without altering walk lengths, we reshape the hexagonal lattice into a "brick" lattice as shown in Figure 3(b);

The complete derivation of d_h is outside the scope of this discussion, so we present an outline:

If $b > a$, $d_h(a,b) \geq a+b$. A walk of length $a+b$ exists, so $d_h(a,b) = a+b$.

If $b \leq a$, we have two sub-cases. If $a+b$ is odd, then $d_h(a,b) \geq 2a-1$. A walk of length $2a-1$ exists, so $d_h(a,b) = 2a-1$. We similarly show for even $a+b$ that $d_h = 2a$.

In summary,

$$d_h(a,b) = \left\{ \begin{array}{cc} a+b, & \text{if } b > a \\ 2a-1, & \text{if } b \leq a, a+b \text{ odd} \\ 2a, & \text{if } b \leq a, a+b \text{ even} \end{array} \right\}.$$

To finish the proof, we show $\frac{d_h}{d} \leq 1.5$ for each case. The algebraic details are outside the scope of this discussion.

4 Lattice Formation

To create a stable hexagonal cell, we divide the population into two types of robots, red and blue. Discriminating between red and blue requires 1 bit, and the robots can be divided into red and blue types by assigning a 50% probability to being one of the colors. Robots of the same color try to maintain a distance of $\sqrt{3}R$ between each other while robots of different colors try to maintain a distance of R between each other. Maintaining these prescribed distances will form the stable hexagonal cell pictured in Figure 4. The nucleus of a hexagonal cell is a triangle with one side of length $\sqrt{3}R$, and two sides of length R. This requires two robots of one color and one of the other. Assuming that we begin with 50% red and 50% blue robots and that three robots are interacting, the probability that two are of the same type and one is of another type is $\frac{3}{4}$: $1 - P(\text{ALL RED}) - P(\text{ALL BLUE}) = 1 - 2\left(\frac{1}{2}\right)^3 = \frac{3}{4}$. Across the entire population, this will generate a hexagonal nucleus with high probability. Our simulation results validate this assumption, the robots readily formed an initial nucleus.

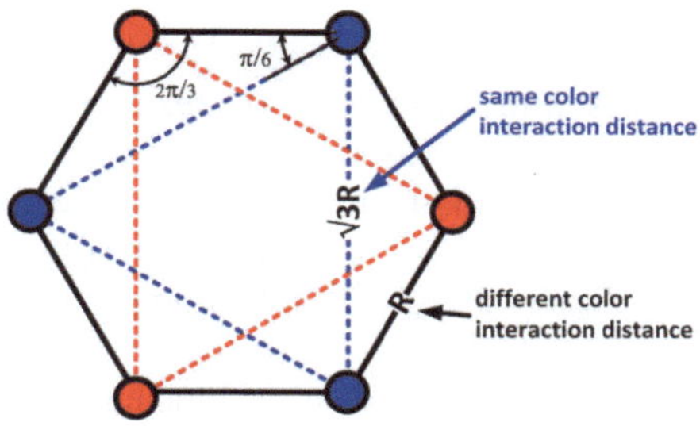

Fig. 4 In the hexagonal cell pictured above, the robots maintain the prescribed distances between each other, making the cell stable

The physicomimetic framework models the force between the robots as a proportionality constant, G, times the product of the masses divided by the distance raised to a power, p. A viscosity term, c, allows the robots to reach a stable state by having their velocity decay to zero. The framework also imposes a maximum force, F_{max}, that the robots can impart on each other to avoid a rapid increase in velocity.

We determined values for constants F_{max}, G, p, and c by searching the parameter space with a genetic algorithm [11, 12]. We used a genetic algorithm to quickly tune parameters to optimize global properties of the lattice. We evolved parameters to minimize global lattice error, which we define later in this paper. Spears et al. tuned their parameters to reduce the number of clusters and to reach a phase transition state, where the robots in a cluster repel each other with more force than the force from the surrounding robots pushing them into a cluster.

4.1 Mass Interactions

To maintain a Euclidean distance R between robots of different types and a distance $\sqrt{3}$R between robots of the same type, we define two stable distances,

$$R_{stable} = \left\{ \begin{array}{ll} R, & \text{for robots of different types} \\ \sqrt{3}R, & \text{for robots of the same type} \end{array} \right\}.$$

By making the force attractive when the distance between the robots, d, is greater than R_{stable}, repulsive when d is less than R_{stable}, and zero when d equals R_{stable}, the robots will move to reach the desired distance between each other. To ensure that the interactions remain local, the force goes to zero for d greater than $1.5R_{stable}$.

The magnitude of the force between two robots of mass m is determined from the equation for the virtual force as defined in Spears et al. [1]:

$$F = \left\{ \begin{array}{ll} \frac{G}{d^p}, & \text{for robots of different types} \\ \frac{G}{(\frac{d}{\sqrt{3}})^p}, & \text{for robots of the same type} \end{array} \right\}.$$

For our simulation, we used $G = 627.30$ and $p = 1.62$.

4.2 Stability

The virtual force approaches infinity as two robots reach a distance where d is very close to zero. When numerical methods are used to integrate an acceleration, resulting from a force that goes to infinity, the velocity can reach a very large value, causing instabilities in the system. Thus, we placed a maximum value on the magnitude of the virtual force, $F_{max} = 50.82$. The total composite virtual force is shown in Figure 5(a).

Additionally, we desire for the robots to lose some energy and eventually settle down into their stable positions. This goal is achieved by introducing a viscosity coefficient, $c = 0.26$, that decreases the magnitude of the velocity.

Changes in the communication radius will not affect the algorithm's stability. The robot's communication radius can fluctuate between R_{stable} and $1.5R_{stable}$, and the robot will still be able to detect the energy well at R_{stable} and settle onto a vertex of the honeycomb lattice.

4.3 Lattice Propagation

We also require that a hexagonal cell, once formed, will give rise to more hexagonal cells and eventually a hexagonal lattice. The potential energy diagram in Figure 5(b)

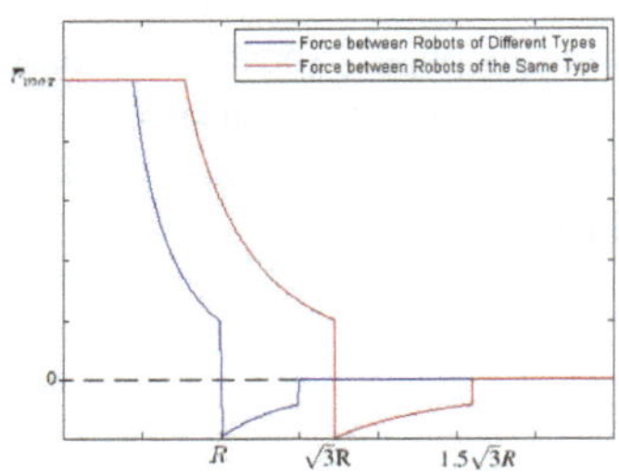

(a) Force Function

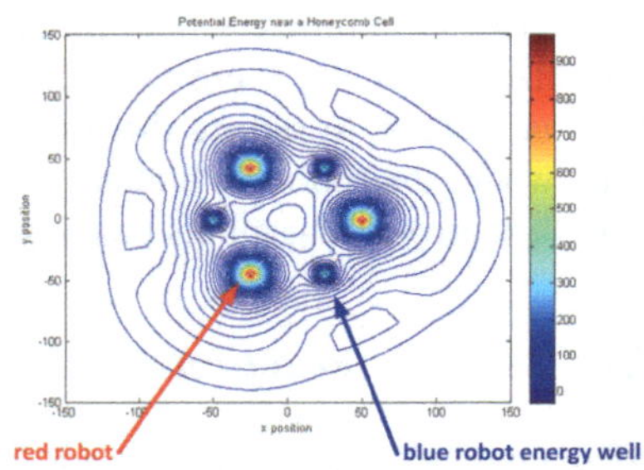

(b) Potential Energy Contours

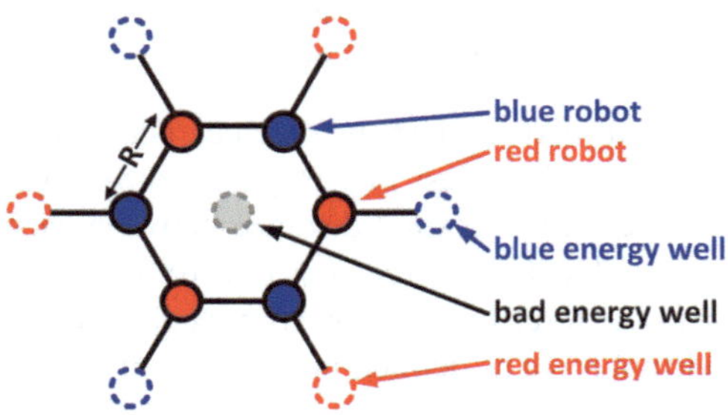

(c) Energy Wells Diagrams

Fig. 5 a) shows the force that two robots impart on each other. b) shows the potential energy wells for incoming red robots. Using symmetry arguments, we can add three more energy wells for blue robots, which are drawn in c).

indicates that there are four stable positions for an incoming red robot in the vicinity of an already assembled hexagonal cell. By symmetry, we can add three energy wells for blue robots, pictured in Figure 5(c). Figure 5(c) also shows that we can assemble significant portions of six adjacent hexagonal cells surrounding the one that has already formed. Thus, a single hexagonal cell acts as a nucleus for the creation of a hexagonal lattice. However, our potential energy field creates a problematic stable position in the center of the hexagonal cell, which causes an *interstitial error*, and causes the hexagonal lattice to become a triangular lattice. This energy well exists for all choice of constants because all forces in this location sum to zero. In order to create a hexagonal lattice, we must design a distributed algorithm to remove a robot caught in this undesirable location.

5 Error Detection and Correction

To remove the robot from the interstitial position at the center of a hexagonal cell, we developed a distributed algorithm that allows each robot to measure its local error and determine if it at the center of a hexagonal cell. If the robot is at the center of the hexagonal cell, it is moved to the exterior of the convex hull surrounding all the robots. The measurement and communication of error requires 32 bits.

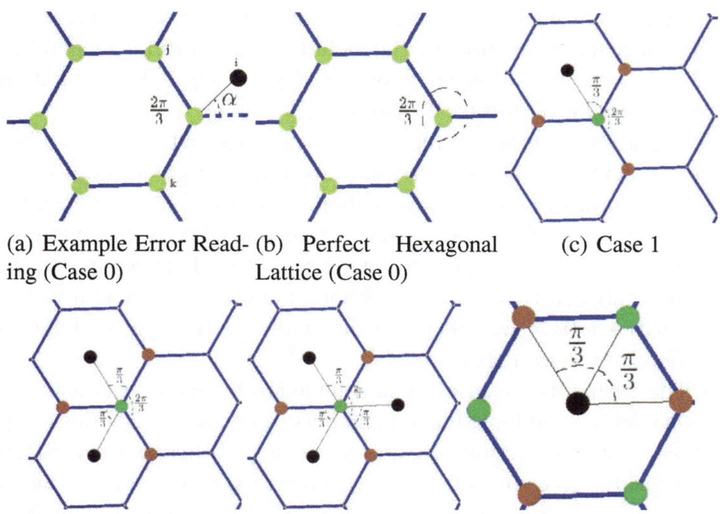

(a) Example Error Reading (Case 0) (b) Perfect Hexagonal Lattice (Case 0) (c) Case 1

(d) Case 2 (e) Case 3 (f) Case 4

Fig. 6 Assuming that the only stable positions are on the vertices of or in the center of a hexagonal cell, the cases above show the possible neighborhoods a robot could find itself in. It can have zero, one, two, or three neighbors in the center of a hexagonal cell.

6 Defining Error

Figure 6(b) show a perfect hexagonal lattice. In this ideal configuration, the angle between any two neighbors, θ_{ij}, is a multiple of $\frac{2\pi}{3}$. We define *local lattice error*, E_{ij}, for a neighbor pair i, j as the smallest difference between θ_{ij} and a multiple of $\frac{2\pi}{3}$: $E_{ij} = min(mod(\theta_{ij}, \frac{2\pi}{3}), \frac{2\pi}{3} - mod(\theta_{ij}, \frac{2\pi}{3}))$. To calculate the total error of robot, we sum the error over all neighbor pairs i, j: $E_{tot} = \sum E_{ij}$. The average error of a robot is defined as the total error divided by the number of neighbor pairs: $E_{avg} = \frac{E_{tot}}{\binom{\Delta}{2}}$, where Δ is the number of neighbors.

6.1 Error Configurations

We simplify this discussion by assuming that most of the robots are at the center of their energy wells, forming a hexagonal lattice. We break the discussion into several cases to illustrate how the error metric detects interstitial lattice defects.

Case 0: Figure 6(b) shows that a robot with all its neighbors positioned on the vertices of a hexagonal lattice will measure θ_{ij} between neighbors i, j that are multiples of $\frac{2\pi}{3}$. This produces a total error $E_{tot} = \sum E_{ij} = 0$.

Case 1: Figure 6(c) shows a robot with three neighbors on the hexagonal lattice and one neighbor in an interstitial position. This will produce an error of $\frac{\pi}{3}$ between the black neighbor and each of the other neighbors, and an error of zero between all other robots. Since there are three θ_{ij} that include the black robot, the total error will be $E_{tot} = 3\left(\frac{\pi}{3}\right) = \pi$. Since there are a total of ${}_4C_2 = 6$ angles between all neighbor pairs, the average error for this robot will be $E_{avg} = \frac{\pi}{6}$

Case 2: Figure 6(d) shows a green robot with three neighbors on a hexagonal lattice and two neighbors in an interstitial position. The green robot will read an error of $\frac{\pi}{3}$ between brown and black neighbors, an error of zero between two black neighbors, and an error of zero between two brown neighbors. Since the robot will measure 6 θ_{ij} between brown and black neighbors, 1 θ_{ij} between two black neighbors, and 3 θ_{ij} between two brown neighbors, the robot will read a total error of $E_{tot} = 6\frac{\pi}{3} + 3(0) + 1(0) = 2\pi$. Since there are a total of 10 angle readings, this total error corresponds to an average error of $E_{avg} = \frac{\pi}{5}$.

Case 3: The green robot in case 3 is surrounded by three brown neighbors that are on a hexagonal lattice and three black neighbors in interstitial positions as pictured in Figure 6(e). The green robot measures an error of $\frac{\pi}{3}$ between brown and black neighbors, measures an error of 0 between two brown neighbors, and measures an error of 0 between two black neighbors. There are a total of 9 θ_{ij} with brown and black neighbors, 3 θ_{ij} between two brown neighbors, and 3 θ_{ij} between two black neighbors. Thus, the total error measured by the robot is $E_{tot} = 9\frac{\pi}{3} + 3(0) + 3(0) = 3\pi$. Since there are a total of 15 ij neighbor pairs, this total error reading corresponds to an average error of $E_{avg} = \frac{\pi}{5}$

Case 4: The black robot in case 4 positions itself in the interstitial position, and it will measure an error of $\frac{\pi}{3}$ between brown and green neighbors, zero error between two brown neighbors, and zero error between two green neighbors. Since there are a total of 9 θ_{ij} between brown and green neighbors, 3 θ_{ij} between two green neighbors, and 3 θ_{ij} between two brown neighbors, the robot will read a total error of $E_{tot} = 9\frac{\pi}{3} + 3(0) + 3(0) = 3\pi$. Since there are a total of 15 angles, this total error reading corresponds to an average error of $E_{avg} = \frac{\pi}{5}$

6.2 *The Error Correction Algorithm*

Figure 7 shows a configuration in which there are many interstitial errors. This is effectively creating a triangular lattice within the hexagonal lattice. With this many errors, there is some boundary between the two lattice types. In such a configuration, our error correction algorithm can start with any interstitial robot that is inside a hexagonal cell that is adjacent to three or more correct hexagonal cells. Removal of this robot would extend the hexagonal lattice. We call this robot a *correction start point*, and have circled such a robot in black in Figure 7.

Figure 7 shows that all robots at the boundary of triangular and hexagonal cells have either case 1 or case 2 neighbors; furthermore, a correction start point has two case 1 neighbors that make an angle of $\frac{\pi}{3}$ with each other; we can discriminate a case 1 robot from a case 2 robot by using the criterion $E_{avg} \leq \frac{\pi}{6}$. Thus, if a robot reads an angle of $\frac{\pi}{3}$ between two neighbors with an average error less than $\frac{\pi}{6}$, it determines itself to be at a correction start point. Having determined itself to be correction start point, it moves to the exterior of the lattice. This process continues indefinitely, removing many of the interstitial errors in the network. Intuitively, we choose robots with case 1 neighbors because the error in a case 1 neighbor is caused entirely by one robot, which could be the robot in question.

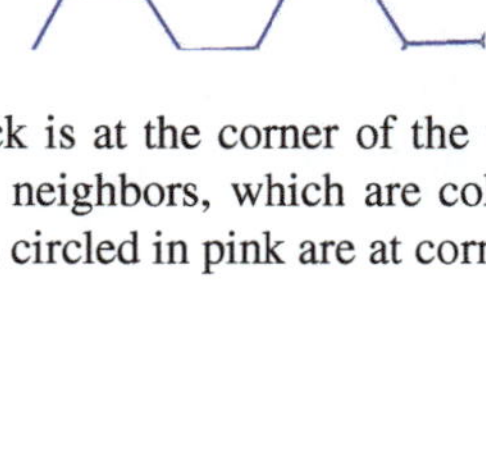

Fig. 7 The robot circled in black is at the corner of the triangular lattice and measures an angle of $\frac{\pi}{3}$ between two case 1 neighbors, which are colored green. Upon removal of the robot circled in black, the robots circled in pink are at corners of the triangular lattice.

7 Simulation Result

Figure 8 shows the position of the robots after running the entire algorithm for 2000 time steps. The hexagonal cells are fairly easy to pick out in both cases. To evaluate the global performance, we define lattice error as: $E_{lat} = \frac{1}{n}\sum_{i=1}^{n} E_{avg,i}$, where n is the number of robots and $E_{avg,i}$ is the E_{avg} corresponding to the ith robot. Figure 9(b)

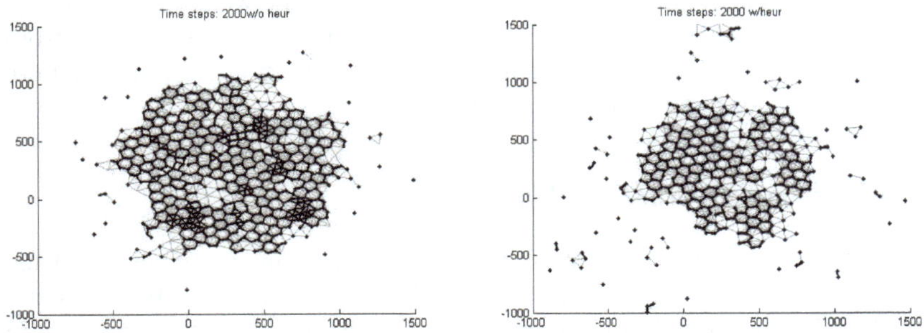

Fig. 8 The diagram on the left shows the position of the robots when the simulation is run without error correction, and the diagram on the right shows the result when the simulation is run with error correction

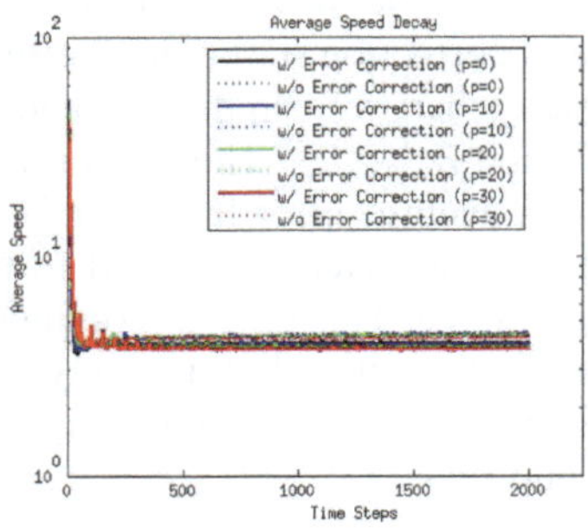

(a) Average Speed Decay

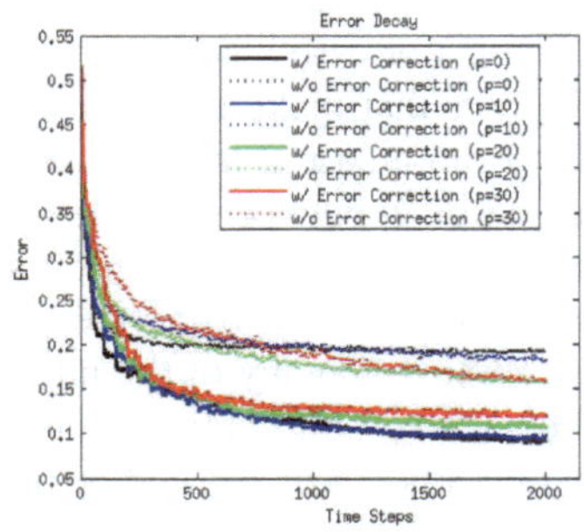

(b) Error Decay

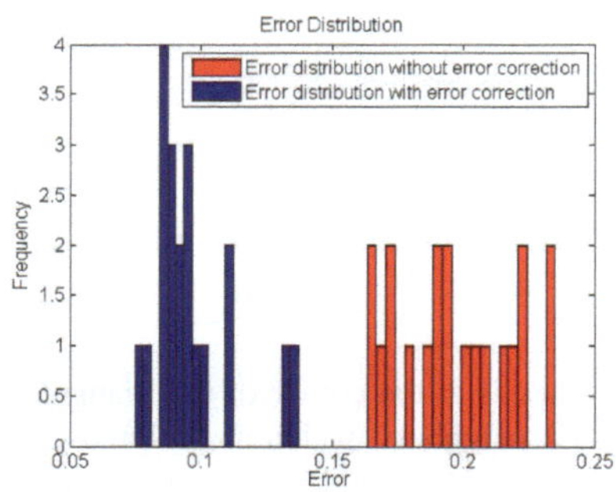

(c) Error Distribution after 2000 Time Steps(p=0)

Fig. 9 In the plot above, we observe decaying speed and error when messages are dropped p percentage of the time. These diagrams show that as time evolves, the simulation with the error correction algorithm decays to a smaller error while also maintaining stability as evidenced by the decreasing average speed. They, also, show that a higher probability of message loss decreases lattice error when the error correction algorithm is not applied. However, with the error correction algorithm, a higher probability of message loss increases lattice error.

shows that for a simulation run with the error correction algorithm, the lattice error decreases up until the 500th time step; whereas, the lattice error for a simulation run without the error correction algorithm ceases to decrease very early in the simulation. Figure 9(c) shows that the error with error correction algorithm is always less than error without the algorithm. Figure 9(a) shows that the error correction algorithm has almost no effect on the average speed, indicating that the error correction algorithm does not cause instabilities in the system.

Figure 9(a) shows that the loss of messages does not cause the average speed to go up, indicating the system remains stable. Figure 9(b) shows that increasing the loss of messages decreases the error of the system run without the error correction algorithm. The decrease in error can be attributed to the loss of stability at the center of the hexagonal cells when one of the robots on the vertices fails to impart a force. However, increasing the loss of messages increases the error of the system run with the error correction algorithm. The increased error can be attributed to an inability to communicate error and fix the lattice.

8 Conclusion and Future Work

Because our algorithm is fully distributed, multiple nuclei start hexagonal lattices simultaneously, ultimately creating crystalline grain boundaries in the final structure. This is unavoidable in our current design, but future implementations could take advantage of symmetry-breaking techniques to break ties at grain boundaries, and have one lattice "rearrange" neighboring grains to produce a single-crystal lattice. Because our current implementation requires communication with neighbors at a range of $\sqrt{3}R$, we cannot form hexagonal cells at the maximum extent of the communication range. We think that it is possible to build a hexagonal lattice using a communication range of R using virtual forces, but it will require a more complicated distributed algorithm to form hexagonal cells, an approach we will explore in future work. Our current algorithm is simple, requires only local network geometry, and is effective. We have calculated fundamental graph properties of the hexagonal lattice. The simulation results show convergence and stable configurations within the first few iterations. Our combination of artificial forces and distributed algorithms produces a high-quality hexagonal lattice efficiently. We have shown the algorithm to work on a 2-dimensional plane. In future work, we hope to place this algorithm on real robots and determine the effect the terrain has on the algorithm.

References

1. Spears, W.M., Spears, D.F., Hamann, J.C., Heil, R.: Autonomous Robots 17(2), 137 (2004)
2. Spears, W., Spears, W., Heil, R., Heil, R., Spears, D., Zarzhitsky, D.: Proceedings of the Third International Joint Conference on Autonomous Agents and Multiagent Systems, AAMAS 2004, pp. 1528–1529 (2004)
3. Gordon-Spears, D., Spears, W.: Lecture Notes in Computer Science, pp. 193–207 (2003)

4. Spears, W.M., Spears, D.F., Heil, R., Kerr, W., Hettiarachchi, S.: An overview of physicomimetics. In: Şahin, E., Spears, W.M. (eds.) Swarm Robotics 2004. LNCS, vol. 3342, pp. 84–97. Springer, Heidelberg (2005)
5. Mullen, R.J., Monekosso, D., Barman, S., Remagnino, P.: Distributed Autonomous Robotic (2010)
6. Martinson, E., Payton, D.: Lattice formation in mobile autonomous sensor arrays. In: Şahin, E., Spears, W.M. (eds.) Swarm Robotics 2004. LNCS, vol. 3342, pp. 98–111. Springer, Heidelberg (2005), http://dx.doi.org/
7. Balch, T., Hybinette, M.: In: Proceedings of the IEEE International Conference on Robotics and Automation, ICRA 2000, vol. 1, pp. 73–80 (2000)
8. Desai, J., Ostrowski, J., Kumar, V.: IEEE Transactions on Robotics and Automation 17(6), 905 (2001)
9. Eikenberry, B., Yakimenko, O., Romano, M.: AIAA Modeling and Simulation Technologies Conference and Exhibit (2006)
10. Hanada, Y., Lee, G., Chong, N.Y.: IEEE Swarm Intelligence Symposium, SIS 2007, pp. 340–347 (2007)
11. Russell, S., Norvig, P.: Artificial Intelligence: A Modern Approach, 2nd edn. Prentice Hall (2003)
12. Goldberg, D.E.: Genetic Algorithms in Search, Optimization and Machine Learning, 1st edn. Addison-Wesley Longman Publishing Co., Inc., Boston (1989)

Direct Policy Search with Variable-Length Genetic Algorithm for Single Beacon Cooperative Path Planning

Tan Yew Teck and Mandar Chitre

Abstract. This paper focuses on Direct Policy Search (DPS) for cooperative path planning of a single beacon vehicle supporting Autonomous Underwater Vehicles (AUVs) performing surveying missions. Due to the lack of availability of GPS signals underwater, the position errors of the AUVs grow with time even though they are equipped with proprioceptive sensors for dead reckoning. One way to minimize this error is to have a moving beacon vehicle with good positioning data transmit its position acoustically from different locations to other AUVs. When the position is received, the AUVs can fuse this data with the range measured from the travel time of acoustic transmission to better estimate their own positions and keep the error bounded. In this work, we address the beacon vehicle's path planning problem which takes into account the position errors being accumulated by the supported survey AUVs. We represent the path planning policy as state-action mapping and employ Variable-Length Genetic Algorithm (VLGA) to automatically discover the number of representative states and their corresponding action mapping. We show the resultant planned paths using the learned policy are able to keep the position errors of the survey AUVs bounded over the mission time.

1 Motivation

Even though marine robotic technologies have matured in recent years, underwater navigation still remains a challenging problem [1]. Due to the lack of availability of GPS signals underwater, AUVs generally rely on the on-board proprioceptive sensors such as compass, Doppler Velocity Log (DVL), and Inertial Navigation System

Tan Yew Teck · Mandar Chitre
ARL, Tropical Marine Science Institute and
Department of Electrical and Computer Engineering,
National University of Singapore,
Singapore 119227
e-mail: {william,mandar}@arl.nus.edu.sg

M.A. Hsieh and G. Chirikjian (eds.), *Distributed Autonomous Robotic Systems*,
Springer Tracts in Advanced Robotics 104,
DOI: 10.1007/978-3-642-55146-8_23,

(INS) for underwater navigation. However, dead reckoning using these sensors suffers unbounded positioning error growth over time. In order to alleviate the problem, methods that involve deploying fixed beacons around the mission area have been reported in the literature. The authors in [2] have developed a low-cost Long Based Line (LBL) navigation system for the AUV while [3] combined data from a DVL and an Utra-Short Based Line (USBL) system to provide superior three-dimensional position estimates to the AUV. Another recent solution uses a GPS Intelligent Buoy (GIB) system which consists of four surface buoys equipped with DGPS receivers and submerged hydrophones for tracking the position of the AUV underwater [4]. Although these systems act as good navigational aids for AUVs, they suffer from a few drawbacks. Firstly, deploying and retrieving these positioning systems require considerable operational effort. Second, they generally operate only at a limited range and are expensive and inflexible. Although the positioning problem can be avoided by having the AUVs surface and obtain a GPS fix, doing this not only costs precious mission time, but may put the AUVs' and the beacons' safety in jeopardy especially around busy shipping channels. Moreover, for some missions the AUVs may be required to be close to the seabed and surfacing during the mission may not be an option.

Recent advances in AUV and underwater communication technology have made inter-vehicle acoustic ranging a viable option for underwater cooperative positioning and localization. The idea of AUV cooperative localization is to have a vehicle with good quality positioning information (beacon vehicle) transmit its position information acoustically to other AUVs (survey AUVs) within its communication range during navigation (Fig. 1(a)). By measuring the propagation delay for the communication signal, the range between the beacon vehicle and the survey AUV can be estimated. Generally, the beacon vehicle is equipped with high accuracy sensors that is able to estimate its position with minimum errors. The range information between the vehicles can then be fused with the data obtained from on-board sensors to reduce the position error during underwater navigation [5, 6].

Fig. 1(b) shows that the error in the survey AUV's position estimate is reduced in the radial direction of the ranging circle centered at the beacon vehicle each time a range estimate becomes available. However, the error in the tangential direction remains unchanged. The key idea underlying the cooperative positioning algorithm for the beacon vehicle is to use the estimated position error ellipse of the survey AUV to plan its own movement. If the beacon vehicle can move to the location where the next range measurement occurs along the direction of the major axis of the error ellipse, the position error of the survey AUV can be minimized.

The idea of cooperative positioning, or localization with moving beacons is not new and has been explored by several researchers [7, 8, 9, 10, 11]. Their work includes observability analysis, algorithms for position determination based on range measurements, and some experimental results. Although all of these authors acknowledge that the relative motion of the beacon vehicle and the survey AUVs is key to having single beacon range-only navigation perform well, the problem of determining the optimal path of the beacon vehicle given the desired path of the survey AUVs has received little attention. For example, the work in [7] assumes a

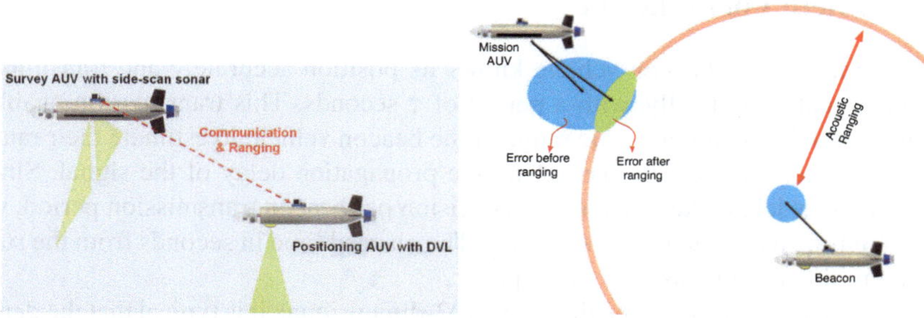

(a) The two AUVs for cooperative positioning. (b) Illustration of error estimates by range measurements.

Fig. 1 Cooperative positioning between the beacon vehicle and the survey AUV. The blue ellipses in (b) represent the position estimation errors for the AUVs before the ranging. The yellow ellipse represents the error after the range data is fused to yield a better position estimate.

circular path for the beacon vehicle, while [8] uses a zig-zag path in the experiments. In [10] the author suggests some maneuvers for the survey AUV while stating that the beacon vehicle would "most likely sprint and drift off side the survey path to force enough relative motion change to fix vehicle position". More recently, for a slightly different application domain, the authors in [12] applied a similar concept in tracking tagged sharks using an AUV. A particle filter algorithm to track the location of the shark and controlling the AUV's position to enable to algorithm converge.

Our previous contributions have focused on path planning for the beacon vehicle using Dynamic Programming (DP) [13] and Markov Decision Processes (MDP) with its policy matrix being learned through the Cross-Entropy method (MDP-CE) [14]. Although managing to achieve some promising results, they require either high computational load or large numbers of manually selected representative states for the policy matrix. In this paper, we further extend the work by approximating the state space in the form of Voronoi Tessellation where the the states are represented by the Voronoi seeds. We then deploy Variable-Length Genetic Algorithm (VLGAP) to automatically discover the optimal number of these states while simultaneously learning their corresponding action mappings.

In what follows, we first formulate the cooperative positioning problem as a path planning problem within the MDP framework in sections 2 and 3. We then describe the DPS algorithm for the MDP using the VLGA in section 4 and validate the performance of the policy learned in cooperative positioning missions in section 5. We discuss our contributions and findings in section 6 and summarize our conclusions in section 7.

2 Problem Formulation

We assume that the beacon vehicle knows its position accurately and transmits a beacon signal periodically, with a period of τ seconds. This transmission enables all survey AUVs within acoustic range of the beacon vehicle to estimate their range from the beacon vehicle by measuring the propagation delay of the signal. Since the beacon vehicle makes a navigation decision per beacon transmission period, we represent time using an index $t \in \{0..T\}$. The elapsed time in seconds from the start of the mission to time instant t is simply $t\tau$.

Although the underwater environment is 3-dimensional, it is typical that the depth for the beacon and survey vehicles is specified in a mission and may not be altered by our path planning algorithm. We therefore represent the position of each vehicle using a 2-dimensional position vector and the direction of travel of each vehicle by a yaw angle. Let $\mathbf{x}_t^B$ be the position and ϕ_t^B be the heading of the beacon vehicle B at time t. Let N be the number of survey AUVs supported by the beacon vehicle. We index the survey AUVs by $j \in \{1..N\}$. Let $\mathbf{x}_t^j$ represent the position of survey AUV j at time t. At every time index t, we have estimates $\hat{R}_t^j$ of the 2-dimensional range (easily estimated from the measured range by taking into account the difference in depths between the vehicles) between the beacon vehicle and each of the survey AUVs. We model the error in range estimation as a zero-mean Gaussian random variable with variance σ^2:

$$\hat{R}_t^j = \mathscr{N}(|\mathbf{x}_t^j - \mathbf{x}_t^B|, \sigma^2) \tag{1}$$

We further model the error in position estimation of the survey AUVs as a 2-dimensional zero-mean Gaussian random variable described by three parameters – the direction θ_t^j of minimum error, the error ε_t^j along direction θ_t^j, and the error $\bar{\varepsilon}_t^j$ in the tangential direction. Just after a range measurement at time $t+1$, the error is minimum along the line joining the beacon and the survey vehicle:

$$\theta_{t+1}^j = \angle(\mathbf{x}_{t+1}^j - \mathbf{x}_{t+1}^B) \tag{2}$$

$$\varepsilon_{t+1}^j = \sigma \tag{3}$$

The range measurement gives no information in the tangential direction and therefore the error grows in that direction. Assuming that the survey AUVs use velocity estimates for dead reckoning, the position error variance in the tangential direction will grow linearly with time:

$$(\bar{\varepsilon}_{t+1}^j)^2 = \frac{(\varepsilon_t^j \bar{\varepsilon}_t^j)^2}{(\varepsilon_t^j \cos \gamma_t^j)^2 + (\bar{\varepsilon}_t^j \sin \gamma_t^j)^2} + \alpha\tau \tag{4}$$

where $\gamma_t^j = \theta_{t+1}^j - \theta_t^j$ and α is the constant of proportionality (determined by the accuracy of the velocity estimate of the survey AUV).

The navigation decision made by the beacon vehicle at each time step t is δ_t^B, the turning angle during the time interval until the next decision. If $\dot{\phi}_{\max}^B$ is the maximum turning rate,

$$|\delta_t^B| \leq \dot{\phi}_{\max}^B \tau \tag{5}$$

If s^B is the speed of the beacon vehicle then the heading and position of the vehicle at time $t+1$ is approximately given by

$$\phi_{t+1}^B = \phi_t^B + \delta_t^B \tag{6}$$

$$\mathbf{x}_{t+1}^B = \mathbf{x}_t^B + \tau s^B \begin{pmatrix} \cos \phi_{t+1}^B \\ \sin \phi_{t+1}^B \end{pmatrix} \tag{7}$$

In order to ensure that the beacon and survey vehicles do not collide but are within transmission range of each other, we require that

$$D_{\min} \leq |\mathbf{x}_{t+1}^j - \mathbf{x}_{t+1}^B| \leq D_{\max} \ \ \forall\, j \tag{8}$$

We assume that the position of each survey AUV is known at the start of the mission with an accuracy of ε_0 in all directions:

$$\varepsilon_0^j = \bar{\varepsilon}_0^j = \varepsilon_0 \tag{9}$$

$$\theta_0^j = 0 \ \ \text{(arbitrary choice)} \tag{10}$$

Given the desired paths $\{\mathbf{x}_t^j \ \forall\, t\}$ of the survey AUVs and the initial position $\mathbf{x}_0^B$ and heading ϕ_0^B of the beacon vehicle, we wish to plan a path for the beacon vehicle such that we minimize the sum-square estimated position error across all survey AUVs for the entire mission duration. The path is fully determined by the sequence of decisions $\{\delta_t^B\}$ made during the mission:

$$\{\delta_t^B\} = \arg\min \sum_{j,t} \left[(\varepsilon_t^j)^2 + (\bar{\varepsilon}_t^j)^2 \right] \tag{11}$$

This naturally translates to the path planning problem for the beacon vehicle which takes into account the errors (both ε_t^j and $\bar{\varepsilon}_t^j$) of the survey AUVs operating within its communication range.

3 MDP Formulation

In this section, we present the formulation of the beacon vehicle's path planning problem within the MDP framework [14]. Generally, an MDP is defined by four main components: the state and action sets, the state transition probability matrix and the reward/cost function. From equation (1), $\hat{R}_t^j$ is the estimated distance between beacon and survey AUV, ϕ_t^{B} represent the beacon vehicle's current bearing at time t and ϕ_{t+1}^j be the survey AUV's bearing at time $t+1$ respectively, our state set is defined as a tuple: $z_t = \{\theta_t^j, \hat{R}_t^j, \phi_t^{\mathrm{B}}, \ \phi_{t+1}^j\}$. Since we assume that ε_{t+1}^j in (3) is a constant, we need to minimize $\bar{\varepsilon}_{t+1}^{j}$ in (4) to obtain (11) for every time step t. This means having γ_t^j in (4) to be as close as possible to 90 deg. Thus, the ability of beacon vehicle B to achieve this with respect to survey AUV j will depend on its knowledge of the components in the state space as well as the actions that it can take. Both the $\hat{R}_t^j$ and θ_t^j can be obtained from the acoustic ranging and communication between the AUVs while ϕ_{t+1}^{j} is usually pre-planned before the mission.

The action a_t is the turning angle from the beacon vehicle's current bearing (ϕ_t^{B}), $|a_t| \leq \dot{\phi}_{\max}^{\mathrm{B}}\tau$. At every time t, after a_t is selected, the corresponding x_{t+1}^{B} can be calculated and the accumulated sum square error can be estimated through (3) and (4). We model this accumulated error as the cost function, C, and we are interested in minimizing this cost over the entire mission path, which is equivalent to solving (11).

Instead of computing the value of being in a state using the state transition probability matrix and value function, we focus our attention on finding a deterministic policy in the form of state-action mapping. Given the beacon vehicle's current bearing, survey AUV's next heading as well as distance and relative angle between the AUVs, the action determines the desired turning angle from the beacon vehicle's current bearing (termed as desired heading in the rest of the paper) so that the position error of the survey AUV can be minimized during the next ranging event.

4 Direct Policy Search Using Variable Length Genetic Algorithm

4.1 State Space Approximation and Action Space Mapping

It is not always easy to design a good policy and predict the value of being in a state based on value function, as it is often computationally infeasible given the limited computational power that an AUV has. In order to alleviate this problem, various approximation techniques have been applied and encouraging results have been reported in the literature [15]. In this section, we describe the approximation technique used to represent the state space in the MDP and employ the evolutionary algorithm to automatically learn the deterministic policy for the beacon vehicle.

We simplify the state space into the form of Voronoi Tessellation where states located within a Voronoi cell are represented by their Representative States (RStates) specified by their Voronoi seeds. Consequently, the path planning policy is the direct mapping of these RStates into the action space as shown in Fig. 2. During cooperative positioning, the beacon vehicle first determines the state using the latest ranging information. It then locates the closest RState in terms of Euclidean distance in the state space. Since each of the RStates is deterministically mapped to a particular action, the decision making using the resultant policy is straightforward. Compared to the previous method [14], this approximation technique greatly reduces both the size of the policy matrix and the computational load of the beacon vehicle.

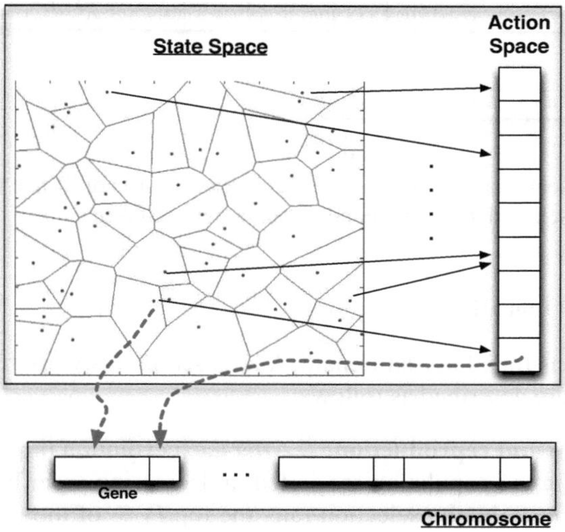

Fig. 2 State-Action space mapping and chromosome representation

4.2 Variable Length Genetic Algorithm

Three important parameters need to be tuned when solving the MDP formulated in section 4.1: the number of RStates to fully represent the entire state space, the locations of each of the RStates and their corresponding action mapping in the action space. To search for the optimal parameters, we use a VLGA with a novel variable-length chromosome representation. The VLGA automatically discovers the number of RStates and their location in the state space, as well as the RState-action mappings for the resultant policy.

4.2.1 Chromosome Representation

The chromosomes are encoded in binary form. Each of the continuous variables in the state and action space is discretized and encoded as a stream of binary numbers. They represent the locations of the state and action within the space domain. Fig. 3 shows an example of the chromosome represented using this scheme. Each of the genes in a chromosome consists of a RState-action pair which represents direct mapping relationship. The length of the chromosomes is variable during the process of evolution and represents the number of RStates for the resulting policy. This representation scheme is important to allow the VLGA to automatically discover the optimal number of the RStates, their locations within the state space, as well as their corresponding action mapping. Since the individual gene encodes the RState's location in the state space and its action mapping, the arrangement of the genes in the chromosome is irrelevant.

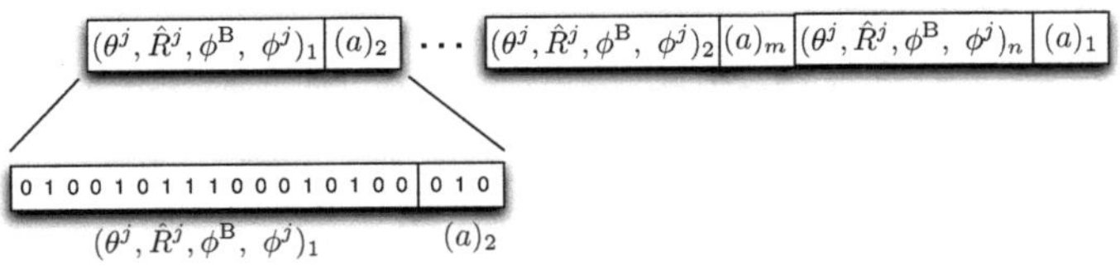

Fig. 3 Gene representation in Chromosome. Each gene consists of RState-action pair; whenever the RState is selected, the corresponding action will be taken.

4.2.2 Genetic Operations

Genetic operations found in traditional GA are used in this work for the process of evolution. They are described as follows:

Elitism selection and reproduction: After each evolution process, the chromosomes in the population are sorted in decreasing order based on their fitness. Let P_s be the selection rate, the top P_s % of the population are selected and reintroduced into the new population. Besides that, the same proportion of new chromosomes are randomly generated and introduced into the new generation. The rest of the population are then randomly reproduced from the pool of best chromosomes. This approach ensure the exploitation of the best found solutions as well as exploration of the new solutions in the new population.

Crossover: Two chromosomes are randomly selected from the population according to the P_c - the crossover rate. One-point crossover is performed between a pair of chromosomes and the new resultant chromosomes are re-introduced into the population. Physically, the crossover operation increases the probability of combining good genes from different parent chromosomes, thus, producing fitter offsprings.

Mutation: Let P_m be the mutation rate. At every generation, P_m chromosomes are chosen from the new population to undergo mutation. In this paper, we applied three different types of mutation operations to the selected sub-population:

- Growth mutation – randomly produces a new gene and appends it to the selected chromosome.
- Shrink mutation – randomly removes a gene from the selected chromosome.
- Flip mutation – applies flipping operation on the genes. The bit is flipped with the probability equal to the mutation rate.

Both the growth and shrink mutation may, hopefully, help to introduce good new genes and remove bad genes from the chromosome. Besides, the flipping mutation aids to maintain the diversity of the new population in searching for an optimal solution.

4.2.3 Fitness Function

The fitness function of the chromosomes are evaluated based on the performance of their encoded policy through Monte Carlo simulation. Detailed descriptions of the simulation are presented in section 4.2.4. Since we are searching for a path planning policy that will minimize the cost function, C, of the MDP described in section 3, the fitness function of the chromosomes is defined as follows:

$$f_i = \frac{1}{C_i} = \frac{1}{\sum_t \left[(\varepsilon_t^{SA})^2 + (\bar{\varepsilon}_t^{SA})^2\right]} \tag{12}$$

where f_i represents the fitness value of the ith chromosome, C_i is the cost incurred from the path planned by the beacon vehicle, which is calculated through the summation of the positioning errors (both the ε_t^{SA} and $\bar{\varepsilon}_t^{SA}$) accumulated by the survey AUV (SA) for a sample survey path of t steps.

4.2.4 Fitness Evaluation through Monte Carlo Simulation

The fitness of each individual offspring is evaluated through Monte Carlo simulation between the beacon vehicle and a survey AUV. During the simulation, a survey path of t steps with lawn mowing pattern is randomly generated to simulate a survey mission. Starting from all the initial states in the state space, the beacon vehicle is deployed and plans its path to support the survey AUV using the encoded policy. Since acoustic ranging information is assumed to be available at each of the t steps, the resultant beacon's path has the same length as the survey path. With both the beacon and survey paths, the sum of the positioning errors (12), which is equivalent to the cost, can be calculated. The same simulation is performed using the policies encoded in all the chromosomes in the population, and the resultant fitnesses are ranked in descending order for the selection operation. Detailed algorithm of the simulation is shown in Algorithm 1.

5 Experimental Results

5.1 Policy Search Setup and Results

Instead of discretizing the map into grid map or graph nodes as is commonly done for the path planning problem of mobile robots [16, 17], we discretized both the state and action space of the beacon vehicle. For the convenience of binary encoding of the chromosome, we discretize the AUVs' bearing and the angle between the AUVs into 32 states spanning from $0 \sim 360$ deg. The distance between the AUVs are discretized into 4 zones: two forbidden zones (less than $D_{\min}$ and more than $D_{\max}$) and two legal zones with each occupying half of the distance in between $D_{\min}$ and $D_{\max}$. Heavy penalty that will contribute to the accumulated errors is given whenever the vehicles are in the forbidden zones. This is necessary to prevent the vehicles from colliding if they are too close together while keeping them within

Algorithm 1. Fitness Evaluation through Monte Carlo Simulation

Require: Z – State Space
Require: Pop – Policies represented by chromosomes in the population
  for all z^s in Z **do**
    Generate a random surveying path with path length of t steps.
    for all p_i in Pop **do**
      Start from the initial state $z_0 = z^s$, set $j = 0$.
      Locate the RState in p_i that is closest to z_0 in terms of Euclidian distance.
      Apply the corresponding action (encoded in the same gene as the selected RState) and generate a new state z_{j+1}. Set $j = j+1$. Repeat until $j = t$.
      Output the total cost (C_{p_i}) of the trajectory $(z_0, z_1, ..., z_t)$.
      Calculate the fitness f_i of the policy p_i.
    end for
  end for
  return f_i of all p_i in Pop.

the communication range. Due to the limitation of the turning radius achievable during navigation, the beacon vehicle's desired turning angle is constrained within [-20,20] deg (obtained from $\tau\dot{\phi}^{\mathrm{B}}_{\mathrm{max}}$ in Table 2(a)) of the vehicle's current bearing and is divided into 8 zones. Detailed parameters setup is shown in Table 1. Table 2 shows the parameters used for the Monte Carlo simulation of the beacon vehicle and the DPS using the VLGA.

Table 1 State and Action Space Discretization

State Space, Z	Number of States	Number of Bits
Beacon vehicle's current bearing	32	5
Surverying AUV's next bearing	32	5
Relatives angle between AUVs	32	5
Distance between AUVs	4	2

Action Space, A	Number of States	Number of Bits
Beacon vehicle's desired turning angle	8	3

The fitness value and the length of the fittest chromosome in each generation of the VLGA are shown in Fig. 4. Even though the length of an individual chromosome in the population was allowed to evolve, it stabilizes at about 220 genes for the fittest chromosome. In some instances during the policy search, we observed that the length of the fittest chromosome dropped (around generation 100, 500 and 700) while their fitness value continue to increase. This shows that the fitness of the chromosome (performance of the policy) does not only depend on the number of the RStates, but also the locations of the RStates and their action mapping.

Table 2 Parameters for Beacon Vehicle and Vlga

(a) Beacon's Parameters

Parameter	Value
τ	10 s
σ	1 m
$\dot{\phi}^{\mathrm{B}}_{\mathrm{max}}$	0.07 rad/s
D_{min}	100 m
D_{max}	1000 m
ε_0	1 m
α	0.1 m^2/s

(b) VLGA Parameters

Parameter	Value
P_s	0.1
P_c	0.6
P_m	0.15
Encoding scheme	Binary
Substring length	20
Population size (*Pop*)	200
Number of generations	1200

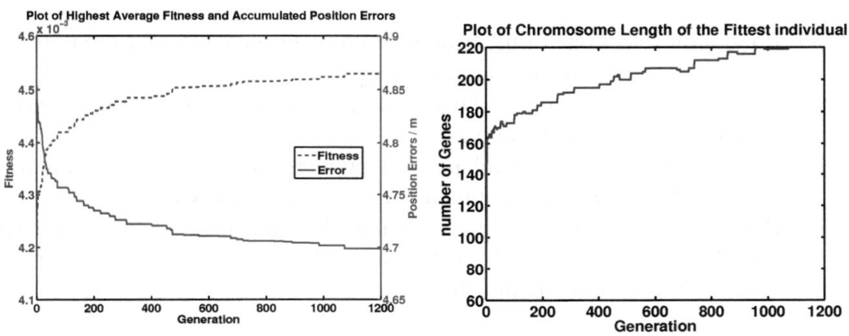

(a) The fitness of the chromosome with the highest fitness value.

(b) The length of the fittest chromosome.

Fig. 4 Result of the VLGA showing the fitness value and the length of the fittest chromosome in each generation

5.2 Cooperative Path Planning Simulations

The fittest chromosome at the end of the VLGA policy search is selected as the cooperative path planning policy for the beacon vehicle. We investigated the performance of the policy in supporting single as well as multiple survey AUVs. The same setups shown in Table 2 (a) were used for the simulations.

5.2.1 Simulation Setup

1. **Supporting Single Survey AUV**
   A survey AUV was given a lawn-mower mission surveying an area of about 500 m by 700 m as shown in Fig. 6(a). The survey AUV's path is pre-planned and shared with the beacon vehicle. All the vehicles are assumed to be moving at the speed of 1.5 m/s and ranging information is available every τ seconds.

The beacon vehicle plans its path iteratively using the policy learned by VLGA until the completion of the mission.

2. **Supporting Multiple Survey AUVs**
In the second simulation scenario, we evaluated the performance of a single beacon AUV supporting 2, 3 and 4 survey AUVs as shown in Fig. 7. Since the policy generates only a desired turning angle with respect to each of the survey AUVs, we get more than one heading commands from the policy after every ranging updates. Choosing one command that favors only one of the AUVs might cause the position error of the other AUVs to grow. Thus, care has to be taken while making the final decision. We studied four different methods to explore the best strategy for the beacon AUV in deciding the desired heading command:

S-1 Randomly select one of the heading commands generated by the policy as the beacon AUV's next desired heading.
S-2 Select the heading command that will favor the survey AUV whose current accumulated error is the highest.
S-3 Select the heading command that will navigate the beacon AUV around the vicinity of the centroid location among the survey AUVs.
S-4 Perform a round-robin selection scheme where the heading commands generated with respect to each of the survey AUVs are selected in a circular order after each ranging updates.

5.2.2 Simulation Results

A simple simulation was performed with a survey AUV moving in a straight line to illustrate the intuition behind the cooperative positioning algorithm (Fig. 5). Starting from the initial position, the beacon vehicle plans its path using the resultant planning policy to support the survey AUV. The simulation results show that, given a straight survey path, the beacon vehicle maneuvered back and forth from the starboard to the port side of the survey AUV to maximize the change of relative aspect when the acoustic range information is exchanged. Also, the resultant paths maneuver the beacon vehicle in the direction of the survey AUV to keep them within the communication range.

Fig. 6(a) shows the resultant cooperative paths planned by the beacon vehicle during the course of supporting a single survey AUV. Even though the beacon vehicle is constrained to navigate within 1 km from the survey AUVs, statistical analysis shows it has "learned" to navigate itself around the proximity of the survey AUVs, in order to increase the chance of achieving maximum change of relative angle with respect to the survey AUVs. The position errors accumulated throughout the mission period are shown in Fig. 6(b). The results are plotted based on the average of 10 simulated runs for the same scenario. The position errors of the survey AUVs are expected to grow unbounded if they rely only on dead reckoning. However, with the ranging information provided by the beacon vehicle at different relatives angles, the errors were kept around $3m \sim 5m$ throughout the mission period.

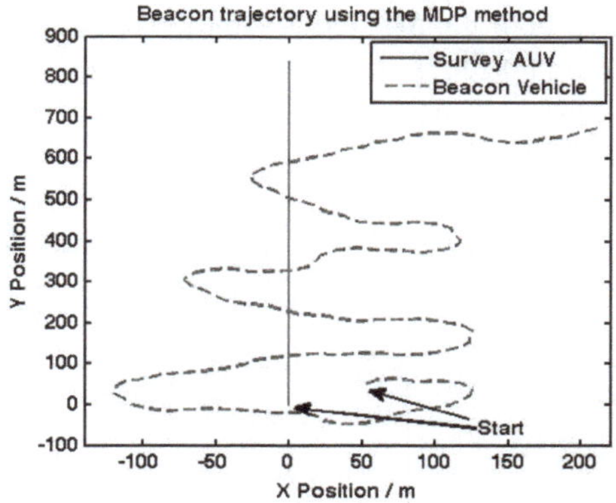

Fig. 5 Simulation result showing the beacon vehicles paths in supporting the survey AUV moving in a straight line

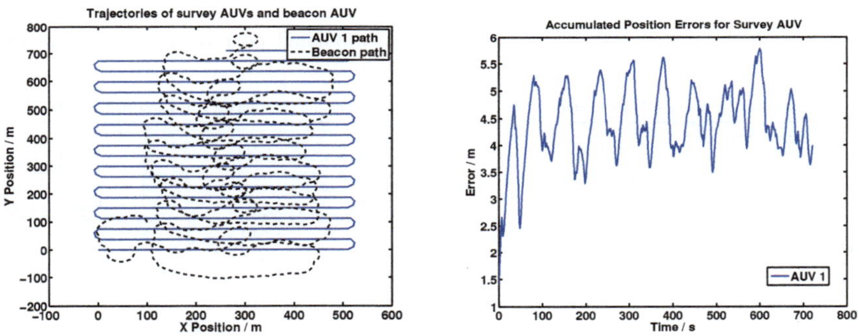

(a) Supporting single survey AUV.

(b) Accumulated position error of survey AUV.

Fig. 6 Simulation results showing the beacon vehicle supporting single survey AUV

The resultant paths planned by a beacon AUV in supporting multiple survey AUVs using the strategy *S-2* are shown in Fig. 7, while the accumulated position errors for the case of supporting 2 survey AUVs is shown in Fig. 7(b). The results from 10 simulated runs using different strategies are summarized in Table 3. Generally, the average Root Mean Square (aRMS) error accumulated by the survey AUVs are kept small within $3m \sim 5m$ across all strategies even though the Maximum (Max) errors varied significantly.

We observed that the performance of *S-2* is slightly better compared to other strategies in both the aRMS and the Max errors. This is due to the fact that the closer the beacon AUV is to the survey AUV team, the chance of achieving the maximum change relative aspect (~ 90deg) with each of the survey AUVs is higher, and consequently, the RMS errors of the survey AUVs can be kept low by acoustic ranging. Not surprisingly, both the *S-1* and *S-4* incurred much higher Max errors especially in the case of supporting 4 survey AUVs, since the decisions were made without considering neither the survey AUV's current accumulated errors nor the distance between the vehicles.

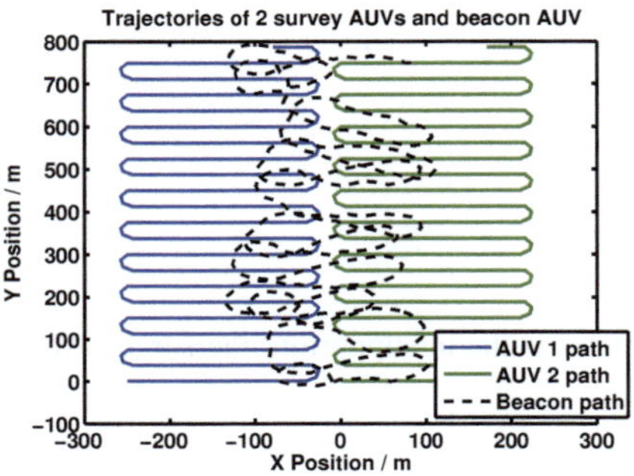

(a) Beacon AUV supporting two survey AUVs.

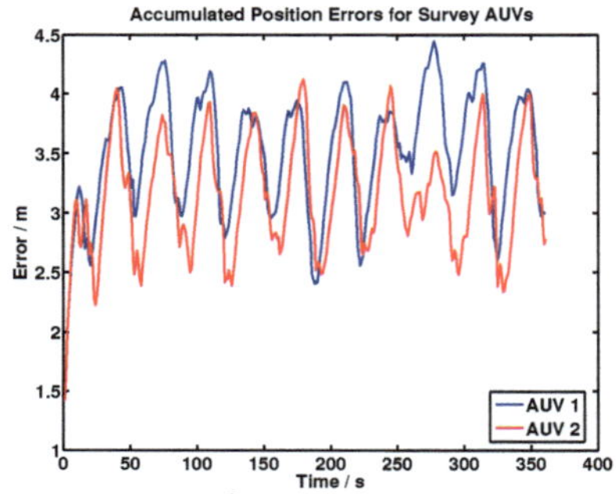

(b) Accumulated position errors of both the survey AUVs.

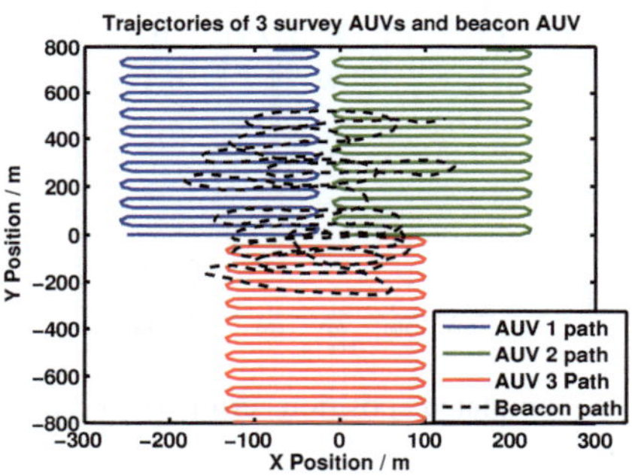

(c) Beacon AUV supporting three survey AUVs.

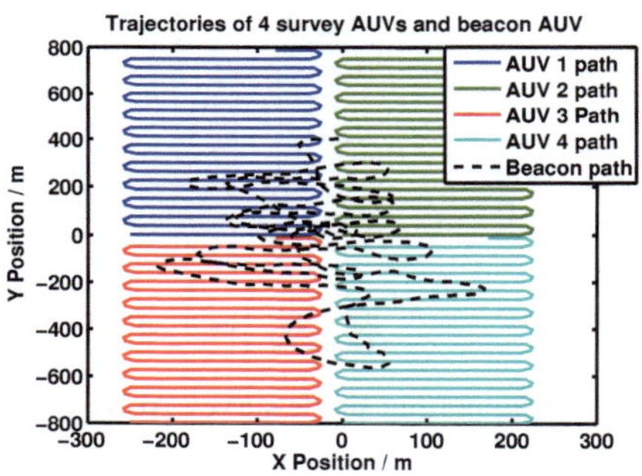

(d) Beacon AUV supporting four survey AUVs.

Fig. 7 Simulation results showing the beacon vehicle supporting multiple survey AUVs using the strategy *S-2*

6 Discussion

The simulation results have demonstrated that the VLGA can be used to automatically discover the optimal number as well as the locations of the RStates that are required to fully represent a multidimensional state space. It is also capable of simultaneously learning the policy in planning cooperative paths for the beacon vehicle. The state space approximation through Voronoi Tessellation has greatly reduced the number of states required for a policy. This not only alleviates the "curse of dimensionality" problem, but also solves the practical issues of applying MDP approach in

Table 3 Simulation Results for Supporting Multiple Survey AUVs

No. Survey AUVs	Strategy							
	S-1		*S-2*		*S-3*		*S-4*	
	Error (m)							
	aRMS	Max	aRMS	Max	aRMS	Max	aRMS	Max
2	4.16	8.07	3.49	6.44	4.39	7.52	4.07	7.51
3	5.19	10.81	4.66	7.72	5.19	11.79	5.37	9.63
4	5.61	22.43	4.60	7.27	5.40	13.67	5.73	23.64

Table 4 Computational Load and Size of Policy Table for DP, MDP-CE and DPS with VLGA

	DP [13]	MDP-CE [14]	DPS with VLGA
Computational Load	$\mathcal{O}(TN_a^{L+1}M)$	$\mathcal{O}(TM)$	$\mathcal{O}(TCM)$
Policy Table (No. of States)	N.A	1119744	7040

autonomous robotic systems due to their limited computational power and memory storage.

The results presented in section 5.2.2 are comparable with the DP approach [13] and MDP-CE approach [14]. Table. 4 showed the comparisons of the computational load and the size of the resultant policy table learned via different approaches. Let L be the number of look-ahead levels, N_a be the action space, M be the number of supported survey AUVs and C be the number of RStates, the computational load of our approach is much lower compared to the DP approach but slightly higher than the MDP-CE method. However, the policy structure of our approach was learned through natural evolution, and its size is much smaller (about 160 times smaller !) compared to the MDP-CE method.

7 Conclusion

We developed a novel method for Direct Policy Search (DPS) for Markov Decision Processes (MDP) using the Variable-Length Genetic Algorithm (VLGA). We demonstrated its capability in discovering the representative states in the state space approximation while simultaneously learning the state-action mapping of a cooperative path planning policy for a beacon vehicle. We showed that the resultant policy is able to plan the path for beacon vehicle so that the position errors of the supported survey AUVs can be kept minimum whenever acoustic ranging information is exchanged. Compared to the previous published approaches, our approach greatly reduces the computational load as well as the size of the policy matrix, yet manages to perform comparatively well in terms of minimizing the survey AUVs' position errors. Future work may include exploring the possibility of online learning given the much simplified policy representation.

References

1. Kinsey, J.C., Eustice, R.M., Whitcomb, L.L.: A survey of underwater vehicle navigation: Recent advances and new challenges. In: IFAC Conference of Manoeuvering and Control of Marine Craft, Lisbon, Portugal (September 2006) (Invited paper)
2. Matos, A., Cruz, N., Martins, A., Lobo Pereira, F.: Development and implementation of a low-cost lbl navigation system for an auv. In: OCEANS 1999 MTS/IEEE. Riding the Crest into the 21st Century, vol. 2, pp. 774–779 (1999)
3. Rigby, P., Pizarro, O., Williams, S.: Towards geo-referenced auv navigation through fusion of usbl and dvl measurements. In: OCEANS 2006, pp. 1–6, 18–21 (2006)
4. Alcocer, A., Oliveira, P., Pascoal, A.: Study and implementation of an ekf gib-based underwater positioning system. Control Engineering Practice 15(6), 689–701 (2007)
5. Rui, G., Chitre, M.: Cooperative positioning using range-only measurements between two AUVs. In: OCEANS 2010 IEEE - Sydney, pp. 1–6 (May 2010)
6. Bahr, A., Leonard, J.J., Fallon, M.F.: Cooperative localization for autonomous underwater vehicles. The International Journal of Robotics Research 28(6), 714–728 (2009)
7. Alleyne, J.C.: Position estimation from range only measurements. Master's thesis, Naval Postgraduate School, Monterey CA (September 2000)
8. Fallon, M.F., Papadopoulos, G., Leonard, J.J., Patrikalakis, N.M.: Cooperative AUV Navigation using a Single Maneuvering Surface Craft. The International Journal of Robotics Research 29(12), 1461–1474 (2010)
9. Song, T.L.: Observability of target tracking with range-only measurements. IEEE Journal of Oceanic Engineering 24, 383–387 (1999)
10. Hartsfiel, J.: Single transponder range only navigation geometry (strong) applied to remus autonomous under water vehicles. Master's thesis. MIT (2005)
11. Gadre, A., Stilwell, D.: Toward underwater navigation based on range measurements from a single location. In: Proceedings of the 2004 IEEE International Conference on Robotics and Automation, ICRA 2004, April-May 1, vol. 5, pp. 4472–4477 (2004)
12. Forney, C., Manii, E., Farris, M., Moline, M., Lowe, C., Clark, C.: Tracking of a tagged leopard shark with an auv: Sensor calibration and state estimation. In: 2012 IEEE International Conference on Robotics and Automation (ICRA), pp. 5315–5321 (May 2012)
13. Chitre, M.: Path planning for cooperative underwater range-only navigation using a single beacon. In: 2010 International Conference on Autonomous and Intelligent Systems (AIS), pp. 1–6 (June 2010)
14. Tan, Y.T., Chitre, M.: Single beacon cooperative path planning using cross-entropy method. In: IEEE/MTS OCEANS, Kona, Hawaii (September 2011)
15. Powell, W.B.: Approximate Dynamic Programming: Solving the Curses of Dimensionality, 2nd edn. Wiley Series in Probability and Statistics. Wiley (2011)
16. Tu, J., Yang, S.: Genetic algorithm based path planning for a mobile robot. In: Proceedings of the IEEE International Conference on Robotics and Automation, ICRA 2003, vol. 1, pp. 1221–1226 (September 2003)
17. Ahn, C.W., Ramakrishna, R.: A genetic algorithm for shortest path routing problem and the sizing of populations. IEEE Transactions on Evolutionary Computation 6, 566–579 (2002)

Optimal Multi-Robot Path Planning with LTL Constraints: Guaranteeing Correctness through Synchronization

Alphan Ulusoy, Stephen L. Smith, and Calin Belta

Abstract. In this paper, we consider the automated planning of optimal paths for a robotic team satisfying a high level mission specification. Each robot in the team is modeled as a weighted transition system where the weights have associated deviation values that capture the non-determinism in the traveling times of the robot during its deployment. The mission is given as a Linear Temporal Logic (LTL) formula over a set of propositions satisfied at the regions of the environment. Additionally, we have an optimizing proposition capturing some particular task that must be repeatedly completed by the team. The goal is to minimize the maximum time between successive satisfying instances of the optimizing proposition while guaranteeing that the mission is satisfied even under non-deterministic traveling times. After computing a set of optimal satisfying paths for the members of the team, we also compute a set of synchronization sequences for each robot to ensure that the LTL formula is never violated during deployment. We implement and experimentally evaluate our method considering a persistent monitoring task in a road network environment.

1 Introduction

Temporal logics [5], such as Linear Temporal Logic (LTL) and Computation Tree Logic (CTL), are extensions of propositional logic that can capture temporal relations. Even though temporal logics have been used in model checking of finite systems [1] for quite some time, they have gained popularity as a means for specifying

Alphan Ulusoy · Calin Belta
Boston University, Boston, MA, USA
e-mail: {alphan, cbelta}@bu.edu

Stephen L. Smith
University of Waterloo, Waterloo, ON, Canada
e-mail: stephen.smith@uwaterloo.ca

M.A. Hsieh and G. Chirikjian (eds.), *Distributed Autonomous Robotic Systems*,
Springer Tracts in Advanced Robotics 104,
DOI: 10.1007/978-3-642-55146-8_24,

complex mission requirements in path planning and control synthesis problems only recently [14, 12, 21]. Existing work on path planning and control synthesis concentrates on LTL specifications for finite state systems, which may be abstractions of their infinite counterparts [14, 17]. Particularly, given the system model and some temporal logic formula, satisfying paths and corresponding control strategies can be computed automatically through a search of the state space for deterministic [9], non-deterministic [15, 17, 12, 10] and probabilistic systems [2, 11, 4].

In [9], the authors propose a method for decentralized motion of multiple robots subject to LTL specifications. Their method, however, results in sub-optimal performance as it requires the robots to travel synchronously, blocking the execution of the mission before each transition until all robots are synchronized. The vehicle routing problem (VRP) [16] and its extensions to more general classes of temporal constraints [7, 8] also deal with finding optimal satisfying paths for a given specification. In [8], the authors consider optimal vehicle routing with metric temporal logic specifications by converting the problem to a mixed integer linear program (MILP). However, their method does not apply to the missions where robots must repeatedly complete some task, as it does not allow for specifications of the form "always eventually". Furthermore, none of these methods are robust to timing errors that can occur during deployment, as they rely on the robots' ability to follow generated trajectories exactly for satisfaction of the mission specification.

In our previous work, we focused on mission specifications given in LTL along with a particular cost function, and proposed an automated method for finding optimal robot paths that satisfy the mission and minimize the cost function for a single robot [13]. Next, we extended this approach to multi-robot teams by utilizing an abstraction based on timed automata [20]. Extending the optimal path planning problem from a single robot to multiple robots is not trivial, as the joint asynchronous motion of all members of the team must be captured in a finite model. Then, we proposed a robust method that could accomodate uncertainties in the traveling times of robots with limited communication capabilities [19]. The methods given in [20] and [19] are actually two extremes: In [20], the robots can follow the generated trajectories exactly and do not communicate at all, while in [19] the robots' traveling times during deployment deviate from those used in planning, and they cannot communicate freely. In this paper, we address the middle between these two extremes: the robots cannot follow the generated trajectories exactly, but they can communicate regardless of their positions in the environment. Thus, after obtaining an optimal satisfying run of the team, we compute synchronization sequences that leverage the communication capabilities of the robots to robustify the planned trajectory against deviations in traveling times.

The main contribution of this paper is threefold. First, we provide an algorithm to capture the joint asynchronous behavior of a team of robots modeled as transition systems in a single transition system. This team transition system is provably more compact than the approach based on timed automata that we previously proposed in [20]. Second, for a satisfying run made up of a finite length prefix and an infinite length cyclic suffix, we propose a synchronization protocol and an algorithm to compute synchronization sequences that guarantee correctness under non-deterministic traveling times that may be observed during deployment. Finally, we provide an automated framework that uses these two methods along with the OPTIMAL-RUN algorithm previously proposed in [13] to solve the multi-robot optimal path planning problem with robustness guarantees.

The rest of the paper is organized as follows: In Sec. 2, we provide some definitions and preliminaries in formal methods. In Sec. 3, we formulate the optimal and robust multi-robot path planning problem and give an outline of our approach. We provide a complete solution to this problem in Sec. 4. In Sec. 5, we present experiments involving a team of robots performing a persistent surveillance mission in a road network environment. In Sec. 6, we conclude with final remarks. Due to page constraints we omit the proofs of all results. The proofs are contained in an extended version available online [18].

2 Preliminaries

In this section, we introduce the notations that we use in the rest of the paper and briefly review some concepts related to automata theory, LTL, and formal verification. For a more rigorous treatment of these topics, we refer the interested reader to [3, 6, 1] and references therein.

For a set Σ, we use $|\Sigma|$, 2^{Σ}, Σ^*, and Σ^{ω} to denote its cardinality, power set, set of finite words, and set of infinite words, respectively. We define $\Sigma^{\infty} = \Sigma^* \cup \Sigma^{\omega}$ and denote the empty string by $\emptyset$.

Definition 1 (Transition System). A (weighted) transition system (TS) is a tuple $\mathbf{T} := (\mathscr{Q}_{\mathrm{T}}, q^0_{\mathrm{T}}, \delta_{\mathrm{T}}, \Pi_{\mathrm{T}}, \mathscr{L}_{\mathrm{T}}, w_{\mathrm{T}})$, where (1) $\mathscr{Q}_{\mathrm{T}}$ is a finite set of states; (2) $q^0_{\mathrm{T}} \in \mathscr{Q}_{\mathrm{T}}$ is the initial state; (3) $\delta_{\mathrm{T}} \subseteq \mathscr{Q}_{\mathrm{T}} \times \mathscr{Q}_{\mathrm{T}}$ is the transition relation; (4) Π_{T} is a finite set of atomic propositions; (5) $\mathscr{L}_{\mathrm{T}} : \mathscr{Q}_{\mathrm{T}} \to 2^{\Pi_{\mathrm{T}}}$ is a map giving the set of atomic propositions satisfied in a state; (6) $w_{\mathrm{T}} : \delta_T \to \mathbb{N}_{>0}$ is a map that assigns a positive integer weight to each transition.

We define a run of $\mathbf{T}$ as an infinite sequence of states $r_{\mathrm{T}} = q^0, q^1, \ldots$ such that $q^0 = q^0_T$, $q^k \in \mathscr{Q}_{\mathrm{T}}$ and $(q^k, q^{k+1}) \in \delta_{\mathrm{T}}$ for all $k \geq 0$. A run generates an infinite word $\omega_{\mathrm{T}} = \mathscr{L}(q^0), \mathscr{L}(q^1), \ldots$ where $\mathscr{L}(q^k)$ is the set of atomic propositions satisfied at state q^k.

In this work, we consider mission specifications expressed as Linear Temporal Logic (LTL) formulas over Π using the standard syntax and semantics defined in [1]. We follow the literal notation for temporal operators $(\mathbf{G}, \mathbf{F}, \mathbf{X}, \mathscr{U})$. We say a run r_{T} satisfies ϕ if and only if the word generated by r_{T} satisfies ϕ.

Definition 2 (Büchi Automaton). A Büchi automaton is a tuple $\mathbf{B} := (\mathscr{Q}_{\mathrm{B}}, \mathscr{Q}^0_{\mathrm{B}}, \Sigma_{\mathrm{B}}, \delta_{\mathrm{B}}, \mathscr{F}_{\mathrm{B}})$, consisting of (1) a finite set of states $\mathscr{Q}_{\mathrm{B}}$; (2) a set of initial states $\mathscr{Q}^0_{\mathrm{B}} \subseteq \mathscr{Q}_{\mathrm{B}}$; (3) an input alphabet Σ_{B}; (4) a non-deterministic transition relation $\delta_{\mathrm{B}} \subseteq \mathscr{Q}_{\mathrm{B}} \times \Sigma_{\mathrm{B}} \times \mathscr{Q}_{\mathrm{B}}$; (5) a set of accepting (final) states $\mathscr{F}_{\mathrm{B}} \subseteq \mathscr{Q}_{\mathrm{B}}$.

A *run* of $\mathbf{B}$ over an input word $\omega = \omega^0, \omega^1, \ldots$ is a sequence $r_{\mathrm{B}} = q^0, q^1, \ldots$, such that $q^0 \in \mathscr{Q}^0_{\mathrm{B}}$, and $(q^k, \omega^k, q^{k+1}) \in \delta_{\mathrm{B}}$, for all $k \geq 0$. A Büchi automaton $\mathbf{B}$ accepts a word over Σ_B if and only if at least one of the corresponding runs intersects with $\mathscr{F}_B$ infinitely many times. For any LTL formula ϕ over a set Π, one can construct a Büchi automaton with input alphabet $\Sigma_B = 2^{\Pi}$ accepting all and only words over 2^{Π} that satisfy ϕ.

Definition 3 (Prefix-Suffix Structure). A prefix of a run is a finite path from an initial state to a state q. A periodic suffix is an infinite run originating at the state q reached by the prefix, and periodically repeating a finite path, which we call the suffix cycle, originating and ending at q. A run is in prefix-suffix form if it consists of a prefix followed by a periodic suffix.

3 Problem Formulation and Approach

In this section we introduce the multi-robot path planning problem with temporal logic constraints for robots with uncertain, but bounded traveling times. Let $\mathscr{E} = (V, \rightarrow_{\mathscr{E}})$ be a directed graph, where V is the set of vertices and $\rightarrow_{\mathscr{E}} \subseteq V \times V$ is the set of edges. We consider $\mathscr{E}$ as the quotient graph of a partitioned environment, where V is the set of labels of the regions and $\rightarrow_{\mathscr{E}}$ is the corresponding adjacency relation.

Consider a team of m robots moving in an environment modeled by $\mathscr{E}$. The motion capabilities of robot $i, i = 1, \ldots, m$ are modeled by a TS $\mathbf{T}_i = (\mathscr{Q}_i, q_i^0, \delta_i, \Pi_i, \mathscr{L}_i, w_i)$, where $\mathscr{Q}_i \subseteq V$; q_i^0 is the initial vertex of robot i; $\delta_i \subseteq \rightarrow_{\mathscr{E}}$ is a relation modeling the capability of robot i to move among the vertices; $\Pi_i \subseteq \Pi$ is the subset of propositions that can be satisfied by robot i; $\mathscr{L}_i$ is a mapping from $\mathscr{Q}_i$ to 2^{Π_i} showing how the propositions are satisfied at vertices; and $w_i(q, q')$ gives the *nominal* time for robot i to go from vertex q to q', which we assume to be a positive integer. However, due to the uncertainty in the traveling times of the robots, the *actual* time it takes for robot i to go from q to q', which we denote by $\tilde{w}_i(q, q')$, is a non-deterministic quantity that lies in the interval $[\underline{\rho_i} w_i(q, q'), \overline{\rho_i} w_i(q, q')]$, where $\underline{\rho_i}, \overline{\rho_i}$ are the predetermined lower and upper *deviation values* of robot i that satisfy $0 < \underline{\rho_i} \leq 1 \leq \overline{\rho_i}$. In this model, robot i travels along the edges of $\mathbf{T}_i$, and spends zero time on the vertices. We also assume that the robots are equipped with motion primitives that allow them to deterministically move from q to q' for each $(q, q') \in \delta_i$, even though the time it takes to reach from q to q' is uncertain. In the following, we use the expression "*in the field*" to refer to the model with uncertain traveling times, and use x and $\tilde{x}$ to denote the *nominal* and *actual* values of some variable x.

We consider the case where the robotic team has a mission in which some particular task must be repeatedly completed and the maximum time between successive completions of this task must be minimized. For instance, in a persistent surveillance mission, the global mission could be *keep gathering data while obeying traffic rules at all times*, and the repeating task could be *gathering data*. For this example, the robots would operate according to the mission specification while ensuring that the maximum time between successive data gatherings is minimized. Consequently, we assume that there is an *optimizing proposition* $\pi \in \Pi$ that corresponds to this repeating task and consider multi-robot missions specified by LTL formulae of the form

$$\phi := \varphi \wedge \mathbf{GF}\pi, \tag{1}$$

where φ can be any LTL formula over Π, and $\mathbf{GF}\pi$ means that the proposition π must be repeatedly satisfied. Our aim is to plan multi-robot paths that satisfy ϕ and minimize the maximum time between successive satisfying instances of π.

To state this problem formally, we assume that each run $r_i = q_i^0, q_i^1, \ldots$ of $\mathbf{T}_i$ (robot i) starts at $t = 0$ and generates a word $\omega_i = \omega_i^0, \omega_i^1, \ldots$ and a corresponding sequence of time instances $\mathbb{T}_i := t_i^0, t_i^1, \ldots$ such that $\omega_i^k = \mathscr{L}_i(q_i^k)$ is satisfied at t_i^k. To define the behavior of the team as a whole, we interpret the sequences $\mathbb{T}_i$ as sets and take the union $\bigcup_{i=1}^m \mathbb{T}_i$ and order this set in an ascending order to obtain the sequence $\mathbb{T} := t^0, t^1, \ldots$. Then, we define $\omega_{team} = \omega_{team}^0, \omega_{team}^1, \ldots$ to be the word generated by the team of robots where ω_{team}^k is the union of all propositions satisfied at t^k. Finally, we define the infinite sequence $\mathbb{T}^\pi = \mathbb{T}^\pi(1), \mathbb{T}^\pi(2), \ldots$ where $\mathbb{T}^\pi(k)$ is the time instance when π is satisfied for the kth time by the team and define the cost function

$$J(\mathbb{T}^\pi) = \limsup_{i \to +\infty} (\mathbb{T}^\pi(i+1) - \mathbb{T}^\pi(i)). \tag{2}$$

The form of the cost function given in Eq. (2) is motivated by persistent surveillance missions, where one is interested in the long-term behavior of the team. Given a sequence $\mathbb{T}^\pi$ corresponding to some run of the team, the cost function in Eq. (2) captures the maximum time between satisfying instances of π once the team behavior reaches a steady-state, which we achieve in finite time as we will discuss in Sec. 4.2. Thus, the problem becomes that of finding an optimal run of the team that satisfies ϕ and minimizes (2). However, the non-determinism in traveling times imposes two additional difficulties which directly follow from Prop. 3.2 in [19]: First, if the traveling times observed during deployment deviate from those used in planning, then there exist missions that will be violated in the field. Second, the worst case performance of the robotic team during deployment in terms of Eq. (2) will be limited by that of a single member.

To guarantee correctness in the field, and limit the deviation of the performance of the team from the planned optimal run during deployment, we propose periodic synchronization of the robots. Using this synchronization protocol, robots synchronize with each other according to pre-computed synchronization sequences $s_i, i = 1, \ldots, m$ as they execute their runs $r_i, i = 1, \ldots, m$ in the field. We can now formulate the problem.

Problem 1. Given a team of m robots modeled as transition systems $\mathbf{T}_i$, $i=1,\ldots,m$, and an LTL formula ϕ over Π in the form (1), synthesize individual runs r_i and synchronization sequences s_i for each robot such that $\mathbb{T}^\pi$ minimizes the cost function (2), and $\tilde{\omega}_{team}$, *i.e.*, the word observed in the field, satisfies ϕ.

Note that our aim in Prob. 1 is to find a run that is optimal under nominal values while ensuring that ϕ is never violated in the field. Since $\tilde{\mathbb{T}}^\pi$, *i.e.*, the sequence of instants at which π is satisfied during deployment, is likely to be sub-optimal, we will also seek to bound the deviation from optimality in the field. As we consider LTL formulas containing $\mathbf{GF}\pi$, this optimization problem is always well-posed.

4 Problem Solution

In this section, we describe each step of our solution to Prob. 1 in detail with the help of a simple illustrative example. We present our experimental results in Sec. 5.

4.1 *Obtaining the Team Transition System*

In [20], we showed that the joint asynchronous behavior of a robotic team modeled as m transition systems $\mathbf{T}_i, i = 1, \ldots, m$ (Def. 1) can be captured using a region automaton. A region automaton, as given in the following definition from [19], is a finite transition system that keeps track of the relative positions of the robots as they move asynchronously in the environment.

Definition 4 (Region Automaton). The region automaton $\mathbf{R}$ is a TS (Def. 1) $\mathbf{R} := (\mathcal{Q}_{\mathrm{R}}, q^0_{\mathrm{R}}, \delta_{\mathrm{R}}, \Pi_{\mathrm{R}}, \mathcal{L}_{\mathrm{R}}, w_{\mathrm{R}})$, where $\mathcal{Q}_{\mathrm{R}}$ is the set of states of the form (q,r) such that q is a tuple of state pairs $(q_1q'_1, \ldots, q_mq'_m)$ where the ith element $q_iq'_i$ is a source-target state pair from $\mathcal{Q}_i$ of $\mathbf{T}_i$ meaning robot i is currently on its way from q_i to q'_i, and r is a tuple of clock values $(x_1, \ldots, x_m)$ where $x_i \in \mathbb{N}$ denotes the time elapsed since robot i left state q_i. $q^0_{\mathrm{R}} \subseteq \mathcal{Q}_{\mathrm{R}}$ is the set of initial states with $r = (0, \ldots, 0)$ and $q = (q^0_1q'_1, \ldots, q^0_mq'_m)$ such that q^0_i is the initial state of $\mathbf{T}_i$ and $(q^0_i, q'_i) \in \delta_i$. δ_{R} is the transition relation such that a transition from (q,r) to (q',r') where the ith state pair $q_iq'_i$ and the ith clock value x_i in (q,r) change to $q'_iq''_i$ and x'_i in (q',r') exists if and only if $(q_i,q'_i),(q'_i,q''_i) \in \delta_i$ for all changed state pairs, $w_i(q_i,q'_i) - x_i$ of all changed state pairs are equal to each other and are strictly smaller than those of unchanged state pairs, and for all changed state pairs, the corresponding x'_i in r' becomes $x'_i = 0$ and all other clock values in r are incremented by $w_i(q_i,q'_i) - x_i$ in r'. $\Pi_{\mathrm{R}} = \cup_{i=1}^m \Pi_i$ is the set of propositions. $\mathcal{L}_{\mathrm{R}} : \mathcal{Q}_{\mathrm{R}} \to 2^{\Pi_{\mathrm{R}}}$ is a map giving the set of atomic propositions satisfied in a state. For a state (q,r), $\mathcal{L}_{\mathrm{R}}((q,r)) = \cup_{i|x_i=0} \mathcal{L}_i(q_i)$. $w_{\mathrm{R}} : \delta_{\mathrm{R}} \to \mathbb{N}_{>0}$ is a map that assigns a positive integer weight to each transition such that $w_{\mathrm{R}}((q,r),(q',r')) = w_i(q_i,q'_i) - x_i$ for each state pair that has changed from $q_iq'_i$ to $q'_iq''_i$ with a corresponding clock value of $x'_i = 0$ in r'.

Example 1. *Fig. 1 illustrates the TS's of two robots that are expected to satisfy the mission* $\phi := \mathbf{G}(\mathrm{p}_1 \Rightarrow \mathbf{X}(\neg \mathrm{p}_1 \,\mathcal{U}\, \mathrm{p}_3)) \wedge \mathbf{GF}\pi$*, where* $\Pi_1 = \{\mathrm{p}_1, \pi\}$*,* $\Pi_2 = \{\mathrm{p}_2, \mathrm{p}_3, \pi\}$*, and* $\Pi = \{\mathrm{p}_1, \mathrm{p}_2, \mathrm{p}_3, \pi\}$*. We also have* $\overline{\rho_1} = \overline{\rho_2} = 1.1$ *and* $\underline{\rho_1} = \underline{\rho_2} = 0.9$*. The region automaton* $\mathbf{R}$ *that models the robots is given in Fig. 2.*

However, as a region automaton encodes the directions of travel of the robots as opposed to their locations, it typically contains redundant states, and thus can typically be reduced to a smaller size. The following example illustrates this fact.

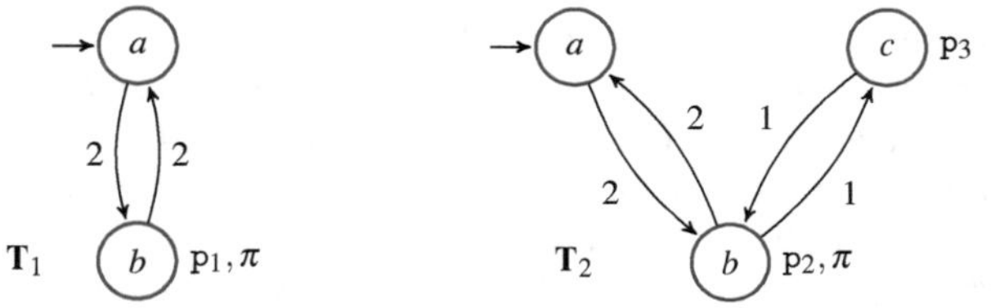

Fig. 1 Transition systems $\mathbf{T}_1$ and $\mathbf{T}_2$ of two robots in an environment with three vertices. The states of the transition systems correspond to vertices $\{a,b,c\}$ and the edges represent the motion capabilities of each robot. The weights of the edges represent the traveling times between any two vertices. The propositions $\mathrm{p}_1, \mathrm{p}_2, \mathrm{p}_3$ and π are shown next to the vertices where they can be satisfied by the robots.

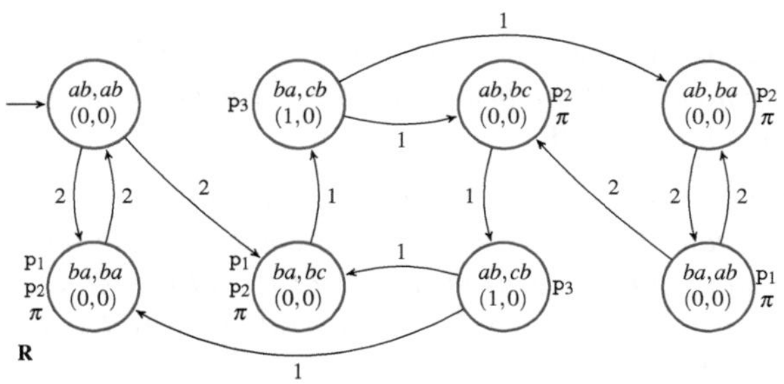

Fig. 2 The finite state region automaton capturing the joint behavior of two robots in 9 states. In a circle representing a state (q,r), the first line is q and the second line is r.

Example 1 Revisited. *State $((ab,bc),(0,0))$ of the region automaton* **R** *given in Fig. 2 is equivalent to the state $((ab,ba),(0,0))$ in the sense that both robots satisfy the same propositions and the positions of both robots are the same at both states, i.e., robot 1 is at a and robot 2 is at b. These two states differ only in the future direction of travel of the second robot, i.e., robot 2 travels towards c in the first state whereas it travels towards a in the second state. This information, however, is redundant as it can be obtained just by looking at the next state of the team in any given run.*

Motivated by this observation, we define a binary relation $\mathscr{R}$ to reduce the region automaton **R** to a smaller team transition system **T**.

Definition 5 (Binary Relation $\mathscr{R}$). Binary relation $\mathscr{R} = \{(s,t) | s \in \mathscr{Q}_{\mathrm{R}}, t \in \mathscr{Q}_{\mathrm{T}}\}$ is a mapping between the states of **R** and **T** that maps a state $s = ((q_1 q_1', \ldots, q_m q_m'), (x_1, \ldots, x_m))$ in $\mathscr{Q}_{\mathrm{R}}$ to a state $t = (t_1, \ldots, t_m)$ in $\mathscr{Q}_{\mathrm{T}}$, where $t_i = q_i$ if $x_i = 0$ and $t_i = q_i q_i' x_i$ if $x_i > 0$. Note that, $x_i = 0$ for at least one $i \in \{1, \ldots, m\}$. We refer to a state $t_i \in \mathscr{Q}_{\mathrm{T}}$ of the form $q_i q_i' x_i$ as a traveling state as it captures the instant where robot i has traveled from q_i to q_i' for x_i time units.

Given a region automaton **R**, we can obtain the corresponding team transition system **T** using the binary relation $\mathscr{R}$ and the following procedure.

Procedure 1 (Obtaining T from R) *Using $\mathscr{R}$ we construct the team transition system* **T** *from the region automaton* **R** *as follows: (1) For each $s \in \mathscr{Q}_{\mathrm{R}}$ we define* *the corresponding $t \in \mathscr{Q}_{\mathrm{T}}$ as given in Def. 5 such that $(s,t) \in \mathscr{R}$. (2) We set $\mathscr{L}_{\mathrm{T}}(t) = \mathscr{L}_{\mathrm{R}}(s)$. Note that, each s that corresponds to a given t has the same set of propositions due to the way* **R** *is constructed (Def. 4) [20]. (3) For each s corresponding to a given t, we define the corresponding transitions originating from t in* **T** *such that $\exists(t,t') \in \delta_{\mathrm{T}} \forall (s,s') \in \delta_{\mathrm{R}}$ where $(s,t) \in \mathscr{R}$ and $(s',t') \in \mathscr{R}$. (4) We mark a state t in $\mathscr{Q}_{\mathrm{T}}$ as the initial state of* **T** *if the corresponding s is an initial state in $\mathscr{Q}_{\mathrm{R}}$. Note that, all states that correspond to a given t are either in q^0_{R} altogether or none of them are in q^0_{R}.*

The following proposition shows that the team transition system **T** obtained using Proc. 1 and the corresponding region automaton **R** are bisimulation equivalent, *i.e.*, there exists a binary relation between the states and the transitions of **R** and **T** such that they behave in the same way [1].

Proposition 1 (Bisimulation Equivalence). *The team transition system* **T** *obtained using Proc. 1 and the region automaton* **R** *are bisimulation equivalent, i.e.,* $\mathbf{R} \sim \mathbf{T}$, *and $\mathscr{R}$ is a bisimulation relation for* **R** *and* **T**.

Example 1 Revisited. *Using $\mathscr{R}$, we construct* **T** *(Fig. 3) that captures the joint asynchronous behavior of the team in 6 states whereas the corresponding region automaton* **R** *had 9 states. A state labeled (a,b) means robot 1 is at region a and robot 2 is at region b, whereas a state labeled $(ba1,c)$ means robot 1 traveled from b to a for 1 time unit and robot 2 is at c.*

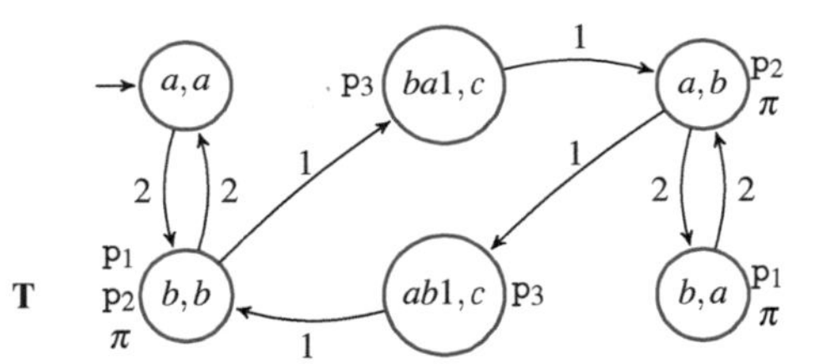

Fig. 3 The team transition system capturing the joint behavior of two robots in 6 states

In [20] we showed that the number of states $|\mathscr{Q}_{\mathrm{R}}|$ of the region automaton **R** that models m robots $\mathbf{T}_i, i = 1,\ldots,m$ is bounded by $(\prod_{i=1}^m |\delta_i|)(\prod_{i=1}^m W_i - \prod_{i=1}^m (W_i - 1))$, where $|\delta_i|$ is the number of transitions in the TS $\mathbf{T}_i$ of robot i and W_i is maximum weight of any transition in $\mathbf{T}_i$. The following proposition provides a bound on the number of states $|\mathscr{Q}_{\mathrm{T}}|$ of **T** and shows that it is indeed significantly smaller than the bound on $|\mathscr{Q}_{\mathrm{R}}|$.

Proposition 2. *The number of states $|\mathscr{Q}_{\mathrm{T}}|$ of* **T** *is bounded by $\prod_{i=1}^m |\mathscr{Q}_i| + (W - 1)\prod_{i=1}^m |\delta_i|$ where W is the largest edge weight in all TS's.*

Finally, we note that the states of **T** correspond to the instants where at least one robot has completed a transition in its individual TS and is currently at a vertex while other robots may still be traveling. Using this fact, one can construct **T** directly by using a depth first search that runs in parallel on the TS's of the individual members of the team as given in Alg. 1. A detailed discussion on Alg. 1 can be found in [18].

Algorithm 1. CONSTRUCT-TEAM-TS

Input: $(\mathbf{T}_1,\ldots,\mathbf{T}_m)$.
Output: Corresponding team transition system $\mathbf{T}$.

$q_{\mathrm{T}}^0 := (q_1^0,\ldots,q_m^0)$, where q_i^0 is the initial state of $\mathbf{T}_i$.
dfsT(q_{T}^0).

Function dfsT(state tuple $q \in \mathscr{Q}_{\mathrm{T}}$)

$q[i]$ is the ith element of state tuple $q \in \mathscr{Q}_{\mathrm{T}}$.
t_i is a transition of $\mathbf{T}_i, i = 1,\ldots,m$, such that $t_i \in \{(q[i],q_i')|(q[i],q_i') \in \delta_i\}$ if $q[i] \in \mathscr{Q}_i$. Else if $q[i] = q_i q_i' x_i$, then $t_i = (q_i, q_i')$.
$T := (t_1,\ldots,t_m)$ is a tuple of such transitions.
$\mathscr{T}$ is the set of all such transition tuples at q.
foreach *transition tuple* $T \in \mathscr{T}$ **do**
  $w \leftarrow$ Shortest time until a robot is at a vertex while the transitions in T are being taken.
  Find the q' that corresponds to this new state of the team using $\mathscr{R}$.
  if $q' \notin \mathscr{Q}_{\mathrm{T}}$ **then**
    Add state q' to $\mathscr{Q}_{\mathrm{T}}$.
    Set $\mathscr{L}_{\mathrm{T}}(q') = \cup_{i|q'[i]\in\mathscr{Q}_i} \mathscr{L}_i(q'[i])$.
    Add (q,q') to δ_{T} with weight w.
    Continue search from q': **dfsT**(q').
  else if $(q,q') \notin \delta_{\mathrm{T}}$ **then**
    Add (q,q') to δ_{T} with weight w.

4.2 Obtaining Optimal Satisfying Runs and Transition Systems with Traveling States

After constructing $\mathbf{T}$ that models the team, we use OPTIMAL-RUN from [13] to obtain an optimal satisfying run $r_{team}^{\star}$ on $\mathbf{T}$ that minimizes the cost function (2) and satisfies ϕ. The optimal run $r_{team}^{\star}$ is always in prefix-suffix form, consisting of a finite sequence of states of $\mathbf{T}$ (prefix), followed by infinite repetitions of another finite sequence of states of $\mathbf{T}$ (suffix) as given in Def. 3.

Example 1 Revisited. *Running Alg.* OPTIMAL-RUN *[13] on* $\mathbf{T}$ *given in Fig. 3 for the formula* $\phi = \mathbf{G}(\mathrm{p}_1 \Rightarrow \mathbf{X}(\neg \mathrm{p}_1 \,\mathscr{U}\, \mathrm{p}_3)) \wedge \mathbf{GF}\pi$ *results in the optimal run with the prefix* $(a,a),(b,b)$ *and the suffix cycle* $(ba1,c),(a,b),(ab1,c),(b,b)$*, which will be repeated indefinitely. The cost as defined in* (2) *is* $J(\mathbb{T}^{\pi}) = 2$.

Since $\mathbf{T}$ captures the asynchronous motion of the robots, the optimal satisfying run $r_{team}^{\star}$ on $\mathbf{T}$ may contain some traveling states which do not appear in the individual TSs $\mathbf{T}_i, i = 1,\ldots,m$ that we started with. But we cannot ignore such traveling states either, as each one of them is a candidate synchronization point for the corresponding robot as we discuss in Sec. 4.3. Instead, we insert those traveling states into the individual TSs so that the robots will be able to synchronize with each other at those points if needed. In the following, we use $q^k[i]$ to denote the ith

element of the kth state tuple in $r^{\star}_{team}$, which is also the state of robot i at that position of $r^{\star}_{team}$. As given in Def. 5, a traveling state of robot i has the form $q_i q'_i x_i$. First, we construct the set $\mathscr{S} = \{(i, q^k[i]) \mid q^k[i] = q_i q'_i x_i \forall k, i\}$ of all traveling states that appear in $r^{\star}_{team}$. Elements of $\mathscr{S}$ are tuples where the second element is a traveling state and the first element gives the transition system this new traveling state will be added to. Next, we construct the set $\mathscr{T} = \{(i, (q^k[i], q^{k+1}[i]), x) \mid ((i, q^k[i]) \in \mathscr{S}) \vee ((i, q^{k+1}[i]) \in \mathscr{S}), x = w_{\mathrm{T}}(q^k, q^{k+1}) \forall k, i\}$ of all transitions that involve any of the traveling states in $r^{\star}_{team}$. Elements of $\mathscr{T}$ are triplets where the second element is a transition, the third element is the weight of this transition, and the first element shows the transition system that this new transition will be added to. Then, we add the traveling states in $\mathscr{S}$ and the transitions in $\mathscr{T}$ to their corresponding transition systems. Finally, using the following definition, we project the optimal satisfying run $r^{\star}_{team}$ down to individual robots $\mathbf{T}_i, i = 1, \ldots, m$ to obtain individual optimal satisfying runs $r^{\star}_i, i = 1, \ldots, m$.

Definition 6 (Projection of a Run on T to T$_i$'s). Given a run r_{team} on $\mathbf{T}$ where $r_{team} = q^0, q^1, \ldots$, we define its projection on $\mathbf{T}_i$ as run $r_i = q^0_i, q^1_i, \ldots$ for all $i = 1, \ldots, m$, such that $q^k_i = q^k[i]$ where $q^k[i]$ is the ith element of tuple q^k.

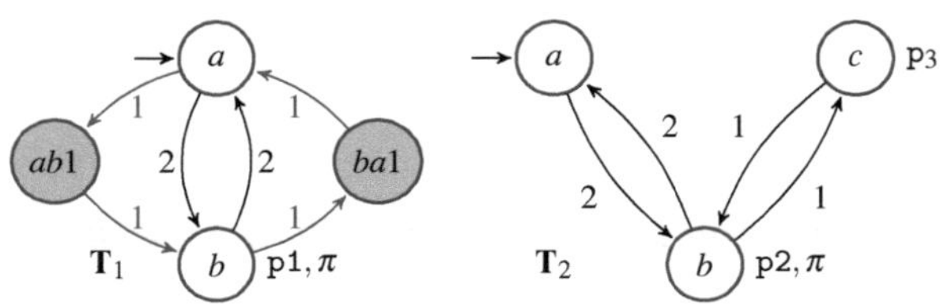

Fig. 4 Transition systems with new traveling states that correspond to the optimal run $r^{\star}_{team}$ that we computed for Ex. 1. The new traveling states and transitions of $\mathbf{T}_1$ are highlighted in red.

Example 1 Revisited. *For this example, we have* $\mathscr{S} = \{(1, ab1), (1, ba1)\}$ *and* $\mathscr{T} = \{(1, (a, ab1), 1), (1, (ab1, b), 1), (1, (b, ba1), 1), (1, (ba1, a), 1)\}$. *Fig. 4 illustrates the corresponding TSs with new traveling states and transitions highlighted in red. Using Def. 6, we obtain the runs of the individual robots as* $r^{\star}_1 = a, b, ba1, a, ab1, b, ba1, a, ab1, \ldots$ *and* $r^{\star}_2 = a, b, c, b, c, b, c, b, c, \ldots$.

4.3 Guaranteeing Correctness through Synchronization and the Optimality Bound

As the robots execute their infinite runs in the field, they synchronize with each other according to the synchronization sequences that we generate using Alg. 2. The synchronization sequence s_i of robot i is an infinite sequence of pairs of sets. The kth element of s_i, denoted by s^k_i, corresponds to the kth element q^k_i of $r^{\star}_i$. Each s^k_i is a tuple of two sets of robots: $s^k_i = (s^k_{i,wait}, s^k_{i,notify})$, where $s^k_{i,wait}$ and $s^k_{i,notify}$ are the *wait-set* and *notify-set* of s^k_i, respectively. The *wait-set* of s^k_i is the set of robots that robot i must wait for at state q_i^k before satisfying its propositions and proceeding to the next state q_i^{k+1} in $r_i^\star$. The *notify-set* of s_i^k is the set of robots that robot i must notify as soon as it reaches state q_i^k. As we discussed earlier in Sec. 4.2, the optimal run $r_{team}^\star$ of the team and the individual optimal runs $r_i^\star, i = 1,\ldots,m$ of the robots are always in prefix-suffix form (Def. 3). Consequently, individual synchronization sequences s_i of the robots are also in prefix-suffix form. A detailed discussion on Alg. 2 can be found in [18].

Algorithm 2. SYNC-SEQ

Input: Individual optimal runs of the robots $\{r_1^\star,\ldots,r_m^\star\}$, Büchi automaton $\mathbf{B}_{\neg\phi}$ that corresponds to $\neg\phi$.

Output: Synchronization sequence for each robot $\{s_1,\ldots,s_m\}$.

$\mathscr{I} = \{1,\ldots,m\}$.
beg $\leftarrow$ beginning of suffix cycle.
end $\leftarrow$ end of suffix cycle.
Initialize each s_i so that all robots wait for and notify each other at every position of their runs.
foreach $k = 0,\ldots,end$ **do**
    foreach $i \in \mathscr{I}$ **do**
        if $k \neq 0$ *and* $k \neq beg$ **then**
            foreach $j \in \mathscr{I} \setminus i$ **do**
                Remove j from $s_{i,wait}^k$.
                Remove i from $s_{j,notify}^k$.
                Construct the TS $\mathbf{W}$ that generates every possible $\tilde{\omega}_{team}$.
                if *the language of* $\mathbf{B}_{\neg\phi} \times \mathbf{W}$ *is not empty* **then**
                    Add j back to $s_{i,wait}^k$.
                    Add i back to $s_{j,notify}^k$.
Rest of each s_i is an infinite repetition of its suffix-cycle, i.e. $s_i^{beg},\ldots,s_i^{end}$.

The following proposition slightly extends the result of Prop. 4.5 in [19] by considering unequal lower and upper deviation values.

Proposition 3. *Suppose that each robot's deviation values are bounded by* $\underline{\rho}$ *and* $\overline{\rho}$ *where* $\overline{\rho} \geq 1 \geq \underline{\rho} > 0$ *(i.e.,* $\underline{\rho_i} \geq \underline{\rho}$ *and* $\overline{\rho_i} \leq \overline{\rho}$ *for each robot i). Let* $J(\mathbb{T}^\pi)$ *be the cost of the planned robot paths and let* $J(\tilde{\mathbb{T}}^\pi)$ *be the actual value of the cost observed during deployment. Then, if the robots use the synchronization sequences generated by Alg. 2, the field value of the cost satisfies* $J(\tilde{\mathbb{T}}^\pi) \leq J(\mathbb{T}^\pi)\overline{\rho} + d_s(\overline{\rho} - \underline{\rho})$ *where* d_s *is the planned duration of the suffix cycle.*

Example 1 Revisited. *Using Alg. 2, we obtain the following individual synchronization sequences:* $s_1 = (\{2\},\{2\}),(\{\},\{\}),(\{2\},\{2\}),(\{\},\{\}),(\{\},\{\}),(\{\},\{\}),\ldots$, *and* $s_2 = (\{1\},\{1\}),(\{\},\{\}),(\{1\},\{1\}),(\{\},\{\}),(\{\},\{\}),(\{\},\{\}),\ldots$. *The elements* *of the kth pair in the synchronization sequences correspond to $s^k_{i,wait}$ and $s^k_{i,notify}$, respectively. Also, from Prop. 3, the field value of the cost function is bounded from above by 3 for $\overline{\rho_1} = \overline{\rho_2} = 1.1$ and $\underline{\rho_1} = \underline{\rho_2} = 0.9$.*

We finally summarize our approach in Alg. 3 and show that this algorithm indeed solves Prob. 1. We discuss the complexity of our approach in Rem. 2.

Proposition 4. *Alg. 3 solves Prob. 1.*

Remark 2 (Computational Complexity). *The main drawback of our approach is its computational complexity, which is exponential in the number of robots (due to generation of the team transition system and the synchronization sequences) and in the length of the LTL formula (due to the conversion to a Büchi automaton). This cost, however, is justified by the globally optimal runs that our approach computes, and in practice, we can solve fairly large problems.*

Algorithm 3. ROBUST-MULTI-ROBOT-OPTIMAL-RUN

Input: m transition systems $\mathbf{T}_i, i = 1, \ldots, m$, corresponding deviation values, and a global LTL specification ϕ of the form (1).

Output: A set of optimal runs $\{r^\star_1, \ldots, r^\star_m\}$ that satisfies ϕ and minimizes (2), a set of synchronization sequences $\{s_1, \ldots, s_m\}$ that guarantees correctness in the field, and the bound on the performance of the team in the field.

Construct the team transition system $\mathbf{T}$ using Alg. 1.
Find an optimal run $r^\star_{team}$ on $\mathbf{T}$ using OPTIMAL-RUN [13].
Insert new traveling states to TSs according to $r^\star_{team}$ (See. Sec. 4.2).
Obtain individual runs $\{r^\star_1, \ldots, r^\star_m\}$ using Def. 6.
Generate synchronization sequences $\{s_1, \ldots, s_m\}$ using Alg. 2.
Find the bound on optimality as given in Prop. 3.

5 Implementation and Case-Study

We implemented Alg. 3 as a python module (available at `http://hyness.bu.edu/lomap/`) and used it to plan optimal satisfying paths and synchronization sequences for the scenario that we consider in this section. Our experimental platform (Fig. 5(a)) is a road network comprising roads, intersections and task locations. Fig. 5(b) illustrates the model that captures the motion of the robots on this platform, where 1 time unit corresponds to 1.574 seconds.

In our experiments, we consider a persistent surveillance task involving two robots with deviation values $\overline{\rho_1} = \overline{\rho_2} = 1.05$ and $\underline{\rho_1} = \underline{\rho_2} = 0.95$. The building in the middle of the platform in Fig. 5(a) is our surveillance target. We define the set of propositions $\Pi = \{$`R1Gather18`, `R1Gather20`, `R2Gather18`, `R2Gather20`, `R1Gather`, `R2Gather`, `R1Upload`, `R2Upload`, `Gather`$\}$ and assign them as $\mathcal{L}_1(18) = \{$`Gather`, `R1Gather18`, `R1Gather`$\}$, $\mathcal{L}_2(18) = \{$`Gather`, `R2Gather18`, `R2Gather`$\}$, $\mathcal{L}_1(20) = \{$`Gather`, `R1Gather20`, `R1Gather`$\}$, $\mathcal{L}_2(20) = \{$`Gather`, `R2Gather20`, `R2Gather`$\}$, $\mathcal{L}_1(22) = \{$`R1Upload`$\}$ and $\mathcal{L}_2(22) = \{$`R2Upload`$\}$.

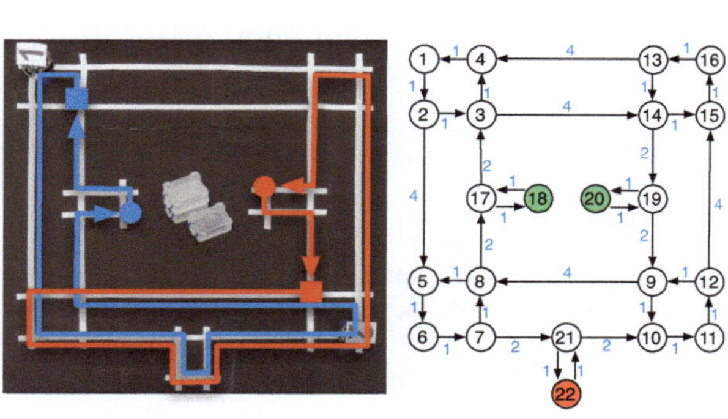

Fig. 5 Left figure shows our experimental platform. The squares and the circles on the trajectories of the robots represent the beginning of the suffix cycle and sync. points, respectively. Right figure illustrates the TS that models the robots. The green and red regions are data gather and upload locations, respectively.

The main objective is to keep gathering data while minimizing the maximum time between successive gathers. We require the robots to gather data in a synchronous manner at data gather locations 18 and 20 while ensuring that they do not gather data at the same place at the same time. We also require the robots to upload their data at upload location 22 before their next data gather. We express these requirements in LTL in the form of (1) as

$$\begin{aligned}\phi = &\mathbf{G}\,(\texttt{R1gather} \Rightarrow \mathbf{X}(\neg\texttt{R1gather}\ \mathcal{U}\ \texttt{R1upload})) \wedge \mathbf{G}\,(\texttt{R2gather} \Rightarrow \\ &\mathbf{X}(\neg\texttt{R2gather}\ \mathcal{U}\ \texttt{R2upload})) \wedge \mathbf{G}\,((\texttt{R1Gather18} \Rightarrow \texttt{R2Gather20}) \wedge \\ &(\texttt{R1gather20} \Rightarrow \texttt{R2gather18}) \wedge (\texttt{R2gather18} \Rightarrow \texttt{R1gather20}) \wedge \\ &(\texttt{R2gather20} \Rightarrow \texttt{R1gather18})) \wedge \mathbf{GF}\,\texttt{Gather},\end{aligned}$$

where `Gather` is set as the optimizing proposition.

Fig. 5(a) illustrates the solution which took our algorithm approximately 20 seconds to compute on an iMac i5 quad-core computer. The planned value of the cost function was 44.072 seconds (28 time units) with an upper bound of 50.683 seconds (32.2 time units) seconds. We deployed our robots in our experimental platform to demonstrate and verify the result. The maximum time between any two successive data uploads was measured to be 48 seconds. The video available at `http://hyness.bu.edu/lomap/dars2012.mov` demonstrates the execution of this run by the robots.

6 Conclusion

In this paper we present an automated method for planning optimal paths for a robotic team subject to a Linear Temporal Logic formula. The robots that we consider have bounded non-deterministic traveling times. We first compute a set of optimal satisfying paths for the members of the team. Then, we compute synchronization sequences for the robots to guarantee correctness during deployment. Our experiments show that our method has practical value in scenarios where the traveling times of the robots during deployment deviate from those used in planning.

Acknowledgements. This work was supported in part by ONR MURI N00014-09-1051, NSF CNS-0834260, and NSERC.

References

1. Baier, C., Katoen, J.P.: Principles of Model Checking. MIT Press (2008)
2. Bianco, A., Alfaro, L.D.: Model Checking of Probabilistic and Nondeterministic Systems. In: Thiagarajan, P.S. (ed.) FSTTCS 1995. LNCS, vol. 1026, pp. 499–513. Springer, Heidelberg (1995)
3. Clarke, E.M., Peled, D., Grumberg, O.: Model checking. MIT Press (1999)
4. Ding, X.C., Smith, S.L., Belta, C., Rus, D.: MDP Optimal Control under Temporal Logic Constraints. In: IEEE Conf. on Decision and Control, Orlando, FL, pp. 532–538 (2011)
5. Emerson, E.A.: Temporal and Modal Logic. In: van Leeuwen, J. (ed.) Handbook of Theoretical Computer Science: Formal Models and Semantics, pp. 995–1072. North-Holland Pub. Co./MIT Press (1990)
6. Hopcroft, J., Motwani, R., Ullman, J.D.: Introduction to Automata Theory, Languages, and Computation. Addison Wesley (2007)
7. Karaman, S., Frazzoli, E.: Complex Mission Optimization for Multiple-UAVs using Linear Temporal Logic. In: American Control Conference, Seattle, WA, pp. 2003–2009 (2008)
8. Karaman, S., Frazzoli, E.: Vehicle Routing Problem with Metric Temporal Logic Specifications. In: IEEE Conf. on Decision and Control, Cancún, México, pp. 3953–3958 (2008)
9. Kloetzer, M., Belta, C.: Automatic Deployment of Distributed Teams of Robots from Temporal Logic Specifications. IEEE Transactions on Robotics 26(1), 48–61 (2010)
10. Kress-Gazit, H., Fainekos, G., Pappas, G.J.: Where's Waldo? Sensor-Based Temporal Logic Motion Planning. In: IEEE Intl. Conf. on Robotics and Automation, pp. 3116–3121 (2007)
11. Kwiatkowska, M., Norman, G., Parker, D.: Probabilistic Symbolic Model Checking with PRISM: A Hybrid Approach. International Journal on Software Tools for Technology Transfer, 52–66 (2002)
12. Kloetzer, M., Belta, C.: Dealing with Non-Determinism in Symbolic Control. In: Egerstedt, M., Mishra, B. (eds.) HSCC 2008. LNCS, vol. 4981, pp. 287–300. Springer, Heidelberg (2008)
13. Smith, S.L., Tůmová, J., Belta, C., Rus, D.: Optimal Path Planning for Surveillance with Temporal Logic Constraints. Intl. Journal of Robotics Research 30(14), 1695–1708 (2011)
14. Tabuada, P., Pappas, G.J.: Linear Time Logic Control of Discrete-Time Linear Systems. IEEE Transactions on Automatic Control 51(12), 1862–1877 (2006)
15. Thomas, W.: Infinite Games and Verification. In: Intl. Conf. on Computer Aided Verification, pp. 58–64 (2002)
16. Toth, P., Vigo, D. (eds.): The Vehicle Routing Problem. Monographs on Discrete Mathematics and Applications. SIAM (2001)
17. Tumova, J., Yordanov, B., Belta, C., Cerna, I., Barnat, J.: A Symbolic Approach to Controlling Piecewise Affine Systems. In: IEEE Conf. on Decision and Control, Atlanta, GA, pp. 4230–4235 (2010)
18. Ulusoy, A., Smith, S.L., Belta, C.: Optimal Multi-Robot Path Planning with LTL Constraints: Guaranteeing Correctness Through Synchronization (2012), `http://arxiv.org/abs/1207.2415`

19. Ulusoy, A., Smith, S.L., Ding, X.C., Belta, C.: Robust Multi-Robot Optimal Path Planning with Temporal Logic Constraints. In: IEEE Intl. Conf. on Robotics and Automation, St. Paul, MN, USA, pp. 4693–4698 (2012)
20. Ulusoy, A., Smith, S.L., Ding, X.C., Belta, C., Rus, D.: Optimal Multi-Robot Path Planning with Temporal Logic Constraints. In: IEEE/RSJ Intl. Conf. on Intelligent Robots & Systems, San Francisco, CA, USA, pp. 3087–3092 (2011)
21. Wongpiromsarn, T., Topcu, U., Murray, R.M.: Receding Horizon Control for Temporal Logic Specifications. In: Hybrid Systems: Computation and Control, Stockholm, Sweden, pp. 101–110 (2010)

Part V

Learning, Adaptation, and Cognition in Many Robot Systems

Evolving Aggregation Behaviors in Multi-Robot Systems with Binary Sensors

Melvin Gauci, Jianing Chen, Tony J. Dodd, and Roderich Groß

Abstract. This paper investigates a non-traditional sensing trade-off in swarm robotics: one in which each robot has a relatively long sensing range, but processes a minimal amount of information. Aggregation is used as a case study, where randomly-placed robots are required to meet at a common location without using environmental cues. The binary sensor used only lets a robot know whether or not there is another robot in its direct line of sight. Simulation results with both a memoryless controller (reactive) and a controller with memory (recurrent) prove that this sensor is enough to achieve error-free aggregation, as long as a sufficient sensing range is provided. The recurrent controller gave better results in simulation, and a post-evaluation with it shows that it is able to aggregate at least 1000 robots into a single cluster consistently. Simulation results also show that, with the recurrent controller, false negative noise on the sensor can speed up the aggregation process. The system has been implemented on 20 physical e-puck robots, and systematic experiments have been performed with both controllers: on average, 86-89% of the robots aggregated into a single cluster within 10 minutes.

1 Introduction

Many studies in swarm robotics have investigated systems where each robot only makes use of localized information. As one may expect, such a restriction often comes at the cost that each robot is required to extract and process a considerable amount of information about its immediate surroundings. For example, each robot may be required to estimate the relative positions of all other robots within some radius. This study aims to investigate an alternative sensing trade-off, which is believed to be potentially useful in a number of ways. In this trade-off, a longer sensing

Melvin Gauci · Jianing Chen · Tony J. Dodd · Roderich Groß
Department of Automatic Control and Systems Engineering,
The University of Sheffield, UK
e-mail: {m.gauci,j.n.chen,t.j.dodd,r.gross}@sheffield.ac.uk

M.A. Hsieh and G. Chirikjian (eds.), *Distributed Autonomous Robotic Systems*,
Springer Tracts in Advanced Robotics 104,
DOI: 10.1007/978-3-642-55146-8_25, © Springer-Verlag Berlin Heidelberg 2014

range is allowed than is normally assumed in swarm robotics; however, each robot only extracts and processes a minimal amount of information per control cycle. One advantage of such a sensing scheme is that, as long as a suitable technology can be used to provide the necessary sensing range, the system is truly scalable, because the amount of information that each robot needs to extract and process does not increase with the number of robots in the swarm. Furthermore, simpler sensing requirements are more likely to be implementable on smaller scale robots, paving the way for nano-scale systems. Finally, using a simple sensing method increases the chance that a controller that performs well in simulation also performs well on the physical system.

The task of aggregation is used as a case study here. Trianni [21] argues that "aggregation is of particular interest since it stands as a prerequisite for other forms of cooperation". Self-organized aggregation is a widely-observed phenomenon in nature. It occurs in a range of organisms, including bacteria, arthropods, fish and mammals [3, 20]. In some cases, self-organized aggregation is aided by environmental heterogeneities, such as areas providing shelter or thermal energy [see 14, 16, and references therein]. However, aggregation can also occur in homogeneous environments [7].

Jeanson et al. [15] investigated aggregation in cockroach larvae, and developed a model of their behavior. The cockroaches were reported to join and leave clusters with probabilities correlated to the sizes of the clusters. Garnier et al. [9] implemented this model as a probabilistic finite-state automaton on 20 Alice robots to achieve aggregation in homogeneous environments. Correll and Martinoli [5] analyzed a similar model and showed that "robots need a minimum combination of communication range and locomotion speed in order to aggregate into a single cluster when using probabilistic aggregation rules". These probabilistic models of aggregation require the agents to obtain estimates of the cluster size or the robot density. For example, Garnier et al. [9] used local infra-red communication to estimate the number of nearby robots.

Ando et al. [1] introduced a deterministic algorithm for achieving aggregation in a group of mobile agents with limited perception in homogeneous environments. Cortés et al. [6] adapted this algorithm and showed that it can be used to achieve aggregation in arbitrarily high dimensions. These algorithms require that the robots initially form a connected visibility graph, and are based on maintaining this graph in every time step. The robots are essentially required to measure the relative positions (distances and angles) to all their neighbors. Gordon et al. [12] relaxed this requirement, such that the robots need only to measure the angles to their neighbors, and not the distances. Although the algorithm was theoretically proven to work, simulation results revealed that "the merging process [was] generally agonizingly slow" [12]. Gennaro and Jadbabie [11] developed a connectivity-maintaining algorithm based on every robot computing the Laplacian matrix of its neighbors. Similar to the work of Ando et al. [1], the algorithm requires the robots to measure the distances and angles to their neighbors.

Dorigo et al. [8] addressed the problem of robotic aggregation by using an evolutionary robotics approach [19]. In their system, the robots can emit a sound and can sense each other using proximity sensors and directional microphones. A neural network controller was evolved and validated in simulation with up to 40 robots. Bahceci and Şahin [2] used a similar setup and investigated the effects of several parameters, such as the number of robots, arena size, and run time.

The problem of aggregating robots with limited information is challenging, because the robots, if not properly controlled, may end up forming separate clusters. This work investigates whether a single bit of information is sufficient to achieve error-free aggregation, and whether memory in the controller is a fundamental requirement.

2 Experimental Setup

2.1 Problem Formulation

N robots are placed in a two-dimensional unbounded, obstacle-free, homogeneous environment, with random positions and orientations. The objective is to bring the robots together at some location in the environment (i.e. aggregate them) via decentralized control. Each robot is equipped with a single binary sensor on some point of its body, which allows it to know whether or not there is another robot in the direct line of sight of the sensor. Formally, the binary sensor gives a positive reading at time t, $I^{(t)} = 1$, if there is a robot in its direct line of sight, and a negative reading, $I^{(t)} = 0$, otherwise. The binary sensor does not provide the distance to the robot being perceived.

2.2 Robotic and Simulation Platforms

The robotic platform that has been used in this study is the e-puck robot [18], shown in Fig. 1(a), which is a miniature, differential wheeled mobile robot that was developed for educational and research purposes. The e-puck is equipped with (among other sensors) a directional camera located at its front. The camera has been used in this study to realize the binary sensor in the physical implementation (see Section 5). The simulations presented here were performed using the open-source Enki library [17], which is used by Webots^{TM} in 2D mode. Enki is capable of modeling the kinematics and the dynamics of rigid bodies in two dimensions, and has a built-in model of the e-puck. In Enki, the body of an e-puck is modeled as a smooth disk of diameter 7.4 cm and mass 152 g. The speeds of the left and the right wheels can be set independently, and the maximum speed of the e-puck is set to 12.8 cm/s, in both the forward and the reverse directions. In simulation, the binary sensor was realized by projecting a line from the robot's front and checking whether it intersects with another robot's body. The length of the control cycle was set to $\Delta = 0.1$ s, both in simulation and in the physical system. In simulation, the physics was updated at a rate of 10 times per control cycle.

(a)

```
if I(t) = 0
    set_speed(s_l^0, s_r^0)
if I(t) = 1
    set_speed(s_l^1, s_r^1)
```

(b)

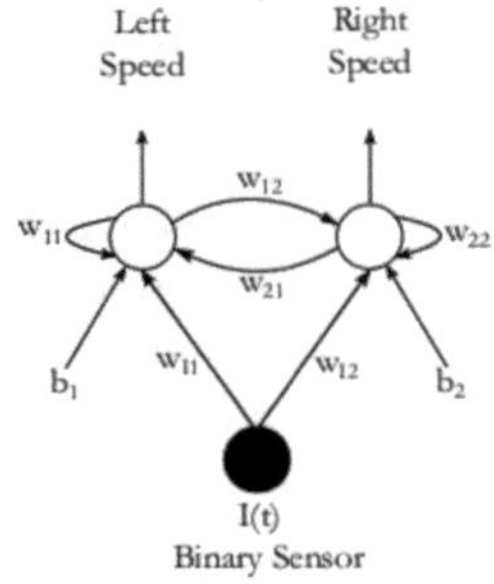

(c)

Fig. 1 (a): An e-puck robot fitted with a black 'skirt' to allow for its visual detection by other robots against a white background. The robot is around 7.4 cm in diameter. (b): Pseudo-code representing the reactive controller. (c): A schematic diagram of the fully-recurrent neural network controller.

2.3 Controllers

Two controllers have been investigated: a reactive controller that does not have any memory, and a recurrent controller with memory. A reactive controller maps all possible sensor readings onto actuation values. For a differential-wheeled robot with a single binary sensor, a reactive controller simply maps each of the two possible sensor readings onto a pair of speeds for the wheels of the robot. Thus, any reactive controller can be represented by 4 parameters (see Fig. 1 (b)): $\mathbf{x} = \left(s_l^0, s_r^0, s_l^1, s_r^1\right)$, $\mathbf{x} \in [-1.0, 1.0]^4$, where s_l^0 is the speed of the left wheel when $I^{(t)} = 0$, and so on, and -1.0 and 1.0 correspond to the maximum backward and forward speeds of a wheel, respectively. The recurrent controller is a fully-recurrent neural network [22] with only two nodes, whose outputs determine the speeds of the left and the right wheels. A schematic diagram of the network is shown in Fig. 1 (c). The network is defined by 8 unbounded real-valued parameters: $\mathbf{x} = (w_{I1}, w_{11}, w_{21}, b_1, w_{I2}, w_{12}, w_{22}, b_2)$, $\mathbf{x} \in \mathbb{R}^8$. The internal states of the neurons, γ_1 and γ_2, are initialized to 0, and are updated according to:

$$\gamma_k^{(t+1)} = w_{Ik} I^{(t)} + w_{1k}\mathrm{sig}\left(\gamma_1^{(t)}\right) + w_{2k}\mathrm{sig}\left(\gamma_2^{(t)}\right) + b_k, \qquad k \in \{1,2\}.$$

where $\mathrm{sig}(\cdot)$ is the standard sigmoidal function, given by $\mathrm{sig}(\cdot) = 1/\left(1+e^{-(\cdot)}\right)$. The speeds of the left and the right wheels are calculated from the states as $s_l^{(t)} = 2\mathrm{sig}\left(\gamma_1^{(t)}\right) - 1$, $s_r^{(t)} = 2\mathrm{sig}\left(\gamma_2^{(t)}\right) - 1$, such that $s_l^{(t)}, s_r^{(t)} \in (-1.0, 1.0)$.

2.4 Evolutionary Algorithm

The aim of the evolutionary algorithm is to synthesize controllers of the forms described in Section 2.3 that give a high aggregation performance. The algorithm used here is based on Classical Evolutionary Programming [23], and is suitable for optimization in continuous, real-valued parameter spaces, $\mathscr{S} \subseteq \mathbb{R}^n$. Its main features are (i) self-adaptation of mutation strengths, and (ii) a stochastic selection method known as q-tournament selection. In this algorithm, an individual can be considered as a 2-tuple: $\mathbf{a} = (\mathbf{x}, \sigma)$, where $\mathbf{x} \in \mathscr{S}$ is a candidate solution (i.e. here, a set of parameters for a controller) and $\sigma \in (0,\infty)^n$ is a vector of mutation strengths. The i-th mutation strength in σ corresponds to the i-th element in $\mathbf{x}$. Each generation g comprises a population of μ individuals, $\mathscr{P}^{(g)} = \left\{\mathbf{a}_1^{(g)}, \mathbf{a}_2^{(g)}, \ldots, \mathbf{a}_\mu^{(g)}\right\}$. In generation $g = 0$, all objective parameters and mutation strengths in $\mathscr{P}^{(0)}$ are initialized according to some distribution. Thereafter, in every generation g, every individual in $\mathscr{P}^{(g)}$ generates a new individual by mutation, to create a mutated population $\mathscr{P}'^{(g)}$ (see Eqs. (1) and (2) in [23]). The population for the next generation, $\mathscr{P}^{(g+1)}$, is selected by q-tournament selection from the combined population $\mathscr{P}^{(g)} \cup \mathscr{P}'^{(g)}$. Each individual $\mathbf{a}_k^{(g)}$ in $\mathscr{P}^{(g)} \cup \mathscr{P}'^{(g)}$, $k \in \{1,2,\ldots,2\mu\}$, competes in a tournament q times. For each tournament, an individual $\mathbf{a}_\chi^{(g)}$, $\chi \neq k$, is chosen at random from $\mathscr{P}^{(g)} \cup \mathscr{P}'^{(g)}$, with replacement. The individuals $\mathbf{a}_k^{(g)}$ and $\mathbf{a}_\chi^{(g)}$ are evaluated using an identical, randomly-generated seed ψ (hence, an identical initial configuration). If $\mathbf{a}_k^{(g)}$ achieves a better fitness than its opponent, its score is increased by one. Therefore, after q tournaments, each individual in $\mathscr{P}^{(g)} \cup \mathscr{P}'^{(g)}$ obtains a score from the set $\{0,1,\ldots,q\}$. The μ individuals with the highest scores are selected to constitute the new population, $\mathscr{P}^{(g+1)}$ (individuals with an identical score have an equal chance of being selected).

2.5 Fitness Evaluation

The aggregation performance, or fitness, of a controller, defined by a vector of parameters, $\mathbf{x}$, is measured by running a simulation with a number of robots employing that controller, and computing their performance as follows: Let $\mathbf{p}_i^{(t)}$, $i \in \{1,2,\ldots,N\}$ represent the positions of the robots at time step t. The average distance to their centroid is given by:

$$d^{(t)} = \frac{1}{N}\sum_{i=1}^{N} \|\mathbf{p}_i^{(t)} - \bar{\mathbf{p}}^{(t)}\|, \tag{1}$$

where $\bar{\mathbf{p}}^{(t)}$ is the centroid of the robots' positions at time step t, computed as $\bar{\mathbf{p}}^{(t)} = \frac{1}{N}\sum_{i=1}^{N} \mathbf{p}_i^{(t)}$, and $\|\cdot\|$ denotes the Euclidean norm. Based on this metric, the fitness of a controller is computed as:

$$F(\mathbf{x}, \psi) = \sum_{t=1}^{T} \frac{1}{\left[d^{(t)}\right]^{\frac{t}{T}}}, \tag{2}$$

where T is the number of time steps for which the simulation is run. Since $d^{(t)}$ is placed in the denominator, a larger value of F signifies a better fitness. This function rewards not only the aggregation metric at the end of the simulation, but also the speed of the aggregation. Through the t/T exponent applied to $d^{(t)}$, a large value of $d^{(t)}$ is increasingly penalized as the simulation progresses. Note that the fitness F is the outcome of a stochastic process using the seed ψ, which determines the initial placement of the robots and the actuation noise.

3 Controller Synthesis and Selection

Two sets of 100 independent evolutionary runs were performed: one set with a reactive controller and one with a recurrent controller. The evolutions were run for 1000 generations, with all object parameters ($\mathbf{x}$) initialized to 0.0 and all mutation strengths (σ) initialized to 1.0. The population size was set to $\mu = 15$ and the tournament selection parameter was set to $q = 5$ (settings as used in [4]). $N = 10$ robots were used for the fitness evaluations of the controllers. Their positions were initialized randomly with a uniform distribution within a square of sides 316.23 cm, for an area of 10000 cm^2 per robot (on average), and their orientations were initialized randomly with a uniform distribution in the range $(-\pi, \pi]$.

Additionally, a grid search was carried out for the reactive controller. A resolution of 21 settings per parameter was used with values between -1.0 and 1.0 in increments of 0.1. Therefore, $21^4 = 194481$ controllers were evaluated. Each controller was evaluated 100 times using Eq. 2 with different initial configurations of robots, with the set of configurations being identical for each controller. The evaluations were done with $N = 10$ robots, and the initialization method was identical to that used in the evolutionary runs, described above. The fitness of each controller was recorded as the mean fitness of the 100 evaluations. Such a search was not possible to perform with the recurrent controller, because this has 8 unbounded parameters, which makes the search space prohibitively large.

Each evolution produced $\mu = 15$ controllers after 1000 generations. In order to extract the best controller found by each evolution, each controller in the final generation was evaluated 100 times with different initial configurations of robots, and the controller with the highest average fitness over the 100 evaluations was selected as the resultant controller of the evolution. Each of the 100 runs for each of the 200 controllers was inspected visually, and it turns out that two distinct behaviors emerged, as shown in Fig. 2(a). The first behavior leads to a compact cluster, while the second behavior leads the robots to form a circle and maintain it. With the reactive controller, 81 evolutionary runs produced controllers leading to a circle configuration, while 19 runs produced controllers leading to a compact cluster configuration. With the recurrent controller, only one evolutionary run out of 100 produced a circle-forming controller. The best reactive controller found by the grid search leads to a

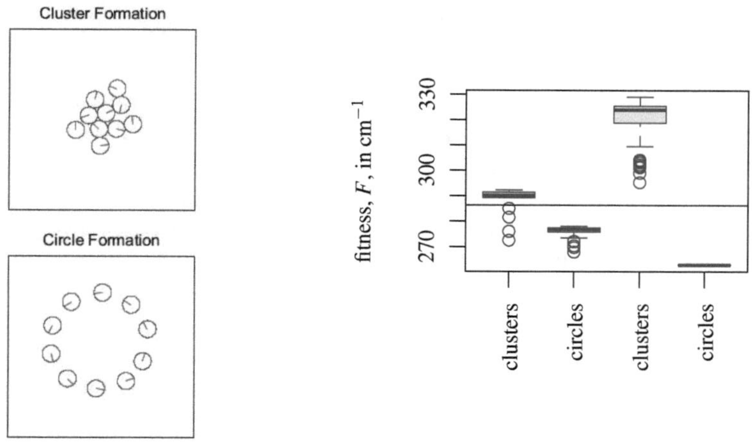

(a) (b)

Fig. 2 (a) Two distinct behaviors emerged from the evolutions: a number of controllers lead to a circle configuration (top), while others lead to a cluster configuration (bottom). (b) This box plot (see Footnote 1) shows the average fitnesses $\bar{F}$ of the 200 evolved controllers, grouped according to (i) the type of controller (reactive or recurrent) and (ii) the type of configuration that the controller leads to. The left (red) two boxes represent reactive controllers, while the right (blue) two boxes represent recurrent controllers. The horizontal green line shows the average fitness of the best controller found by the grid search. Higher values indicate a better fitness. The expected value of F for randomly-placed robots that do not move throughout the simulation is 185.562 cm^{-1}. The maximum 'possible' value of F is 417.514 cm^{-1}, corresponding to robots that are in the most compact configuration possible [see 13] from start to finish.

compact cluster configuration. Fig. 2(b) shows a box plot[1] with the average fitnesses $\bar{F}$ of the 200 evolved controllers, grouped according to (i) the type of controller (reactive or recurrent) and (ii) the type of configuration that the controller leads to. The circle-forming controllers achieved a significantly lower fitness than the compact cluster-forming controllers. This is because the fitness function in Eq. 2 was specifically designed to reward compactness. The circle-forming recurrent controller has a worse fitness than the worst circle-forming reactive controller, as the circle forms over a longer time. In terms of compact-cluster-forming controllers, it is clear from Fig. 2(b) that the recurrent controllers lead to significantly higher fitnesses than the reactive controllers. In fact, the worst compact cluster-forming recurrent controller evolved has a higher fitness than the best reactive controller evolved. The fitness of the best reactive controller located by the grid search (green line in Fig. 2(b)) is

[1] The boxplots presented here are all as follows. The line inside the box represents the median of the data. The edges of the box represent the lower and the upper quartiles (25-th and 75-th percentiles) of the data, while the whiskers represent the lowest and the highest data points that are within 1.5 times the inter-quartile range from the lower and the upper quartiles, respectively. Circles represent outliers.

lower than the fitness of the best reactive controller found by an evolution. This is because of the limited resolution that had to be used in the grid search, whereas the evolutionary algorithm has a practically infinite resolution.

The circle forming controllers yield an interesting behavior. Preliminary experiments show that they scale well with the number of robots (as tested with 100 robots in simulation), and are in principle also implementable on real robots. However, the next sections will investigate the cluster-forming controllers, which outperform the circle-forming controllers in terms of compactness. An investigation of the circle forming controllers will be left as future work.

4 Post-evaluations with the Best Controller

During the controller synthesis stage, the controller evaluations were limited to $N = 10$ robots, in order to keep the computation time within reasonable limits. The best synthesized (recurrent) controller was chosen for post-evaluations with 100 robots. In the following, 100 robots were initialized within a virtual square of sides 1000 cm, such that the area per robot (on average) is 10000 cm^2, identical to that used for controller synthesis.

The effect of the sensing range: As the robots are initialized within a virtual square of sides 1000 cm, a sensing range of $\sqrt{1000^2 + 1000^2} = 1414$ cm can be considered as practically unlimited. This will be denoted by δ_∞. In order to investigate the effect of the sensing range on aggregation performance, simulations with different proportions of the sensing range δ to δ_∞ were performed; 100 simulations for each $\delta/\delta_\infty = \{0.0, 0.1, \ldots, 1.0\}$. For each simulation, the value of $1/d^{(t)}$ (see Eq. 1) after 1800 s was recorded (hence a higher value signifies more compactness). Fig. 3(a) shows a box plot for all the simulations. The performance is virtually unaffected as δ/δ_∞ is reduced from 1.0 to 0.3. As δ/δ_∞ is reduced to 0.2, the median performance drops instantly, almost to the value of robots that do not move during the simulation (lower magenta line in Fig. 3(a)).

The effect of sensing noise: Two types of noise on the binary sensor were considered. With false positive noise, the binary sensor erroneously gives the wrong reading with probability p when no robot should be observed, while it always gives the correct reading when a robot should be observed. Conversely, with false negative noise, the binary sensor never indicates the presence of a robot when there is none, but when there is a robot, it fails to detect it with probability p. False positive noise was found to be detrimental to aggregation performance. However, false negative noise was found to speed up the aggregation process. In order to investigate this, 100 simulations were run for each of 10 values of p: $\{0.0, 0.1, \ldots, 0.9\}$. Each simulation was stopped when all the robots were aggregated in a single cluster[2], and

[2] Two robots are said to be neighbors at time step t if the distance between their peripheries is less than some threshold, τ. The value of τ used here is equal to the diameter of the robots, i.e. $\tau = 7.4$ cm. A depth-first search algorithm is used to find the cluster profile at time step t, where a cluster is defined as a set of robots that form a connected graph (i.e. every two robots within the set are either neighbors, or connected by a chain of neighboring robots).

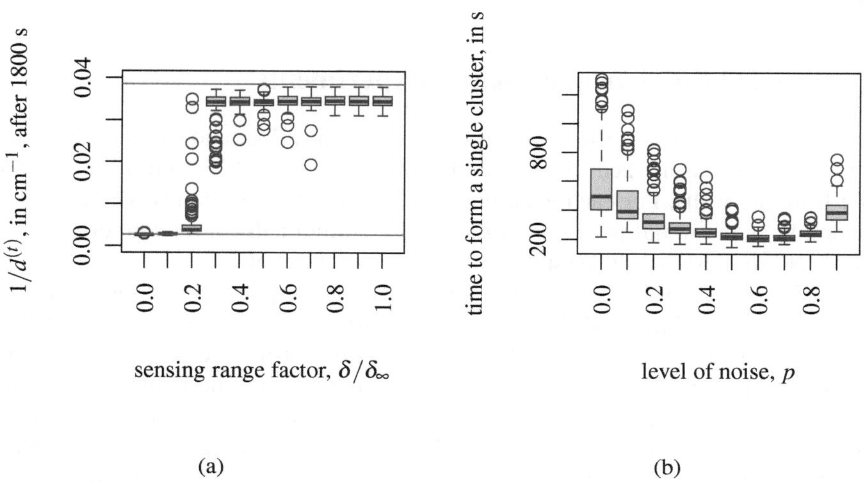

Fig. 3 (a): This box plot shows how the aggregation performance with $N = 100$ robots is affected by the range of the binary sensor. The horizontal axis shows the sensing range, δ, as a proportion of the maximum, effectively infinite, sensing range, δ_∞, while the vertical axis shows a measure of aggregation, $1/d^{(t)}$, as defined in Eq. 1. Each box represents 100 values of $1/d^{(t)}$ obtained from 100 runs. The lower magenta line shows the value of $1/d^{(t)}$ for robots that do not move. The upper magenta line shows the maximum value of $1/d^{(t)}$ obtainable with 100 robots, corresponding to the optimal packing [see 13]. (b): This box plot shows how the time taken by $N = 100$ robots to form a single cluster is affected by the level of false negative noise on the binary sensor. Each box represents 100 values of time obtained from 100 runs.

the time taken was recorded. Fig. 3 (b) shows a box plot of these times for false negative noise. The plot shows a clear 'bowl' shape, and it is striking that the optimum (fastest aggregation) occurs somewhere between $p = 0.6$ and 0.7.

From a qualitative visual analysis of the aggregation dynamics with the recurrent controller, with and without noise, it turns out that when there is no noise, the robots quickly form two to four large clusters. These clusters then take a long time to move as whole units and merge with each other. On the other hand, when there is false negative noise on the binary sensors, the robots do not form large clusters, but rather smaller clusters of only a few robots each. These clusters then are able to move as units more quickly than the larger clusters that form in the case with no noise, and therefore, aggregation happens faster.

Scalability: In order to investigate the scalability of aggregation with binary sensors, the best synthesized controller was tested with increasing numbers of robots. 100 simulations with different initial configurations of robots were performed for each value of $N \in \{100, 200, \ldots, 1000\}$. In each trial, the robots formed a single cluster, meaning that the controller is capable of achieving consistent, error free aggregation with at least 1000 robots.

5 Physical Implementation and Experiments

A square arena of sides 250 cm was used, having a white floor and enclosed by white walls. Its floor was marked with a grid of 9×9 points, spaced 25 cm from each other and from the walls. For each experiment, 20 out of these 81 points where chosen randomly as the initial positions of the robots. Additionally, a random orientation from {North, East, South, West} was chosen for each robot.

The binary sensor has been implemented using the e-puck's directional camera. The robots were fitted with black 'skirts' in order to make them distinguishable against the white arena. The middle column of pixels from the camera's image is sub-sampled so as to obtain 15 equally-spaced pixels from the bottom of the image to its half-way height. The gray value of these 15 pixels is compared against a threshold, which was empirically set to $2/3$ of the maximum possible value (pure white). If one or more of the pixels has a gray value below this threshold, the sensor gives a positive reading. The implemented sensor has been found to provide reliable readings up to a range of around 150 cm.

Two sets of 30 trials each were performed with $N = 20$ robots; one set with the best-found reactive controller, and one set with the best-found recurrent controller. The robots were started by issuing an infra-red signal from a remote control, and stopped automatically after 10 minutes. Each trial was recorded by an overhead camera, and all of the videos can be found in the online supplementary material

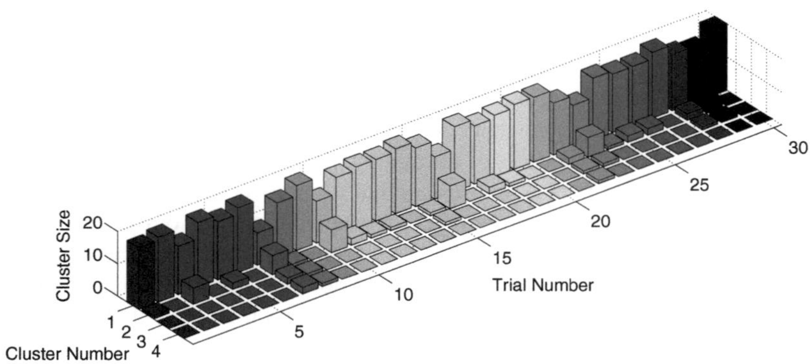

Fig. 4 This plot shows the final configurations of all the 30 trials performed with the reactive controller. The bar at the back shows the number of robots in the largest cluster, the bar in front of it shows the number of robots in the second-largest cluster, and so on. A robot that is not within 7.4 cm of any other robot is considered as a cluster in itself. Across all the trials, there were at most 4 clusters in the final configuration.

[10]. Throughout the 60 trials, 9 robots had a mechanical or electrical problem, and stopped moving. Whenever this happened, an infra-red signal was re-issued to the robot in case it might start again (normally-operating robots ignored this signal). Otherwise, the stationary robot was left in the arena for the rest of the trial. No other mechanical or electrical problems were observed. The cluster configuration (see Footnote 2 for a definition) of the robots at the end of each trial was examined. In the trials with the recurrent controller, the mean size (over the 30 trials) of the largest cluster in the final configuration was 17.10, meaning that 85.50% of the robots were aggregated in a single cluster. In the trials with the reactive controller, this value was 17.77, or 88.83%. These results are not significantly different (Mann-Whitney test performed with a $p = 0.05$ threshold). However, with the recurrent controller, substantially more robots became stuck at the arena walls than with the reactive controller (49 times against 15 times, out of a possible total of 600 each). Fig. 4 shows the cluster sizes in the final configurations of the 30 trials with the reactive controller.

6 Conclusion

This paper has investigated the usefulness of an alternative sensing trade-off for swarm robotic systems: one in which the robots have a longer sensing range than is normally assumed, but process only a minimal amount of information. This idea has been implemented to solve the problem of decentralized robot aggregation in a homogeneous environment using a sensor that provides each robot with a single bit of information per control cycle. The simplicity of the sensor does not come at the cost of a complex controller: in fact, even the simplest possible memoryless controller has been shown to be effective. To the best of the authors' knowledge, this is the simplest sensor/controller combination capable of aggregating robots in a single location. The system was implemented on 20 physical e-puck robots, and two sets of 30 trials were conducted, one set with a memoryless controller, and one set with a recurrent controller. Both controllers led to a good aggregation performance within 10 minutes. In the future, it is intended to extend the work performed to more complex environments (e.g. with obstacles), and to test the proposed sensing trade-off on other, more challenging swarming tasks.

Acknowledgements. The research work disclosed in this publication is funded by the Marie Curie European Reintegration Grant within the 7th European Community Framework Programme (grant no. PERG07-GA-2010-267354). M. Gauci acknowledges support by the Strategic Educational Pathways Scholarship (Malta). The scholarship is part-financed by the European Union – European Social Fund (ESF) under Operational Programme II – Cohesion Policy 2007–2013, "Empowering People for More Jobs and a Better Quality of Life".

References

[1] Ando, H., Oasa, Y., Suzuki, I., Yamashita, M.: Distributed memoryless point convergence algorithm for mobile robots with limited visibility. IEEE Trans. Robotic. Autom. 15(5), 818–828 (1999)
[2] Bahceci, E., Şahin, E.: Evolving aggregation behaviors for swarm robotic systems: a systematic case study. In: Proc. 2005 IEEE Swarm Intelligence Symposium, pp. 333–340 (2005)
[3] Camazine, S., Franks, N.R., Sneyd, J., Bonabeau, E., Deneubourg, J.-L., Theraulaz, G.: Self-Organization in Biological Systems. Princeton University Press, Princeton (2001)
[4] Chellapilla, K., Fogel, D.B.: Evolving an expert checkers playing program without human expertise. IEEE. Trans. Evolut. Comput. 5(4), 422–488 (2001)
[5] Correll, N., Martinoli, A.: Modeling and designing self-organized aggregation in a swarm of miniature robots. Int. J. Robot. Res. 30(5), 615–626 (2011)
[6] Cortés, J., Martinez, S., Bullo, F.: Robust rendezvous for mobile autonomous agents via proximity graphs in arbitrary dimensions. IEEE Trans. Automat. Contr. 51(8), 1289–1298 (2006)
[7] Deneubourg, J.-L., Gregoire, J.-C., Le Fort, E.: Kinetics of larval gregarious behavior in the bark beetle Dendroctonus micans (Coleoptera: Scolytidae). J. Insect. Behav. 3, 169–182 (1990)
[8] Dorigo, M., Trianni, V., Şahin, E., Groß, R., Labella, T.H., Baldassarre, G., Nolfi, S., Deneubourg, J.-L., Mondada, F., Floreano, D., Gambardella, L.: Evolving self-organizing behaviors for a swarm-bot. Auton. Robot. 17, 223–245 (2004)
[9] Garnier, S., Jost, C., Gautrais, J., Asadpour, M., Caprari, G., Jeanson, R., Grimal, A., Theraulaz, G.: The embodiment of cockroach aggregation behavior in a group of micro-robots. Artif. Life 14(4), 387–408 (2008)
[10] Gauci, M., Chen, J., Dodd, T.J., Groß, R.: Online supplementary material (2012), `http://naturalrobotics.group.shef.ac.uk/supp/2012-003/`
[11] De Gennaro, M.C., Jadbabie, A.: Decentralized control of connectivity for multi-agent systems. In: Proc. 45th IEEE Conf. Decision and Control, pp. 3628–3633 (2006)
[12] Gordon, N., Wagner, I.A., Bruckstein, A.M.: Gathering multiple robotic a(ge)nts with limited sensing capabilities. In: Dorigo, M., Birattari, M., Blum, C., Gambardella, L.M., Mondada, F., Stützle, T. (eds.) ANTS 2004. LNCS, vol. 3172, pp. 142–153. Springer, Heidelberg (2004)
[13] Graham, R.L., Sloane, N.J.A.: Penny-packing and two-dimensional codes. Dicrete Comput. Geom. 5, 1–11 (1990)
[14] Halloy, J., Sempo, G., Caprari, G., Rivault, C., Asadpour, M., Tâche, F., Saïd, I., Durier, V., Canonge, S., Amé, J.M., Detrain, C., Correll, N., Martinoli, A., Mondada, F., Siegwart, R., Deneubourg, J.L.: Social integration of robots into groups of cockroaches to control self-organized choices. Science 318, 1155–1158 (2007)
[15] Jeanson, R., Rivault, C., Deneubourg, J.-L., Blanco, S., Fournier, R., Jost, C., Theraulaz, G.: Self-organized aggregation in cockroaches. Anim. Behav. 69(1), 169–180 (2005)
[16] Kernbach, S., Thenius, R., Kernbach, O., Schmickl, T.: Re-embodiment of honeybee aggregation behavior in an artificial micro-robotic system. Adapt. Behav. 17(3), 237–259 (2009)
[17] Magnenat, S., Waibel, M., Beyeler, A.: Enki: The fast 2d robot simulator (2011), `http://home.gna.org/enki/`

[18] Mondada, F., Bonani, M., Raemy, X., Pugh, J., Canci, C., Klaptocz, A., Magnenat, S., Zufferey, J.-C., Floreano, D., Martinoli, A.: The e-puck, a robot designed for education in engineering. In: Proceedings of the 9th Conference on Autonomous Robot Systems and Competitions, vol. 1, pp. 59–65 (2009)
[19] Nolfi, S., Floreano, D.: Evolutionary Robotics: The Biology, Intelligence, and Technology of Self-Organizing Machines. The MIT Press, Cambridge (2000)
[20] Parrish, J.K., Edelstein-Keshet, L.: Complexity, pattern, and evolutionary trade-offs in animal aggregation. Science 284(5411), 99–101 (1999)
[21] Trianni, V.: Evolutionary Swarm Robotics. SCI, vol. 108. Springer, Berlin (2008)
[22] Williams, R.J., Zipser, D.: A learning algorithm for continually running fully recurrent neural networks. Neural. Comput. 1(2), 270–280 (1989)
[23] Yao, X., Liu, Y., Lin, G.: Evolutionary programming made faster. IEEE Trans. Evolut. Comput. 3(2), 82–102 (1999)

Robot Teams: Sharing Visual Memories

Raphael Grech, Francisco Flórez-Revuelta, Dorothy N. Monekosso, and Paolo Remagnino

Abstract. This paper presents, the Growing Neural Gas (GNG), an unsupervised learning algorithm, which allows a team of robots to memorize scenes and collectively create a general understanding of the environment that is easily understood and referenced by humans is presented. Each robot will have its own memory represented by a graph with nodes encoding the visual information of a video stream as a limited set of representative images. GNG are self-organizing neural networks that can dynamically adapt their reference vectors and topology. Frames are sequentially processed by the GNG, automatically generating nodes, establishing connections between them and creating clusters dynamically. We mainly focus on creating a robot team learning mechanism to achieve a distributed system of robots automatically sharing acquired knowledge with others available within the area. This is done using keyframes representing clusters within the robot memory.

1 Introduction

There are situations where tasks cannot be carried out by a single robot. When these situations arise, tasks are carried out with multiple robot systems (MRS) to collaborate and work together as a team in order to achieve the required goal. One such task is that of surveying and patrolling large areas. MRS can accomplish tasks that no single robot can accomplish by itself, since ultimately a single robot, no matter how capable, is spatially limited [3]. When using a MRS, each robot may be

Raphael Grech · Francisco Flórez-Revuelta · Paolo Remagnino
Faculty of Science, Engineering and Computing, Kingston University, London, UK
e-mail: {R.Grech,F.Florez,P.Remagnino}@kingston.ac.uk

Dorothy N. Monekosso
Faculty of Computing and Engineering, University of Ulster, UK and Visiting Academic at Kingston University, London, UK
e-mail: dn.monekosso@ulster.ac.uk

M.A. Hsieh and G. Chirikjian (eds.), *Distributed Autonomous Robotic Systems*, Springer Tracts in Advanced Robotics 104,
DOI: 10.1007/978-3-642-55146-8_26, © Springer-Verlag Berlin Heidelberg 2014

designed for a different task, and the required goal is achieved with proper collaboration by the robots. This would provide a more generic structure as the robots would be able to reconfigure themselves as required. This highlights the importance of proper communication for effective teamwork. Furthermore, using several robots introduces redundancy. Teams of robots therefore can be expected to be more fault-tolerant than a single robot. The merging of overlapping information coming from the robot team can help compensate for sensor uncertainty [2]. Extensive work has been carried out on robot mapping of the environment [23, 5, 1, 9]. Mapping alone however does not give information about the environment itself. Whilst the focus of robot mapping is building a geometrical map of the environment, our aim is merging meaningful visual information to the geometrical map, by creating a topological map of the environment. We therefore are of the idea that similar to humans, only salient images, which we call keyframes, are stored. This strategy requires less processing and storage power but is still able to maintain all the relevant information. As an example, office blocks generally have the same geometrical layout, but the contents may vary between floors. Assuming limited resources on the robot, an efficient way for the robot to learn the main scene differences and memorize them is needed. We propose a method to automatically create a visual memory for robots equipped with cameras surveying, monitoring or searching an area of interest. We wish to encode the visual information into a limited set of representative images on-line and with limited computational over-head. The idea behind our approach is to provide a flexible graphical representation of visual memory to be subsequently used for the semantic description of a captured scene.

Various video segmentation methods exist in literature [4, 20, 22, 17] and some are extensively reviewed [15, 16, 8, 19], however these generally assume that the video is stored and can be post-processed. Some algorithms can also be very computationally intensive. Ngan and Li in [19] highlight four main challenges in image/video segmentation. The first challenge is how to effectively bridge the semantic gap between low-level and high-level features. The second is how to yield accurate segmentation and how to extract accurate masks. The third challenge is that of working in real time without compromising accuracy and the fourth is the need to develop appropriate validation and evaluation approaches, by providing a common database and by developing an evaluation technique. Ngan and Li state that most evaluation methods in the current literature are based on the computation of the scores between the ground truth mask and the segmented result. It is reasonable but not sufficient to address the segmentation quality. Gao et al. in [19] state that the usage of machine learning techniques has proven to be a robust methodology for semantic scene analysis and understanding. The main characteristic of learning-based approaches is their ability to adjust their internal structure according to input and respective desired output data pairs in order to approximate the relations implicit in the provided (training) data, thus elegantly simulating a reasoning process.

Although the Growing Neural Gas (GNG) was originally designed for offline training [14], this work extends the ideas suggested in [10] for building visual memories of video streams to a multi-robot scenario. Several robots can be used to create a visual memory of the environments in a faster and more efficient manner. Our intention is to generate a graph of small size as a reduced representation of the environment that could be easily shared among robots or a distributed set of computational nodes and easy to grow in cooperation. The robot's visual memories are incrementally built using the GNG self-organizing model. The GNG was chosen because it provides flexibility and portability, by dynamically building a representative graph of the input space, in this case a video or scene video sequence. The main contribution of this paper is an efficient method for learning and memorizing in real time an environment from a sequential input video stream in a very concise and compact manner using a team of robots. By real time we mean that the sampling input frequency of the environment scene being high enough for the GNG visual memory to reflect the salient views of the robot. This work is primarily intended for robot understanding of their own environment and possibly localization.

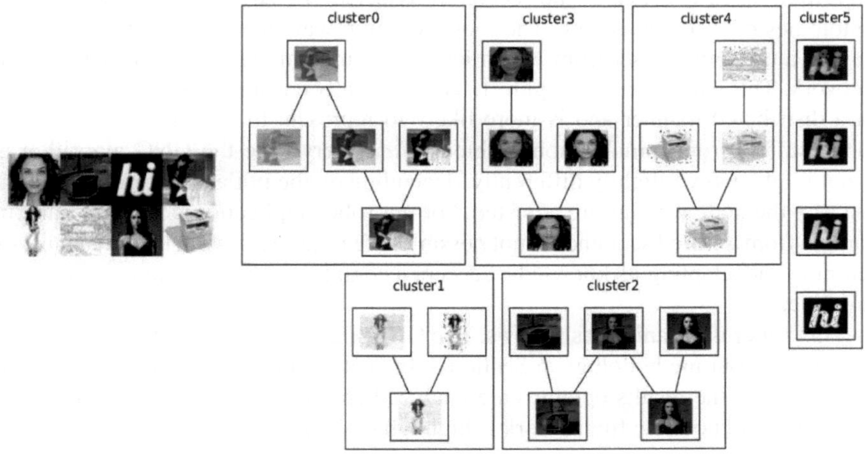

Fig. 1 Training Images (Left) and Generated GNG Clustered Nodes (Right)

Figure 1 illustrates an example of how GNG structures network topology. In this example, a fixed set of training images and a maximum number of nodes have been used. After a random iteration of the trained images illustrated on the left of Figure 1, GNG creates the links between the nodes and the clusters, which are shown on the right hand side of Figure 1. One can note that some of the images were merged together. This is mainly due to the feature vector used. In our case the feature vector is the grayscale pixel value, so images having similar grayscale distributions tend to cluster together. The more defined the input feature vector the better the classification is expected to be. It might be argued that there could be better features one can pass to the GNG rather than the scaled grayscale images. Unless the selected features are highly descriptive, and since only these features will be memorized, there is no way to revert to the original image. This would make it impossible for a human to understand straight away what the robot has memorized. Another possible option could be that of feeding in a higher semantic level to the GNG. This however necessitates the robot to be able to recognize what it sees. This could work well in an environment where the contents are somewhat expected, however it would fail miserably in a total alien environment.

GNG was proved to be superior to existing unsupervised methods, such as self-organising Kohonen maps, K-means, growing cell structures [6, 12, 13]. Florez et al. [6] conclude that networks having an evolving topology adapt better to the input manifold than any network with pre-established topologies. In a GNG graph, nodes can be disconnected while the network is evolving, creating a separation between uncorrelated memories. The number of nodes need not be fixed a priori, since they are incrementally added during execution. Insertion of new nodes ceases when a set of user defined performance criteria is met, or alternatively the maximum network size is reached. The algorithm iteratively learns to identify similarities of input data and classifies them into clusters. GNG is much more than a simple clustering algorithm since it provides the means to associate visual memories and to build ontologies of visual concepts. The most common way of training GNG is that of having a training dataset from which items are randomly selected and fed into the network. This can be seen in Figure 1. This generally ensures that the GNG evolves in a distributed manner and is more likely to represent the input data accurately. GNG suffers from initialization problems, *i.e.*, every time the GNG algorithm is run it might evolve slightly differently, depending on the initial seeding and also on the way the node weights are adjusted. For our robot application, feeding a random sample from a stored sequence is not possible. We require a system that is capable of learning and adapting its knowledge, accept a continuous video stream, and process it online.

This paper is organized as follows. The GNG algorithm is presented in Section 2 followed by our methodology in Section 3. In Section 4 the experiments carried out together with the results obtained are analyzed and discussed. Finally conclusions are drawn and possible future work is highlighted in Section 5.

2 The GNG Algorithm

GNG was originally introduced by Fritzke [7], as an unsupervised learning technique where no prior training is needed. The system starts with two linked nodes; new nodes are inserted at every fixed number of input cycles up until the maximum number of allowed nodes is reached. Connections between nodes are also inserted and removed adapting the network topology. Moreover, nodes which are disconnected are removed thus allowing for new nodes to be inserted in a better position within the topological map. This results in a network having a topological structure composed of N nodes in Y clusters connected by edges closely reflecting the topology of the feature distribution. The GNG algorithm operates as shown in Algorithm 1. The GNG network is specified as:

- A set N of nodes (neurons). Each node $k \in N$ has its associated reference vector $\mathbf{w}_k$ belonging to the input space (80×60 grayscale images).

- A set of edges (connections) between pairs of nodes. These connections are not weighted and its purpose is to define the topological structure. An edge ageing scheme is used to remove connections that are invalid due to the adaptation of the node during the learning process.

Algorithm 1. GNG Algorithm

Set two nodes containing random values, age edge = 0, error = 0

while (Stopping Criterion = **false**) **do**
  - capture an input image vector **x**
  - from all nodes, find winning node s_1 and second best node s_2
  - increase the age of all the edges from s_1 to its topological neighbours
  - update the error of s_1
  - move s_1 and its neighbors towards **x**

  if (s_1 and s_2 are connected by an edge) **then**
    - set the age of the edge to 0.
  else
    - create an edge between them.
  end if

  if edges are older than age threshold **then**
    - remove edges
  end if
  - remove isolated neurons

  if (current iteration is a multiple of λ) **and** (maximum node count = **false**) **then**
    - find node u with largest error.
    for all neighbors of u **do**
      find node v with largest error
    end for
    - insert a new node r between u and v
    - create edges between u and r, and v and r
    - remove edge between u and v
    - decrease the error variables of u and v
    - set the error of node r
  end if
  - decrease error value of all nodes
end while

Due to the sequential nature of the robot's visual data acquisition the GNG was adapted for our application as follows:

- During testing it was noted that the Best Matching Unit (BMU) will be the same one for a number of consecutive frames which are very similar. In our case this is good, however we do not want to over train. If this happens one possible option could be to skip input frames.

- We also need to learn fast. The BMU is made to converge to the actual input very fast by adding a large proportion of the error between the input and the BMU. This basically sets the BMU to the input image.
- The second BMU (for the same frame) is slightly adjusted. This allows for more information storage within the same cluster rather than having several nodes with the same value within the cluster.
- A new node is inserted at a relatively fast rate (e.g. every other iteration). Thus allowing for a large number of nodes to be used from early age and new nodes are inserted soon after isolated nodes are killed.

3 Methodology

We want each robot in the team to memorize its own area and at the same time share some of its acquired knowledge with its peers in an efficient and compact way. As it happens with human memory, details of a scene are retained and memorized images can be blurred or somewhat unclear. However, sufficient information is retained to recall relevant information from memory about a part of the scene [11]. Likewise, our proposed algorithm does not produce a perfect photographic memory, but rather retains image representations, which contain meaningful information about the explored environment. In our method each node consists of an 80×60 pixel grayscale grid representing an evolving memory image. One basic way of having a global understanding of the environment is to have each robot surveying its own area, generating their own set of clusters and then feeding them into another learning network to create a common central memory. This however has some disadvantages. Aside from the fact there is reliance on a centralized system, robots would only know their area and would be totally unaware of what other robots are experiencing in other areas. This means that if one robot dies (e.g. run dry on battery power) all the information obtained from the robot would be lost, unless it has already been provided to the central system. We want to have a distributed system to complement the above idea so that if one robot dies along the process some of its most relevant information will be retained. We therefore suggest having several robots each with its own visual memory within a distributed environment. Robots are to memorize what they see and also accept incoming visual information from neighboring robots, diagrammatically shown in Figure 2. We take inspiration from the island model genetic algorithm [24]. This methods revolves around the concept of migration where each island (in our case a robot) will periodically exchange a portion of its population (nodes) with other islands. Each robot will start generating clusters of similar images within its own visual memory. The average of this cluster is then calculated and a single image is produced thus generating one image per cluster (keyframe). One of these generated images will then be selected at random and shared with the other robots. With our suggested method we can have both a distributed and centralized system working together. The distributed system consists of robots which memorize mostly their environment with some influence from other robots and the centralized

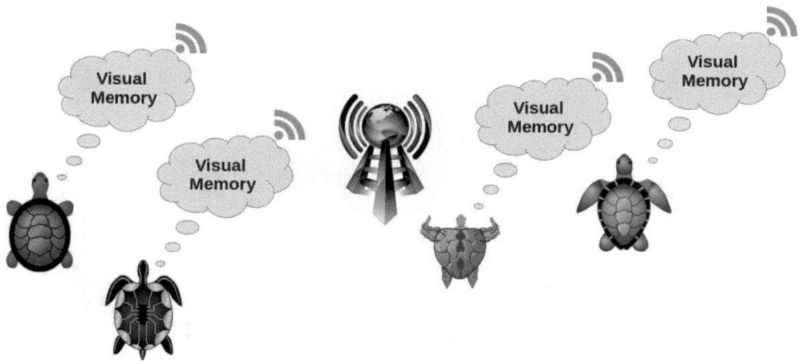

Fig. 2 Concept behind the distributed visual memory

system to have a general understanding of all the environments being monitored. One way to do this in a distributed manner is to have a "visually impaired" robot together with the other "normal" robots. This means that this robot would only receive inputs coming from other robots and share its own clusters based on these inputs. Referring to Figure 3 the noise cluster 5 is most likely generated by the blind robot itself since in the first iteration it could only share noise as it had seen nothing before. At times this cluster can evolve into meaningful images but at others, as in this case, it might linger in the robot network.

The way GNG is used in this paper bears some resemblance to well-known video annotation and segmentation methods [19] and [21]. These methods generally look for specific changes in the video segment such as scene change or cross fading between scenes. They also use more complex segmentation algorithms such as graph-cuts and eigen-based methods. The main limitation of such methods is that they cannot segment and annotate in realtime, as the whole video has to be processed to create a reliable segmentation. Our choice provides several advantages when implemented for a robot. Storing and transmitting a video stream requires a large amount of memory and a high bandwidth, usually scarce on a robot. Each robot will operate using the procedure presented in Algorithm 2.

In the proposed algorithm each robot is the "expert" of its area, however it will have enough information from the other robots to know what else is in the surroundings. Figure 4 shows a summary of how cluster sharing happens between robots. Robots 1 and 2 generate an average cluster (keyframe) image which is shared with Robot 3. Robot 3 will generate its own keyframe which in turn will be shared with the other robots. One can note that the outcome fidelity is reduced. This is to be expected as observed in humans. When person A and person B say something to person C, it is highly unlikely that person C will relay accurate information to person D. This would mean that in the case of a general scene identification query sent to all the robots, the one with the original data (the expert) is more likely to respond with the best match. Robot learning is similar to that of a child. Initially its

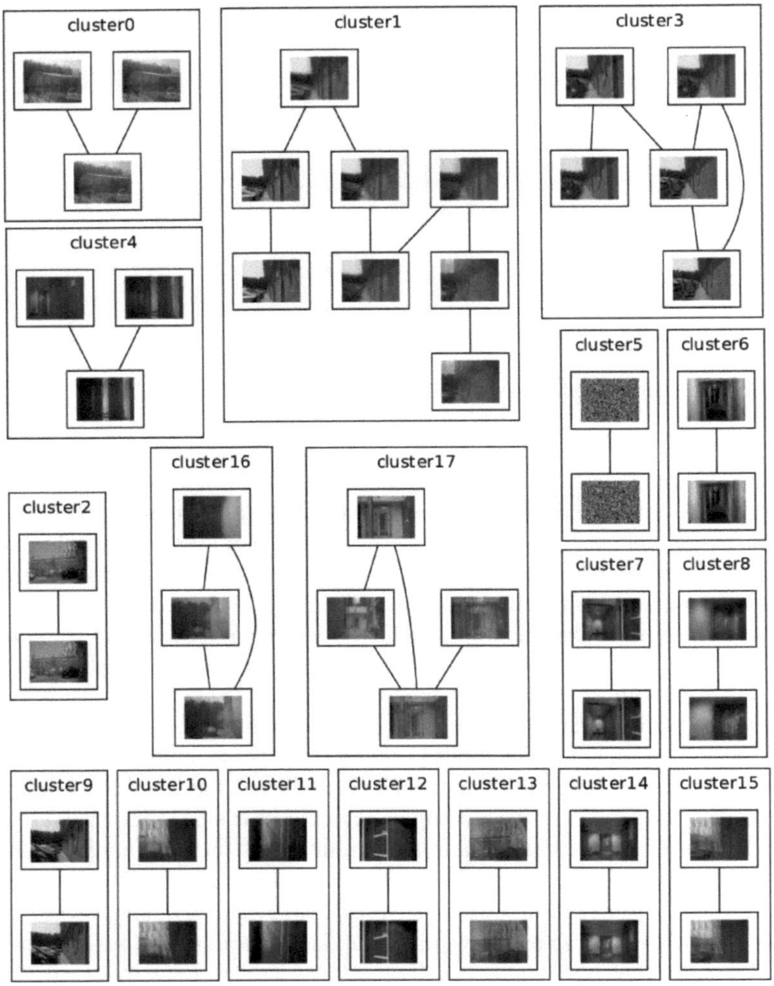

Fig. 3 Visually Impaired Robot - This robot only memorizes what other robots share

knowledge will be "blurred". As time passes, its clusters will be better defined and therefore the will start sharing "better" information. Same applies to when it has to share its knowledge. A child will have unclear or "blurred" concepts which will become more clear by time. The main advantage is that of initializing seeds within the peer robots with information about scenes which were not previously seen by that robot. This seed would allow robots to learn a new environment faster, if it happens to be the similar to one already visited and shared by other robots.

Algorithm 2. Robot Learning and Sharing Procedure

while (Stopping Criterion = **false**) **do**
- Capture image from the environment
if (broadcast image from other robots = **true**) **then**
- Accept broadcast image
end if
- Scale and convert captured images to grayscale
- Feed images into memory learning algorithm {learning algorithm starts generating its own clusters}
if (specified or random time elapsed = **true**) **then**
- select a cluster at random and broadcast its average image (keyframe)
end if
end while

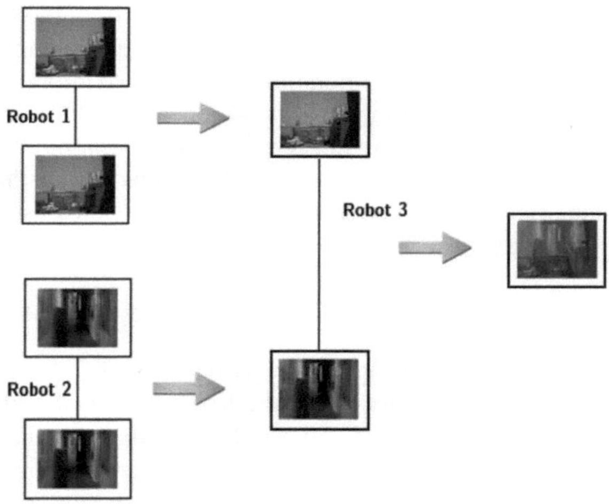

Fig. 4 Cluster Sharing

4 Experimental Analysis and Results

In our study, we have mainly analyzed how individual learning evolves with the change in number of robots and image injection rate from other robots. The implementation was developed in ROS (Robot Operating System). ROS is a robot-specific middle-layer solution for distributed computation and message passing. It allows easy integration of sensor drivers and data processing components including both off-the-shelf and in-house components. The distributed nature of ROS allows each independent component to function with some degree of independence and facilitates extensibility [18]. The main reasons for using ROS are that we can have multiple robot instances running as separate nodes / threads, it can be implemented on real robots, the input to the robot learning algorithm can come from any visual capturing device available on the robot / ROS network and robots can join or leave at any point. This also caters to the likely case that robots die along the way. For our experiments only video inputs were required. The video streams were captured using a digital camera and fed into the ROS environment as a camera node to which each robot subscribed using its memory node. A range between two and five robots were used, each having a different input video sequence all of same length (2500 frames). Robot 1 and 2 surveyed two different corridors, Robot 3 was moving along an outside passageway between two buildings. Robot 4 was moving outside along a pavement in a car park and Robot 5 was on a road leading to the car park. The GNG parameters and the maximum number of nodes was kept fixed, the reason being that we want to analyze the effect of varying the number of robots and the frequency of sharing. The maximum number of nodes in the GNG was set to 50 and a new node was inserted every 5 iterations. The best matching unit coefficient was set to 0.95 and that of its neighbors to 0.001. The maximum edge age was set to 2 and for every iteration the error of each node is decreased by a factor of 0.005. For each set of robots (2, 3, 4, 5) a different sharing frequency was used (one image shared by each robot every 5, 15, 25, 35, 45, 55 iterations). This led to a total of 84 different graphs: 12 for 2 robots, 18 for 3 robots, 24 for 4 robots and 30 for 5 robots. Each graph was checked for which nodes in the visual memory were not from the robot's input but rather from the common pool by manually comparing it to the ground truth data. The percentage sharing between robots was then calculated using

$$\frac{100}{n}\sum_{i=1}^{n}\frac{x(i)}{z(i)}$$

where $x(i)$ is the number of nodes within the visual memory of robot i not originating from the onboard camera and $z(i)$ is the total number of nodes within that network. The outcome was plotted in Figure 5 showing the percentage of memory originating from the other robots (y-axis) versus frame sharing frequency (x-axis).

The higher the sharing frequency between the robots the higher the percentage of shared memory between the robots. A monotonic curve with negative gradient could therefore be assumed. Given a number of robots and a desired percentage of shared memory to be stored, the frame sharing frequency should be set accordingly. In order to find the best fitting curve a 3 robot case scenario was used. The sharing frequency was varied from 2 up to 40 in steps of 2. Various curves where fitted and their R-squared value noted. Out of all the curves fitted, the best match was that using a polynomial of order 3, followed by a log, then power, exponential, and finally the least accurate being the straight line approximation. When varying the number of robots, the best overall performance was given by the log curve with an R-squared value of over 0.8 for the 2 and 5 robot case and a 0.6 for the 3 robot case. The 4 robot case performed at 0.2, still higher than the straight line approximation. Due to the initialization process of the GNG and the random nature of cluster image selection to feed other robots, the content of each visual memory will be different (i.e. not repeatable). This means that Figure 5 cannot be reproduced exactly for every run, however the general negative trend still holds. If the frequency of sharing is low robots will tend to learn only their environment with low influence from the

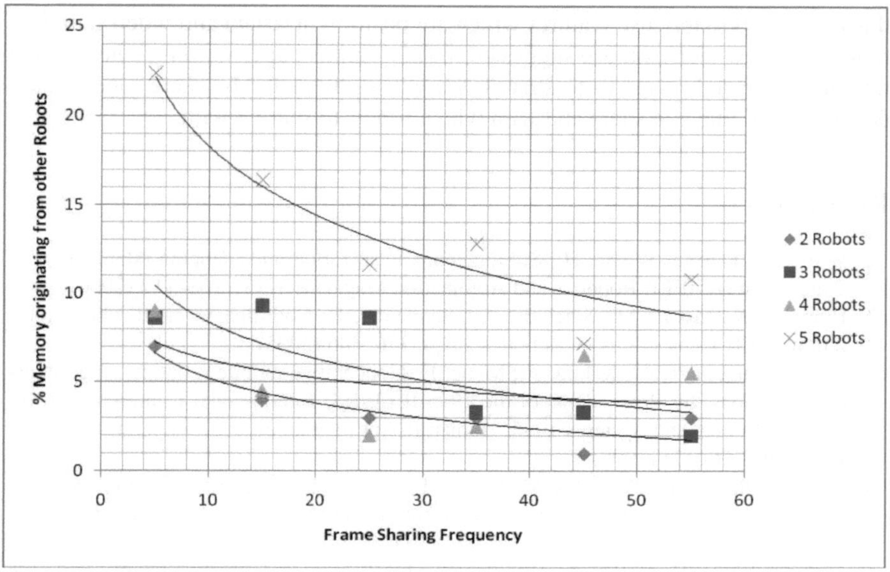

Fig. 5 Percentage memory sharing (y-axis) vs Frame sharing frequency (x-axis) for various team sizes.

neighboring robots. If on the other hand the frequency of sharing is too high then it might be that some robots could be overwhelmed with information coming from other robots and end up learning what others are memorizing rather than building a memory of their environment. As the results tend to indicate, this situations becomes more acute as the number of robots increases.

5 Conclusion

In this paper we provided a method where a team of robots efficiently creates a distributed visual memory. This is implemented by learning and memorizing the robots' environment in real time from sequential input video streams into a flexible graphical representation using a Growing Neural Gas (GNG) network. We tested the system on various raw video streams coming from the robots. Experimental results show that the proposed method suits its intended application and a very concise yet meaningful representation of input data is obtained. We saw that as the sharing frequency between the robots increases, the higher the percentage of shared memory between the robots, however a good balance between the number of robots available and how much information they share is required so as not to overwhelm the robots with external information. In our system we feed in the scaled down raw images. If the movement of the robot is not smooth, sequence frames capturing the same scene will generate a different Euclidean distance. This was noted during the experiments and this tends to generate multiple clusters of the same scene. We intend to look into this next and possibly use features which are position invariant and video stabilization techniques such as those suggested in [25]. In this paper we only considered a randomly selected image to be shared. We will study methods where robots could decide which image to share and with which robots to share it with.

References

1. Bailey, T., Durrant-Whyte, H.: Simultaneous localization and mapping (slam): Part ii: State of the art. IEEE Robotics and Automation Magazine 13(3), 108–117 (2006)
2. Burgard, W., Moors, M., Stachniss, C., Schneider, F.: Coordinated multi-robot exploration. IEEE Transactions on Robotics 21(3), 376–386 (2005)
3. Cao, Y., Fukunaga, A., Kahng, A., Meng, F.: Cooperative mobile robotics: antecedents and directions. In: Proceedings of the 1995 IEEE/RSJ International Conference on Intelligent Robots and Systems 1995, Human Robot Interaction and Cooperative Robots, vol. 1, pp. 226–234 (1995)
4. Chiu, P., Girgensohn, A., Polak, W., Rieffel, E., Wilcox, L.: A genetic algorithm for video segmentation and summarization. In: IEEE International Conference on Multimedia and Expo, IEEE, New York (2000)
5. Durrant-Whyte, H., Bailey, T.: Simultaneous localisation and mapping (slam): Part i the essential algorithms. IEEE Robotics and Automation Magazine 13(2), 99–108 (2006)
6. Florez, F., Garcia, J., Garcia, J., Hernández, A.: Representing 2d objects. comparison of several self-organizing networks. In: 3th WSES Conference on Neural Networks ancl Applications, Interlaken, pp. 69–72 (2002)
7. Fritzke, B.: A growing neural gas network learns topologies. In: Tesauro, G., Touretzky, D.S., Leen, T.K. (eds.) Advances in Neural Information Processing Systems 7, pp. 625–632. MIT Press, Cambridge (1995)
8. Gao, W., Tian, Y., Huang, T., Yang, Q.: Vlogging: A survey of videoblogging technology on the web. ACM Comput. Surv. 42, 15:1–15:57 (2010)
9. Gil, A., Mozos, O., Ballesta, M., Reinoso, O.: A comparative evaluation of interest point detectors and local descriptors for visual slam. Machine Vision and Applications 21, 905–920 (2010)
10. Grech, R., Monekosso, D., Remagnino, P.: Building visual memories of video streams. Electronics Letters 48(9), 487–488 (2012)
11. Green, M.: Eyewitness memory is unreliable, `http://www.visualexpert.com/Resources/eyewitnessmemory.html` (last accessed on March 10, 2012)
12. Heinke, D., Hamker, F.: Comparing neural networks: a benchmark on growing neural gas, growing cell structures, and fuzzy artmap. IEEE Transactions on Neural Networks 9(6), 1279–1291 (1998)
13. Holmström, J.: Growing Neural Gas Experiments with GNG, GNG with Utility and Supervised GNG. Master's thesis, Uppsala University (2002)
14. Kirstein, S., Wersing, H., Körner, E.: A biologically motivated visual memory architecture for online learning of objects. Neural Networks 21, 65–77 (2008)
15. Koprinska, I., Carrato, S.: Temporal video segmentation: A survey. Signal Processing: Image Communication 16(5), 477–500 (2001)
16. Lefèvre, S., Holler, J., Vincent, N.: A review of real-time segmentation of uncompressed video sequences for content-based search and retrieval. Real-Time Imaging 9, 73–98 (2003)

17. Lezama, J., Alahari, K., Sivic, J., Laptev, I.: Track to the future: Spatio-temporal video segmentation with long-range motion cues. In: 2011 IEEE Conference on Computer Vision and Pattern Recognition (CVPR), pp. 3369–3376 (2011)
18. Meger, D., Muja, M., Helmer, S., Gupta, A., Gamroth, C., Hoffman, T., Baumann, M., Southey, T., Fazli, P., Wohlkinger, W., Viswanathan, P., Little, J., Lowe, D., Orwell, J.: Curious george: An integrated visual search platform. In: 2010 Canadian Conference on Computer and Robot Vision (CRV), pp. 107–114 (2010)
19. Ngan, K.N., Li, H.: Video Segmentation and Its Applications, 1st edn. Springer, New York (2011)
20. Porter, S.: Video segmentation and indexing using motion estimation. Ph.D. thesis, University of Bristol (2004)
21. Siagian, C., Itti, L.: Storing and recalling information for vision localization. In: ICRA, pp. 1848–1855. IEEE (2008)
22. Song, X., Fan, G.: Selecting salient frames for spatiotemporal video modeling and segmentation. IEEE Transactions on Image Processing 16(12), 3035–3046 (2007)
23. Thrun, S.: Robotic mapping: A survey. In: Exploring Artificial Intelligence in the New Millenium. Morgan Kaufmann (2002)
24. Whitley, D., Rana, S., Heckendorn, R.B.: The island model genetic algorithm: On separability, population size and convergence. Journal of Computing and Information Technology 7, 33–47 (1998)
25. Yang, J., Schonfeld, D., Mohamed, M.: Robust video stabilization based on particle filter tracking of projected camera motion. IEEE Trans. Cir. and Sys. for Video Technol. 19(7), 945–954 (2009)

Distributed Particle Swarm Optimization for Limited Time Adaptation in Autonomous Robots

Ezequiel Di Mario and Alcherio Martinoli

Abstract. Evaluative techniques offer a tremendous potential for on-line controller design. However, when the optimization space is large and the performance metric is noisy, the time needed to properly evaluate candidate solutions becomes prohibitively large and, as a consequence, the overall adaptation process becomes extremely time consuming. Distributing the adaptation process reduces the required time and increases robustness to failure of individual agents. In this paper, we analyze the role of the four algorithmic parameters that determine the total evaluation time in a distributed implementation of a Particle Swarm Optimization algorithm. For a multi-robot obstacle avoidance case study, we explore in simulation the lower boundaries of these parameters with the goal of reducing the total evaluation time so that it is feasible to implement the adaptation process within a limited amount of time determined by the robots' energy autonomy. We show that each parameter has a different impact on the final fitness and propose some guidelines for choosing these parameters for real robot implementations.

1 Introduction

Human design of high-performing robotic controllers is not a trivial task for a number of reasons. In the first place, even the simplest of modern robots have a large number of sensors and actuators, which implies a large number of control parameters to optimize. Secondly, real systems often present discontinuities and nonlinearities, making it difficult to apply well-understood linear control techniques. Finally, when porting the designed controller to real robots there might be an unexpected

Ezequiel Di Mario · Alcherio Martinoli
Distributed Intelligent Systems and Algorithms Laboratory, School of Architecture, Civil and Environmental Engineering, École Polytechnique Fédérale de Lausanne,
1015 Lausanne, Switzerland
e-mail: {ezequiel.dimario,alcherio.martinoli}@epfl.ch

M.A. Hsieh and G. Chirikjian (eds.), *Distributed Autonomous Robotic Systems*,
Springer Tracts in Advanced Robotics 104,
DOI: 10.1007/978-3-642-55146-8_27,

performance drop due to a number of factors such as imperfections in fabrication, changes in the environment, or modeling inaccuracies.

Machine-learning techniques are an alternative to human-guided design that can address some of the previously mentioned challenges. They can automatically synthesize robotic controllers in large search spaces, coping with discontinuities and nonlinearities, and find innovative solutions not foreseen by human designers by working with a pool of potentially very diverse candidate solutions. Furthermore, the learning process can be implemented fully on-board, enabling automatic adaptation to the underlying hardware and environment.

However, the main drawback of working with a pool of candidate solutions is the amount of time needed to evaluate all candidates, which is substantially larger than that required to generate them. Moreover, due to several sources of uncertainty, such as sensor noise, manufacturing tolerances, or lack of strict coordination in multi-robot settings, it may be necessary to re-evaluate some solutions to gather sufficient statistics for meaningful adaptation. Because of these two reasons, the adaptation process is considered an expensive optimization problem.

Implementing the adaptation process in a distributed fashion brings two distinct advantages. Firstly, it reduces the required evaluation time through parallelization. Secondly, it increases robustness by avoiding a critical point of failure, which is of particular interest in real robot implementations.

Thus, the goal of this paper is to analyze how different algorithmic parameters in a distributed implementation affect the total evaluation time and resulting fitness. We aim to reduce the total evaluation time such that it is feasible to implement the adaptation process within the limits of the robots' energy autonomy without renouncing the benefits of a population-based learning algorithm.

2 Related Work

Particle Swarm Optimization (PSO) is a relatively new metaheuristic originally introduced by Kennedy and Eberhart [8]. PSO is inspired by the movement of flocks of birds and schools of fish, and represents a set of candidate solutions as a swarm of particles moving in a multi-dimensional space. Particles can recall at which position of the search space they obtained their best performance and also the position of the best performing particle in a pre-established neighborhood.

Because of its simplicity and versatility, PSO has been used in a wide range of applications such as antenna design, communication networks, finance, power systems, and scheduling. Within the robotics domain, popular topics are robotic search, path planning, and odor source localization [13].

PSO is well suited for distributed/decentralized implementations due to its distinct individual and social components and the use of the neighborhood concept. Most of the work on distributed implementations has been focused on benchmark functions running on computational clusters [1,3,16]. Implementations with mobile robots are mostly applied to odor source localization [9,17], and robotic search [6], where, as opposed to optimizing a set of control parameters for the task at hand, the particles' position is usually directly matched to the robots' position in the arena. Thus, the search is conducted in two dimensions and with few or even only one local extrema. For these reasons, even though robotic search is a challenging practical task, it does not represent a complex optimization problem.

An example of a more challenging on-line optimization problem is the work of Floreano and Mondada [5], who used Genetic Algorithms to optimize the weights of an artificial neural network controller. The task was to navigate a path and avoid obstacles with a tethered mobile robot. Even though the population manager and other resource-intensive tasks were carried out on a dedicated off-board computer, this study was still able to show the advantages of evaluation with hardware in the loop. For example, the evolved direction of motion was a result of the interplay between the robot morphology (higher density of proximity sensors facing forward) and the environment in which it was deployed. It is worth noting that the experiment required 67 hours of total evaluation time, and it would require the same time to recreate it nowadays since the limit was not imposed by computational capabilities but rather by the wall-clock time needed to gather enough information on the quality of the candidate solutions.

Most of the research on optimization in noisy environments has focused on evolutionary algorithms [7]. The performance of PSO under noise has not been extensively studied. Parsopoulos and Vrahatis showed that standard PSO was able to cope with noisy and continuously changing environments, and even suggested that noise may help to avoid local minima [12]. Pan et al. [11] proposed a PSO variation based on statistical tests to select particles, but was only applied to benchmark functions with added gaussian noise.

Pugh et al. showed that PSO could outperform Genetic Algorithms on benchmark functions and for certain scenarios of limited-time learning under the presence of noise [14, 15]. Pugh also showed that PSO can perform satisfactorily with low population sizes, a result that is of particular interest for multi-robot implementations because a smaller number of robots can be used while leaving the optimization process robust to connectivity issues between the robots.

3 Materials and Methods

This paper is focused on a case study of obstacle avoidance, a basic behavior in robotics. Robots navigate autonomously in a square arena of 1 m^2 in which walls and other robots are the only obstacles. We use the same metric of performance as Floreano and Mondada [5], which consists of three factors, all normalized to the interval [0, 1] (Eq. 1).

$$F = V \cdot (1 - \sqrt{\Delta v}) \cdot (1 - i) \tag{1}$$

V is the average wheel speed, Δv the wheel speed difference, and i the proximity sensor activation value of the most active sensor. Each factor is calculated at each time step and then averaged for the total number of time steps in the evaluation period. This function rewards robots that move quickly, turn as little as possible, and spend as little time as possible near obstacles.

We chose the obstacle avoidance task because it is scalable in the number of robots, requires basic sensors and actuators that are available in most mobile robots, and the chosen performance metric can be fully evaluated with on-board resources. Thus, it can serve as a benchmark for testing distributed learning algorithms with real robots in the same way that standard benchmark functions are used in numerical optimization.

Our experimental platform is the Khepera III mobile robot, a differential wheeled vehicle with a diameter of 12 cm. The Khepera III is equipped with nine infra-red sensors as well as five ultrasound sensors for short and medium range obstacle detection. Our experiments were carried out in simulation using Webots [10], a realistic, physics-based, submicroscopic simulator that models dynamical effects such as friction and inertia. In this context, by submicroscopic we mean that it provides a higher level of detail than usual microscopic models, faithfully reproducing intra-robot modules (e.g., individual sensors and actuators).

The controller architecture is an artificial neural network of two units with sigmoidal activation functions, and a single output per unit to control the two motors. Each neuron has 12 input connections: the 9 infrared sensors, a connection to a constant bias speed, a recurrent connection from its own output, and a lateral connection from the other neuron's output, resulting in 24 weight parameters in total.

The adaptation algorithm is the noise-resistant variation of PSO introduced by Pugh et al. [15], which operates by re-evaluating personal best positions and aggregating them with the previous evaluations (in our case a regular average performed at each iteration of the algorithm). The movement of particle i in dimension j depends on three components: the velocity at the previous step weighted by an inertia coefficient w, a randomized attraction to its personal best $x^*_{i,j}$ weighted by w_p, and a randomized attraction to the neighborhood's best $x^*_{i',j}$ weighted by w_n (Eq. 2). $rand()$ is a random number drawn from a uniform distribution between 0 and 1.

$$v_{i,j} = w \cdot v_{i,j} + w_p \cdot rand() \cdot (x^*_{i,j} - x_{i,j}) + w_n \cdot rand() \cdot (x^*_{i',j} - x_{i,j}) \tag{2}$$

The neighborhood presents a ring topology with one neighbor on each side. Particles' positions and velocities are initialized randomly with a uniform distribution in the [-20, 20] interval, and their maximum velocity is also limited to that interval. The robots' pose (position and orientation in the arena) is randomized at the beginning of each evaluation. At the end of each optimization run, the best solution is tested with 40 evaluations of 30 s, and the final performance is the average of these final evaluations.

The total evaluation time for PSO depends on four factors: population size (N_p), individual candidate evaluation time (t_e), number of iterations of the algorithm (N_i), and number of re-evaluations of the personal best position associated with each candidate solution within the same iteration (N_{re}), as shown in Eq. 3.

$$t_{tot} = t_e \cdot N_p \cdot N_i \cdot (N_{re} + 1) \tag{3}$$

If we assume that there is a fixed upper limit for the total evaluation time, an increase in any of the parameters would result in a proportional decrease in the rest.

In a parallelized or distributed implementation, fitness evaluations are distributed among N_{rob} robots, and the wall-clock time t_{wc} required to evaluate candidate solutions is reduced (Eq. 4).

$$t_{wc} = t_e \cdot \left\lceil \frac{N_p}{N_{rob}} \right\rceil \cdot N_i \cdot (N_{re} + 1) \tag{4}$$

It is worth noting that the number of robots is not necessarily the same as the population size. In fact, the choice of the population size depends on the dimension of the search space and the complexity of the task, while the choice of number of robots is based on more experimental considerations (e.g., availability of robots, targeted number of robots needed for a specific mission in a given environment). Thus, a robot may have several particles to evaluate within the same candidate solution pool as opposed to only one.

A full optimization of the algorithmic parameters to minimize the total evaluation time would be a computationally very expensive problem, due to the large number of candidate configurations, the combination of continuous and discrete parameters, the large variation between runs, and the possible existence of local optima. Thus, our approach is to analyze each parameter individually, taking into account its impact on the final performance as compared to a full-time adaptation baseline.

Our baseline set of parameters, based on the work of Pugh et al [14], is shown in Table 1. This set of parameters amounts to a total evaluation time of approximately 417 hours if carried out on a single robot, what we refer to as full-time adaptation.

To complement our robotic case study and add more generality, we also perform runs on four traditional mono and multi-modal benchmark functions without noise: the sphere, Rosenbrock's, Rastrigin's, and Griewank's, defined in Eq. 5. The baseline parameters for the algorithm ran on benchmark functions are also shown in Table 1.

$$\begin{aligned}
f_1(\mathbf{x}) &= \sum_{i=1}^{D} x_i^2 \\
f_2(\mathbf{x}) &= \sum_{i=1}^{D-1} [(1 - x_i^2) + 100(x_{i+1} - x_i^2)^2] \\
f_3(\mathbf{x}) &= 10D + \sum_{i=1}^{D} [x_i^2 - 10\cos(2\pi x_i)] \\
f_4(\mathbf{x}) &= 1 + \frac{1}{4000}\sum_{i=1}^{D} x_i^2 - \prod_{i=1}^{D} \cos\left(\frac{x_i}{\sqrt{i}}\right)
\end{aligned} \tag{5}$$

Table 1 Algorithmic parameter values

Parameter	Obstacle Avoidance	Benchmark functions
Population size N_p	100	100
Iterations N_i	50	500
Evaluation span t_e	150 s	-
Re-evaluations N_{re}	1	0
Personal weight w_p	2.0	2.0
Neighborhood weight w_n	2.0	2.0
Dimension D	24	24
Inertia w	0.8	0.6
V_{max}	20	5.12

4 Results and Discussion

The results of this paper are presented as follows: Section 4.1 introduces the discussion with a comparison of the fitness as a function of the total evaluation time. In Sections 4.2 to 4.5 we analyze each of the four previously mentioned algorithmic parameters and propose guidelines for setting their values. Finally, in Section 4.6 we apply the proposed guidelines to reduce the total evaluation time and compare the results with the full-time adaptation.

4.1 Parameter Comparison

In the first place, we started from the total evaluation time baseline of 417 h and reduced N_p, N_i, and t_e individually to 5, 5, and 5 s respectively, while keeping the other two parameters at their baseline values, plotting the three curves in the same graph for better comparisons. We performed 100 independent runs for each set of parameter values and with 1, 2, 4, and 8 robots. When multiple robots were considered, all of them were learning in parallel. All runs were performed in a 1x1 m arena unless noted otherwise. The resulting fitness can be observed in Figure 1. In all cases, reducing the evaluation span t_e has the least impact on the resulting fitness, followed by N_p and N_i. When comparing the same total evaluation time across different numbers of robots, it can be noted that as the number of robots increases, the arena becomes more crowded and obstacle avoidance becomes harder, thus causing the average fitness to decrease. Also, performances are noisier (see larger error bars in lower right corner) and therefore there is less impact of a decreased N_p or N_i (flatter profile than with 1-2 robots). The following sections present a more detailed analysis of each individual parameter.

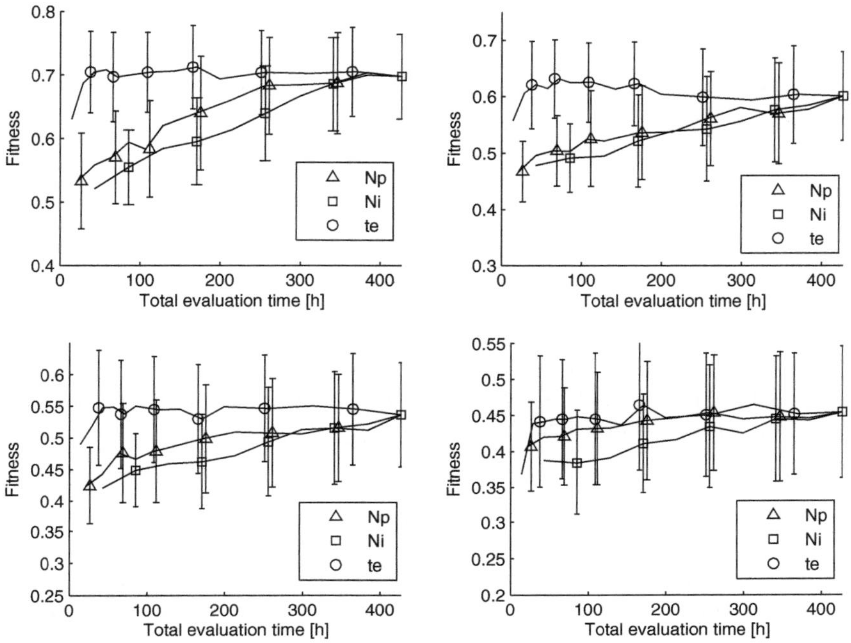

Fig. 1 Fitness as a function of total evaluation time for 1, 2, 4, and 8 robots respectively. Each curve represents the reduction of one individual parameter (population size, number of iterations, and evaluation span) with the others held constant.

4.2 Evaluation Span

To analyze the effect of the evaluation span, we reduced t_e from 150s to 5s, with 20s steps in the higher range and 5s steps in the low range. We did 100 runs for each value, and plotted the mean and standard deviation in Figure 2, left.

The mean fitness remains fairly constant for $30\,s < t_e < 150\,s$ for all number of robots. In fact, the difference in fitness between 150s and 30s is not statistically significant in all cases (Mann-Whitney U test, 5% significance level). For $t_e < 30\,s$ the fitness starts to decrease, although at different rates for different numbers of robots. In particular, for 8 robots t_e can be reduced to 10 s without a major change in fitness, which suggests that a crowded arena may allow for shorter evaluation spans due to more frequent collisions, and thus more opportunities to learn to avoid them. It is interesting to note that reducing the evaluation span does not seem to increase the fitness variation between runs.

The evaluation span parameter depends on the task and the environment. With the goal of trying to explain the lower limits for our task, we varied the evaluation span for several arena sizes using one robot (Figure 2, right). In this case, the point where performance starts to drop occurs at longer evaluation spans for larger arena sizes (15 and 25 seconds for 4 and 8 meters respectively). We suspect this point is

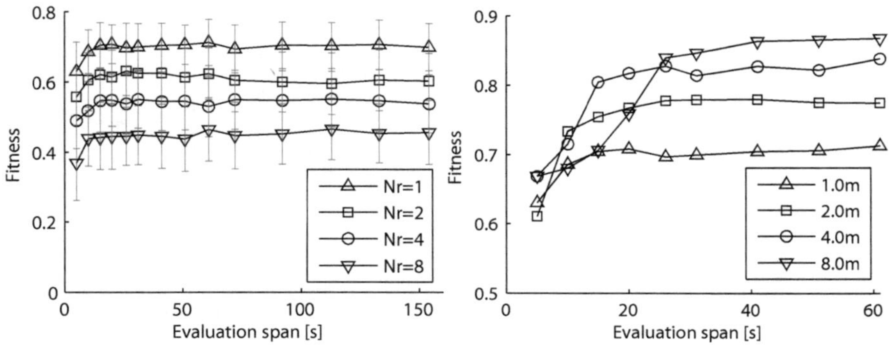

Fig. 2 Left: Mean fitness for different evaluation span values and number of robots in a 1x1 m arena. Error bars represent one standard deviation. Right: Mean fitness for different evaluation span values and arena sizes

related to the minimum time it takes to have at least one collision. In fact, if the robot moves in a straight line at a maximum speed of 0.25 m/s, it takes 16 and 32 seconds to cross one side of an arena of 4 and 8 meters respectively. Therefore, the robot speed and environment size can be used as guidelines to choose an evaluation span. We suggest that, in general, the minimum evaluation span should guarantee at least one interaction of the robot with other components of the environment relevant to the task at hand.

4.3 *Re-evaluations*

We then compared the performance of noise-resistant PSO with standard PSO to determine if re-evaluations improve performance in limited time scenarios. For any given set of parameters, noise-resistant PSO takes twice as much evaluation time as standard PSO due to the personal best re-evaluations. In order to perform a fair comparison, if we remove re-evaluations we need to double one of the other parameters to keep total evaluation time constant. We thus compared four alternatives: re-evaluations, doubled iterations, doubled evaluation span, and doubled population size. We performed 100 runs for each algorithmic variant and plotted the final fitness in Figure 3.

In the single robot case, noise-resistant PSO performed significantly worse than standard PSO with doubled iterations and doubled population size. However, as the number of robots is increased, the relative performance of noise-resistant PSO improves: for 2 robots there is no significant difference, and for 4 and 8 robots noise-resistant PSO significantly outperforms standard PSO with doubled iterations and doubled population size. It is worth noting that there is no significant difference between doubling population size and number of iterations for all number of robots, and that doubling the evaluation span performs significantly worse in all cases except for 8 robots.

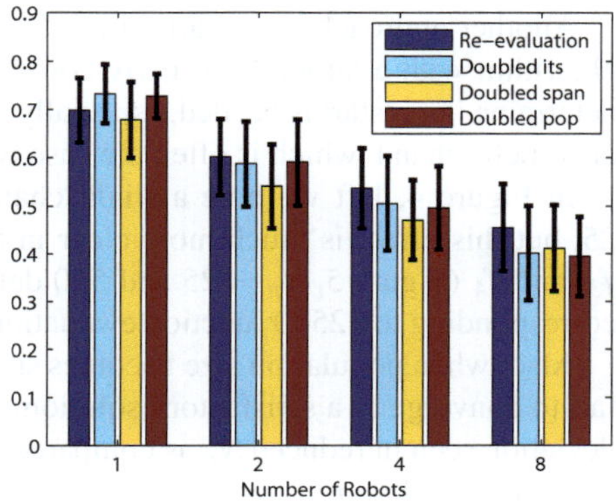

Fig. 3 Average fitness and standard deviation for noise-resistant PSO, standard PSO with doubled number of iterations, standard PSO with doubled evaluation span, and standard PSO with doubled population size

These results suggest that the decision to invest time in re-evaluations depends on the amount of uncertainty in fitness evaluations. In fact, as the arena becomes more crowded, there is more uncertainty in fitness evaluations, which depend both on the performance of other robots (avoiding other robots is easier if other robots are also trying to avoid you) and on initial conditions such as position in the arena (a robot is more likely to be trapped against a corner in a crowded arena).

Re-evaluations may also reduce the effect of heterogeneities on solution sharing in multi-robot evaluations, as shared solutions are re-evaluated at each iteration and thus can be dropped if they do not perform as well as they had done on other robots. The final advantage of re-evaluations can be seen in the case of dynamic environments, where a previously found good solution may no longer be valid after a certain amount of time. Thus, re-evaluations seem to be a good recommendation in general for multi-robot learning scenarios.

4.4 Population Size

Both for our robotic case study and the benchmark functions, the population size was reduced from 100 to 30 in steps of 10, and from 30 to 5 in steps of 5, in order to obtain more data points in what we expected to be an interesting region, while keeping all the other parameters the same as in the baseline. We did 100 independent runs for each value, results are shown in Figure 4, left and Figure 5.

It is clear from figures 1 and 5 that, at least with our baseline parameters, reducing the population size is better in terms of mean fitness than reducing the number of iterations, both for obstacle avoidance and for all benchmark functions[1]. Now the question that arises is how low should we set the population size? While there is no clear consensus in PSO literature [2], there are a few guidelines based on the dimension D of the search space such as $N_p = D$ or $N_p = 10 + 2\sqrt{D}$ [2].

[1] Note that in benchmark functions lower fitness values mean better performance

[2] This last formula is used in Standard PSO 2006, an effort to define a PSO standard published online in Particle Swarm Central `http://www.particleswarm.info`.

Another approach is to start with a fixed value such as $N_p = 40$ and restart the algorithm with a larger N_p if early convergence is noticed. However, it is hard to determine if a restart is needed, especially when the maximum feasible fitness is not clear beforehand, which is often the case when learning robotic behaviors.

In Figure 4, left we note a slight change in the fitness slope at around $N_p = 25$, but this effect is much more clear in the case of the benchmark functions f_2, f_3, and f_4 (Figure 5, $N_p = 25$ and 500 default iterations, as mentioned in Table 1, corresponding to 12500 function evaluations).

Also, when population size becomes small, more outliers appear due to runs that fail to converge to a satisfactory solution. This can be noted in the higher standard deviation seen in reduced N_p as compared to reduced N_i with the same total evaluation time (see Figure 1).

Thus, because of higher fitness, lower variance, the possibility to distribute particles among robots, and the impracticality of the restart process, we prefer to err on the side of larger population sizes, and we suggest the following guideline:

$$N_p = \max(D, N_{rob}) \tag{6}$$

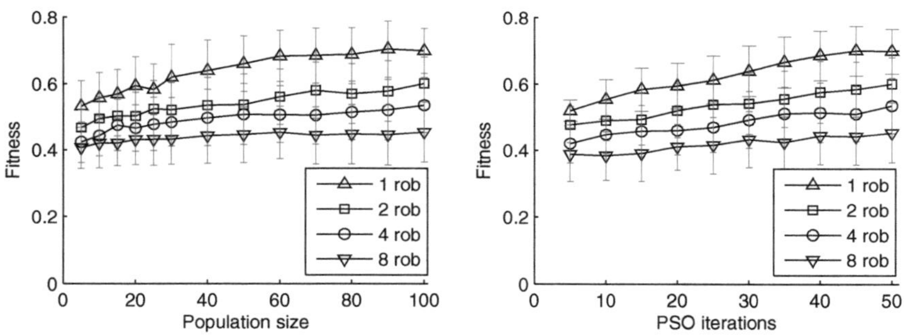

Fig. 4 Left: Mean fitness for different population size values. Right: Mean fitness for different number of iterations. Error bars represent one standard deviation.

4.5 PSO Iterations

For our robotic case study, the number of iterations was reduced from 50 to 5 in steps of 5, again keeping all the other parameters the same as in the baseline. We did 100 independent runs for each parameter value, results are shown in Figure 4, right. The chosen benchmark functions traditionally use larger values of N_i, so we chose 500 as a baseline and reduced it to 50 in steps of 50 (see Figure 5).

We observed a nearly linear performance drop for obstacle avoidance and on benchmark functions f_3 and f_4. For f_1 and f_2, the behavior of N_i was similar to that of N_p, but with a worse fitness overall.

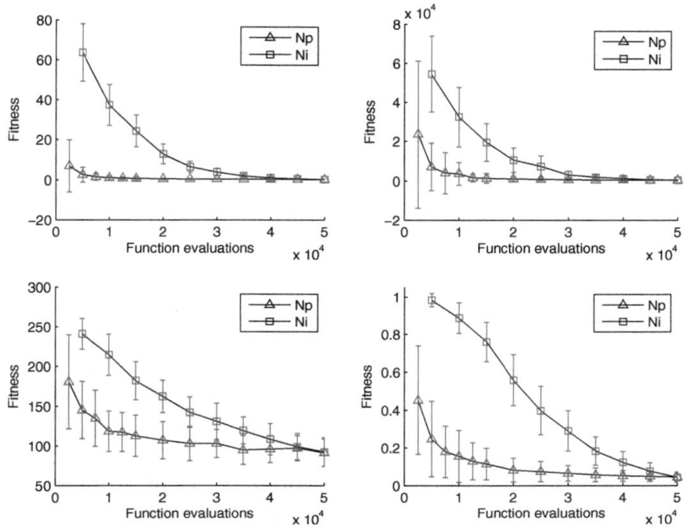

Fig. 5 Fitness as a function of number of evaluations for benchmark functions f1, f2, f3, and f4. Each curve represents the reduction of one individual parameter (population size and number of iterations) with the other held constant.

Given that N_i is the easiest parameter to adjust on the fly, this parameter seems suitable for trade-offs between performance and available learning time. That is, for a fixed available time t_{wc}, we suggest using the previous guidelines to determine the 3 other parameters and allocate all remaining time to N_i using Eq. 7, which is derived from Eq. 4.

$$N_i = \frac{t_{wc}}{t_e \cdot \left\lceil \frac{N_p}{N_{rob}} \right\rceil \cdot (N_{re} + 1)} \tag{7}$$

4.6 Limited Time Adaptation

For our limited time adaptation runs, we set a maximum total evaluation time of 8 h, which if distributed among 8 robots results in a wall-clock time of 1 h, about one third of the battery autonomy of our robots[3]. Following our proposed guidelines, we used $N_p = 24$, $t_e = 20$ s, $N_{re} = 1$, and $N_i = 30$, and run the adaptation process in simulation for $N_{rob} = \{1,2,4,8\}$ (see Figure 6).

The fitness difference between full-time and limited time adaptation is 17%, 17%, 14%, and 9% for 1, 2, 4, and 8 robots respectively. These values are relatively low considering the evaluation time was reduced more than 52 times. More importantly,

[3] Our autonomy is lower than that specified by the manufacturer due to additional modules such as an active tracking turret and a Wi-Fi card.

both limited and full-time adaptation converged to the same obstacle avoidance strategy, regardless of the number of robots: going back and forth between walls in a straight line, reversing the direction of motion every time an obstacle was detected. We verified the sameness of the solution strategies by analyzing the trajectories described by the robots, focusing on the step length and angle distributions as described in [4].

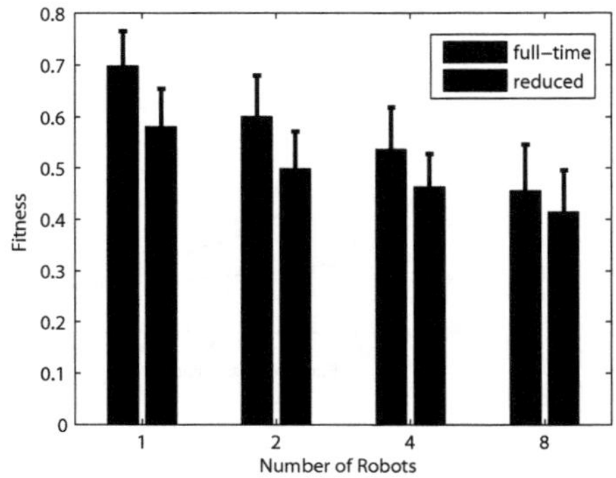

Fig. 6 Average fitness and standard deviation for full-time and limited-time adaptation

5 Conclusion

We analyzed the effect of the four PSO algorithmic parameters that determine total evaluation time for a case study of multi-robot limited time learning of an obstacle avoidance behavior. Each parameter was varied independently, and based on the resulting fitness we proposed guidelines to choose these parameters for the case of multi-robot distributed learning. To add more generality to our guidelines, we ran analogous tests on benchmark functions traditionally used for numerical optimization problems.

For the population size parameter, we suggested to use at least the dimension of the search space D, and to use the number of robots if it is greater than D to take advantage of parallel evaluations and increased robustness. We proposed using the robot speed and environment size as guidelines to choose an evaluation span that guarantees at least one interaction of the robot with other components of the environment relevant to the task at hand. Due to the inherent uncertainty in controller evaluations when using more than one robot, we showed that re-evaluations have a positive impact in multi-robot learning scenarios. The last parameter, the number of iterations, can be adjusted to fit the total evaluation time available.

By applying our guidelines, we were able to reduce the total adaptation time to an amount which can be easily completed without fully depleting the batteries of the individual robots. This resulted in a maximum quantitative performance drop of 17% but with no observable difference in the qualitative behaviors of the solutions. Even though we employed the PSO algorithm, we believe that the evaluation time issues and the guidelines presented in this paper are not limited to PSO and are relevant to population-based multi-robot learning in general.

Our next step is to validate these results with real robots. In particular, we intend to study the effect of asynchronous updates and dynamic neighborhood topologies on multi-robot learning. In the future, we hope to extend our analysis to tasks of increasing complexity, requiring a higher degree of coordination between robots. We are also interested in exploring PSO variations and other population-based algorithms that can be applied to limited-time distributed learning. Our final goal is to devise a set of general guidelines for fast, robust adaptation of high-performing robotic controllers.

Acknowledgements. This research was supported by the Swiss National Science Foundation through the National Centre of Competence in Research Robotics.

References

1. Akat, S.B., Gazi, V.: Decentralized asynchronous particle swarm optimization. In: IEEE Swarm Intelligence Symposium (2008), doi:10.1109/SIS.2008.4668304
2. Bratton, D., Kennedy, J.: Defining a Standard for Particle Swarm Optimization. In: IEEE Swarm Intelligence Symposium, pp. 120–127 (2007)
3. Chang, J., Chu, S., Roddick, J.: A parallel particle swarm optimization algorithm with communication strategies. Journal of Information Science, 809–818 (2005)
4. Di Mario, E., Mermoud, G., Mastrangeli, M., Martinoli, A.: A trajectory-based calibration method for stochastic motion models. In: IEEE/RSJ International Conference on Intelligent Robots and Systems, pp. 4341–4347 (2011)
5. Floreano, D., Mondada, F.: Evolution of homing navigation in a real mobile robot. IEEE Transactions on Systems, Man, and Cybernetics, Part B: Cybernetics 26(3), 396–407 (1996)
6. Hereford, J., Siebold, M.: Using the particle swarm optimization algorithm for robotic search applications. In: IEEE Swarm Intelligence Symposium, pp. 53–59 (2007)
7. Jin, Y., Branke, J.: Evolutionary Optimization in Uncertain Environments A Survey. IEEE Transactions on Evolutionary Computation 9(3), 303–317 (2005)
8. Kennedy, J., Eberhart, R.: Particle swarm optimization. In: IEEE International Conference on Neural Networks, vol. 4, pp. 1942–1948 (1995)
9. Marques, L., Nunes, U., Almeida, A.T.: Particle swarm-based olfactory guided search. Autonomous Robots 20(3), 277–287 (2006)
10. Michel, O.: Webots: Professional Mobile Robot Simulation. Advanced Robotic Systems 1(1), 39–42 (2004)
11. Pan, H., Wang, L., Liu, B.: Particle swarm optimization for function optimization in noisy environment. Applied Mathematics and Computation 181(2), 908–919 (2006)
12. Parsopoulos, K.E., Vrahatis, M.N.: Particle Swarm Optimizer in Noisy and Continuously Changing Environments. In: Hamza, M.H. (ed.) Artificial Intelligence and Soft Computing, pp. 289–294. IASTED/ACTA Press (2001)
13. Poli, R.: Analysis of the publications on the applications of particle swarm optimisation. Journal of Artificial Evolution and Applications 2008(2), 1–10 (2008)

14. Pugh, J., Martinoli, A.: Distributed scalable multi-robot learning using particle swarm optimization. Swarm Intelligence 3(3), 203–222 (2009)
15. Pugh, J., Zhang, Y., Martinoli, A.: Particle swarm optimization for unsupervised robotic learning. In: IEEE Swarm Intelligence Symposium, pp. 92–99 (2005)
16. Rada-Vilela, J., Zhang, M., Seah, W.: Random Asynchronous PSO. In: The 5th International Conference on Automation, Robotics and Applications, pp. 220–225 (2011)
17. Turduev, M., Atas, Y.: Cooperative Chemical Concentration Map Building Using Decentralized Asynchronous Particle Swarm Optimization Based Search by Mobile Robots. In: IEEE/RSJ International Conference on Intelligent Robots and Systems, pp. 4175–4180 (2010)

A Multi-Robot Cognitive Sharing System Using Audio and Video Sensors

Daniel McGibney, Ryo Morioka, Kosuke Sekiyama,
Hiro Mukai, and Toshio Fukuda

Abstract. In this paper, we present a multi-robot system that integrates a vision-based navigation system with a non-speech-based audio system for the purpose of sorting objects. Here we use two separate robotic systems each utilizing a different sensor type: video and audio. We propose using vision-based sensors with audio-based sensors because a single sensor such as a video sensor may not understand the entire environment. Additionally, robots can come to inaccurate conclusions based on their observations so we increase the accuracy of the system by interpreting the input from multiple robots using cognitive sharing. Therefore, cooperation of multiple robots using multiple sensors provides a better understanding of the environment. The results show the effectiveness of the multi-robot cognitive sharing system.

1 Introduction

Generally, robots find relevant information from objects using many different types of sensors and methods to get a better understanding of their environment and surroundings. In addition to using multiple sensors, information sharing among robots can also lead to the robots understanding their environment better.

Research has been done to classify and localize non-speech-based sounds. One notable workshop, the Classification of Events, Activities and Relationships

Daniel McGibney · Hiro Mukai
Washington University in St. Louis, St. Louis, MO 63130, USA
e-mail: {dpm2,mukai}@wustl.edu

Ryo Morioka · Kosuke Sekiyama · Toshio Fukuda
Nagoya University, Nagoya, 464-8603, Japan
e-mail: morioka@robo.mein.nagoya-u.ac.jp,
{sekiyama,fukuda}@mein.nagoya-u.ac.jp

M.A. Hsieh and G. Chirikjian (eds.), *Distributed Autonomous Robotic Systems*,
Springer Tracts in Advanced Robotics 104,
DOI: 10.1007/978-3-642-55146-8_28,

(CLEAR) 2007 workshop [10], had several teams compare results using different non-speech classification systems. There were 12 acoustic events of interest in the data sets. Examples of the acoustic events are steps, keys jingling, coughing and laughing. One result from the CLEAR 2007 workshop was a survey of sound classifiers. Some non-speech-based sound classification systems are described in [3], [18], [19], and [20]. The classification of sounds is also addressed by pattern recognition as in [14] and [4]. However, another important function that can be accomplished through the use of robots with audio sensors is to approximate the location of a sound. In [1], sound localization methods are discussed and compared.

In addition to analysis of audio, there is a large amount of research devoted to interpreting images and video. The openCV library has become a popular source of information where images and video are analyzed through various methods. In [2], the libraries are explained in detail. Additionally, the ARToolKit has also become a popular set for analyzing video, discussed in great length in [7].

Cognitive sharing involves identifying an object or event and sharing it with the other robots in the system. In [17], objects are identified and the information is shared among robots. The shared features are visual-based features including color, contour, and circularity using objects such as ball, rectangular prism, etc. Sharing of vision features among robots allows the robotic system to better understand the observed object. Cognitive sharing plays a large role in multi-robot cooperative tasks. In [13] and [12] tasks were simplified by using an RFID marker on the target. Although, it is not always necessary to have a marker, as in [6], since an environment can provide landmarks. In [6], a description of how cognitive sharing is handled by utilizing the relation between landmarks and an object. Robot cooperation through sharing is an area of research that will become more essential as the number of robots that play a role in everyday living grows.

In this paper, we extend the work in [8] and [17] and use cognitive sharing as a means to share the information from the audio-based system and the visual-based system. The combined system uses both audio-based and visual-based systems, and using cognitive sharing perform a cooperative multi-robot task.

The remaining sections of this paper are organized as follows. In the next section, section 2, we describe the robots' task. The following two sections, 3 and 4, describe the design of the audio-based system and the vision-based system respectively. Section 5 describes the multi-robot experiment and its results. In the final section, we make conclusions about the experiments.

2 Task Description

In our research, we integrated an audio-based system with a visual-based monitoring navigation system in order to perform a task. The audio-visual navigation system, which combines visual-based and audio-based robots, is designed to collectively detect and locate an object that enters the system, determine the object type, and transport the object to the appropriate sorting area.

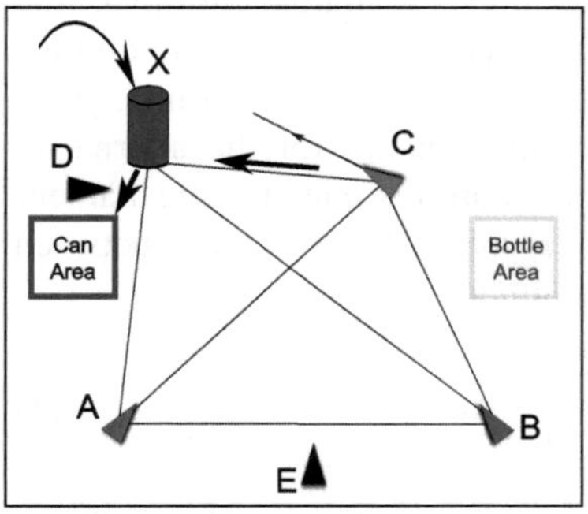

Label	Sensor	Function
A	Video	Monitoring robot
B	Video	Monitoring robot
C	Video	Carrier robot
D	Kinect	Audio robot
E	Kinect	Audio robot
X	(none)	Object

Fig. 1 The set up of the visual-based navigation system

In our experiment, five independent mobile robots work together to perform the task of multi-robot navigation as depicted in Fig. 1. First, an object, a blue can or a yellow bottle, is thrown into the experiment space. Once the object makes a sound, the audio robots (robots D and E) detect and locate the object. Then the audio robots classify the object based on the sound the object makes. Then the information is transmitted to visual-based robots A and B, which use their visual sensors to navigate robot C to the object. Then robot C grasps the object with its robotic arm and the visual robots search for the area to take the object. Finally, robot C takes the object to the appropriate sorting location.

3 Audio-Based System

The sharing of audio information is made possible by the individual audio robots which segment the audio signal, calculate features of those segments, select some features of interest, and classify the feature sets.

3.1 Segmentation and Features

After data collection, the system processes the audio data by segmenting it into smaller subsets from which features are calculated. The segmentation method chosen for the feature generation process was the Hamming window, a popular choice that is used in [14]. For the classification system in this paper we chose cepstrum filter bank coefficients, Mel-Frequency Cepstral Coefficients (MFCC) and Discrete Cosine Transform (DCT) coefficients. The aforementioned feature sets are a popular choice in speech based audio and are discussed in detail in [9].

To find the usefulness of the aforementioned features, the mutual entropy of the features was used as a measure of the discriminative capability. In order to do this, we estimate the probability distribution of a feature for a particular sound, $p(y)$. Then we estimate the probability distribution of that feature for the alternative sounds, $q(y)$. Both of these probability distributions are estimated using Parzen-window density estimation as in [15]. Next, a measure of the "distance" between the two distributions is given by the equation for mutual entropy:

$$D(p\|q) = \int_{-\infty}^{\infty} p(y) \log \frac{p(y)}{q(y)} dx. \tag{1}$$

If p_{ij} is the estimated probability distribution of the feature i and sound j, and q_{ij} is the estimated probability distribution of the feature i and the remaining sounds not including j, then we can find the mutual entropy

$$d_{ij} = D(p_{ij}\|q_{ij}) \tag{2}$$

for each unique sound type. Since mutual entropy can be thought of as a distance between two distributions, the resulting mutual entropies can be used as a means of finding features that discriminates different sounds. In particular, we define

$$d_{i,\max} = \max_j d_{ij} \tag{3}$$

as a means of evaluating the usefulness of a particular feature. Mutual entropy was also used in [18] and [19] as a means of finding the most useful features.

3.2 Classification

In order to classify a sound signal, we compare the sound signal with another sound signal to get a measure of similarity by using Dynamic Time Warping (DTW). From a sound signal, we segment the file and generate features for every segment. The result is a feature matrix A where

$$A = [a_1, a_2, \ldots, a_M] \tag{4}$$

with $a_1, a_2, \ldots, a_M$ representing M feature vectors. Similarly we define a feature matrix B from a separate sound signal with feature vectors $b_1, b_2, \ldots, b_M$ as

$$B = [b_1, b_2, \ldots, b_M]. \tag{5}$$

From these two different sets of features, we can calculate a similarity measure:

$$S_{ij} = \frac{a_i^T b_j}{\|a_i\| \|b_j\|}, \tag{6}$$

with indices i and j of the feature vectors from $1,2,\ldots,M$. Using this measure, we note that similar a_i and b_j result in $S_{ij} \rightarrow 1$. For the DTW process, we set

$$D_{ij} = 1 - S_{ij} \tag{7}$$

for each i and j. Then we recalculate for each matrix element using

$$D_{ij} = D_{ij} + \min(D_{i-1,j}, D_{i,j-1}, D_{i-1,j-1}) \tag{8}$$

recursively from $i = j = 1$ to $i = j = M$. Next, we take as the similarity measure

$$C = D_{M,M} \tag{9}$$

which denotes greater similarity when C is smaller. Using this similarity measure, we can classify feature matrix F as more similar to A or more similar to B.

3.3 *Localization*

Identifying the location of objects was done using two independent robots, both equipped with the Kinect sensor. During a short interval, the acoustic angle can be calculated. A built-in function of the Kinect software development kit calculates this angle using time difference of arrival with Kinect's four-microphone array. Using two Kinects at different locations allows us to calculate an angle from each Kinect. With both angles and the locations of the Kinects known, a simple calculation determines the location the sound originated from.

3.4 *Audio Robot Operation*

The audio-based system by which the two robots interact and share to classify and localize objects from audio information, brings together the auditory knowledge of the system. For this system, we use two audio robots both equipped with the Kinect for localization and classification of the sound.

The operation of the audio robots provides the time, acoustic angle, and object type to the audio-based system. Both robot programs begin by sampling the audio. When the sampled audio reaches an SNR threshold, an object is detected, and each robot begins to observe the audio data. While observing the audio data, each audio robot calculates its acoustic angle. The robots observe the angles and the audio independently for 8 intervals of 125 ms. After the observation, each robot produces angles which are used to calculate the coordinates of the sound. Also after the observation, each robot produces audio output of the sound, which is then classified. The classification process consists of segmenting, calculating features, and classifying the sounds. Fig. 2 further describes the operation of the audio robot.

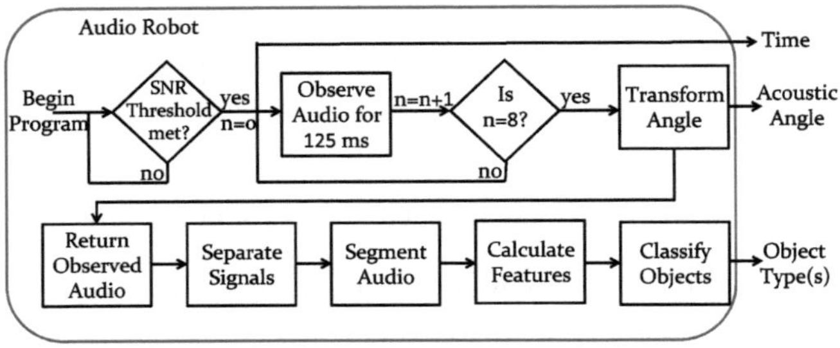

Fig. 2 Block diagram of operation for each audio robot

3.5 Cognitive Sharing of Audio Information

Several possible outcomes can occur with our system since the audio robots function independently. It is possible that after a sound is produced, neither robot detects the sound, only one robot detects the sound, or both robots detect the sound.

If neither robot detects the sound, it is a failure of the audio system. It is therefore advantageous to choose a low value for the SNR threshold to prevent such failures.

If only one robot makes a detection, the robot system classifies the observation. The classification vector $\mathbf{C}^i$ from robot i consists of elements C_j from (9) for $j = 1,2,...,N$ where N is the number of unique classifications:

$$\mathbf{C}^i = [C_1, C_2, ..., C_N]. \tag{10}$$

The vector $\mathbf{C}^i$ could possibly indicate that the sound was silence and therefore the detection was unnecessary. However, if the sound was not classified as silence, the detection is reported to the other robots in the system.

If a robot detects a sound within time Δt of another audio robot detection, the detections are assumed to be the same event. It is therefore the case that we have $\mathbf{C}^1$ from robot 1 and $\mathbf{C}^2$ from robot 2. Here we use the summation of the classification vectors, $\mathbf{C}^M$ as a multi-robot classification measure:

$$\mathbf{C}^M = \mathbf{C}^1 + \mathbf{C}^2. \tag{11}$$

We use the index of the minimum value of $\mathbf{C}^M$ as the object type since the lower the value of C means greater similarity.

3.6 Results

The classification methodology was tested using 10 objects and silence in a quiet office-type setting similar to where the data was collected. We used 30 sound data signals for each object, 20 of which were used for training and 10 for testing. Using these classification methods we were able to correctly classify the objects 92% of

Fig. 3 The 10 objects classified in the system; 4 plastic bottles, 4 metal cans, and 2 cardboard objects are displayed from left to right

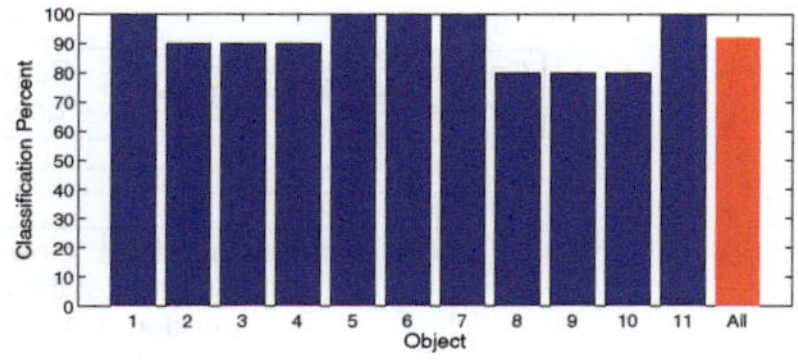

Fig. 4 The percent of correctly classified objects by the object type; the object number in Fig. 4 corresponds to the order number of the object in Fig. 3

the time. Fig. 3 displays the 10 objects, and Fig. 4 graphs the percent of correctly classified objects in the test data set by object type where the 11th object is silence.

The localization technique was tested using 30 trials in which an object was thrown into an experiment space of 1.524 meters by 1.524 meters. The robots were able to identify the location within 0.3 meters 80% of the time.

To test the audio-based system, it was desired that the audio robots could detect, identify, and localize the object of interest when an object was thrown into the reception area. The objects were not always detected which required further tweaking of the SNR threshold. Next, we used a different floor for the test data so that classification would be more difficult. The results using a single robot were 80% correct classification, but, when there was an agreement that an object was detected and the vector $\mathbf{C}^M$ was used, the classification accuracy improved to 87%. However, it should be noted that when only one audio robot makes a detection, the localization method cannot provide the localization of the object.

4 Visual-Based System

4.1 Visual Features

A visual sensor can provide a lot of information, such as the angle between a camera and an easily recognized marker. Once a marker is identified, we use the information to get a transformation matrix from the marker. The transformation matrix consists of a component of rotation and a component of translation. From the transformation matrix, we can calculate the angle between the camera and the marker.

Another popular feature type that the visual-based system uses is color. If we can find the contour of an object and identify the color, we can sort different objects based on those features. In order to get the color feature, we take the input from the image, the image is given a binary mask, the image is converted to grayscale, the colors are smoothed, and finally contour extraction is performed on the image. The resulting information gives us the number of colors, the position, and area of the color in the image. The color feature extraction process is shown in Fig. 5.

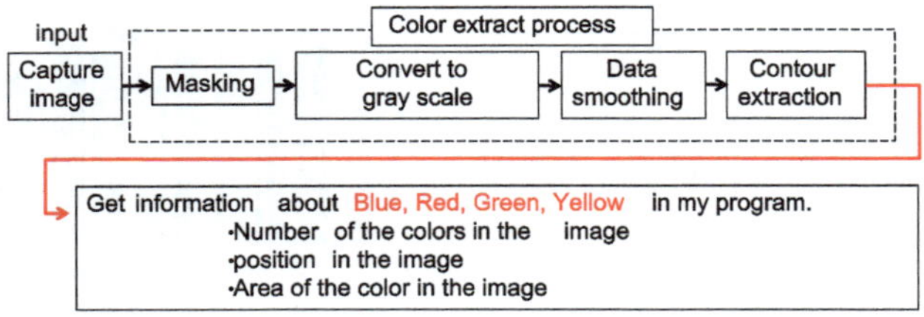

Fig. 5 Depiction of color extraction process

4.2 *Cooperative Vision-Based Navigation*

In order to navigate the carrier robot to the object, the monitoring robots observe the pan angles θ_{BC} and θ_{CA} from the location of robot C. The monitoring robots share their observations with the carrier robot. From these observations, the carrier robot calculates θ_{AB}, **a**, and **b**. Additionally, the angle Φ is observed by the carrier robot. The navigation vector, **n**, is given by

$$\mathbf{n} = \mathbf{a} + \mathbf{b}. \tag{12}$$

Next, the carrier robot calculates the yaw angle, δ, resulting from the navigation vector. Then, the carrier robot adjusts so that it is facing at angle δ. Once the aim of the robot is adjusted, the robot simply moves straight for a short duration. When the carrier robot finishes moving, the robots repeat the process of collecting observations and calculating the yaw angle. If the carrier robot detects an object within its grasp, then the navigation process is complete. The visual-based navigation system is depicted in Fig. 6.

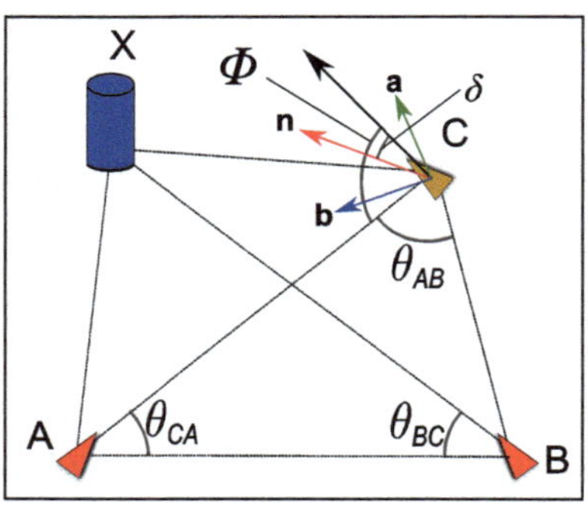

Notation	Meaning
θ_{BC}	Pan angle given by A
θ_{CA}	Pan angle given by B
θ_{AB}	$180^o - \theta_{BC} - \theta_{CA}$
Φ	Angle given by C
a	Direction vector given by A
b	Direction vector given by B
n	Navigation vector
δ	Yaw angle

Fig. 6 The set up of the visual-based navigation system

5 The Audio-Visual System

Here we present experiments on the audio-visual system which shows the usefulness of using multiple sensor types and sharing the information.

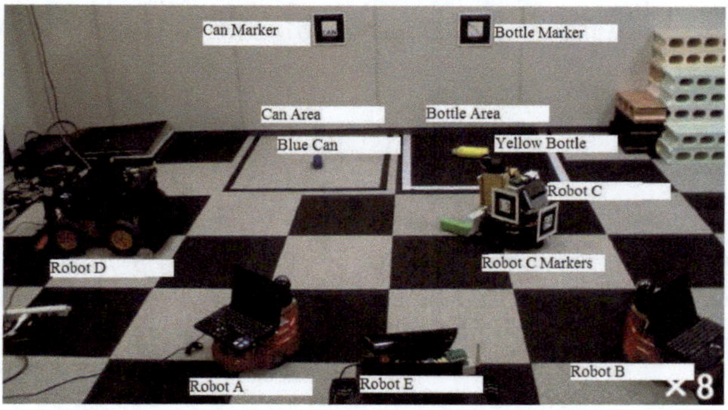

Fig. 7 Actual experiment space

5.1 Experiment Design

The experiments were designed to add benefit to the sorting process at a recycling center making use of two distinct objects, a bottle and a can. Sorting objects is an essential task for recycling valuable resources. We design our experiments to test a robot's ability to identify objects and to locate the object using only sound data. Fig. 7 shows the actual experiment space. To add the usefulness of vision-based features, we color the can blue and the bottle yellow. Therefore our system relies upon both the color of the objects and the sounds produced.

5.2 Results

The visual-based navigation system was able to perform its function as desired. First, we placed an object into the center of the experiment space and activated the system. The system was able to locate, move towards, and grasp the object using visual input. Once the object was within the carrier robot's grasp, the robot was then able to place the object in its proper sorting area.

The combined audio-visual navigation system was also tested, and was able to work efficiently most of the time. The initial classification and localization data was given to the system via the audio robots. The initial localization data was able to reduce some of the time that the system required to find the object, however, the classification only added marginal benefit because of the high accuracy of the visual-based system. Fig. 8 shows the experiment during major system events. In Fig. 8a, a yellow bottle makes a sound. The sound is received by the audio robots which process the sound and output the location and object type. The observations from the audio robots are brought together and interpreted by the vision monitoring robots. The vision monitoring robots send the location of the bottle to the carrier

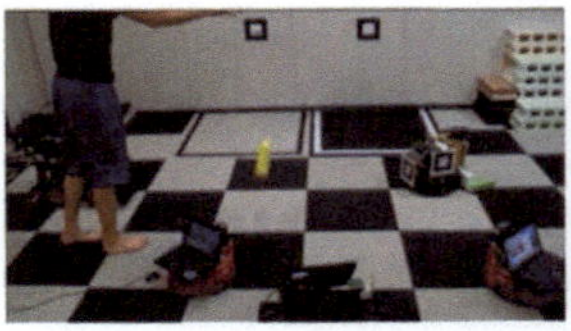

(**a**) Time t=0.2 seconds

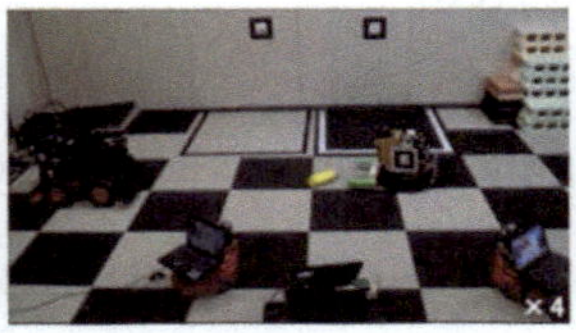

(**b**) Time t=6 seconds

(**c**) Time t=12 seconds

(**d**) Time t=14 seconds

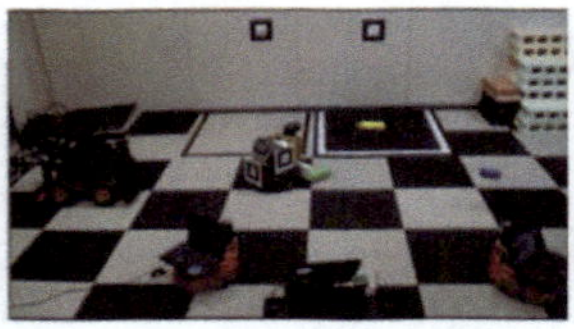

(**e**) Time t=15 seconds

(**f**) Time t=26 seconds

(**g**) Time t=32 seconds

(**h**) Time t=34.8 seconds

Fig. 8 The experiment space at different times

robot. Then, as shown in Fig. 8b, the carrier robot moves toward the bottle based on the inputs of the monitoring robots and the carrier robot. Fig. 8c shows the bottle being grasped by the carrier robot. Next, in Fig. 8d the bottle is taken to the bottle area. Then the system process is repeated for the next object. In Fig. 8e, the blue can is dropped into the reception area. In Fig. 8f, the carrier robot moves toward the

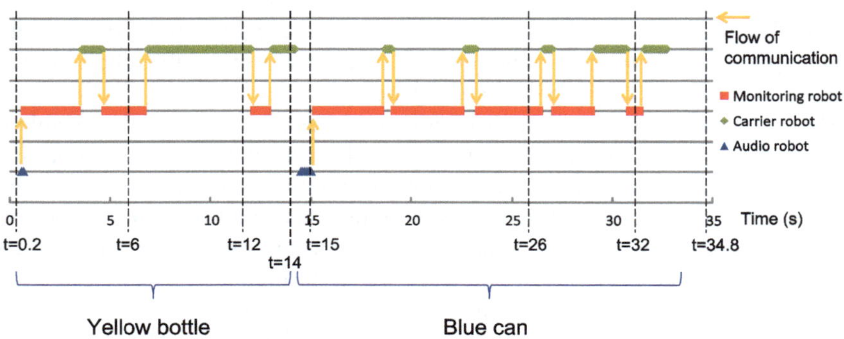

Fig. 9 Operation time graph

can. Fig. 8g shows the can being grasped by the carrier robot. Next, in Fig. 8h the can is taken to the can area.

We display an operation time graph in Fig. 9 that indicates which robot is in operation throughout the span of the experiment. The snapshots from Fig. 8 correspond to the times indicated by the dashed lines in Fig. 9.

6 Conclusion

In this paper, we presented a multi-robot system that uses cognitive sharing of information from both visual and auditory sensors. The goal of this design was to provide a novel idea for sharing features in a multi-robot system. The method consists of two independent multi-robot systems, one which uses only audio input and the other which uses only visual input. Separately, the two independent systems serve different purposes. We were able to combine the two systems in a way that allowed both systems to work cooperatively. Sharing methods can help a system perform a task better than a single robot. In this system we integrated two independent systems that shared their results for a common task. Robot sharing and cooperation can be used in several other important areas such as the operation of an unmanned autonomous aircraft or the operation of a self-driving automobile. Both instances focus on a single robot but the action taken by these robots depends upon input from surroundings which is often from other similar robots.

Acknowledgements. Part of this research is supported by MEXT/JSPS KAKENHI Grant Number 22500174 and the National Science Foundation (NSF) under grant number DGE-0538541. This document was created with help from Seema Dahlheimer at the Engineering Communication Center in Washington University in St. Louis.

References

1. Badali, A., Valin, J., Michaud, F., Aarabi, P.: Evaluating Real-time Audio Localization Algorithms for Artificial Audition in Robotics. In: Proceedings of the IEEE/RSJ International Conference on Intelligent Robots and Systems (IROS), pp. 2033–2038 (2009)
2. Bradski, G.R., Kaehler, A.: Learning OpenCV: Computer vision with the OpenCV library. O'Reilly Press, Cambridge (2008)
3. Cowling, M.: Non-speech environmental sound classification system for autonomous surveillance. Ph.D. Thesis, Griffith University, Gold Coast Campus, School of Information Technology (2004)
4. Duda, R.O., Hart, P.E., Stork, D.G.: Pattern Classification. Wiley Interscience (2004)
5. Ellis, D.: Dynamic Time Warp (DTW) in Matlab (2003), `http://www.ee.columbia.edu/dpwe/resources/matlab/dtw/`
6. Gohring, D., Homann, J.: Multi Robot Object Tracking and Self Localization Using Visual Percept Relations. In: Proceedings of IEEE/RSJ International Conference of Intelligent Robots and Systems, pp. 31–36 (2006)
7. Hashimoto, T.: Birth in the real world 3D Character! Introduction to Programming ARToolKit Augmented Reality. ASCII Media Works (2008)

8. McGibney, D., Umeda, T., Sekiyama, K., Mukai, H., Fukuda, T.: Cooperative Distributed Object Classification for Multiple Robots with Audio Features. In: IEEE International Symposium on Micro-NanoMechatronics and Human Science, pp.134–139 (2011)
9. Slaney, M.: Auditory Toolbox. Interval Research Corporation, version 2 (1998), https://engineering.purdue.edu/~malcolm/interval/1998-010/AuditoryToolboxTechReport.pdf
10. Stiefelhagen, R., Bernardin, K., Bowers, R., Garofolo, J., Mostefa, D., Soundararajan, P.: The CLEAR 2006 Evaluation. In: Stiefelhagen, R., Garofolo, J.S. (eds.) CLEAR 2006. LNCS, vol. 4122, pp. 1–44. Springer, Heidelberg (2007)
11. Stiefelhagen, R., Bernardin, K., Bowers, R., Rose, R.T., Michel, M., Garofolo, J.S.: The CLEAR 2007 Evaluation. In: Stiefelhagen, R., Bowers, R., Fiscus, J.G. (eds.) RT 2007 and CLEAR 2007. LNCS, vol. 4625, pp. 3–34. Springer, Heidelberg (2008)
12. Takahashi, J., Sekiyama, K., Fukuda, T.: Cooperative Object Tracking with Mobile Robotic Sensor Network. Distributed Autonomous Robotic Systems 8, 51–62 (2008)
13. Tan, K.G., Wasif, A.R., Tan, C.P.: Objects Tracking Utilizing Square Grid Rfid Reader Antenna Network. Journal of Electromagnetic Waves and Applications 22(12), 27–38 (2008)
14. Theodoridis, S., Koutroumbas, K.: Pattern Recognition. Academic Press (2009)
15. Theodoridis, S., Pikrakis, A., Koutroumbas, K., Cavouras, D.: Introduction to pattern recognition: a MATLAB approach. Academic Press (2010)
16. Turetsky, R., Ellis, D.: Ground-truth transcriptions of real music from force-aligned MIDI syntheses. In: 4th International Conference on Music Information Retrieval, pp. 135–141 (2003)
17. Umeda, T., Sekiyama, K., Fukuda, T.: Vision-Based Object Tracking by Multi-Robots. Journal of Robotics and Mechatronics 24(3), 531–539 (2012)
18. Zhou, X., Zhuang, X., Liu, M., Tang, H., Hasegawa-Johnson, M., Huang, T.: HMM-based Acoustic Event Detection with AdaBoost Feature Selection. In: Stiefelhagen, R., Bowers, R., Fiscus, J.G. (eds.) RT 2007 and CLEAR 2007. LNCS, vol. 4625, pp. 345–353. Springer, Heidelberg (2008)
19. Zhuang, X., Zhou, X., Hasegawa-Johnson, M.A., Huang, T.S.: Real-world acoustic event detection. Pattern Recognition Letters 31(12), 1543–1551 (2010)
20. Zieger, C.: An HMM Based System for Acoustic Event Detection. In: Stiefelhagen, R., Bowers, R., Fiscus, J.G. (eds.) RT 2007 and CLEAR 2007. LNCS, vol. 4625, pp. 338–344. Springer, Heidelberg (2008)

Conditional Random Fields for Behavior Recognition of Autonomous Underwater Vehicles

Michael Novitzky, Charles Pippin, Thomas R. Collins,
Tucker R. Balch, and Michael E. West

Abstract. This paper focuses on multi-robot teams working cooperatively in an underwater application. Multi-robot teams working cooperatively to perform multiple tasks simultaneously have the potential to be more robust to failure and efficient when compared to single robot solutions. One key to more effective interaction is the ability to identify the behavior of other agents. However, the underwater environment presents specific challenges to teammate behavior identification. Current decentralized collaboration approaches, such as auction-based methods, degrade in poor communication environments. Sensor information regarding teammates can be leveraged to perform behavior recognition and task-assignment in the absence of communication. This work illustrates the use of Conditional Random Fields (CRFs) to perform behavior recognition as an approach to task monitoring in the absence of robust communication in a challenging underwater environment. In order to demonstrate the feasibility of performing behavior recognition of an AUV in the underwater domain, we use trajectory based techniques for model generation and behavior discrimination in experiments using simulated trajectories and real sonar data. Results are presented with comparison of a CRF method to one using Hidden Markov Models.

1 Introduction

Multi-robot teams, in comparison with single robot solutions, can offer solutions that are more economical, robust to failure, and more efficient than single robot

Michael Novitzky · Tucker R. Balch
College of Computing, Georgia Institute of Technology, 801 Atlantic Drive,
Atlanta, Georgia 30332
e-mail: misha@gatech.edu, tucker@cc.gatech.edu

Charles Pippin · Thomas R. Collins · Michael E. West
Georgia Tech Research Institute, 430 N. 10th St. Atlanta, GA 30332
e-mail: first.last@gtri.gatech.edu

M.A. Hsieh and G. Chirikjian (eds.), *Distributed Autonomous Robotic Systems*,
Springer Tracts in Advanced Robotics 104,
DOI: 10.1007/978-3-642-55146-8_29,

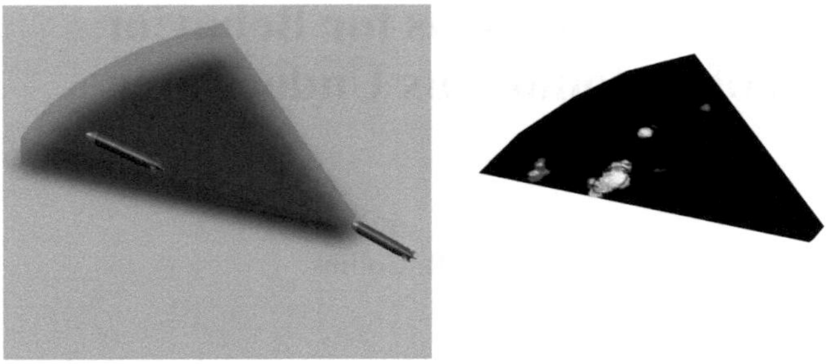

Fig. 1 An AUV using its forward-looking sonar to *Track & Trail* a leader AUV in order to perform behavior recognition

solutions [8, 6]. A team of robots can work on tasks in parallel, perform distributed sensing and operate in multiple locations at once. Furthermore, multiple robots add redundancy to the system. Unfortunately, a tradeoff is that these teams must communicate and work together with the added uncertainty regarding the behaviors of robots. For instance, a team member may have trouble cooperating due to communication errors, because they are busy performing other tasks, or have conflicting goals [2]. Many different methods for performing distributed cooperation exist, including centralized optimization algorithms and game theoretic techniques. A centralized method requires at least one agent or a home base to make task/role assignments. Although this may be optimal when communication links are reliable, its efficacy degenerates with intermittent communication, and a central point of failure makes the whole system come to a halt. Thus, a decentralized approach is much more viable as it is more robust to failures of communication. Auction-based algorithms generally have low communication requirements (where agents coordinate tasks through bid messages). Therefore, they are well suited to environments with communication constraints. Auctions can perform computations in parallel and the methods take advantage of the local information known to each agent [7, 9].

However, this method can still degrade in overall efficiency as communication deteriorates [14]. Such poor communication environments are encountered by autonomous underwater vehicles (AUVs) as acoustic transmissions suffer from surface reflections, bottom reflections, ambient noise, and noise sources within the water column, such as emissions from other vessels. Sotzing and Lane [15] have demonstrated that using teammate prediction improves overall performance of a cooperative AUV system. Such a system still needs communication to be of relatively good quality, as without a sufficient amount of communication the system degrades as predictions accrue error over time without correction from teammate communication.

The ultimate purpose of this research is to create a system that can efficiently operate with as little explicit communication as possible as this is the type of environment our own AUV will encounter [18]. We envision a system similar to that proposed by Novitzky [11] which will utilize auction-based methods along with prediction of teammate tasks during periods of poor communication. If the confidence in a prediction of a teammate's task is low, then an AUV can perform prediction verification through behavior recognition, as suggested in [3]. Additionally, this handles the situation where an AUV may have been assigned a task only to discover another agent already performing the task but not communicating.

2 Related Work

This work focuses on behavior recognition of autonomous mobile robots, specifically in the underwater domain. Baxter et al. [3] performed behavior recognition using HMMs on post-mission analysis of self-localization provided by an AUV. The post-mission analysis converted GPS pose trajectories to low-level actions such as *track-west* and *left u-turn east*. The main drawback of this method is that it claimed to be agnostic to the environment yet still required the use of cardinal direction, which in itself is still somewhat constraining to the compass orientation within an environment. The authors improved upon their discretization methods in [4] where they also enhanced HMMs to deal with behaviors of variable length. They began with AUV location information from simulated sonar data. These trajectories were fed into maneuver recognition algorithm capable of identifying an AUV's actions such as *straight* and *veer-left* thus making it more environmentally agnostic. While the authors were searching for top-level goals such as *mine-countermeasure* (MCM), *mine-countermeasure inspection* (MCMI), and *point inspection* (PI), they further divided the top level goals into sub-goals which included *dive*, *track*, *right u-turn*, and *left u-turn* along with *inspection*. Their results also included that top-level goals are achieved via the AUV performing sub-goal behaviors.

More recently, Novitzky et al. [12] performed exploratory work using HMMs to discriminate a small number of robot behaviors. The authors successfully applied their technique to a very limited amount of real data. Their data consisted of collected trajectories gathered while unmanned aerial vehicles (UAVs) performed search and track behaviors of surface targets and UUV trajectories collected via a forward-looking sonar. However, their data consisted of only a few behaviors and at most a handful of tracks for each behavior.

Of specific importance to this work is that performed by Vail et al. [17] in which the authors compared the accuracy of CRFs and HMMs for activity recognition on robot systems. Their chosen domain was simulated robot *Tag*. In their simulation, two robots were passively moving from waypoint to waypoint while a third was the *Seeker* searching for a robot to *Tag*. As part of the analysis of CRFs and HMMs, the authors tested the accuracy with different observations such as raw positions only, including velocities, and chasing features. The authors also examined the effect of incorporating features which violate the independence assumptions between observations. The results showed that a discriminatively trained CRF performed as well as or better than an HMM in their robot *Tag* domain.

Vail and Veloso [16] used CRFs for multi-robot domains. The authors experimented with two approaches to feature selection: grafting, and l_1 regularization. They applied these methods to data recorded during RoboCup soccer small-size

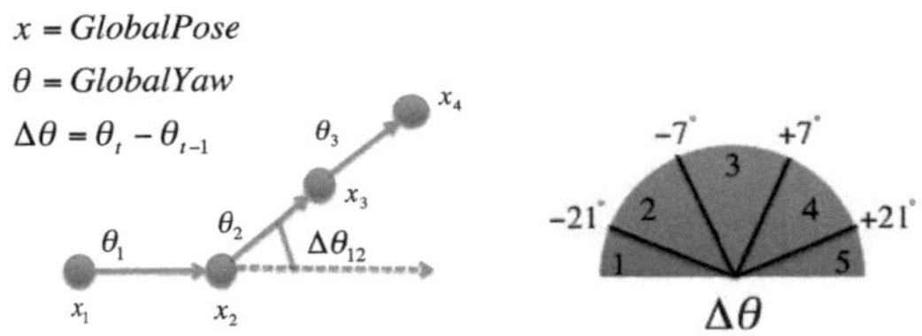

(a) Location to $\Delta\theta$. (b) $\Delta\theta$ discretization.

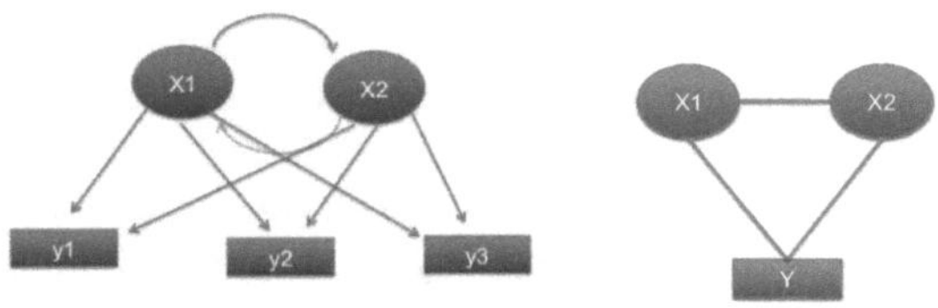

(c) Hidden Markov Model (HMM) (d) Conditional Random Field

Fig. 2 In (a) an AUV's location over time is used to determine its global yaw. The change in global yaw from one time step to the next is encoded as an integer value which represents a given range, as seen in (b). The two behavior recognition methods are Hidden Markov Models and Conditional Random Fields, as seen in (c), and (d) respectively.

league games. The goal of their work was to create a classifier that can provide useful information to robots that are playing against a team whose roles are being classified. They found that using feature selection can dramatically reduce the number of features required by CRFs to achieve error rates that are close to or identical to the error rate achieved by the model with its full complement of features. Reducing the number of features dramatically speeds up online classification and training.

Unlike previous behavior recognition work in the AUV domain, this work does not rely on trajectories provided through post-mission analysis nor only through simulation. Furthermore it performs behavior recognition of an AUV through the use of a simple discretization method, resulting in only one feature, on both simulated trajectories and actual sonar data comparing the results of a method using a CRF and a method using HMMs.

3 Trajectory Discretization

The encoding method used is agnostic to any environment. The only measurement required is the location $x = (x,y)$ coordinates of an AUV in a fixed 2D plane, as seen in Fig. 2a. The motion model of the AUV is assumed to be non-holonomic and always moving with a forward motion similar to a tricycle model. The yaw of the AUV is calculated from the vector of motion from one time-step to the next.

$$\Delta x_{(t-1,t)} = x_t - x_{t-1} \tag{1}$$

$$\theta_t = \arctan(\Delta x_{(t-1,t)}) \tag{2}$$

$$\Delta\theta_t = \theta_t - \theta_{t-1} \tag{3}$$

The encoding used in this research is the change in yaw between time steps. Possible changes in yaw are discretized according to bins. Each bin corresponds to a range of values. Bin 3, for example, represents a change in yaw between -7 and 7 degrees. As seen in Fig. 2b, an AUV moving straight ahead is observed as having a 0° change in yaw and thus encoded as a *3* while one turning by $-15°$ is encoded as a *2*. A series of these encodings are combined into a trajectory string for input into the Hidden Markov Model (HMM) or the Conditional Random Field (CRF).

4 Discrimination Methods

In general, observations are labeled as $Y = \{y_1,...,y_T\}$ and states are labeled as $X = \{x_1,..,x_T\}$, where the index represents successive time steps. In our domain y_t contains an integer value of the change in yaw of the AUV, described above. In the HMM method each hidden state x may not have an explicit definition. In the CRF method the labels x_t are drawn from one of the three behaviors.

4.1 Hidden Markov Model

In this research each behavior is modeled using a separate Hidden Markov Model (HMM). Each HMM is first trained on example trajectories of a specific behavior. The trained HMM is then given test trajectories to determine the log-likelihood that the test trajectory was generated by that behavior.

4.1.1 Training

The Hidden Markov Model (HMM), as seen in Fig. 2c, is composed of hidden states and observable states [13]. In Fig. 2c the hidden states are labeled with $x_1...x_n$ while the observation states are labeled $y_1...y_n$. A random process can be in any one of the hidden states and can emit any one of the observable states. In this work the observable states consist of the labeled changes in yaw, $\Delta\theta$. The number of hidden states are empirically determined. An HMM must learn the transition probabilities between hidden states and the probabilities that a hidden state may produce an observation. The Baum-Welch algorithm estimates the maximum likelihood of the parameters when given a corpus of training data.

4.1.2 Testing

Testing an HMM trained on a behavior is produced by the forward algorithm. An HMM can be used to determine the negative log-likelihood that a test trajectory instance was produced by the behavior it was trained upon [13]. A trial consists of an instance of a behavior trajectory being tested against each possible HMM. At each trial the HMM producing the maximum negative log-likelihood is determined as the representative behavior of the trial. If the representative behavior matches the true test instance label then it is logged as a positive identification. The accuracy of each trained HMM is the number of positive identifications over the entire corpus of similarly labeled instances.

4.2 Conditional Random Field

As seen in Fig. 2d, Conditional random fields (CRFs) are undirected graphical models for structured classification [10]. CRFs are built from a vector of weights and a vector of features. Features take the form $f_i(t, x_{t-1}, x_t, Y)$ where i is an index into the feature vector f and t is an offset into the sequence, x_{t-1} and x_t are values of the label pair at time $t-1$ and t, respectively. Y represents the entire observation sequence across all values of t.

4.2.1 Training

Training of CRFs is performed by finding a weight vector w^* that maximizes the conditional log-likelihood of labeled training data:

$$l(X|Y;w) = w^T f(t, x_{t-1}, x_t, Y) - log(Z_Y) \tag{4}$$

$$w^* = \arg\max_y l(X|Y;w) \tag{5}$$

4.2.2 Testing

The conditional probability of a label sequence given an observation sequence is computed from the weighted sum of the features as:

$$P(X|Y) = \frac{1}{Z_Y} \prod_{t=1}^{T} exp(w^T f(t, x_{t-1}, x_t, Y)) \tag{6}$$

$$Z_Y = \sum_{X'} \prod_{t=1}^{T} exp(w^T f(t, x'_{t-1}, x'_t, Y)) \tag{7}$$

The most likely behavior label x is assigned to each observation for each test instance presented to the trained CRF.

5 Experiments

5.1 Stationary Observer

The experiments were first performed using trajectory data gathered through simulation and then using a stationary forward-looking sonar. In order to test our method

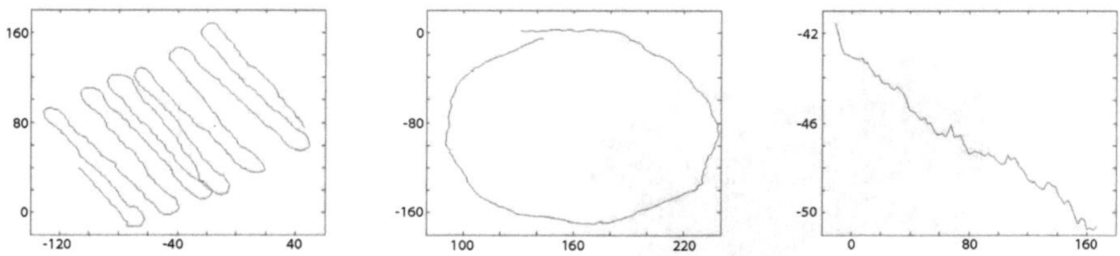

(a) Noisy *SearchPattern* (b) Noisy *Loiter* (c) Noisy *GoToWaypoint*

Fig. 3 Noisy versions of the template trajectories are depicted in (a), (b),and (c)

with two AUVs, trajectory data was gathered in simulation with one *Tracking & Trailing* a leader vehicle.

5.1.1 Simulation

MOOS-IvP is used to generate the simulated trajectory data [5]. The behaviors *GoToWaypoint*, *Loiter*, and *SearchPattern* are run within iMarineSim and viewed through pMarineViewer. The locations of the AUVs are recorded as each behavior is executed, providing a template trajectory. In order to create more realistic results, the template trajectories undergo rotation and translation transformations and are injected with Gaussian noise. Variations of each behavior trajectory template are created as they undergo a random assignment of transformations, including changes in rotation and translation along with an injection of cumulative Gaussian noise with random assignments of standard deviation, as seen in 3a, 3b, and 3c. This will demonstrate that our methods are agnostic to the environment as they are robust to rotations and translations and environmental noise. The global change in yaw for these experiments is discretized into seven bins with a spread of four degrees per bin.

5.1.2 Real Sonar Data

For our real sonar data experiments, a surrogate vehicle called the YellowRay ROV is used instead of the Yellowfin AUV due to space limitations in our testing tank[1]. The testing tank is 7.62 meters deep, 7.62 meters wide, and 10.36 meters long. The YellowRay is a Video Ray ROV [1] which has been modified to act as a viable surrogate of the Yellowfin. This includes the addition of Yellowfin subsystems such as the WHOI acoustic micro-modem and a BlueView forward looking sonar, as seen in Fig. 4a. The experiments were conducted with a BlueView forward-looking sonar positioned statically in a corner while it recorded the location of a human piloted YellowRay ROV. Throughout the experiment the YellowRay ranged between 1 to 10 meters from the BlueView sonar. For these experiments the perception algorithm

[1] Acoustic Water Tank Facility, Woodruff School of Mechanical Engineering, Georgia Tech

(a) YellowRay ROV

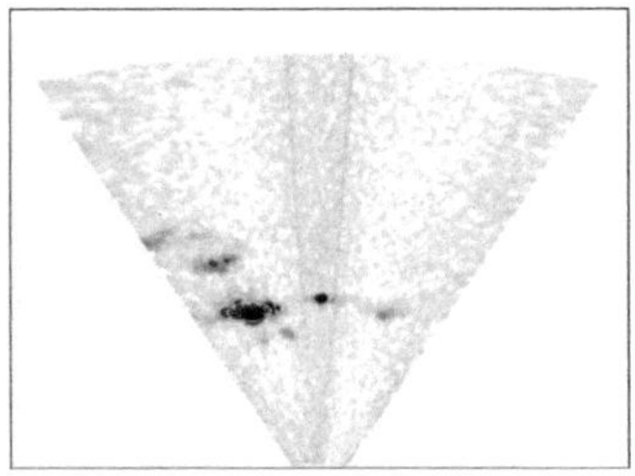

(b) Inverted Sonar Image

Fig. 4 The YellowRay ROV is seen in 4a. An inverted sonar image of the YellowRay ROV along with false positive noise is seen in 4b. The largest object is assumed to be the target ROV while the smaller objects are noise.

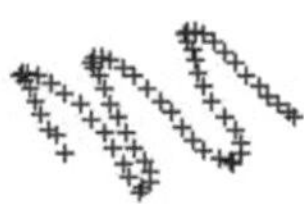

(a) Sonar *SearchPattern*

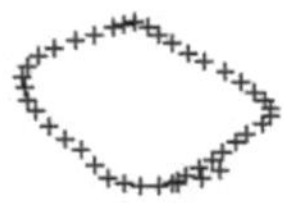

(b) Sonar *Loiter*

(c) Sonar *GoToWaypoint*

Fig. 5 Real trajectories captured by a BlueView forward-looking sonar of an AUV performing *SearchPattern*, *Loiter*, and *GoToWaypoint* are seen in (a), (b), and (c), respectively

makes the simplifying assumptions that there is only one relevant object in the scene, the YellowRay, and that it will always be in the FOV of the sonar. The YellowRay operators were asked to perform multiple runs of three behaviors, *GoToWaypoint*, *Loiter*, and *SearchPattern*.

The BlueView forward-looking sonar provides an image with intensity values corresponding to the acoustic response of a surface, as seen in Fig. 4b. The more intense a pixel, the more likely that an object exists at that location. In order to smooth the ROV's trajectory, only every *ith* frame is used. This reduces the number of outliers significantly as the sonar data is extremely noisy. Edges are found in the sonar image which are used to create contours. The contour with the largest area is assumed to be the ROV, as we assume that only the ROV is in the image and the smaller blobs are noise. The API of the BlueView sonar then produces the range and bearing of the center pixel relative to the sonar itself. Range and bearing is then converted to x and y coordinates to produce trajectories, as seen in Fig. 5. In this form, discretization can take place converting location to global yaw then to change in yaw as described above. In this experiment, the global change in yaw is discretized into five bins with a spread of four degrees each.

5.2 Track and Trail

In order to test our methods with a non-stationary observer, testing was performed on simulated data with one AUV performing *Track & Trail* of a leader performing a behavior. As in the experiments above, the MOOS-IvP simulator is used to generate template trajectories of a leading AUV performing *GoToWaypoint*, *Loiter*, and *SearchPattern* while an observing AUV performs *Track & Trail*.

In order to use the change in yaw method of encoding, the template trajectories of each vehicle are used to produce the pose (x, y, θ) of the trailing vehicle along with range and bearing to the lead vehicle. Using this information allows the trailing AUV to reconstruct the leading AUV's trajectory which will be discretized for use in behavior recognition. In order to create more realistic results, the original measurements of the trailing AUV's location (x, y, θ) along with range and bearing to the leader AUV are injected with Gaussian noise, similar to those seen in 3a, 3b, and 3c. This more accurately represents the uncertainty an AUV will have of its own location and the uncertainty of the location of the target AUV present in sonar data. In this experiment, the global change in yaw is discretized into seven bins with a spread of four degrees each.

6 Results

The results of three different sources of data are analyzed. The accuracy of the Hidden Markov Model (HMM) and Conditional Random Field (CRF) methods are considered for each data source.

6.1 *Stationary Observer*

6.1.1 Simulation

As seen in Table 1, each method was trained on a specific behavior using a corpus of 600 instances of trajectories generated by that behavior. A total of 400 trajectories from each behavior, for a total of 1200 instances, were presented to both methods. As is seen in Table 1, the HMMs performed well for *SearchPattern*, *Loiter*, and *GoToWaypoint* as they were able to accurately discriminate trials by 100%, 99.25%, and 96%, respectively. The CRF performed better for *SearchPattern*, *Loiter*, and *GoToWaypoint* as it was able to discriminate the behaviors with 100% accuracy. The HMM method had the most difficulty in discriminating *GoToWaypoint* as it recognized 16 instances of that behavior as *SearchPattern*.

Table 1 Accuracy of Simulated Stationary Behavior Recognition

Behavior	Training	Testing	HMM	CRF
SearchPattern	600	400	100%	100%
Loiter	600	400	99.25%	100%
GoToWaypoint	600	400	96%	100%

Table 2 Confusion matrix for the HMM method applied to simulated stationary data

		SearchPattern	*Loiter*	*GoToWaypoint*
HMM	*SearchPattern*	400	0	0
	Loiter	0	367	3
	GoToWaypoint	16	0	384

6.1.2 Sonar Data

As seen in Table 3, each method was trained on real sonar data while an ROV performed a specific behavior using a corpus of 21 instances for *SearchPattern*, 23 instances for *Loiter*, and 14 instances for *GoToWaypoint*. A total of 38 instances were presented to both methods for testing. The HMM discrimination method had the best accuracy of 100%, 68.75% and 100%, respectively. The CRF performed worse than the HMM method with discrimination of *SearchPattern*, *Loiter*, and *GoToWaypoint* with accuracy of 75%, 68.75%, and 80%, respectively. The HMM method only had false positives with five instances of *Loiter* being identified as *SearchPattern*, as seen in Table 4. The CRF method suffered similarly to the HMM method in discriminating *Loiter* as *SearchPattern*. Additionally, the CRF method identified one instance of *SearchPattern* as *Loiter* and two instances as *GoToWaypoint*. The CRF method's best performance on the real sonar data was in discriminating *GoToWaypoint* as it only mis-identified two instances as *Loiter*.

Table 3 Accuracy of Sonar Behavior Recognition

Behavior	Training	Testing	HMM	CRF
SearchPattern	21	12	100%	75%
Loiter	23	16	68.75%	68.75%
GoToWaypoint	14	10	100%	80%

Table 4 Confusion matrices for the CRF and HMM methods applied to stationary sonar data

		SearchPattern	*Loiter*	*GoToWaypoint*
HMM	*SearchPattern*	12	0	0
	Loiter	5	11	0
	GoToWaypoint	0	0	10
CRF	*SearchPattern*	9	1	2
	Loiter	5	11	0
	GoToWaypoint	0	2	8

6.2 Track and Trail

As seen in Table 5, using the change in yaw of the leading vehicle as a discretization method resulted in sufficient accuracy. The results are from inserting Gaussian noise with a standard deviation of 0.75 on the location (x, y, θ) of the trailing vehicle, range and bearing to the leader. The HMM discrimination method had accuracy of 97.25% with *SearchPattern*, 94.75% with *Loiter*, and 95.25% with *GoToWaypoint*. The CRF discrimination method had the best accuracy of discrimination of *SearchPattern*, *Loiter*, and *GoToWaypoint* with accuracy of 99.50%, 99.75%, and 99.75%, respectively. As seen in Table 6, the CRF method only had at worst two mis-discriminations of *SearchPattern* versus the HMM method which had a best case of only 11 mis-discriminations.

Table 5 Accuracy of Simulated *Track & Trail* Behavior Recognition

Behavior	Training	Testing	HMM	CRF
SearchPattern	600	400	97.25 %	99.50%
Loiter	600	400	94.75 %	99.75%
GoToWaypoint	600	400	95.25 %	99.75%

Table 6 Confusion matrices for CRF and HMM methods applied to the *Track & Trail* data

		SearchPattern	*Loiter*	*GoToWaypoint*
HMM	*SearchPattern*	389	0	11
	Loiter	1	379	20
	GoToWaypoint	13	6	381
CRF	*SearchPattern*	398	2	0
	Loiter	1	399	0
	GoToWaypoint	0	1	399

7 Conclusion

The work presented here demonstrates the feasibility of performing behavior recognition of an AUV in situ. In general, using Hidden Markov Models (HMM) resulted in sufficient performance. Using the Conditional Random Field method resulted in better performance than the HMM method when there was ample training data available, as in the simulated stationary observer data or the simulated *Track & Trail* data. However, the CRF method performed poorly in discrimination of the real sonar data. Due to the small sample size of real sonar data it may be an indication of under training the CRF. However, all the methods struggled discriminating the *Loiter* behavior in the real sonar data set. It is possible that the sonar-captured *Loiter* behavior should be further separated into *left* and *right Loiter* as that could be the reason for the methods performing poorly. This is in contrast to the simulated *Loiter* trajectories which only performed them in the left direction.

Discretization parameters for the change in yaw observations play a crucial role in the success of behavior recognition in these experiments. For example, an experiment that discretizes the global change in yaw with five bins each with a spread of four degrees has a limited resolution. Any change in yaw greater than six or less than negative six degrees is placed into bins one and five, respectively. Thus, if a crucial distinction between two behaviors occurs beyond these terminal edge bins they will not be properly discriminated. The discretization parameters were empirically determined for each experiment. While simple, global changes of yaw as an observation method is very susceptible to changes in speed. An alternative may be to describe higher level primitives such as straight, left turn, and right turn as observations for our behavior recognition methods.

Future work includes further investigation of discretization and behavior recognition methods along with a larger data set. It may be possible that each behavior recognition method requires different change in yaw discrimination parameters for improved accuracy. Alternatively, higher level motion primitives may increase accuracy and robustness. Alternative recognition methods may be more appropriate. Handwriting recognition has similarities with behavior recognition and may yield improved methods. Recognition should be verified with more behaviors than the ones used in these experiments, as they are a small sample representation. The next step is to obtain a larger corpus of real data. To truly test the feasibility of behavior recognition of one AUV *Tracking & Trailing* a leader experiments should be performed with real AUVs while performing recognition in real-time.

Acknowledgements. The authors thank their colleagues Paul Robinette and Andrew Melim along with the faculty and staff of the Georgia Tech Mechanical Engineering Acoustic Tank for their valuable comments and aiding in data gathering. This work was conducted through internal research and development funding by the Georgia Tech Research Institute (GTRI).

References

1. Videoray (2011), `http://www.videoray.com/`
2. Arkin, R.C.: Behavior-Based Robotics, ch. 9. MIT Press, Cambridge (1998)
3. Baxter, R., Lane, D., Petillot, Y.: Behaviour recognition for spatially unconstrained unmanned vehicles. Proceedings of the IJCAI 9, 1–8 (2009)
4. Baxter, R.H., Lane, D.M., Petillot, Y.: Recognising agent behaviour during variable length activities. In: Proceedings of The 19th European Conference on Artificial Intelligence (2010)
5. Benjamin, M., Schmidt, H., Newman, P., Leonard, J.: Nested autonomy for unmanned marine vehicles with moos-ivp. Journal of Field Robotics 27(6), 834–875 (2010)
6. Cao, Y., Fukunaga, A., Kahng, A.: Cooperative mobile robotics: Antecedents and directions. Autonomous Robots 4(1), 7–27 (1997)
7. Dias, M., Stentz, A.: A free market architecture for distributed control of a multirobot system. In: 6th International Conference on Intelligent Autonomous Systems (IAS-6), pp. 115–122 (2000)
8. Dudek, G., Jenkin, M., Milios, E., Wilkes, D.: A taxonomy for multi-agent robotics. Autonomous Robots 3(4), 375–397 (1996)
9. Gerkey, B., Mataric, M.: Sold!: Auction methods for multirobot coordination. IEEE Transactions on Robotics and Automation 18(5), 758–768 (2002)
10. Lafferty, J., McCallum, A., Pereira, F.: Conditional random fields: Probabilistic models for segmenting and labeling sequence data (2001)
11. Novitzky, M.: Improvement of multi-auv cooperation through teammate verification. In: Workshops at the Twenty-Fifth AAAI Conference on Artificial Intelligence (2011)

12. Novitzky, M., Pippin, C., Balch, T., Collins, T., West, E.: Behavior recognition of an auv using a forward-looking sonar. In: Workshops at the Robotics: Science and Systems, Los Angeles, CA (2011)
13. Rabiner, L.: A tutorial on hidden Markov models and selected applications in speech recognition. Proceedings of the IEEE 77(2), 257–286 (1989)
14. Sariel, S., Balch, T., Erdogan, N.: Naval mine countermeasure missions. IEEE Robotics & Automation Magazine 15(1), 45–52 (2008)
15. Sotzing, C., Lane, D.: Improving the coordination efficiency of limited-communication multi-autonomus underwater vehicle operations using a multiagent architecture. Journal of Field Robotics 27(4), 412–429 (2010)
16. Vail, D., Veloso, M.: Feature selection for activity recognition in multi-robot domains. In: Proceedings of AAAI, vol. 2008 (2008)
17. Vail, D., Veloso, M., Lafferty, J.: Conditional random fields for activity recognition. In: Proceedings of the 6th International Joint Conference on Autonomous Agents and Multiagent Systems, pp. 1–8. ACM (2007)
18. West, M., Novitzky, M., Varnell, J., Melim, A., Sequin, E., Toler, T., Collins, T., Bogle, J.: Design and development of the yellowfin uuv for homogenous collaborative missions (2010)

Imitation Learning and Behavior Generation in a Robot Team

Huan Tan

Abstract. In this paper, we propose a method to apply robotic imitation learning in robot teams. In our method, behavior primitives with task-relevant information are defined as the basic units for robots to complete a task. Each behavior has its own task-relevant affordances. The learned behavior primitives format TAEM based behavior libraries and are stored in a database for robots to share. The motivation of learning, conducted by robots, is goal-oriented and strongly related to the given task. Given a task, a robot analyzes the environment and searches the behavior library to find suitable behaviors to generate a behavior sequence to complete the task. If it thinks that it cannot complete this task, this robot requests other robots for assistance or request a human teacher to demonstrate the required behaviors. The newly learned behaviors will be added into the existing behavior library. We also develop inhibiting properties for robots to evaluate the current behaviors, which enables robots to request collaborations from other robots. The experimental results show the validity our proposed method.

1 Introduction

Robotic imitation learning has been considered as a powerful class of tools for transferring behaviors and skills between robots and human teachers [1]. These imitation learning methods can be divided into two types [4]: one is to teach robots to generate similar movement trajectories [7]; the other is to teach robots to learn behavior sequences which consist of several behavior primitives [6]. Robotic imitation learning has been shown to be effective method for enabling robots to learn knowledge and skills rapidly from human teachers and complete similar tasks in slightly different situations [2].

However, we cannot expect that a robot can complete all the tasks by itself. If we only apply teaching, learning, and generation on a single robot, since: 1) it is

---

Huan Tan
Vanderbilt University

M.A. Hsieh and G. Chirikjian (eds.), *Distributed Autonomous Robotic Systems*,
Springer Tracts in Advanced Robotics 104,
DOI: 10.1007/978-3-642-55146-8_30,

impossible to expect that we can teach one robot everything that it needs to know to complete all tasks; 2) a robot cannot complete some tasks due to its physical limitations; 3) sometimes, it is necessary to use several robots to complete tasks through cooperation. Therefore, transferring of learned behaviors in a robot team and cooperating among team members are necessary.

The motivation of this paper is to design a robotic imitation learning strategy for a robotic team and address the three issues described in the former paragraph. In order to realize imitation learning in a robotic team, the most important issue is to design shared behavior libraries, so that robots can obtain the knowledge and skills from other team members and update its own local behavior library. At the behavior generation stage of the imitation learning, robots need to analyze the relationship between behaviors in a required behavior sequence. If necessary, several robots should cooperate to complete a task.

The rest of this paper is organized as follows: Section 2 explains the system design including the overall system framework design and the imitation learning framework on an individual robot; Section 3 describes the experiments carried out on a real robot and in a simulation environment and the experimental results; Section 4 discusses the experimental results and our proposed system; and Section 5 summarizes this paper.

2 System Design

The overall system design is divided into two parts: one is an imitation learning framework on an individual robot (Tan et al., 2012a); the other is a team-based imitation learning and generation framework in a robotic team.

2.1 Overall Learning Framework

In our method, in task-relevant situations, the generation of the behaviors in a robotic team should be related to the inner behavior library of an individual robot and the shared libraries among all the robot members. In task-relevant situations, a robot should first use learned skills (normally represented as behaviors) in its own library to try to complete the tasks. If the robot has required behaviors, it can use

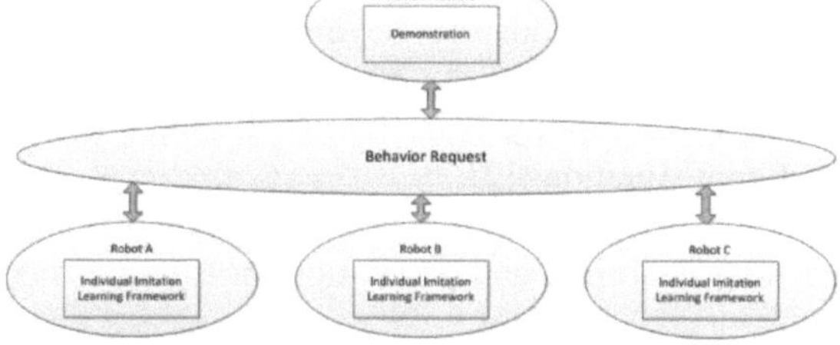

Fig. 1 Overall System Framework

them to complete the goal-driven task. If it fails to do so, the robot should try to find suitable behaviors from other robots. If it still cannot find a suitable behavior to complete the task, a human teacher will be required to demonstrate how to do it.

Fig.1 displays the overall imitation learning framework of a robot team. Each robot can request behaviors by sending signals to other robots, and the human teachers can demonstrate the required behaviors to robots when necessary.

2.1.1 Behavior Library

In our design, behaviors are considered as raw data of movement trajectories for completing tasks. Two important requirements are necessary to describe a behavior: one is a regression model of describing the movement trajectory and the other is its task-related affordances which describe what tasks can be applied to. In addition, one should be able to easily update and modify the designed behavior libraries.

In our system, we used TAEMS [9] as the basis of our behavior library design. TAEMS is a multi-agent framework, which is domain independent. Fig.2 displays a behavior graph designed using TAEMS. The leaves, which are shown as ellipses, are basic behaviors. The root and sub-roots are shown by circles, are complex task-relevant behaviors and consist of basic behaviors. The links could be considered as the logic operations. For example, the node with its children connected using an AND means that this task must be completed by sequentially executing the behaviors of the children, and the node with its children connected using an OR means that this task can be completed by selecting one child. In this figure, T represents the Task, and B represents the Behavior.

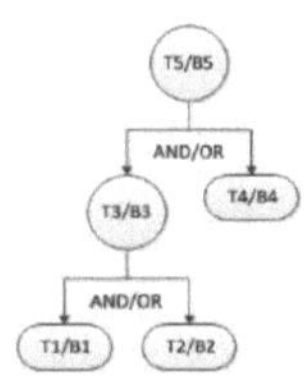

Fig. 2 An Example of the Behavior Tree using TAEMS

Behavior is task-relevant, and each behavior is used to complete a task. In Fig.2 (assume all the children are connected by AND), task 1 is related to behavior 1, and task 2 is related to behavior 2. Task 3 is more complex than behavior 1 and behavior 2, and is related to behavior 3. From Fig.2, behavior 3 consists of behavior 1 and behavior 2. If the robot wants to complete task 3, it needs to execute behavior 1 and behavior 2 subsequently. Task 4 is related to behavior 4 as shown. Task 5 is related to behavior 5 which consists of behavior 3 and behavior 4. This tree architecture could help the robot find the required behavior to complete a task.

We can use a table to describe each behavior as shown in Fig.3. Each behavior has been assigned a unique ID and a name as shown in Fig.3. A goal-driven task is described as: $Goal(Parameters)$. Using the adaptive generation method described later in this paper, robots can generate similar motion trajectories to complete tasks with different parameters.

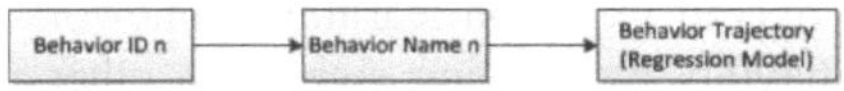

Fig. 3 Representation of a Single Behavior

The behaviors in the behavior library should not only be used by its father nodes, but also be used by the adjacent root nodes. Therefore, we extended the behavior graph in Fig.2 to construct the behavior graph as shown in Fig.4.

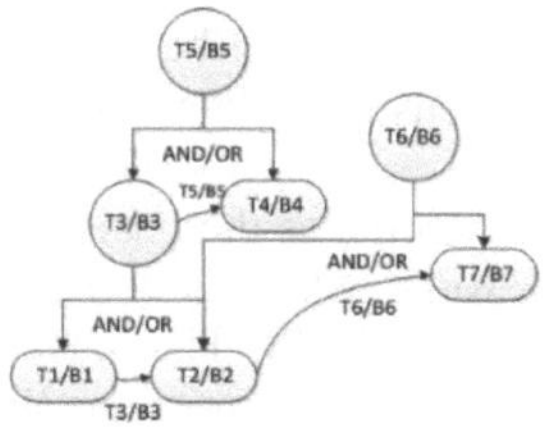

Fig. 4 A Behavior Library with Transitions

The links between the nodes and the leaves describes the root-children relationships. In order to make the temporal order clear among the children, we add the transitions among the children as shown in Fig.4. If a robot wants to use Behavior 6 to complete the TASK 6, it should execute Behavior 1, Behavior 2, and Behavior 7 consequently. The labeled transitions reflect which root node the children belong to. Using this method, there could be multiple transitions between children nodes that belong to different root nodes.

Behaviors are executed in a temporal way, which means 1) the robot can only execute one behavior at a time period; 2) adjacent behaviors in a behavior sequence should be temporally compatible. For example, the robot is asked to grasp an object in the environment and move it to a designated place, and it is impossible for the robot to grasp another object before it releases the object in its grasp. In a learned behavior sequence, if such temporal incompatibility happens, the former behaviors can inhibit the latter behaviors.

Assume that Behavior 1 is the Reaching, Behavior 2 is the grasping, Behavior 4 is the Moving, and Behavior 7 is the Lifting in Fig.5. The robot needs to execute Behavior 5 and Behavior 6 to complete a task. In Behavior 5, the robot reaches an object, grasps it and moves it to a position. In Behavior 6, the robot reaches for another object, and needs to grasp it. In this situation, there is an object already in its hand after the robot executes Behavior 5, so it cannot grasp another object. Thus Behavior 6 is inhibited unless the robot releases the already grasped object. If Behavior 6 is inhibited by Behavior 5, Behavior 6 is drawn using dotted lines. The dotted lines also mean that this behavior should be executed by another robot.

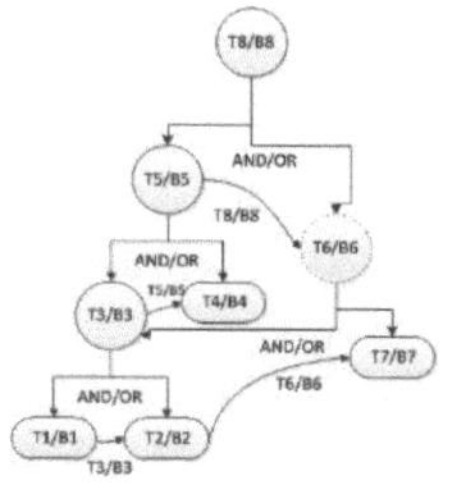

Fig. 5 Inhibiting Behavior

When Robot A encounters Robot B, it requests assistance from Robot B, and shares its current behavior library with Robot B. The behavior library in Robot B is described in an opposite way as shown in Fig.6. The Behavior is drawn using dotted lines, and it means that it should wait for Robot A to execute this behavior.

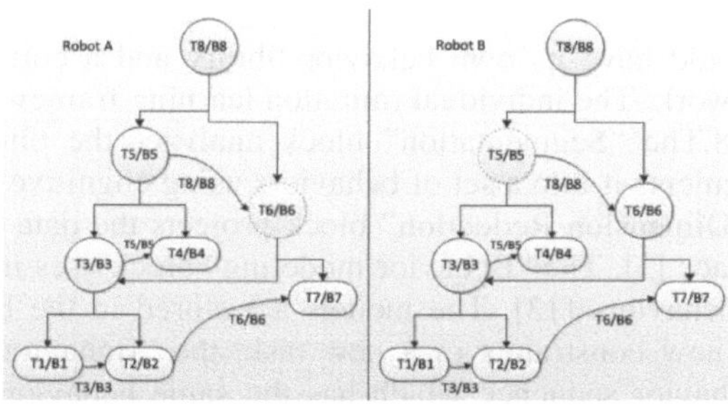

Fig. 6 Inhibiting Behaviors in Two Robots

2.1.2 Behavior Search and Behavior Sequence Generation

Given a task, the robot needs to find whether it has suitable behaviors or not. In order to simplify the task analysis, we assume that the tasks are given in behavior sequences. Then the robot only needs to find corresponding behaviors in its behavior library. The search is conducted using Depth-First Search (DFS) [5]. The reason for using this method is to keep the temporal relationships among the children. When a required behavior is found as a node A in the behavior library, robots can still use DPS to return all the leaves of node A. A behavior sequence is generated by putting all the returned leaves in a sequence using DFS.

2.1.3 Behavior Update

When robot A cannot find a suitable behavior in in its own behavior library, it first requests other robotic team members to share their behavior libraries. If one of the team members has the required behaviors, it will send a behavior graph to robot A. The root node of the behavior graph is the required behavior which may be composed of several basic behaviors. Then robot A uses the received behavior graph to update its own library as shown in Fig.7.

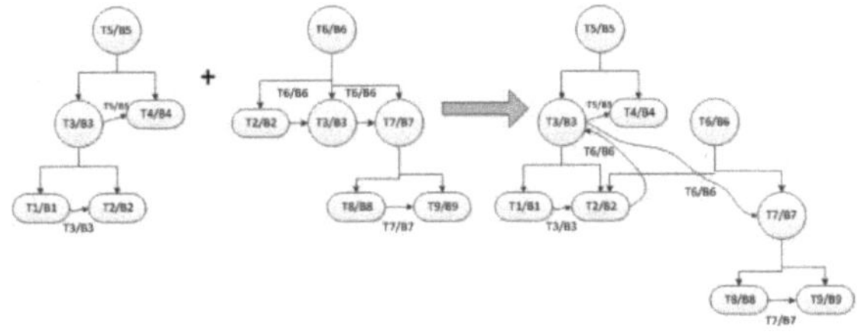

Fig. 7 Behavior Update

If all the robotic members in this team do not have the required behaviors or skills, robot A needs to ask a human teacher to demonstrate how to complete this task and imitation learning by the robot is initiated.

2.2 Individual Imitation Learning Framework

Each robot should have its own behavior library and a corresponding imitation learning framework. The individual imitation learning framework on each robot is shown in Fig.8.The "Segmentation" block analyzes the observed behavior sequence and segments it into a set of behaviors using cognitive segmentation method [11]. The "Dimension Reduction" block projects the data from the data space to the latent space [3]. The "Behavior modeling" block uses mathematical models to record the behaviors [13]. The models are stored in the block "Behavior Library". Given new constraints in a new task, the "Behavior Generation" block generates a behavior sequence which has the same behavior descriptions as the demonstration. The "Reconstruction" block projects the data from the latent space to the joint space and sends the generated points to the actuator to execute [15]. If the robot cannot find suitable behaviors to complete the task, the "Demonstration Request/Behavior Request" block requests the demonstration from human teachers or ask other robotic team members to provide suitable behaviors. In some situations, the "Dimension Reduction" and "Reconstruction" block are not necessary.

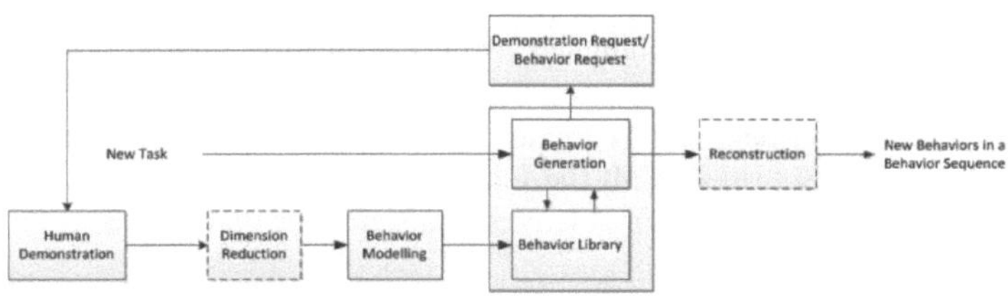

Fig. 8 Individual Imitation Learning Framework

2.2.1 Behavior Modeling

Each behavior has its related movement trajectories. In this paper, we choose to use Gaussian Process (GP) [10] to model the movement trajectory. Assume the N data points have the following probabilistic distribution,

$$p(\overrightarrow{z_e}|\overrightarrow{x_e}) = (\overrightarrow{z_e}|\overrightarrow{x_e}, \beta^{-1}\boldsymbol{I}) \tag{1}$$

where $\boldsymbol{x_e}$ is the target value and $\boldsymbol{z_e}$ is the predicted value. We assume $p(\boldsymbol{X}) = (\boldsymbol{X}|0, \boldsymbol{C_N})$, where the covariance matrix $C(n,m) = k(\boldsymbol{t_n}, \boldsymbol{t_m}) + \beta^{-1}\delta_{nm}$, where $k(t_n, t_m)$ is the kernel function. Normally,

$$k(t_n, t_m) = \theta_0 exp\left\{-\frac{\theta_1}{2}\|\boldsymbol{t_n} - \boldsymbol{t_m}\|^2\right\} + \theta_2 + \theta_3 \boldsymbol{t_n}^T \boldsymbol{t_m} \tag{2}$$

and $\boldsymbol{t_n}$ and $\boldsymbol{t_m}$ are considered as the temporal information in the demonstration. When $\boldsymbol{t}_e$ is given as an enquiry point and GP is used to calculate the corresponding data value $\boldsymbol{z}_e$.

$$p(\{\overrightarrow{\boldsymbol{X}}, \boldsymbol{z}_e\}) = (\{\overrightarrow{\boldsymbol{X}}, \boldsymbol{z}_e\}|0, \boldsymbol{C}_{N+1}) \tag{3}$$

The covariance matrix is:

$$\boldsymbol{C}_{N+1} = \begin{pmatrix} \boldsymbol{C}_N & \overrightarrow{\boldsymbol{k}} \\ \overrightarrow{\boldsymbol{k}}^T & c \end{pmatrix}, \tag{4}$$

where $k = k(\boldsymbol{t_n}, \boldsymbol{t_e})$ for n=1,2,…, N. Using Baye's rule, $\boldsymbol{z_e}$ is given by:

$$\boldsymbol{z}_e = \overrightarrow{\boldsymbol{k}}^T \boldsymbol{C}_N^{-1} \overrightarrow{\boldsymbol{X}}. \tag{5}$$

2.2.2 Generation

The Dynamic Movement Primitives (DMP) [7, 8] algorithm is configured as:

$$\tau\dot{z} = \alpha_z(\beta_z(g - y) - z), \tag{6}$$

$$\tau\dot{y} = z + f. \tag{7}$$

It can be considered a second-order attractor modulated by a non-linear function which represents the dynamics of the demonstration. In (6) and (7) g is the goal state, z is the internal state, f is calculated to record the dynamic of the demonstration and to guarantee convergence of the new generated trajectories, y is the position generated by the DMP differential equations, and $\dot{y}$ is the generated velocity correspondingly. Additionally, α_z, β_z, and τ are the constants in this equation. In this work, f is a GP regression model. Using the DMP method, robots can generate motion trajectories which are similar to the demonstrated motion trajectories and have different goal-states and avoid obstacles in the environment [15].

3 Experimental Results

We designed four experiments to validate our system. The first experiment is carried out on a real humanoid robot, named ISAC, and the second and the third experiment are carried out in the simulation environment due to the limitations of the hardware in our lab.

3.1 Experiment 1

The objective of this experiment is to verify that the robot can generate trajectories, which are similar to the movement trajectory in the demonstration, to reach different locations. The movement trajectory is generated using DMP.

The left picture of Fig.9 shows a human teacher demonstrating how to reach an object on a table in front of ISAC. The middle graph of Fig.9 shows the movement trajectory in a demonstration. In order to visualize the trajectories, we compute the trajectory in the Cartesian space. The first row of the middle graph displays the positions in X, Y, and Z axis, and the second row shows the corresponding velocities. The goal is to reach (520, -70, -300) in the Cartesian space. The right graph of Fig.9 shows the generated trajectories when the robot is tasked to reach $(500, 0, -300)$.

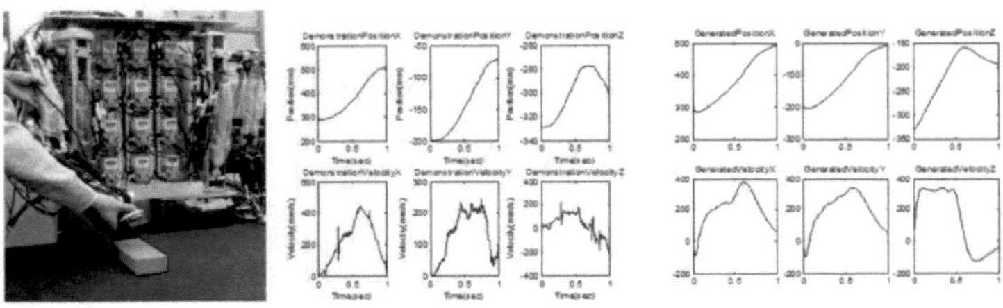

Fig. 9 Imitation Learning Results

We normalized the generated trajectory and demonstrated trajectory and compute the distance between the corresponding points on the trajectory using the following equation:

$$d = \frac{\sum_{i=1}^{N} \|Gen\_tra(i) - Demo\_tra(i)\|_2}{N} \tag{8}$$

where $\|\cdot\|_2$ is the norm-2 Euclidean distance.

The measured distance value is 0.3013, which means the trajectories are close. By comparing the middle picture and the right picture of Fig.9, we can see that the robots can generate a movement trajectory similar to the demonstration. By observing the ending point of the generated trajectory, we can see that the final value is not exactly the same as the desired destination value. But the error is around 1 cm and is small. While this is not good for precise manipulation tasks, it is acceptable for most general tasks that do not require much precision.

3.2 Experiment 2

In this experiment, robot A is tasked to remove an object in the environment. This "remove" behavior is composed of "reach", "grasp", and "lift". Assuming the robot only has the knowledge of "reach" and "grasp", it should acquire the behavioral information from other robots or ask a human teacher to demonstrate how to lift an object.

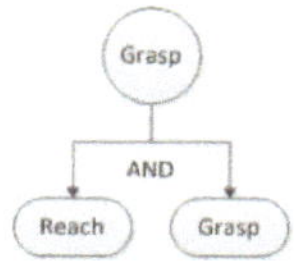

Fig. 10 Behavior Library

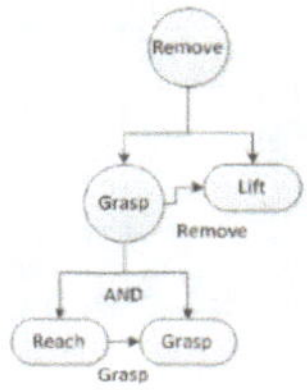

Fig. 11 Update Behavior Library

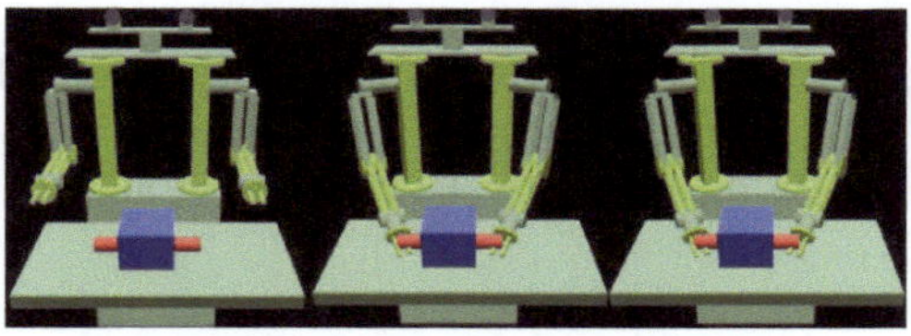

Fig. 12 Simulation Results for Experiment 2

Fig. 10 shows the behavior library of robot A. Fig.11 shows the updated behavior library when robot A requested sharing of behavioral information from robot B. Fig.12 shows the simulation result where robot A successfully removes the object in the environment.

3.3 Experiment 3

In this experiment, robot A is tasked to first remove a big box on the table, and then remove a small box from under the big box. The "remove" behavior is composed of "reach", "grasp", and "lift". Since robot A does not release the big box, it cannot remove the small box. Therefore, the two "remove" behaviors are incompatible. Robot A requests robot B to assist it to complete this task. Fig.13 displays the corresponding behavior libraries in each robot. Fig.14 displays that the robot A and B successfully removes the two boxes.

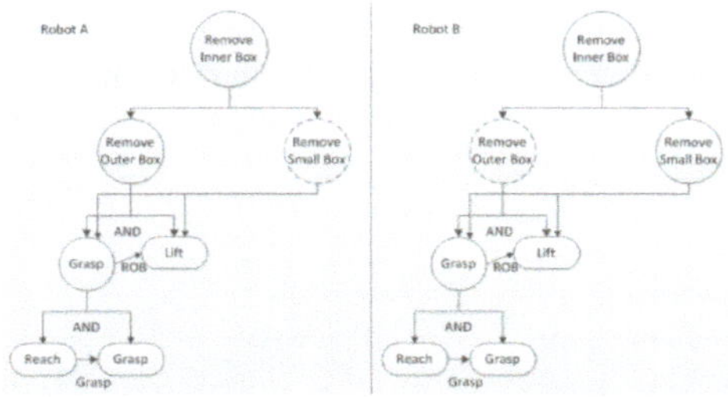

Fig. 13 Behavior Libraries

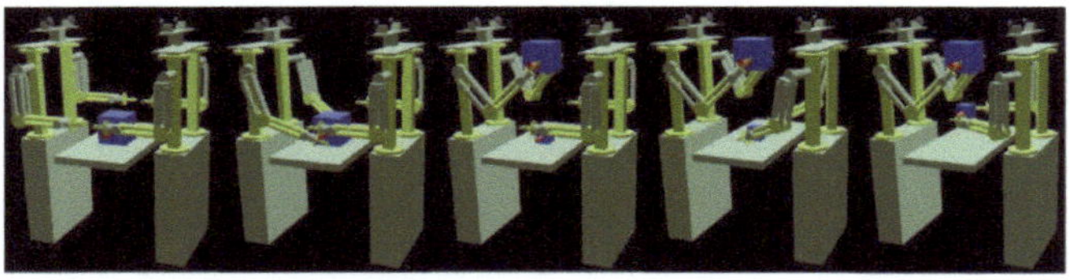

Fig. 14 Simulation Results for Experiment 3

3.4 *Experiment 4*

In this experiment, robot A is required to lift the table in front of it. Lifting a table requires both two sides of the table to be grasped. Fig.15 shows corresponding behavior libraries in each robot. Fig.16 shows robots A and B successfully lifting the table.

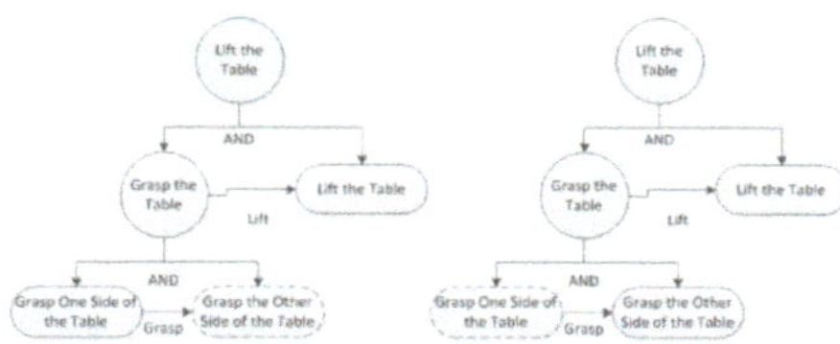

Fig. 15 Behavior Libraries

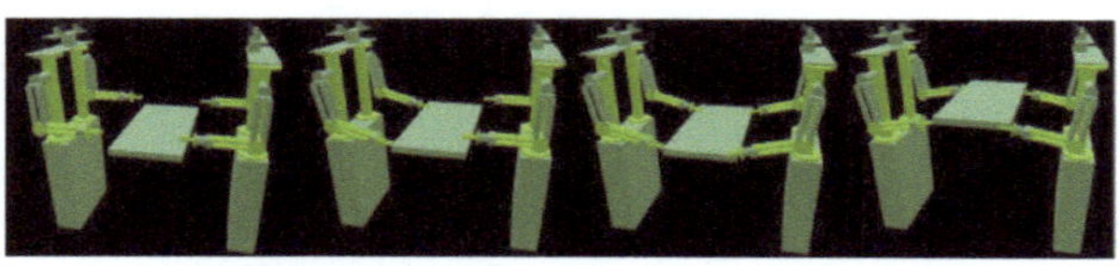

Fig. 16 Simulation Results for Experiment 4

4 Discussion and Future Work

The experimental results and simulation results demonstrate that our method is effective. Our system could be considered as a typical hierarchical learning framework. The sub-systems on individual robots are typical imitation learning processes. This was achieved by modifying existing popular frameworks to enable robots to use this framework to communicate with other robots and interactively learn from human teachers. The necessity of learning from human teachers or obtaining behaviors from other robots is based on the evaluation of the behavior libraries of the robot. When a robot considers that it is necessary to obtain extra knowledge from other robots or humans, this imitation learning framework is similar to teamwork.

Since the robots are able to share knowledge between each other, cooperation becomes possible. The designed behavior libraries not only enable robots to update their knowledge base, but also can reflect the cooperative relationships among robots. In our method, the behavior inhibiting properties can tell a robot whether it can executive this behavior or not. If the answer is no, it naturally requests help from other robots.

In this paper, we propose a prototype for applying imitation learning in a robotic team, however significant work remains. At the behavior generation stage, the robot needs to make a decision to learn or obtain knowledge from human teachers or from other robots respectively. However, the decision making process is not robust. In the future, we need to implement probabilistic strategies which take into consideration the dynamic environment. Another approach is to incorporate cognitive methods in our system.

An additional challenge is how the robot can use current knowledge to develop new behaviors to complete new tasks, which is called behavior evolution. Incorporating this feature will require the quantitative evaluation of behaviors.

5 Conclusion

This paper proposes a method for implementing robotic imitation learning in a robot team. The major contribution of this paper is: 1) we developed a hierarchy framework for robotic teams to learn the behaviors and skills from human teachers; 2) we applied TAEMS to describe learned behaviors for robots. 3) we adapted the TAEMS method, which enables robots to share and update its own local libraries through learning and sharing. The experimental results and simulation results reflect the success of our approach.

References

1. Atkeson, C., Schaal, S.: Robot Learning from Demonstration. In: Fisher Jr., D.H. (ed.) The Fourteenth International Conference on Machine Learning, pp. 11–73. Morgan Kaufmann (1997)
2. Billard, A., Calinon, S., Dillmann, R., Schaal, S.: Robot programming by demonstration. In: Siciliano, B., Khatib, O. (eds.) Handbook of Robotics. Springer, New York (2007)
3. Bitzer, S., Vijayakumar, S.: Latent Spaces for Dynamic Movement Primitives. In: The 2009 IEEE-RAS International Conference on Humanoid Robots, pp. 574–581 (2009)

4. Calinon, S., Guenter, F., Billard, A.: On learning, representing, and generalizing a task in a humanoid robot. IEEE Transactions on Systems, Man, and Cybernetics, Part B 37, 286–298 (2007)
5. Cormen, T.H., Leiserson, C.E., Rivest, R.L., Stein, C.: Introduction to algorithms. MIT Press (2001)
6. Dillmann, R., Rogalla, O., Ehrenmann, M., Zollner, R., Bordegoni, M.: Learning robot behaviour and skills based on human demonstration and advice: the machine learning paradigm. In: Ninth International Symposium of Robotics Research (ISRR 1999), Citeseer, Snowbird, UT, USA, vol. 9, pp. 229–238 (2000)
7. Ijspeert, A., Nakanishi, J., Schaal, S.: Learning attractor landscapes for learning motor primitives. In: Becker, S., Thrun, S., Obermayer, K. (eds.) Advances in Neural Information Processing Systems, vol. 15, pp. 1547–1554. MIT Press (2003)
8. Ijspeert, A., Nakanishi, J., Schaal, S.: Movement imitation with nonlinear dynamical systems in humanoid robots. In: 2002 IEEE International Conference on Robotics and Automation, Citeseer, Washington, DC, USA, vol. 2, pp. 1398–1403 (2002)
9. Lesser, V., Decker, K., Wagner, T., Carver, N., Garvey, A., Horling, B., et al.: Evolution of the GPGP/TAEMS domain-independent coordination framework. Autonomous Agents and Multi-agent Systems 9, 87–143 (2004)
10. Rasmussen, C.: Gaussian processes in machine learning. In: Advanced Lectures on Machine Learning, pp. 63–71. MIT Press, Cambridge (2004)
11. Tan, H.: Implementation of a Framework for Imitation Learning on a Humanoid Robot using a Cognitive Architecture. In: Zaier, R. (ed.) The Future of Humanoid Robots: Research and Applications, pp. 189–210. InTech (2012)
12. Tan, H., Du, Q., Wu, N.: A Framework for Cognitive Robots to Learn Behaviors through Imitation and Interaction with Humans. In: 2012 IEEE International Multi-Disciplinary Conference on Cognitive Methods in Situation Awareness and Decision Support, New Orleans, USA, pp. 235–238 (2012a)
13. Tan, H., Du, Q., Wu, N.: Robots Learn Writing. Journal of Robotics (2012b)
14. Tan, H., Erdemir, E., Kawamura, K., Du, Q.: A Potential Field Method-based Extension of the Dynamic Movement Primitive Algorithm for Imitation Learning with Obstacle Avoidance. In: 2011 IEEE International Conference on Mechatronics and Automation, Beijing, China, pp. 525–530 (2011)
15. Tan, H., Kawamura, K.: A Computational Framework for Integrating Robotic Exploration and Human Demonstration in Imitation Learning. In: 2011 IEEE International Conference on System, Man and Cybernetics, Anchorage, AK, USA, pp. 2501–2506 (2011)

Supervised Learning in Robotic Swarms: From Training Samples to Emergent Behavior

Gregory Vorobyev, Andrew Vardy, and Wolfgang Banzhaf

Abstract. Emergent behavior in swarm robotic systems is key to obtaining complex behavior by a group of relatively simple agents. The question is how to design the individual behaviors of agents in such a way that the desired global behavior emerges. Different approaches have been proposed to solve this problem: from biologically inspired probabilistic behavioral models to evolutionary techniques. In some situations, however, creating a complex probabilistic model of the behavior or developing a proper setup for an evolutionary process can be challenging. In this paper we propose a new method, based on supervised learning on a relatively small number of training samples. We apply our method to the well-known clustering problem and show that this approach yields the desired global clustering behavior.

1 Introduction

Emergent behavior in swarm robotic systems has been a subject of extensive research for the last two decades [1, 2]. A robotic swarm, provided that it has been designed in a particular way, can produce global behavior, that is often conceptually more complex than the behaviors of the individual agents. For example, a group of robots can collect scattered objects into a single cluster, although each individual robot follows a simple pick-up-and-deposit procedure without any explicit knowledge of where the cluster has to be formed [3, 4, 5]. This phenomenon is widely known in biology. For example, ants exhibit an astonishing degree of collaboration and coordination in wars against other ant colonies and even other animals, with attacks and retreats and, in the worst case, evacuation of the queen and larvae carried out by the whole ant colony as if it were controlled by someone who is always aware of the current situation in the world [6]. Honeybees have been observed

Gregory Vorobyev · Andrew Vardy · Wolfgang Banzhaf
Memorial University of Newfoundland, St. John's, NL A1B 3X5
e-mail: {gvorobyev,av,banzhaf}@mun.ca

M.A. Hsieh and G. Chirikjian (eds.), *Distributed Autonomous Robotic Systems*, Springer Tracts in Advanced Robotics 104,
DOI: 10.1007/978-3-642-55146-8_31, © Springer-Verlag Berlin Heidelberg 2014

as they select a new place as home: this process is truly democratic and involves "voting", i.e., collective decision making, in which many individuals participate [7]. Due to the degree of self-organization, coordination and unity exhibited by these groups of animals, some scientists consider these groups as superorganisms [8].

The emergence of global behavior is a result of actions taken by individuals; thus, relatively simple behavioral patterns followed by the individuals eventually produce the more "intelligent" behavior of the swarm. The question is therefore how to design individual behaviors in such a way that they will construct the basis from which the desired global swarm behavior will emerge [9]. One way to do this is to take inspiration from biology. For example, clustering behavior is observed in ants as they collect their dead peers into piles [6], or in honeybees distributing pollen into cells in their hives [7]. The results of experiments conducted by biologists have led to probabilistic models of individual behavior [3]. The feasibility of these models is further confirmed by observing the global swarm behavior in simulated (or real) robotic systems and comparing it with what is observed in nature [10]. Another approach exploits evolutionary techniques. In this methodology, individual behaviors evolve in such a way that the global behavior is improved [11, 12, 13, 14].

While both methods of designing individual behaviors have proved to be successful, they have their own issues. In the first case, a swarm designer needs to have feasible behavioral models, which may not be available (for example, if a behavior which is desired for the swarm has not been observed in nature). In the other case, the problem of a proper set-up for the evolutionary process arises (for example, which parameters of the behavior are subject to evolution and which are not); moreover, computational costs are typically high for the evolutionary approach.

In this paper, we propose an alternative simple method of designing the individual behaviors of agents for the clustering problem. In our approach, the designer considers a small number of characteristic situations that an agent might encounter. While it is hard to predict each possible configuration of the environment in which the agent may find itself and to generate a corresponding rule for this situation, it is much easier to accomplish this task if the number of situations being considered is relatively small (in our work, only 4). Yet, as we demonstrate in this paper, such a small number of training samples is sufficient for the agents to learn the task of clustering in such a way that the swarm starts to produce the desired behavior. We conduct experiments to test our approach in a custom 3D simulator with a realistic physics engine, and we show that our agents are capable of accomplishing the clustering task without any explicit probabilistic models embedded into them.

The rest of the paper is organized as follows. In Section 2, a short review of the relevant work is given. Section 3 describes the methodology used to solve the behavioral design problem. In Section 4, we conduct experiments and discuss the results. Finally, conclusions and future work are given in Section 5.

2 Related Work

In one of the pioneering works in the area of swarm robotics, Deneubourg et al. consider a generic sorting problem, where a swarm of robots collect objects (pucks) of different types into homogeneous clusters [4]. Each agent moves randomly between cells in a grid-based environment. Whenever the agent encounters a puck (i.e., enters a cell with a puck), it decides whether or not to pick it up. This decision is based upon how many objects of the same type this agent has encountered in the recent past. This information is stored in a short-term memory which is represented as an array of 10 items, for instance 00AAA0B00A (4 pucks of type A, 1 puck of type B, and 5 empty cells encountered during the last 10 time steps). The more pucks of a given type the agent remembers from his recent experience, the less is the probability of picking up a puck of that type. Further on, if an agent carries a puck and enters an empty cell, it decides whether or not to put the puck down. Intuitively, the more pucks of the same type as the puck which is being carried the agent has encountered in the recent past, the larger is the probability of depositing this puck. Ultimately, these simple rules result in a global sorting behavior of the swarm. There is no communication between agents or centralized control in the swarm: agents effectively are not aware of each other and act completely independently.

The idea of creating probabilistic behavioral controllers similar to what was proposed by Deneubourg et. al. has been applied to many different problems [1, 2]. For example, the task of collective aggregation has been solved by a group of cockroach-like robots with probabilistic controllers [15, 16]. The robots move randomly and stop with a certain probability which is a function of the number of other robots in immediate proximity (note that this mechanism is very similar to what has been used by Deneubourg et. al., although the task is slightly different). Thus, the behavior `Stop` is activated with a certain probability P_{stop}. In [17], the aggregation task is accomplished by robots with 4 atomic behaviors: `ObstacleAvoidance`, `Repel`, `Wait`, and `Approach`, with the last three organized into a probabilistic finite-state automaton. A robot approaches the largest group of robots with a probability P_{return}, waits for a random period of time, and then runs away from it with the probability P_{leave}. The results of this work demonstrate that the best performance is achieved with the $P_{return} = P_{leave} = 1$; in this case, the probabilistic behavior is reduced to deterministic, or procedural, behavior.

Deterministic behaviors of swarm agents have also been systematically studied in [9], where 6 basic behaviors are presented and tested: `Aggregation`, `Homing`, `CollisionAvoidance`, `Following`, `Dispersion`, and `Flocking`. Each behavior is a simple procedure; for example, `CollisionAvoidance` can be summarized as *"If there is another robot on the right, turn left; otherwise, turn right"*. In more recent work, [18], relatively simple deterministic behaviors of the agents have been applied to the chain formation problem.

In [5], similar deterministic rules have been embedded into a subsumption architecture to solve the clustering problem. For example, if an obstacle is detected in front of the robot, the `ObstacleAvoidance` behavior is activated. Different behaviors are activated depending on certain conditions. Similar experiments have been conducted in [19]. However, the condition checks used to trigger the behaviors in this work are "encoded" as the weights of neural connections going from the sensors rather than hard-coded procedural boolean expressions. The neural networks approach for activating behaviors based on a certain perception snapshot has been more explicitly used in [12] for the aggregation problem. In [20], the similar approach has been applied to the problem of self-assembly in a swarm-bot.

The impact of different parameters of atomic behaviors on the overall performance is commonly estimated through systematic experiments in these works. For example, in case of probabilistic controllers, the parameters that are subject to testing may include probabilistic thresholds, such as P_{leave} or P_{return} [17]. Such tests proved to be important, because it is tricky to predict which values for these parameters will be optimal for each particular experimental configuration. If the number of parameters is large, the designer's task becomes even more challenging.

Evolutionary techniques have been introduced in swarm robotics as another approach to designing individual behaviors. In [12] the aggregation problem was solved by evolving the weights of a perceptron using a fitness function which computes the average distance from a robot to the largest group. This work has been further improved in [11], where the authors deduced general rules for selecting evolutionary parameters in the swarm design problem. In the most recent work, [13], the evolutionary approach has been applied in swarm robotics to obtain emergent self-organizing behavior inspired by the collective flashing behavior of fireflies.

While evolutionary algorithms allow to avoid difficulties with fine-tuning parameters of the individual behaviors, they raise new issues. For example, as it is stressed in [13], the designer of a robotic swarm should determine which behavioral parameters are fixed and which are subject to evolution. The most difficult part, however, is probably the fitness function. Fitness functions, like those used in [12], tend to require some global knowledge (for example, distance between robots), which sometimes could hardly be obtained. Finally, the computational costs are usually large for evolutionary algorithms: for example, in [13], 500 experimental trials have been executed to evolve the individual behaviors.

In this paper, we present an alternative approach to designing the individual behaviors with application to the clustering problem. Our agent's controller is based on a neural network, which is similar to the networks described in [19] and [12]. However, we do not use hard-coded neural weights (as in, e.g., [19]), and we do not use evolutionary algorithms to evolve the weights (as in, e.g., [12]). Rather, we consider a set of 4 training samples. Each sample represents a perceptual snapshot. From a set of 3 behaviors - `BackUpAndTurn`, `Turn`, and `MoveStraightAhead` - we select the most suitable. For example, if a robot "sees" a large number of pucks in front of it, it should activate `BackUpAndTurn`. We show that this approach, being extremely simple and easy to follow, yields the desired clustering behavior.

3 Methodology

In this work we revisit the clustering problem, in which agents collect initially scattered pucks into a single pile. Similar to [21], our agents have no specialized grippers to manipulate the pucks. Rather, the agents push the pucks with a plow.

As mentioned above, a neural network is a central part of the agent's architecture in our work. We use a simple single-layer perceptron, with 2 inputs and 3 outputs. Each time step, sensory data is used as an input to this neural network. The output of the neural network is then interpreted as a code of the basic behavior to activate.

In the rest of this section, we discuss what kind of sensory data we use, describe the basic behaviors, and explain how sensory data is normalized to be fed into the neural network and how the network's output is interpreted. Finally, we describe the samples used to train the neural network.

3.1 Perception Areas

We assume that a robot has a sensor that can detect pucks and their position relative to it. It can be implemented in physical robotic systems with a calibrated camera, some image processing, and exploiting some knowledge about the geometry of the local environment (i.e. that the floor is planar) [23].

Sensory data in our work are the number of pucks in perception areas. Two such areas are provided for an agent (see Fig. 1). The `Central` area is important for detecting clusters of pucks in the immediate vicinity in front of the agent. The size of the `Central` area is approximately 6x6 puck diameters. The `Exploration` area stretches forward and is used for detecting pucks that are relatively far from the robot. The size of the `Exploration` area is approximately 4x25 puck diameters.

The input fed to the neural network reflects the number of pucks in the perception areas. This input, however, must be normalized within the range $[0, 1]$ (which is conventional for neural networks). The normalization is done by dividing the actual number of pucks in the area by the maximum number of pucks for that area. Thus, the **relative density** of pucks in an area is calculated. The question is then how to define the maximum numbers for the perception areas.

The neural signal from the `Central` region is saturated at 1 (is maximized) when the number of pucks in this region, assuming that they are uniformly

Fig. 1 Perception areas of an agent (blue). 1. The `Central` area is directly in front of the agent. 2. The `Exploration` area stretches ahead.

distributed, is large enough to form a single cluster[1]. In our setup, this number (further referred to as `MaxCentral`) is equal to 16. Experiments have shown that if the `MaxCentral` parameter is chosen to be significantly lower, for example, 8, performance of the swarm is unsatisfactory[2].

For the `Exploration` area the interpretation of the maximum number is different. This area is intended to serve for searching pucks, and not for detecting clusters. Hence, it is only important whether there is at least one puck in this area or not. The maximum number for the `Exploration` area is therefore 1.

A saturating linear function is used to normalize the number of pucks for both perception regions. Thus, if the number of pucks in either of the regions is larger than the corresponding maximum number, the input to the associated neuron will be saturated at 1.

3.2 Basic Behaviors

Three behaviors are available for agents. `BackUpAndTurn` behavior is inspired by work in [4], where it has been proved to be efficient and the most important for the clustering behavior. `Turn` behavior is used for locating pucks. Finally, `MoveStraightAhead` behavior, as the name implies, moves an agent forward. The description of the behaviors is given in Table 1.

Table 1 Description of the behaviors

`BackUpAndTurn`	Back up for N time units, then turn at random angle*.
`Turn`	Turn at a small angle.
`MoveStraightAhead`	Move straight ahead.

*To implement this behavior, the procedure `Update`, which is called each time step and which is responsible for collecting data, feeding it into the neural network, getting the output and activating behaviors, needs an additional condition. If `BackUpAndTurn` behavior is currently being executed, then `Update` immediately returns without referring to the neural network. This is repeated until N times units have passed. After that, the `Update` procedure is executed normally.

The output of the network is an array of three floating point numbers. The methodology "winner-takes-all" is used to convert this array to an array of three

[1] We define clusters as follows. Suppose that at time t we have a graph with P vertices, where P is the number of pucks in the experiment. There is a one-to-one correspondence between the set of vertices and the set of pucks; i.e., each vertex i is a mathematical representation of a corresponding puck p_i. An *edge* between vertices a and b in this graph exists if and only if the $d_t(p_a, p_b) < h$, where $d_t(x, y)$ is the distance between x and y at time t and h is a distance threshold. In our work, we define h as one diameter of a puck. Each connected component in this graph will then represent a cluster of pucks.

[2] Further discussion of this question is given in Section 4.

integers: the maximum element of the array is rounded to 1, while two other elements are rounded to 0. The interpretation of the resulting integer array is given in Table 2.

Table 2 Interpretation of the output.

`[1, 0, 0]`	Activate `BackUpAndTurn` behavior.
`[0, 1, 0]`	Activate `Turn` behavior.
`[0, 0, 1]`	Activate `MoveStraightAhead` behavior.

3.2.1 Obstacle Avoidance

The obstacle avoidance behavior, which is used in many works (for example, [5, 9, 17, 19]; see Section 2) is not explicitly present here. Rather, to avoid collisions between robots we use `BackUpAndTurn` behavior: in the same manner as we avoid disturbing large clusters, we avoid collisions between robots. This way, we do not require an additional behavior for obstacle avoidance; thus, we reduce the complexity of learning.

Therefore, another robot should be perceived in a similar way as a large cluster. To achieve that, we maximize (make it equal to 1) the input from the `Central` area if a robot is detected within this region.

3.3 Network Training

The sketch of the neural network used in this paper is presented in Fig. 3.

It may be hard to develop a probabilistic behavioral model even for such a simple task as clustering. However, it is relatively simple to define qualitative rules. The short summary of these rules, partially inspired by works discussed in Section 2, is as follows.

1. *If there is a cluster directly in front of an agent, it should back up and turn (to avoid disturbing the cluster).*
2. *If there is no cluster directly in front of an agent, it should always move to the puck in the* `Exploration` *area (this would help to bring some of the pucks the agent is plowing, if any, to that puck).*
3. *If there are no pucks in the* `Exploration` *area, an agent should turn around until it finds at least one. (Extrusions on the sides of the plow will help the agent keep the pucks that it has plowed.)*

Given this summary, we define characteristic situations that an agent may encounter (i.e., a training set) and provide a "solution" for each of these situations (see Fig. 2 and Fig. 4). We then use the backpropagation learning method to train the

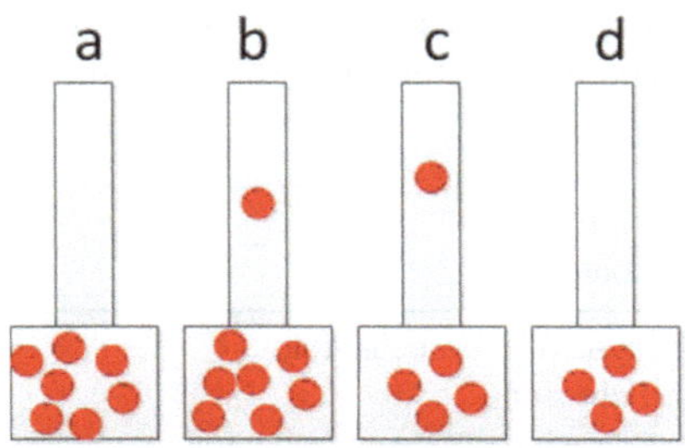

Fig. 2 The sketch of the training set. The lower rectangle is the `Central` area, the rectangle above is the `Exploration` area. Red circles are pucks. In situations ***a*** and ***b***, `BackUpAndTurn` behavior should be activated, because the relative density of pucks in the `Central` area is high. In ***c***, `MoveStraightAhead` should be activated: this will increase the number of pucks in the `Central` area; therefore, a larger cluster will be created. Finally, in ***d*** the agent should activate `Turn` behavior in attempt to find a puck in the `Exploration` area.

neural network until it starts producing correct output for all training samples[3]. It appears that although we considered only 4 possible situations from more than 3,200 possible combinations (0.00125%) of the numbers of pucks in the perception areas[4], the neural network interpolates to recognize all other situations correctly. The proposed method is somewhat similar to linear support vector machines (SVMs) used in supervised learning and statistical analysis [22], though we do not follow this approach directly.

4 Experiments

Experiments have been conducted using a custom simulator written in C#. The simulator uses the MOGRE engine[5] for 3D visualization and the MogreNewt engine[6] for modeling realistic physics in the simulation. The process of clustering observed in the simulator can be seen at Fig. 5.

[3] Since we are using a single-layer perceptron, the backpropagation is effectively reduced to the delta rule. However, in a more general case, neural networks with hidden layers can be used; therefore, the backpropagation method will be needed.

[4] In theory, the `Central` area can accommodate as many as 32 pucks (not 36, although its size is 6x6 puck diameters; this is because the plow extrusions are located within the `Central` area, leaving less space for pucks), provided that they are clustered in an extremely tight manner; the `Exploration` area in a similar way provides space for about 100 pucks. Thefore, the total number of combinations of pucks in the perception areas is $N = 32 \times 100 = 3200$.

[5] http://www.ogre3d.org/tikiwiki/MOGRE

[6] http://www.ogre3d.org/tikiwiki/MogreNewt

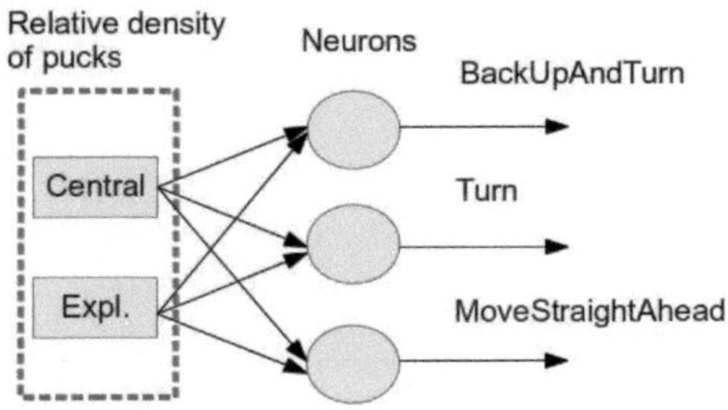

Fig. 3 The neural network. The inputs (in picture, bounded by a dashed line) are normalized relative densities of pucks in perception areas. (The normalization is made using saturating linear function, see Sect. 3.1.) The output is an array of floating-point numbers; we then apply the 'winner-takes-all' methodology to obtain one of the codes from Table 2. The operation of the neural network is described by the following equation: $\mathbf{a} = logsig(\mathbf{Wp} + \mathbf{b})$, where $logsig(n) = \frac{1}{1+e^{-n}}$, $\mathbf{a}$ is the output vector $(a_1, a_2, a_3)^T$, $\mathbf{W}$ is the weight matrix 3×2, $\mathbf{p}$ is the input vector $(p_1, p_2)^T$, and b is the bias vector $(b_1, b_2)^T$.

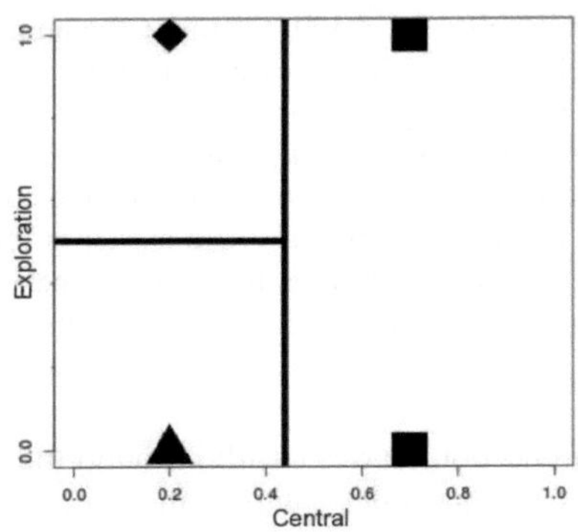

Fig. 4 Target outputs for the training set. The inputs for the neural network are coordinates of the points; the output is denoted by symbols, which are interpreted as follows: ■ - `BackUpAndTurn`, ▲ - `Turn`, ♦ - `MoveStraightAhead`. Black solid lines are decision boundaries obtained by the application of the delta rule. It can be seen that the given patterns are linearly separable. Note that this particular problem could be solved by using only two neurons (one for each decision boundary) with saturating linear transfer functions. The output of the network will then be the binary code of a behavior. In this paper, however, we prefer to use three neurons (one for each behavior) with log-sigmoid transfer functions; therefore, the output of the network is an array of floating-point numbers which we interpret as the 'confidence' of the network in a given behavior.

Initially, 100 pucks are scattered randomly (using the uniform distribution) over a squared area with sides of 40 puck diameters. Five agents[7] start in random positions

[7] The number of agents has been chosen more or less arbitrarily; most works in the clustering problem, including [4], [5], [19], and [23] tend to choose the number of robots in the range from 1 to 10, so we concluded 5 to be the most typical value.

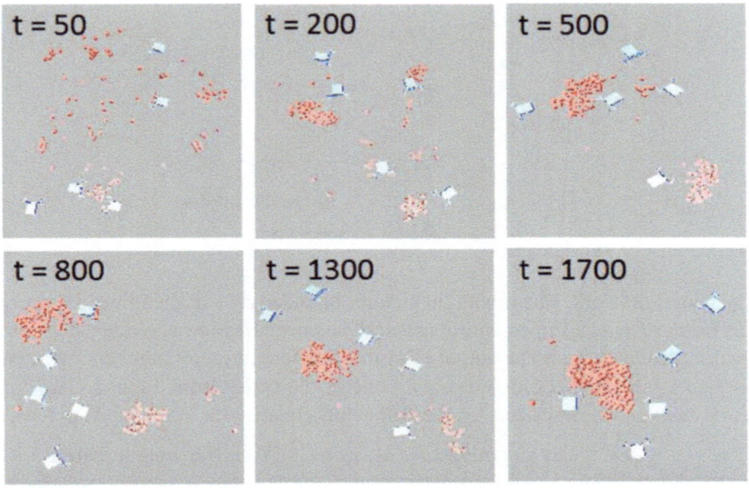

Fig. 5 Simulation of the clustering problem in process. Several phases of the process can be distinguished. In the first stage ($0 < t < 500$, t is time), the number of clusters is quickly decreasing and a few large clusters (two in this case; sometimes it may be three or even four clusters) are formed. In the second stage ($500 < t < 1700$), the smaller cluster is gradually destroyed; pucks from the second cluster are delivered to the bigger cluster. (This process, however, is statistical; we have observed situations when, on the contrary, the bigger cluster was destroyed.) In the final phase ($t > 1700$), a single cluster is formed. A few pucks may be removed from the cluster from time to time, but they are then returned back. (All time intervals are given in simulation time units and may vary from trial to trial.)

within this area. There are no boundaries (walls) around the simulation area. Note, however, that agents tend to keep pucks within the initial area, due to the `Turn` behavior.

To measure the performance, we use the metrics from [5]: the average size of a cluster and the size of the largest cluster. Statistics are collected every 10 time units of the simulation. In Section 3.1 we have mentioned that we predefine the maximum number of pucks for perception areas (for `Central` area this parameter is named `MaxCentral`). These numbers are the only tunable parameters for the learning mechanism. We can change these parameters by effectively reprogramming a robot (i.e., modifying the "software"). All other parameters, such as robot and puck sizes, or the number and the sizes of perception areas, most likely cannot be adjusted without affecting the "hardware" (we consider the "hardware" parameters to be given as is). The efficiency of the proposed "software" can be estimated by using trials with different values for the "software" parameters. Thus, we conduct experiments with different values for the `MaxCentral` parameter to determine the performance of the proposed learning mechanism. Results, averaged for 10 runs for each value of the `MaxCentral` parameter, can be seen in Fig. 6 and Fig. 7.

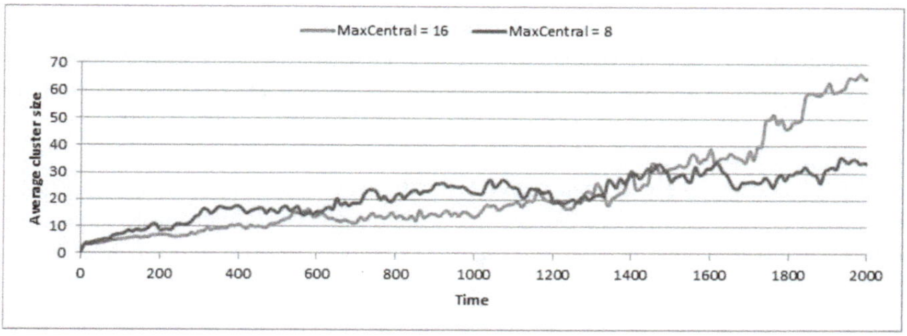

Fig. 6 Average cluster size for different values of the `MaxCentral` parameter. With `MaxCentral` = 16, the average of 33 (1 main cluster with 98 pucks and 2 clusters with a single puck in each), 50 (1 main cluster with 99 pucks and a separated puck) or 100 is typical in the final stage of the process, since pucks are occasionally removed from the cluster (but are quickly returned back). With `MaxCentral` = 8, the average cluster increases slower.

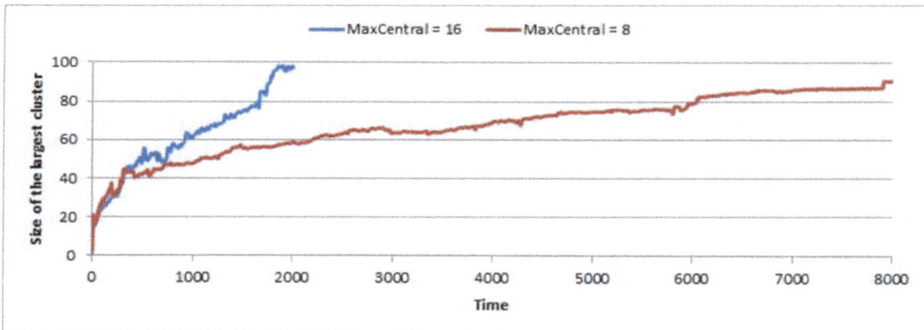

Fig. 7 Size of the largest cluster for the different values of the `MaxCentral` parameter. The horizontal axis (time) has been extended in this figure. Experiments with `MaxCentral` = 16 were stopped once a cluster of 100 pucks was formed. Experiments with `MaxCentral` = 8 took approximately 4 times more time to converge to a single cluster. However, in a few experiments with `MaxCentral` = 8 the convergence to a single cluster was not achieved even after 8000 time units.

The `MaxCentral` parameter has proved to be critical for performance. If this parameter is chosen to be too low (for example, 8), the `BackUpAndTurn` behavior is often triggered prematurely. In this case, agents are highly unlikely to destroy smaller clusters in order to push a few pucks to bigger clusters; thus, the size of the largest cluster and the average cluster size grow slowly (see Fig. 7; the size of the largest cluster is growing approximately 4 times slower than for the value for `MaxCentral` equal to 16). When the `MaxCentral` parameter is high enough (for example, 16), agents are allowed to "steal" pucks from one cluster to deliver them to another one; therefore, the size of the largest cluster and the average cluster size grow faster.

It is interesting that in the first 1000 times steps the performance yielded by the smaller parameter value is better (see Fig. 6). This is because agents guided by a smaller `MaxCentral` parameter tend to form many clusters of a relatively small size (10-20 pucks), whereas "greedier" agents (with `MaxCentral` = 16) start creating bigger clusters from the very beginning, and smaller clusters which are occasionally formed are likely to be destroyed shortly. However, once several big clusters have been formed, the average cluster size starts to grow relatively fast, whereas in the first case it is growing more slowly. We conclude that the tradeoff here is between convergence to smaller, bigger clusters at the expense of increased time.

It appears that the probability of removing pucks from a cluster is a function of its size. The only way to remove pucks from a big cluster is to follow a tangent line to the cluster and to plow some pucks from its skirt. Due to the `Turn` behavior, which is likely to be activated afterwards, the pucks are then either returned back to the cluster (if there are no other piles of pucks outside), or pushed to another cluster. If the cluster is relatively large, then the chance that the first puck that will appear in the `Exploration` area will belong to the same cluster is large; the pucks will thus be returned to the cluster. If the cluster is relatively small, there is a high chance that a significant part of it (or even the entire cluster) would be removed without activating `BackUpAndTurn` behavior. Hence, the smaller the cluster is, the bigger is the probability to remove pucks from it. This probabilistic process is functionally similar to what has been developed by Deneubourg et al. [4]. However, no explicit probabilistic rules are present in our system. The global probabilistic behavior emerges based on the geometrical shapes of the robots and pucks, and the individual behaviors provided by neural-based controllers. More detailed theoretical derivations related to this subject can be found in [24].

5 Conclusions

Emergent behavior is a key to using a swarm of simple cheap robots for solving complex tasks without centralized control. Different approaches have been proposed to designing the agents' behavior in such a way that the desired global swarm behavior emerges. However, in some situations these approaches may turn out to be inapplicable or inefficient. In this work, we have presented a simple method to design a swarm behavior based on supervised learning of a small number of samples representing situations that an agent may encounter in the world of clustering.

While obtaining global probabilistic behavior which is functionally similar to what has been described in [4] and subsequent works, we avoid creating explicit probabilistic models of the individual behavior of the agents. Our approach is extreme in its simplicity. We use no specialized grippers for the agents; the number of behaviors is limited to three; the neural network is a single-layer perceptron. Yet, such a simple approach yields the desired result: the swarm of agents accomplishes the clustering task.

With increasing complexity of sensory input and the number of behaviors the training will most probably become more complicated. However, the basic idea of swarm robotics implies an extreme minimalism of robots [23]; the number of sensory inputs and behaviors is expected to be relatively small. If it is difficult to distinguish "characteristic", or "boundary" situations from the set of possible sensory inputs, the process of supervised learning can be made iterative: add one sample after another, until the produced behavior becomes acceptable. With this assumption, we think that the proposed approach may be efficient for other tasks studied in swarm robotics.

Acknowledgements. W.B. acknowledges funding from NSERC under the Discovery Grant Program RGPIN 283304-07 and RGPIN 283304-12.

References

1. Bayindir, L., Sahin, E.: A Review of Studies in Swarm Robotics. Turkish Journal of Electrical Engineering 15, 115–147 (2007)
2. Yogeswaran, M.: Swarm Robotics: An Extensive Research Review. In: Advanced Knowledge Application in Practice, pp. 259–278 (2010)
3. Bonabeau, E., Dorigo, M., Theraulaz, G.: Swarm Intelligence: From Natural to Artificial Systems. Oxford University Press (1999)
4. Deneubourg, J.L., Goss, S., Franks, N., Sendova-Franks, A., Detrain, C., Chretien, L.: The dynamics of collective sorting robot-like ants and ant-like robots. In: First International Conference on the Simulation of Adaptive Behaviour, pp. 356–363 (1991)
5. Beckers, R., Holland, O.: From local actions to global tasks: Stigmergy and collective robotics. In: Artificial Life, vol. IV, pp. 181–189 (1994)
6. Holldobler, B., Wilson, E.O.: Journey to the Ants: A Story of Scientific Exploration. Belknap Press of Harvard University Press (1994)
7. Seeley, T.D.: Honeybee democracy. Princeton University Press (2010)
8. Holldobler, B., Wilson, E.O.: The Superorganism: The Beauty, Elegance and Strangeness of Insect Societies. W. W. Norton & Company, Inc (2009)
9. Mataric, M.: Designing emergent behaviors: From Local Interactions to Collective Intelligence. In: From Animals to Animats 2. Proceedings of the Second International Conference of Simulation Adaptive Behavior, pp. 432–441 (1992)
10. Martinoli, A., Ijspeert, A.: A Probabilistic Model For Understanding And Comparing Collective Aggregation Mechanisms. In: Floreano, D., Mondada, F. (eds.) ECAL 1999. LNCS, vol. 1674, pp. 575–584. Springer, Heidelberg (1999)
11. Bahceci, E., Sahin, E.: Evolving Aggregation Behaviors For Swarm Robotic Systems: A Systematic Case Study. Technical report METU-CENG-TR-2005-03, Middle East Technical University, Turkey (2005)
12. Trianni, V., Groß, R., Labella, T.H., Şahin, E., Dorigo, M.: Evolving Aggregation Behaviors in a Swarm of Robots. In: Banzhaf, W., Ziegler, J., Christaller, T., Dittrich, P., Kim, J.T. (eds.) ECAL 2003. LNCS (LNAI), vol. 2801, pp. 865–874. Springer, Heidelberg (2003)
13. Trianni, V., Nolfi, S.: Engineering The Evolution of Self-Organizing Behaviors In Swarm Robotics: A Case Study. Artificial Life 17(3), 183–202 (2011)
14. Trianni, V.: Evolutionary Swarm Robotics: Evolving Self-Organising Behaviours in Groups of Autonomous Robots. SCI, vol. 108. Springer, Heidelberg (2008)
15. Jeanson, R., Rivault, C., Deneubourg, J., Blancos, S., Fourniers, R., Jost, C., Theraulaz, G.: Self-Organized Aggregation in Cockroaches. Animal Behaviour 69, 169–180 (2005)

16. Garnier, S., Jost, C., Jeanson, R., Gautrais, J., Asadpour, M., Caprari, G., Theraulaz, G.: Collective Decision-Making by a Group of Cockroach-Like Robots. In: Proceedings of the IEEE Swarm Intelligence Symposium, pp. 233–240 (2005)
17. Soysal, O., Sahin, E.: Probabilistic Aggregation Strategies in Swarm Robotic Systems. In: Proceedings of the IEEE Swarm Intelligence Symposium, pp. 325–332 (2005)
18. Nouyan, S., Dorigo, M.: Chain Formation in a Swarm of Robots. Technical report TR/IRIDIA/2004-18, IRIDIA - University Libre de Bruxelles, Belgium (2004)
19. Martinoli, A., Mondada, F.: Collective and cooperative group behaviours: Biologically inspired experiments in robotics. In: Proceedings of the Fourth International Symposium on Experimental Robotics ISER-95. LNCIS, vol. 223, pp. 1–10. Springer, Heidelberg (1997)
20. Gross, R., Bonani, M., Mondada, F., Dorigo, M.: Autonomous Self-assembly in a Swarmbot. In: Proceedings of the Third International Symposium on Autonomous Minirobots for Research and Edutainment, pp. 314–322 (2006)
21. Parker, C., Zhang, H.: Blind bulldozing: multiple robot nest construction. In: Proceedings of the International Conference on Robots and Systems, pp. 2010–2015 (2003)
22. Christianini, N., Shawe-Taylor, J.: An Introduction To Support Vector Machines and Other Kernel-Based Learning Methods. Cambridge University Press (2003)
23. Vardy, A.: Accelerated patch sorting by a robotic swarm. In: Canadian Conference on Robot Vision (2012)
24. Kazadi, S., Abdul-Khaliq, A., Goodman, R.: On the convergence of puck clustering systems. Robotics and Autonomous Systems 38, 93–117 (2002)

Author Index

MIX
Papier aus verantwortungsvollen Quellen
Paper from responsible sources
FSC® C105338

If you have any concerns about our products,
you can contact us on
ProductSafety@springernature.com

In case Publisher is established outside the EU,
the EU authorized representative is:
Springer Nature Customer Service Center GmbH
Europaplatz 3, 69115 Heidelberg, Germany

Printed by Libri Plureos GmbH
in Hamburg, Germany